1st Conference on Traditional and Renewable Energy Sources: Perspectives and Paradigms for the 21st Century (TRESP 2021)

E3S Web of Conferences Volume 250 (2021)

Prague, Czech Republic
22 – 23 January 2021

Editor:

Wadim Strielkowski

ISBN: 978-1-7138-2870-9

Printed from e-media with permission by:

Curran Associates, Inc.
57 Morehouse Lane
Red Hook, NY 12571

Some format issues inherent in the e-media version may also appear in this print version.

For additional information, please contact EDP Sciences – Web of Conferences
at the address below.

EDP Sciences – Web of Conferences
17, Avenue du Hoggar
Parc d'Activité de Courtabœuf
BP 112
F-91944 Les Ulis Cedex A
France

Phone: +33 (0) 1 69 18 75 75
Fax: +33 (0) 1 69 28 84 91

contact-edps@webofconferences.org

Additional copies of this publication are available from:

Curran Associates, Inc.
57 Morehouse Lane
Red Hook, NY 12571 USA
Phone: 845-758-0400
Fax: 845-758-2633
Email: curran@proceedings.com
Web: www.proceedings.com

TABLE OF CONTENTS

Preface

It is my utmost pleasure to present this collection of scholarly articles of the international scientific conference "1st Conference on Traditional and Renewable Energy Sources: perspectives and paradigms for the 21st century (TRESP 2021)". The conference was held at the Prague Institute for Qualification Enhancement on the 22nd and 23rd of January 2021 in Prague, Czech Republic.

TRESP 2021 conference consisted of the eight main topics (tracks):

- Energy perspectives for the 21st century
- Traditional energy sources
- Renewable energy sources
- Sustainable development and energy markets
- Energy and earth sciences
- Legal regulation of the digital economy and digital relations on energy markets
- Energy research and social sciences
- Smart grids and future energy systems

The 1st International Conference on Traditional and Renewable Energy Sources brought together researchers and stakeholders from the fields of energy economics and policy, sustainable development, renewable and traditional energy sources, earth sciences, energy research and social sciences, as well as policy-makers responsible for the creation and maintenance of the sets of rules for the future sustainable development and growth.

The conference offered a wide range of keynote speeches and presentations and resulted in a fruitful discussion and exchange of opinions on the perspectives and paradigms for the traditional and renewable energy sources in the 21st century.

I would like to express my high appreciation to the organizers, members of the Conference Scientific Committee and Conference Publication Committee as well as the reviewers of TRESP 2021 for their hard work. Without their participation, it would be impossible to prepare and to hold

TRESP 2021 successfully as well as to ensure high quality of papers in the conference proceedings. I would also like to express my gratitude to the unbelievable number of participants for contributing their excellent research results to our conference. Special acknowledgements go to the EDP Sciences team. I am looking forward at meeting more potential contributors and speakers in our upcoming conferences.

On behalf of the TRESP 2021 Organizing Committee

Prof. Wadim Strielkowski, Ph.D.
Prague Institute for Qualification Enhancement
and
Centre for Energy Studies, Prague Business School

Statement of Peer review

In submitting conference proceedings to *Web of Conferences*, the editors of the proceedings certify to the Publisher that

1. They adhere to its **Policy on Publishing Integrity** in order to safeguard good scientific practice in publishing.
2. All articles have been subjected to peer review administered by the proceedings editors.
3. Reviews have been conducted by expert referees, who have been requested to provide unbiased and constructive comments aimed, whenever possible, at improving the work.
4. Proceedings editors have taken all reasonable steps to ensure the quality of the materials they publish and their decision to accept or reject a paper for publication has been based only on the merits of the work and the relevance to the journal.

Title, date and place of the conference

1st Conference on Traditional and Renewable Energy Sources: perspectives and paradigms for the 21st century (TRESP 2021), 22nd-23rd of January 2021, Prague (Czech Republic)

Proceedings editor(s):

Prof. Wadim Strielkowski, Director of the Prague Institute for Qualification Enhancement and Professor of Prague Business School (Prague Czech Republic)

Date and editor's signature

30.03.2021

Digital transformation of the energy sector: a case of Russia

Galina Chebotareva[1 *]

[1]Academic Department of Energy and Industrial Enterprises Management Systems, Graduate School of Economics and Management, Ural Federal University, Mira str. 19, 620002 Yekaterinburg, Russian Federation

Abstract. The article outlines the main stages of the digital transformation of the energy sector of Russia since 2014. The methodological framework of the summary is composed of the key federal laws and industrial regulations that formed the foundation of the digitalization of the Russian energy sector. The study revealed two focus segments for Russian energy transformation. The first one deals with the development of the methodological framework for digital transformation (departmental project "Digital Energy"). The second one works towards the development of new technical solutions, their active field testing and elimination of emerging administrative barriers (EnergyNet road map, "Unified Technical Policy - Power Supply Reliability" project). It has been established that technological development trends in the Russian energy sector generally match those in the rest of the world (smart energy, distributed generation, renewable energy, consumer services etc.). Private projects that are being implemented as part of these initiatives prove highly efficient technically and economically and are attracting wide attention in the global energy market.

1 Introduction

The traditional structure of the energy system of Russia is currently out of date. The average rate of depreciation for fixed assets is 70 to 80% depending on the specialization of the energy business [1, 2], while the asset renewal rate is usually not so high. New groundbreaking technologies that are emerging in traditional generation as well as in alternative energy should be put in use. At the same time, the development of the energy market should take into account constantly growing demand for electric power and meet qualitatively new demands of society (environmental friendliness, availability, mobility, digital quality) [3-6]. This is only possible with the start of new investment cycles that are aimed not only at replacing obsolete energy capacity, but also at the adoption of new digital technologies at energy facilities [7-9].

As a result, the transformation of the national energy sector that is gaining pace, new laws and regulations that are being drafted, national, sectoral and ministerial programs that have been approved for implementation and emerging technical solutions and administrative initiatives call for systematization of a large amount of incoming information. It is, therefore,

* Corresponding author: g.s.chebotareva@urfu.ru

the objective of this study to produce an overview of digital transformation methods and practices that are currently in use in the Russian energy sector. This article draws upon up-to-date laws and regulations addressing the digitalization of the economy and industries, road maps, project profiles, and other official sources documenting the implementation of digital projects.

The practice-related subject matter and objective of the article determine the following structure of the work. Part 2 describes the key stages of the digital development of the Russian economy, including the energy sector. Part 3 looks at the tangible results of the implementation of the "EnergyNet" program with regard to four types of projects. Part 4 investigates the methodological side of the "Digital Energy" project. Part 5 presents the current practical achievements of the "Unified Technical Policy - Power Supply Reliability" project. In the Conclusions part of the article, the main results are summarized and technological trends that accompany the digital transformation of the Russian energy sector are described.

2 Main stages of Russian energy sector digitalization

A strategy towards sweeping digitalization of the Russian economy was formally announced on 4 December 2014 during the Russian president's *annual address to the Federal Assembly* [10]. In the address, Vladimir Putin sent the goal of "creating a system of long-term forecasts feeding a strategy of choosing suitable technologies in the next 10 to 15 years". The ultimate goal of the initiative is to ensure national security, improve quality of life, and promote industries operating in a new technological environment. *The National Technology Initiative* (NTI) took shape by mid-2015 [11, 12] and was officially approved in 2016 [13] to run till 2035. The initiative comprises 8 road maps and around 54 projects of the NTI Fund and 578 projects of the Foundation for Assistance to Small Innovative Enterprises [14]. The NTI's road maps include "Neuronet", "Technet", "Healthnet" and a few others. Their implementation should forge new markets and create conditions for Russia's global technological leadership.

It is within the framework of the NTI that the *"EnergyNet" action plan* for the digitalization of the national energy sector has been implemented since 2016-2017. The mission of this action plan is largely in line with the NTI, but is customized to the needs of the energy sector. The ultimate goal of EnergyNet is to secure the global technological leadership of Russian companies in new energy markets through forward-looking development of digital transformation practices for the national electric power infrastructure. The specific goals of the project [15, 16] are associated with:

- Penetration of new energy markets;
- Development/adoption of promising smart energy technologies and services;
- Import of technologies under development from "target countries" (e.g. BRICS member states);
- Energy asset modernization;
- Training of personnel for the successful development of the energy market;
- Refinement of legislation and elimination of administrative barriers in the course of the transition to a new energy paradigm;
- Breaking the constraints of infrastructure by expanding a chain of research centers, laboratories, and experimental sites;
- Overcoming funding constraints, primarily with the help of private investments;
- Fulfilling the challenges of economic development facing the energy sector.

The key segments of the "EnergyNet" road map are "reliable and flexible distribution networks", "smart distributed energy", and "consumer services".

In 2018, the decree of the Russian President [17] set the key targets for the *Russian Federation national program "Digital Economy"* [18, 19] through to 2024:

a) a three-fold increase in domestic expenditures on the development of the digital economy;

b) creation of sustainable and safe information and telecom infrastructure for high-speed transmission, processing and storage of large data volumes that would be accessible to all organizations and households (access to the internet, 5G network rollout, data protection);

c) predominant use of domestic software by government agencies, local governments and organizations;

d) boosting the efficiencies of the key industries;

e) training of highly qualified personnel for the digital economy.

This decree [17] also sets a number of tasks directly applicable to the development of digital energy:

- transformation of priority industries, including energy infrastructure, through introduction of digital technology and platform solutions;
- guaranteed provision of affordable electric power, including: a) implementation of smart grid management systems based on digital technologies, b) development of centralized energy systems, c) development of distributed generation, renewable energy, including remote and isolated areas.

As part of this federal program, an *industry-specific project called "Digital Energy"* was formed. Its agenda was approved at the end of 2018 with the purpose of supporting the digital development of power engineering, the oil and gas industry and coal mining [20]. The project is aimed at systematizing the experience of the implementation of digital technologies, at ensuring clear goals for the digitalization of the energy sector and setting basic requirements and criteria for solutions being implemented, and at digital transformation monitoring. Cooperation between the ministry, the expert community and businesses is an effective tool for the implementation of the project. The part of the project that is universal throughout the industry envisages the creation of conditions for designing and developing digital services and solutions in specific branches of the fuel and energy sector.

There is, however, a need for a specific approach to the digitalization of the electric power industry. For this reason, the Energy Ministry of Russia initiated in 2019 a separate *ministerial project* called "*Unified Technical Policy - Power Supply Reliability*". It is aimed at creating conditions for the implementation of risk-oriented industrial energy asset management by refining the legal and technical regulatory framework of the electric power industry [21].

On 9 June 2020, the new *"Energy Strategy of Russia through to 2035"* was approved [22]. It is this document that sets the goal of "the digital transformation and "smartization" of the energy and fuel industries" [22] and makes references to the NTI EnergyNet road map and sector-specific projects of the Energy Ministry. The projected outcomes of the strategy implementation are [22]:

- creation of a system for control, coordination and monitoring of the fuel and energy sector transformation;
- integration of digital technology into state governance and oversight activities in the fuel and energy industries;
- implementation of pilot projects to implement digital technology and platform solutions tailored to the sector;
- a transition to risk-oriented industrial asset management on the basis of digital technology;
- creation of smart electricity metering systems etc.

3 Current results of "EnergyNet" road map implementation

All projects that are being implemented as part of the EnergyNet program are divided into four groups [23]:

- Flagship projects: major and backbone projects (supported by [13]);
- Comprehensive projects: pilot projects aimed at testing technical, economic, organizational and regulatory aspects of new business practices in the field of power engineering;
- Technology projects that are aimed at overcoming the technological barriers of the "EnergyNet" road map (under implementation in cooperation with the Foundation for Assistance to Small Innovative Enterprises);
- Regulatory projects that are aimed at refining legislation and eliminating administrative barriers [24].

Table 1. "EnergyNet" flagship projects

Project	Year	Status	Contents
Digital PDA - YantarEnergo	2018	Completed	- Commercial technology prototypes for digitalization of a grid company - Technology trials in pilot zones - Verification of potential technical and economic effects and target performance indicators of a model grid company - Initiatives to adjust regulatory and technical documentation in order to replicate project results
Internet of Energy architecture	2018	Completed	- Development of decentralized energy system - Application of smart distributed control to be carried out through energy transactions - Adherence to the principles of transaction, intelligence, sustainability and flexibility
Energozapas	2018	In progress	- Creation of lifted weight storage plant (with capacity above 300 MWh)
Topaz	2018	In progress	- Creation of portable power source that is capable of continuously generating electricity from organic fuel at temperatures ranging from minus 40 to plus 50 degrees Celsius (with capacity of 3,000 watt-hour per kg)
«∀Platform	2019	In progress	- Smart distributed energy - Provision to consumers of high-quality low-cost mechanisms of accessing the full range of capabilities and functions of energy sector actors
CableWalker	2019	In progress	- Pilotless transport platform, inspection and maintenance modules, software and automated expert system of defect detection - Detailed monitoring and technical maintenance of transmission wires, condition-based repairs - Precision investment in specific trouble areas and extension of service life - Reducing the human factor; reducing failure occurrence

The structure of the EnergyNet projects and their contents are briefly presented in Tables 1 [25-30] and 2 [31].

Table 2. Some "EnergyNet" technology projects being implemented since 2020

Initiating company	Contents
Iserv	Digital platform for automation of processes of transportation and selling of energy and utilities
ANSystems	Decision Support System in the energy sector on the basis open digital ontologies with the capability of crowdsourced structure building and editing
Optispark	Solar-blind filters for inspection of overhead power lines
Technotronics	Battery monitoring and predictive analytics system
ELIOT	Platform for electricity bill calculation and payment on the basis of smart contracts
DonEnergoMash	Simulation modeling, creation of management algorithms and software for automated control over MicroGrid smart energy system

In 2020, a total of eight technology projects received support; in 2019 – five; in 2018 – 15; in 2017 – 15; in 2016 – 21 [31].

Comprehensive projects within the "EnergyNet" framework are being implemented in six areas [32]:

- active energy complexes;
- demand-side management aggregators;
- power supply to isolated and remote areas;
- use of energy storage systems;
- digital distribution electrical networks;
- user services.

Regulatory projects "EnergyNet" are distributed among five segments [33]:

- Development of active energy complexes;
- Development of the electrical energy storage systems market;
- Development of the demand-side management (DSM) market on the basis of DSM aggregators;
- Digital grids – improving the quality of electrical power supply services;
- Concessions – ensuring return on investment in smart energy.

The above projects have already won recognition in the domestic and international markets. For example, the "CableWalker" project received exceptional attention at the international trade show Middle East Electricity [34]. Trial runs of its pilot projects are being conducted by Russian company Rosseti Urala and by Saudi Electricity Company.

4 Current results of "Digital Energy" project

The departmental project Digital Energy implies a more methodical approach as compared to other programs. This is the deciding factor in its current results [20]:

- Creation of the council for digital transformation of the branches of the energy industry.

The purpose of the council is to work out a unified stance and key decisions regarding the digital transformation of the entire fuel and energy sector. Its creation was the first step towards the formation of an all-encompassing system for coordination and monitoring of digital transformation.

- Creation of centers of competence in the electric power industry, the oil and gas industry and coal mining.

They mainly serve to determine priority areas of action and technologies for the industries, to identify regulatory and technological barriers to implementation, as well as to create digital development road maps for the industries. The centers of competences were established by leading companies of the fuel and energy sector in cooperation with the Energy Ministry of Russia.

- Drafting of the core document for the digital development of the Russian energy sector – the Concept of the Digital Transformation of the Fuel and Energy Sector for the mid-term (through to 2024) and long-term (through to 2035) period. The concept contains scenarios, forecasts and priority action areas in terms of the digitalization of the sector.

Applied projects of energy digitalization are being implemented as part of another departmental project. They are described in the following section.

5 Current results of the "Unified Technical Policy - Power Supply Reliability" project

This departmental project envisages the implementation of 25 pilot projects at various energy companies before the end of 2024. The shared goal of the projects is to refine the legal and technical regulatory framework for the adoption of risk-oriented industrial energy asset management.

One of the projects that aimed to design "A digital model of a power grid and electricity delivery points" [35] in compliance with national standards was implemented by MRSK South (a subsidiary of Rosseti). As an outcome, trials were conducted to test the collection of input data for calculating the reliability of services being provided by utilities on the basis of the Industrial Internet. In the future, this solution could enable accurate and fast assessment of reliability metrics from the consumers' perspective and make it possible to improve operations and technology control and situational management processes in power networks.

A pilot project implemented by Kubanenergo (also a Rosseti subsidiary) [36] created a tool that uses an approved method [37] for automated calculation of the probability and impacts of a failure and technical risks. Further development of intelligent systems built upon algorithms that generate forecasts on the basis of measurements done by an automated monitoring and diagnosis system would make it possible to get a more precise list of priority equipment for technical operations.

A joint project by the System Operator of the United Energy System and RusHydro resulted in the launch of a digital remotely controlled load shedding system for ten hydroelectric power plants. The system is operated by control centers. The project was a step forward in the deployment of digital remote equipment control and work schedule management at energy facilities [38].

In addition, Rosseti's Mobile Gas-Turbine Power Stations group presented a pilot project aiming to create a prototype system for predictive analysis of energy generating equipment using intelligent data processing approaches [39]. As part of the project, models and methods were developed for predictive analysis of the main elements and units, detection of performance deviations, prediction of actual equipment status.

6 Conclusions

The modern world is witnessing the beginning of the fourth energy transition that will go through stages of the digitalization of the energy sector, including widespread use of unconventional renewable energy sources, a bigger role of distributed generation, the development and improvement of energy accumulation and storage systems. The objective of the new energy transition is to enhance energy security, energy efficiency; to make energy more accessible and greener; to make renewable energy competitive as innovations in the market reach critical mass.

The Russian economy, too, is actively involved in step-by-step digitalization of industries, including the fuel and energy sector. To achieve that, not only relevant regulatory documents are being drafted, but also targeted incentives are being proposed to facilitate the creation and adoption of digital technologies. For example, the EnergyNet action plan and the departmental project "Unified Technical Policy – Power Supply Reliability" were developed. They serve as umbrella projects for the design of new models of digitalization of grid companies, the development of decentralized energy systems, the creation of efficient energy storage systems, portable power sources etc. Russian technical solutions are sought after in the international energy market. Being a product of synergy of the state, businesses, the innovation community and science, such projects ensure the maximum system effect of digitalization.

According to experts, there is a number of global digital trends that will be actively implemented in the near future: digitalization of infrastructure; decentralization of power generation; intelligent control and engineering; the increasing appeal of the sector to private investors; dismantlement of regulatory barriers for business; introduction of a new generation of Impact Investing economic technologies; creation of new opportunities for end consumers of energy, service organizations, etc.

Acknowledgement

The work was supported by Grant of the President of the Russian Federation (MK-4549.2021.2).

References

1. B.E. Murmansky, K.E. Aronson, Y.M. Brodov, Journal of Physics: Conference Series, **891**, 012279 (2017)
2. L.D. Gitelman, B.E. Ratnikov, *Economics and business in the electric power industry [Ekonomika i biznes v elektroenergetike]* (2013)
3. V. Panchenko, Y. Harust, Y. Us, O. Korobets, V. Pavlyk, Marketing and Management of Innovations, **1**, 256-264 (2020)
4. Y. Wu, Y. Wu, J.M. Guerrero, J.C. Vasquez, International Journal of Electrical Power and Energy Systems, **126,** 106593 (2021)
5. E. Lisin I. Lebedev E. Sukhareva I. Komarov, International Economics Letters, **3(3)**, 105-114 (2014)
6. G. Chebotareva, W. Strielkowski, D. Streimikiene, Journal of Cleaner Production, **269**, 122110 (2020)
7. L. El Iysaouy, N.E. El Idrissi, M. Tvaronavičienė, M. Lahbabi, A. Oumnad, Insights into Regional Development, **1(3)**, 259-271 (2019)
8. R. Carmichael, R. Gross, R. Hanna, A. Rhodes, T. Green, Renewable and Sustainable Energy Reviews, **139,** 110701 (2021)
9. O. Konova I. Komarov E. Lisin, Czech Journal of Social Sciences, Business and Economics, **1(1)**, 101-109 (2012)
10. Message of the President of the Russian Federation to the Federal Assembly [Poslanie Prezidenta RF Federal'nomu Sobraniyu], *http://www.consultant.ru/document/cons_doc_LAW_171774* (2014)

11. National Technology Initiative: Russia's Leadership in Global Technology Markets by 2035 [Nacional'naya tekhnologicheskaya iniciativa: liderstvo Rossii na global'nyh tekhnologicheskih rynkah k 2035 godu], *https://www.rvc.ru/press-service/news/company/41396* (2015)
12. National Technology Initiative. Dossier [Nacional'naya tekhnologicheskaya iniciativa. Dos'e], *https://tass.ru/info/4421323* (2017)
13. Government Resolution "On the Implementation of the National Technology Initiative" [Postanovlenie pravitel'stva "O realizacii Nacional'noj tekhnologicheskoj iniciativy"], *https://base.garant.ru/71380666* (2016)
14. Register of NTI projects [Reestr proektov NTI], *https://nti2035.ru/upload/353016.2.project_register_NTI_317.pdf* (2020)
15. Roadmap "EnergyNet" [Dorozhnaya karta "EnergyNet"], *https://nti2035.ru/documents/docs/DK_energynet.pdf* (2016)
16. Regulatory roadmap "EnergyNet" [Normativnaya dorozhnaya karta "EnergyNet"], https://nti2035.ru/documents/docs/NDK_energynet_2020.pdf (2020)
17. Decree of the President of the Russian Federation No. 204 "On National Goals and Strategic Objectives of the Development of the Russian Federation for the period up to 2024" [Ukaz Prezidenta Rossijskoj Federacii 204 "O nacional'nyh celyah i strategicheskih zadachah razvitiya Rossijskoj Federacii na period do 2024 goda"], *https://www.primorsky.ru/authorities/executive-agencies/departments/departament-proektov/Ukaz%20204/Указ%20Президента%20РФ%20от%2007.05.2018%20г.%20№%20204.pdf* (2018)
18. National Program Digital Economy [Nacional'naya programma Cifrovaya ekonomika], *https://digital.ac.gov.ru* (2018)
19. Passport of the national program "Digital Economy of the Russian Federation" [Pasport nacional'noj programmy "Cifrovaya ekonomika Rossijskoj Federacii"], *http://government.ru/info/35568* (2019)
20. Departmental project "Digital Energy" [Vedomstvennyj proekt "Cifrovaya energetika"], *https://in.minenergo.gov.ru/energynet/docs/Ведомственный%20проект%20Цифровая%20энергетика%20-%20прил%201.pdf* (2018)
21. Departmental project of the Ministry of Energy "Unified Technical Policy – Power Supply Reliability" [Vedomstvennyj proekt Ministerstva energetiki "Edinaya tekhnicheskaya politika – nadezhnost' elektrosnabzheniya"], *https://www.digital-energy.ru/trends/analytics/projects/unified-technical-policy* (2019)
22. Russia's Energy Strategy for the period up to 2035 [Energeticheskaya strategiya Rossii na period do 2035 goda], *http://static.government.ru/media/files/w4sigFOiDjGVDYT4IgsApssm6mZRb7wx.pdf* (2020)
23. Project "EnergyNet" [Proekt "EnergyNet"], *https://energynet.ru (*2016)
24. Order of the Government of the Russian Federation 1526-r "On Approval of the Action Plan ("Road Map") to improve Legislation and Eliminate Administrative Barriers in order to ensure the implementation of the National Technology Initiative in the direction of "Energinet" [Rasporyazhenie Pravitel'stva RF 1526-r "Ob utverzhdenii plana meropriyatij ("dorozhnoj karty") po sovershenstvovaniyu zakonodatel'stva i ustraneniyu administrativnyh bar'erov v celyah obespecheniya realizacii Nacional'noj tekhnologicheskoj iniciativy po napravleniyu "Enerdzhinet"], http://www.consultant.ru/document/cons_doc_LAW_354841 (2020)
25. Flagship projects of NTI "Energinet" [Flagmanskie proekty NTI "Enerdzhinet"], *https://energynet.ru/projects#!/tab/171646271-1* (2021)

26. Project "Architecture of the Internet of energy" [Proekt "Arhitektura Interneta energii"], https://idea-go.tech (2021)
27. Project "Energozapas" [Proekt "Energozapas], *http://energozapas.ru* (2021)
28. Project "Topaz" [Proekt "Topaz"], *https://topaz.center/* (2021)
29. Project "A-Platform" [Proekt "A-Platforma"] (2021), https://a-platform.ru/
30. Project "CableWalker" [Proekt "Kanatohod"], *https://drive.google.com/file/d/1P8Eo3QruTDTDJ8W0IP6Ht22QPIULpdVu/view* (2021)
31. Technological projects of NTI "Energinet" [Tekhnologicheskie proekty NTI "Enerdzhinet"], https://energynet.ru/Projects#!/tab/171646271-3 (2021)
32. Comprehensive projects of NTI "Energinet" [Kompleksnye proekty NTI "Enerdzhinet"] (2021), *https://energynet.ru/Projects#!/tab/171646271-2* (2021)
33. Regulatory projects of NTI "Energinet" [Regulyatornye proekty NTI "Enerdzhinet"], *https://energynet.ru/Projects#!/tab/171646271-4* (2021)
34. The project "CableWalker" conquers the world [Proekt "Kanatohod" pokoryaet mir], https://urfu.ru/ru/news/23003/ (2018)
35. The Ministry of Energy approved a pilot project to determine the reliability of energy supply indicators based on the digital network topology [V Minenergo byl odobren pilotnyj proekt po opredeleniyu pokazatelej nadezhnosti energosnabzheniya po cifrovoj topologii seti], *https://digitalsubstation.com/blog/2020/03/08/v-minenergo-byl-odobren-pilot-po-opredeleniyu-pokazatelej-nadezhnosti-energosnabzheniya-po-tsifrovoj-topologii-seti* (2021)
36. The Ministry of Energy considered a pilot project in the field of predictive analysis of equipment condition [V Minenergo rassmotreli pilotnyj proekt v oblasti prediktivnogo analiza sostoyaniya oborudovaniya], *https://digitalsubstation.com/blog/2020/04/09/v-minenergo-rassmotreli-pilotnyj-proekt-prediktivnogo-analiza-sostoyaniya-oborudovaniya* (2021)
37. Order of the Ministry of Energy of the Russian Federation 123 "On Approval of Methodological Guidelines for Calculating the Probability of Failure of a Functional Unit and a Unit of Basic Technological Equipment and Assessing the Consequences of such a Failure" [Prikaz Ministerstva energetiki RF 123 "Ob utverzhdenii metodicheskih ukazanij po raschetu veroyatnosti otkaza funkcional'nogo uzla i edinicy osnovnogo tekhnologicheskogo oborudovaniya i ocenki posledstvij takogo otkaza"], *https://www.garant.ru/products/ipo/prime/doc/72113770/* (2019)
38. The System operator suggests using remote control technologies in thermal generation [Sistemnyj operator predlagaet primenyat' tekhnologii distancionnogo upravleniya v teplovoj generacii], *https://www.eprussia.ru/news/base/2020/2151473.html* (2020)
39. Digital predictive analysis of the operation of power generation equipment [Cifrovaya sistema prediktivnogo analiza raboty generiruyushchego oborudovaniya], *https://energypolicy.ru/czifrovaya-sistema-prediktivnogo-anal/energetika/2020/13/10* (2020)

Energy and industrial tourism: a specific niche on the tourism market

Natalya Alekseeva[1*], *Katarína* Hercegová[2]

[1] Volga Region State University of Service, Gagarina str. 4, 445017 Togliatti, Russian Federation
[2] Prague Business School, Werichova str. 1145/29, 15200 Prague, Czech Republic

Abstract. Energy tourism belongs to the not-so-well-researched fields of tourism. Being a part of the industrial tourism or a special interest tourism, this type of tourism includes visits to the energy facilities and locations such as factories, mines, power stations and renewable energy sites. This paper describes the niche of the energy tourism within the tourism market and assesses its potential. We show that sometimes the energy tourism might intervene with dark tourism (for example, in the case of the tourist visits to the site of the Chernobyl nuclear power station). In addition, we show that this type of tourism is closely correlated with the public and social acceptance of traditional and renewable energy facilities. We conclude that this might be a promising and emerging type of tourism that will likely grow due to the ongoing industrialization and expenditure of energy-generating facilities envisaged for meeting the growing demand for energy all around the world.

1 Introduction

Nowadays, sustainable tourism is a fast-growing industry that seeks to protect tourism industry objectives and reduce the impact on the environment and local culture [1, 2]. Demographic changes, such as an ageing population and growing children, are making themselves felt in the special-interest tourism (SIT) market [3]. In addition, digital tourism is also becoming an important issue, especially nowadays when the travel is disrupted by the COVID-19 pandemic [4, 5]. Tourists can visit some unusual and extraordinary places while using web-based interfaces and programmes, reducing the need for traditional travel and the use of fossil fuels and other forms of energy. This reduces the travel costs and environmental impact of the tourism industry [6, 7].

There are many types of special tourist interests, but tourism of special interest will continue to gain in importance as tourists become people who want to experience and experience traditional mass holidays instead of traditional "mass holidays" [8]. A special tourist interest can take the form of a special event, such as a visit to a national park or even a trip to the local community. Another example are the Antarctic tourism products, followed by the general management problems and problems faced by small tourism businesses in Antarctica and beyond. In order to study the special-interest tourism in general and energy and industrial tourism in particular, we must first discuss the nature of tourism and its

* Corresponding author: natali0174@mail.ru

variants. Travel of special interest is travel for people who go somewhere because they have a particular interest that can be pursued in a particular region or destination [9].

With regard to the above, one can take a look at energy tourism which is often considered a form of tourism of particular interest, but it is not the only form [10]. In general, demand in any SIT is expected to rise as tourist groups shrink. The adaptations already implemented by the travel and tourism industry allow for an expansion of the SIT and allow for the overall planning and development of travel experiences in these hubs.

In the Figure 1 that follows, one can see the photo of the Cheboksary Dam on Volga river which is a part of the Volga-Kama Cascade and the only hydro power station in the world that has a highway built over its roof. The dam is a local tourist attraction visited by thousands of people every year [11].

Fig.1. Cheboksary Dam on Volga river in Russia

With the remarkable increase in responsible travel decisions, the demand for tourism with special interest (such as energy tourism) is likely to increase. In fact, more and more individual travellers are being accompanied on their personal journey to various destinations that are less travelled and are not marked by mass tourism. They are run by local experts, such as experts in the field of energy tourism and energy saving. The World Travel and Tourism Council said that members of the tourism industry have improved their carbon efficiency by almost 20% in the last decade. With the aim of promoting sustainable tourism, the World Tourism Organisation (WTO) and the International Energy Agency (IEA) announced this year the launch of a new sustainable tourism strategy for the 2030s [12]. The development of the energy tourism as a special niche tourism opens up the possibility of focusing on sustainable energy and energy - efficient travel options such as renewable energy represented by the wind, hydro and solar energy [13]. There are also challenges for destinations that have to cope with a sudden and unplanned influx of visitors, such as hotels, restaurants, hotels and hotels.

The development of specialised types of tourism must be properly planned to ensure the long-term viability of the destination and the sustainability of its economy, and to maximise the benefits of this objective through sustainable local economic activity [14, 15]. The opportunities to start a business in the tourism sector are plentiful and can often start with minimal investment. The establishment of a tourism company is particularly attractive as these companies are usually locally owned and offer more economic advantages than chain stores. If the goal is to generate more external revenue for the local economy, it is a viable strategy to expand, retain and establish businesses that manufacture, sell and provide services to tourists. The expansion, retention and creation of tourist businesses can also be a viable strategy, especially when the company manufactures, sells or provides services for products. Young people with a lot of energy can start bicycle taxis, become fishing guides, take tourists on kayaking adventures or start restaurants. Improved waste management should help tourism companies save money, create more jobs and make the destination more attractive. Sustainable tourism also refers to tourist activities that are and can be sustained indefinitely. Therefore, specific forms of tourism are not the only form of sustainable tourism, but rather they can strive to be more sustainable. This means that ecotourism should strive to protect the environment through the use of renewable energy sources such as wind, solar, water, geothermal and hydroelectric power [16]. This is embedded into the Sustainable Development Agenda 2030 which goals are closely directed to tourism (and energy tourism) development.

Tourism trends of particular interest are driven primarily by the increasing demand for energy tourism in the countries such as the USA, Canada, Australia, and New Zealand. Such developments and initiatives will have a significant impact on the emerging tourism of particular interest in these countries and in other parts of the world. Particular interests also promote the development of renewable energies as a key element of their tourism strategy. Some researchers predict that energy tourism as a part of adventure and industrial tourism will emerge as the fastest growing niche tourism [17]. Within this context, it is useful to define the place of energy tourism withing tourism studies, as a special part of the special-interest tourism: it can be placed in the intersection of the adventure tourism and industrial tourism (see Figure 1).

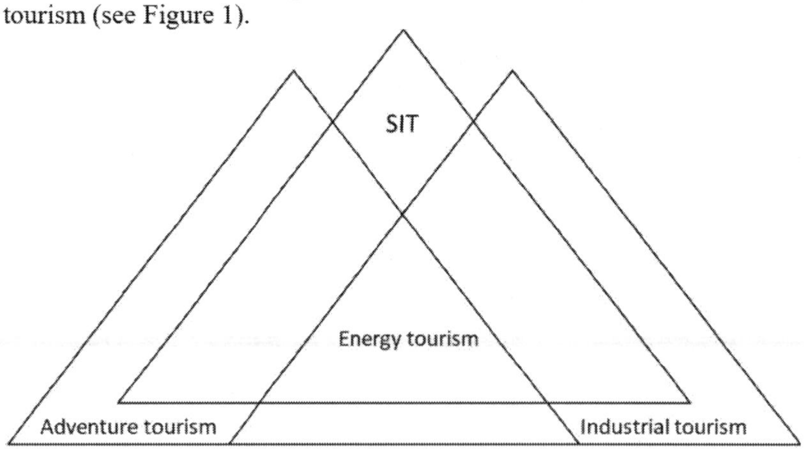

Fig.2. The place of energy tourism within the tourism industry

Moreover, tourism will by nature constitute the bulk of the tourism market in the coming years, with a steady growth rate after the massive vaccination that would help to eradicate the COVID-19 pandemic for good.

Therefore, one can say that energy tourism as tourism of particular interest would definitely have some impact on the tourism industry and the world economy. The tourism industry is one of the founders of plans to protect and preserve the environment and has played a decisive role in drawing attention to important environmental issues. Indeed, it is appropriate to promote the promotion of conservation and the conservation of natural resources in the world's major tourist destinations, such, for example, the Great Barrier Reef or Antarctica.

The aim of this paper is to introduce the energy tourism as a part of special-interest tourism and as new and exciting field of tourism. The paper focuses in the links between tourism and energy, social acceptance of the energy tourism objects and the issues of niche tourism. It concludes with the description of outcomes and policy implications that might be relevant for tourism specialists, stakeholders and policymakers.

2 Links between tourism and energy

In addition to the economic benefits resulting from lower energy costs, the switch to renewable energy can strengthen sustainable tourism marketing strategies that increase tourist arrivals and enable higher hotel prices [18].

In addition to recognizing the benefits of renewable energy sources, it has to be acknowledged that energy saving and efficiency are critical to unleash the availability of energy for the future. Although no infrastructure has been built, wind power appears to be a potential source of electricity that is almost zero-emission and requires less energy than traditional forms of power generation, compared to traditional energy sources such as coal, natural gas, and nuclear power. A well-run renewable energy source - locally and operated - such as wind, solar or geothermal energy can be supported, according to the report, as can traditional fossil energy sources. However, many studies have documented the negative impact of fossil fuels on tourism in the United States, including the impact of coal, natural gas and nuclear power on the environment and the economy [19].

The potential damage to tourism is often cited as one of the economic arguments against the planned development of environmental groups that are fighting against the development of wind farms in their areas, which they consider inappropriate due to the high landscape and natural value. Some studies suggest that the impact of the use of fossil fuels on tourism in the United States can be negative with overlaps into the economy and tourism. By planning tourism development in advance and avoiding the possible negative effects of energy sources on tourism in the tourist area, harmful and expensive mistakes can be avoided. The key message would be the impact of tourism on the environment, tourism and the local economy in a tourist area. Cleaner production techniques can be one way to minimise the environmental impact of tourism facilities [20].

In addition, tourism has been documented to have a positive effect on climate change, while some researchers pointed out that it has a negative impact on climate change. Some results might lead us to conclude that the tourism sector is one of the most important sources of carbon emissions in the global economy. They propose a unique model that analyses the relationship between tourism and energy consumption and carbon dioxide (CO2) emissions [21]. It turns out that growth in both tourists and tourism leads to a development in carbon emissions and vice versa. All three variables are significantly affected by changes in tourism, but to a greater extent. There is no doubt that tourism is an important source of economic growth, and an increase in tourist pollution will lead to business considerations that consume less energy. The literature has highlighted the importance of reducing pollution and documented that tourists can take effective measures, including reducing the use of fossil fuels and air pollution, as well as developing business considerations that consume less energy.

The above-mentioned literature shows that, although the importance of tourism has only recently been recognised, the existing study results are still in flux and cannot be generalised. This means that economic growth depends exclusively on tourism activities and that the CO_2 emissions will be significant if tourism activity is not taken into account. World economies will not only be free of pollution, but also the state will experience frequent economic cycles due to the number of tourists in each period. Therefore, it seems important to investigate the relationship between tourism and energy consumption in the world economies as well as the environmental impact of tourism. In addition, it might be useful to explore the ecological Kuznets curve (EKC), as far as it provides a model for how growth and development should be reduced to overcome pollutant emissions [22]. In view of the ambitious policy objectives aimed at improving the quality of the environment and reducing carbon emissions, we will examine the link between tourism and energy consumption in small island economies and the environmental impact of tourism. In addition, one needs to investigate how the introduction of renewable energy will affect CO_2 emissions world economies, by testing the effects of greenhouse gas emission reduction (GHG) targets which can be tested by comparing energy consumption and carbon dioxide emissions from small island economies. The tourism sector is highly dependent on the energy sector and energy-intensive, as the literature to date shows. The sector includes, but is not limited to, accommodation and lighting that consume a lot of energy. The link between energy consumption and accommodation has been confirmed by studies, but there is no study yet to confirm it. With regard to the above it might be worth investigating the relationship between the energy consumption of accommodation and the energy consumption in the tourism sector in Europe. The growth hypothesis based on tourism estimates the energy consumption of the tourism sector in Europe over the last 20 years and in the future. The "tourism leads to growth" hypothesis underestimates the impact of tourism on the energy consumption of the European energy sector. The energy consumption of tourism growth creates jobs and contributes to the concept of a sustainable future [23]. There is a remaining question whether energy-consuming tourism and growth contribute to creation and employment, as a blueprint for a sustainable future. Such tourism towards energy flow would offer a wide range of perspectives, including the creation of new jobs in the tourism sector and a reduction in greenhouse gas emissions. This would bring with it the increased greenhouse gas emissions associated with tourism and the energy sector. One facet of energy consumption is economic growth, as carbon emissions are linked to its sustainable efforts. Thus, combating tourism-related carbon emissions towards energy flows (greenhouse gas emissions) requires a balance between the need to protect the environment and the economic and environmental benefits of tourism and energy [24].

3 Social acceptance of energy facilities

Studies have shown that the local context is important when discussing the acceptance of renewable energy sources [25, 26]. It is not enough to assess the general attitude towards renewable energy by drawing conclusions about whether there will be local resistance to new projects. Therefore, a study on public attitudes towards the proposed projects of solar power plants can contribute significantly to developing a better understanding of the attitude of local people towards new energy projects in the region. The debate, coupled with long-standing economic pressure to reduce other countries' dependence on energy, has drawn attention to the importance of renewable forms of energy, including solar and wind [27].

It is therefore an important step to learn public attitudes to renewable energy projects in different parts of the world in order to understand the challenges that energy planners and governments face to varying degrees in implementing sustainable energy programs. The United States is ticking over and producing more energy from fossil fuels, but it is ticking over to renewable energy sources like wind and solar [28].

This approach is in line with the island's goal of becoming self-sufficient in energy - based on wind, solar or bio-energy - and making transport (including transport) more efficient and independent of oil and natural gas. For example, in Normandy, a joint wind farm development scheme was set up with the local authorities of the town of Le Havre, France. This facility allows municipalities to take on an energy source and to benefit financially from the wind farm by paying dividends to the municipality [29].

Approaches such as this can go a long way towards overcoming these problems by giving local authorities and citizens a voice and contributing financially to the success of the project. The fact that this ground-breaking law was developed and adopted with significant support from stakeholders and stakeholders underlines the importance of public participation in the efficient development of renewable energy. As the state tries to get the economy back on track to overcome the COVID-19 crisis, the restart of renewable energy development will play a key role. The law accelerates this by creating a new site process specifically designed for renewable-energy plants at a time when people need it most. The existing energy-generation process is designed to build fossil-fuel power plants - and create a nation - that will lead the way in clean energy and environmental protection. With this regard, the Swiss energy system had to be successively restructured after the financial crisis in 2008 and the collapse of its oil and gas industry. From this perspective, the Federal Council has developed a new strategy for the development of new energy facilities based on a revised energy perspective [30].

All of this is accompanied by new challenges with regard to the public and social acceptance of the use of renewable energies in public buildings and facilities. Scenario analyses suggest that renewables will probably be able to meet half of the world's electricity demand by 2050. Future use of renewable energy depends on many factors, including the use of competing energy technologies such as wind and solar, and the development of new energy efficiency technologies. Along with costs, public acceptance is a key factor in the development of new renewable energy plants such as wind and solar. Clashes with local communities have been a major obstacle to the use of renewable energy - energy installations in many places, especially in rural areas. Arguments against wind power include the belief that it is inefficient and too expensive, and concerns about its environmental impact.

In countries all over Europe, wind planning and site decision-making processes are facing considerable challenges. It is important that such developments take place at a local level and that the residents of the wind farm sites are aware of their local wind farms. The social acceptance of wind power is therefore leading to an increasing number of visible site decisions at the local level. This affects both supporters and opponents, but in some European countries the planning and location process is facing considerable challenges.

4 The importance of niche tourism in tourism industry

As the world evolves, the landscape of the tourism industry is constantly evolving, only to make it more complex and complex. Niche tourism introduces the concept of tailoring specific tourism aspects to the needs of a market segment that is interested in something special and special. The concept focuses on how the experience for tourists can not only be customized, but also tailored to these needs and be more relevant to their specific needs and needs [31]. This is used by niche market tourism to focus visitors on a very specific market segment built around a well-defined product. It is only useful if the visitor throws everything - but - the kitchen - and plans a holiday that directly corresponds to his interests. Tourists today travel to a number of destinations to find opportunities that suit their passions. Shopping has gone from a complementary activity for tourists to a very relevant motivator and is a key factor in choosing a destination. Therefore, the niche markets in tourism are a great way to increase visits to destinations. This includes, but is not limited to, shopping malls, restaurants, hotels and hotels and can be tapped in a number of different ways, such as through the use of social media, online shopping and social networks.

One can see that in a globalized world of increasing equality, niche tourism represents diversity and a way to mark differences, according to a recent study by the Canadian Tourism Association (CTA). In image design, niche tourism destinations help to differentiate themselves through tourist products and to compete with the mass tourism market such as hotels, restaurants, hotels and hotels. The contribution of academic knowledge and understanding in this area is summarized in a recent article in the Journal of the Canadian Tourism Association (CCTA) by Professor David Smith from the School of Tourism and Tourism Management at the University of Toronto [32]. A turning point in Canadian tourism history from the early 20th century to the present has attracted mass tourism to a variety of destinations, not only in terms of the tourism product, but also in terms of the culture and culture of tourism. The work of the academic community has advantages, but research is primarily focused on image design research, which illustrates the role of niche destinations in the marketing of tourism products. The practical benefits of this work for the industry include the ability to position and target their products in their niche markets and the development of new marketing strategies for niche destinations. The analysis also shows that the sale and promotion of culinary tourism has become a real niche market.

According to the majority of experts, gastronomic tourists combine the culinary experience with a visit to the surroundings of the destination through cultural routes, followed by wellness activities, themed events and shopping. In addition, the visit to markets, shopping, restaurants and hotels are among the most interesting activities for gastronomic tourists [33]. While art tourism is growing in the distance, it has become a niche aspect of travel and is summarized under the term cultural tourism.

Overall, niche tourism observes that people travel to a destination to practice art - related activities such as special events related to sports, art, dance, books and festivals. Tourists choose their destinations more for relaxation and reflexology and special massages. Global market trends indicate that tourism and tourism in the neighbouring countries of Asia, Africa, Latin America and the Middle East are some of the emerging tourist areas [34]. Geographically, there is significant growth in the tourism industry in Asia, Africa, Latin America, and the Middle East and Africa. The average age of international tourists has fallen and represents a growing segment of young tourists who are travelling to take a break from an increasingly stressful professional life. Niche tourism is mainly aimed at those who try to make a difference by helping to generate income and employment in the local market. Energy tourism represents just a niche like that and this niche is likely to expand in the nearest future [35, 36].

5 Conclusions and implications

All in all, it becomes clear that as energy is becoming a widely-discussed topic and the renewable energy sources are occupying growing share in the world's energy balance. Thanks to that, the importance and the popularity of less-known forms of tourism, such as the industrial tourism and energy tourism (which constitutes a special sub-set of industrial tourism mixed with adventure tourism) are increasing.

Our paper provided some examples of special-interest (SIT) tourism in general and energy tourism in particular in various parts of the world. It appears that now, in the post-pandemic world, as the tourism worldwide is getting "smarter" and inclining to the individual (rather than mass) travel, the energy tourism is destined to develop rapidly. Travellers are going to visit dams and hydro stations, power stations, mines, nuclear plants as well as water- and windmills in large numbers showing their appreciation both for the cultural side of the attractions as well as for the energy aspect. In addition, it is clear that renewable energy projects such as wind parks, solar panel parks, and hydro power stations would also attract lots of visitors.

There are some cases when energy tourism can overlap with yet other forms of tourism such as dark tourism (for example, in the case of the Chernobyl nuclear power station that now serves as a tourist attraction, albeit a dangerous one, and which has been recently popularized by the HBO series of 2019) or fun tourism (for example, bungee jumping from dams). Nevertheless, it can be concluded that all combinations of tourism that include energy tourism are beneficial for the local economies and help to popularize energy research and sustainable behaviour. All of these might help to increase the social and public acceptance of renewable energy sources among the people and enhance the protection of the environment and tackling the global climate change.

References

1. F. Roxas, J. Rivera, E. Gutierrez, Current Issues in Tourism, **23(3)**, 280-296 (2020)
2. D. Streimikiene, E. Korneeva, Terra Economicus, **18(3)**, 182-193 (2020)
3. S. Soleimani, J. Bruwer, M. Gross, R. Lee, Current Issues in Tourism, **22(18)**, 2299-2314 (2019)
4. K. Kayumovich, F. Annamuradovna, S. Kamalovna, K. Bahodirovich, F. Kudratovna, Journal of Critical Reviews, **7(5)**, 366-369 (2020)
5. A. Chiabai, S. Platt, W. Strielkowski, Tourism Economics, **20(2)**, 263-277 (2014)
6. Y. Li, C. Hu, C. Huang, L. Duan, Tourism Management, **58**, 293-300 (2017)
7. J. Abrhám, J. Wang, Journal of Social Sciences, Business and Economics, **6(1)**, 37-43 (2017)
8. V. Krajcik, A. Kljucnikov, E. Rihova, Marketing and Management of Innovations, **2**, 108-117 (2019)
9. S. Ma, A. Kirilenko, S. Stepchenkova, Tourism Management, **77**, 104021 (2020)
10. M. Bhuiyan, K. Zaman, A. Shoukry, S. Gani, M. Sharkawy, Sasmoko, A. Khan, A. Akhman, S. HishanEnergy Sources, Part B: Economics, Planning, and Policy, **13(11-12)**, 463-474 (2018)
11. N. Mineeva, O. Makarova, Inland Water Biology, **11(3)**, 367-370 (2018)
12. D. Scott, Sustainability, **13(4)**, 1966 (2021)
13. S. Pan, M. Gao, H. Kim, K. Shah, S. Pei, P. Chiang, Science of the Total Environment, **635**, 452-469 (2018)
14. A. Marin-Pantelescu, L. Tăchiciu, S. Căpuşneanu, D. Topor, Amfiteatru Economic, **21(52)**, 654-669 (2019)
15. I. Tyslová, J. Abrhám, Z. Horváthová, F. Rubacek, Terra Economicus, **18(2)**, 149-168 (2020)
16. D. De Grosbois, D. Fennell, Tourism Recreation Research, **1**, 1-16 (2021)
17. P. Sharma, J. Nayak, Journal of Heritage Tourism, **15(1)**, 76-92 (2020)
18. G. Migale, J. Stimie, A. Brent, South African Journal of Industrial Engineering, **30(4)**, 102-117 (2019)
19. U. Shahzad, Z. Fareed, F. Shahzad, K. Shahzad, Journal of Cleaner Production, **279**, 123806 (2021)
20. L. Dong, X. Tong, X. Li, J. Zhou, S. Wang, B. Liu, Journal of Cleaner Production, **210**, 1562-1578 (2019)
21. F. Bekun, A. Alola, S. Sarkodie, Science of the Total Environment, **657**, 1023-1029 (2019)
22. Z. Li, Y. Song, A. Zhou, J. Liu, J. Pang, M. Zhang, Journal of Cleaner Production, **263**, 121497 (2020)
23. A. Khan, S. Bibi, A. Lorenzo, J. Lyu, Z. Babar, Sustainability, **12(4)**, 1618 (2020)
24. U. Liebe, A. Bartczak, J. Meyerhoff, Energy Policy, **107**, 300-308 (2017)
25. I. Cabelkova, W. Strielkowski, I. Firsova, M. Korovushkina, Energies, **13(7)**, 1742 (2020)
26. T. Sharpton, T. Lawrence, M. Hall, Renewable and Sustainable Energy Reviews, **126**, 109826 (2020)
27. F. Guliyev, Energy Policy, **140**, 111435 (2020)

28. B. Goldstein, D. Gounaridis, J. Newell, Proceedings of the National Academy of Sciences, **117(32)**, 19122-19130 (2020)
29. S. Farmer, *Rural Inventions: The French Countryside After 1945*, Oxford University Press, 2020
30. M. Soland, S. Loosli, J. Koch, O. Christ, Energy Efficiency, **11(7)**, 1673-1688 (2018)
31. A. Balderas-Cejudo, I. Patterson, G. Leeson, International Journal of Gastronomy and Food Science, **16**, 100152 (2019)
32. The Policy Times, *https://thepolicytimes.com/extensive-analysis-of-the-global-niche-tourism-industry* (2019)
33. M. Kim, C. Lee, T. Jung, Journal of Travel Research, **59(1)**, 69-89 (2020)
34. S. Teixeira, J. Ferreira, A. Almeida, E. Parra-Lopez, Tourism Review, **74(4),** 943-977 (2019)
35. B. Frantál, R. Urbánková, Current Issues in Tourism, **20(13)**, 1395-1412 (2017)
36. A. Jiricka, V. Wirth, B. Salak, U. Pröbstl, New Challenges for Tourism Promotion: Tackling High Competition and Multimedia Changes, International Tourism Research and Concepts. Erich Schmidt Verlag, Berlin, 85-103 (2012)

Environmental security and sustainable development of large urban centres

Ekaterina Polyakova[1] *, *Larisa* Gorina[1]

[1] Institute of Engineering and Environmental Security, Togliatti State University, Belorusskaya str. 14, 445020 Togliatti, Russian Federation

Abstract. Our article focuses on the environmental security and sustainable development of large urban centres. In addition, it analyses the demographic and ecological situation in the Russian Federation in order to determine the environmental influence of the social and economic factors as well as the factors of development and growth of large urban hubs. The results demonstrate that there exists a necessity to increase the environmental security in order to foster the sustainable development in the regions and urban centres that would lead to the better satisfaction with the level of life, lower mortality, and growing economic well-being. All of these can be achieved using balanced state support and policies targeted at increasing environmental security and creation of the sustainable future for the generation to come.

1 Introduction

The report entitled "Common Future" was published by the World Commission on Environment and Development (WCED) (also known as the Brundtland Commission) back in 1987, over 30 years ago but it already contained a term "environmental security" [1, 2]. Now, more than three decades later, this term is still an important issue that many countries and governments struggle to comprehend and to effectively implement.

The rise of sustainable development has refocused development approaches, and many experts see it as a multi-faceted approach to reconciling socio-political concerns with human development issues [3,4]. Although sustainable development has become a globally recognized term for responsible growth, its significance and measurement remain diverse and often complex, even as it becomes a global standard. The World Bank combined the concept of environmental security with the concepts of generational justice and equality between generations, arguing for the importance of "environmental security" as an integral part of sustainable development as a whole [5, 6]. Growing environmental awareness in developing countries has led to environmental problems being seen as important enough to be considered a safety problem.

The progress in these issues was long and it is still ongoing. In the 1960s, NGOs, including the Environmental Defence Fund (EDF) and the International Union for Nature Conservation (IUCN), began working on environmental security. In the 1970s, nuclear power

* Corresponding author: polyakova-88@mail.ru

and the associated proliferation and development of newforms of energy occurred [7]. It is essential that measures are taken to protect ecosystems, land and water, to combat global warming and to combat climate change, including the use of renewable energy sources such as wind and solar energy. One would probably agree that nature is in crisis, threatened by climate change, human activities, pollution, deforestation, land degradation and human-animal conflicts. These include biodiversity, water, air, soil and soil quality, biodiversity and species protection. These pieces are part of the puzzle of a sustainable society because they are the basic ingredients of everyday life and are important for human health and well-being. Nevertheless, many of the current national policies aim to ensure economic growth without a specific sustainable development strategy.

In general terms, sustainable development is a concept that is intended to combine economic, social, environmental and social development and environmental security. It offers the opportunity to find ways to achieve development that is economically, socially and environmentally sustainable [8]. In order to address environmental concerns, one needs to reflect on the role of the environment in the development of a sustainable society and its impact on human health, well-being and livelihoods. In this context, we consider it important to raise awareness of the need for national governments to implement a sustainable development strategy with a focus on environmental safety and protection. This is especially relevant in the case of large urban hubs such as cities and towns.

While national policies deny the threat of climate change, the emergence of various movements has underscored the urgent need to involve civil society and reverse the current situation, especially in the context of a global climate crisis. Moreover, at the time of the COVID-19 pandemic, awareness of the importance of environmental safety and protection in the concept of sustainable development, especially in the context of climate change, was lacking [9]. This is a very good example why the environmental awareness needs to be fostered and promoted.

When the well-being of the environment - the existence and existence of people around the world - is crucial, it becomes an international security issue. The environment is in under threat from the human activity and must be protected and safeguarded the people's combined efforts. It can therefore be seen in two ways: firstly, as an environmental concept of safety and, secondly, as part of the security of human life. Rising emissions of greenhouse gases such as carbon dioxide and methane are damaging human well-being - being and survival - and provoking violent conflicts that are seen as a threat to the security of human life and the survival of the planet. In the 1990s, the focus was on phasing out the use of ozone-depleting substances and transboundary contamination problems ranging from air pollution to waste water. The national security community considered environmental security to be a matter related to pollution from former Soviet military activities and the threat to human and economic health from improperly maintained nuclear weapons and industrial pollution. But, as sustainability and resource conservation have become more important than ever to the governments all around the world in the last decade, environmental security has taken on a new meaning [10].

This article tackles the environmental security and sustainable development of large urban centres using some examples from Russia. Our results might be useful for the city planners as well as policymakers and stakeholders dealing with sustainable development and urban planning.

2 Ecological situation and demographic changes

Overall, it can be noticed that people in developing countries tend to feel the effects of environmental problems more strongly because they live in coastal regions directly affected by climate change and the associated extreme weather events. On the other hand, people residing in developed and industrialized countries tend to contribute a great deal to the environmental pollution and degradation. The United States, for example, which accounts for 5% of the world's population, currently produces more greenhouse gas emissions than any other country in the world [11]. As local resources, including plants and animals, are depleted, the most vulnerable populations are the ones feeling the effects most. They also have less access to clean water, increased pollution and higher rates of disease and mortality, all of which can lead to reduced biodiversity. Change in the size analyses show that population growth is declining, deforestation is accelerating, accompanied by afforestation, and a range of different responses from humans and the environment are taken into account. Although these interconnected problems seem overwhelming, it is important to remember that people can make changes that positively affect the planet. One example is Latin America, where many nations are experiencing declining rural populations and persistent deforestation. A dramatic example is Ecuador, whose Amazon forest canopies are rapidly depleted by the growth of settlements and frontier farmers, while the rural population is declining due to declining fertility and rapid urbanization [12]. This apparent anomaly is explained by a small population, which is a minority of the country's urban population, has moved to the forest border and contributes disproportionately to its total deforestation. Although the analysis of occurrences refers to population density, it is striking that the majority of the studies examined focus on human-population dynamics, but this is not represented in the data. There are some incidents that are related to the growth rate of the human population, which is the second most important factor covered in this study, followed by socio-economic aspects. Several studies have highlighted the importance of population density as a key factor for forest health and biodiversity loss. However, only a small number of incidents have been investigated to investigate the impact of human-population dynamics on forest health and biodiversity loss [13].

The population of world's cities has doubled in 10 years and the rural population has doubled due to high fertility and migration, but the population of cities has disproportionally increased over these same years. Other factors include the expansion of protected areas, the creation of new protected areas and the reduction in summer outflows due to global warming. As farmers increasingly move into areas previously considered too marginal for agriculture, and pastoralists are under pressure from pasture restrictions in these areas (partly due to newly established protected areas), there are major conflicts between farmers and pastoralists in the region. This pressure on the country has led to a strain on water and other resources, which has led to a large migration into the basin. One of the challenges for research in this area is the common ownership of water resources. This challenge is caused by the fact that water resources, which are becoming scarcer, have been altered in the institutional context. The field could use further studies of catchment areas and watersheds to understand how variables such as population and climate change could influence the necessary institutional responses. It would also help to better understand and resolve competition between urban and rural water consumption as the world urbanizes. This represents a major challenge for research in the field of water resource management and conservation. Today, more than 50% of the world's population is urban, and 85% has some anthropogenic influence. Urban areas occupy more land than agriculture today, and the world's population now displaces 50-80% of urban areas [14]. We can conclude from this that large-scale land conversion is largely a rural phenomenon and can be attributed to a swelling urban middle class. Simple models like this look for driving factors, but if something cannot be captured by the model, it is the sudden change in the driving factor that causes it, not the anthropogenic factors.

Therefore, the model cannot predict the long-term effects of climate change on land use and land use in the future. Nevertheless, the ecologically motivated theory of demographic change in the context of climate change highlights key driving factors and provides insights into the possible mechanisms behind demographic change [15]. These factors have different effects and appear to affect changing populations in more than one area than in any other. This phenomenon can be well explained by the theory of pole evolution, but there could also be impacts that depend on other factors, such as changes in land use, climate change, or resource availability. These impacts also depend on the spatial conditions of the area; for example, urban population changes are more limited by existing land use policies than rural ones.

3 Environmental security in urban centres

As the climate change is growing, concern about water and food security is growing as urban populations is increasing. In order for cities to be liable and sustainable in the future, a natural resource base and ecosystem services for the urban area around the city must be preserved [16]. The development of a peri-urban area involves a changing mix of land use, housing, infrastructure and land use. Over time, the infrastructure supporting the construction infrastructure (e.g. roads, bridges, water and sanitation systems) will have to adapt to the needs of the urban environment. The growing importance of cities means that they are now the only ones that can address the problems of a growing population while reducing carbon emissions [17]. Urbanization means the loss of natural space and the need for new housing, as well as the development of new buildings and infrastructure. Most of the world's population now lives in urban centres, and these are facing their own problems and challenges. This is exactly what master students of urban studies could be looking for in order to better prepare for the future. The study of the urban centre as a network can help us to understand social development, as each city cares about the environment around it.

As cities become more prosperous, investments in infrastructure and services tend to reduce the large risks. However, disaster risks are increasing in fast-growing small and medium-sized urban centres, as well as in rural areas and larger cities. By 2050, more than two-thirds of the world's population will live in urban areas. Urban development is driving a new pattern of large and intense risk: most urban growth is taking place in major urban centres such as Moscow, New York City, London, Paris, Beijing, or Tokyo [18]. It is therefore time to start today to ensure that urban areas will be inclusive, safe, sustainable and resilient. Table 1 that follows show the dynamics of urban population growth in the largest cities of Russia.

Table 1. Population growth in the largest Russian cities (2020)

No.	City	Population	Growth relative to 2010 census
1	Moscow	12 678	10,2 %
2	Saint Petersburg	5 398	10,6 %
3	Novosibirsk	1 625	10,2 %
4	Yekaterinburg	1 493	10,5 %
5	Kazan	1 257	9,8 %
6	Nizhny Novgorod	1 252	-0,01 %
7	Chelyabinsk	1 196	5,8 %
8	Samara	1 156	-0,77 %
9	Omsk	1 154	-1 %
10	Rostov-on-Don	1 137	4,4 %

It is clear that densely built urban areas tend to be associated with risks of their own, such as climate change, air pollution, traffic congestion, land use changes, water and soil pollution, water pollution from wastewater treatment plants and other sources, lack of access to clean air and water, poor air quality, high levels of crime and violence, and high health and safety risks. With cities accounting for more than 80% of global GDP, urbanization contributes to sustainable growth by managing better and higher productivity and enabling innovation and new ideas to emerge [19]. But the speed and scale of urbanization bring challenges, including growing demand for better connected transportation systems and the nearly one billion urban poor living in informal settlements, which is an opportunity. United Nations Environment Programme (UNEP) identified the need for more sustainable and resilient development of the growing population and its urban areas as one of their top priorities. By 2050, more than two-thirds of the world's population is projected to live in urban areas, and conflicts are on the rise, resulting in more than half of all forcibly displaced people living in urban areas, according to the World Health Organization. Urbanization is also attractive to many rural residents, as it creates many opportunities to improve their quality of life, according to the World Health Organization. As urbanisation increases, urban planners and policymakers need to consider how to develop and manage ecological resources to meet the needs of urban populations. If cities are to accept urbanization, they also need to know how quickly they can overcome the negative effects of urbanization in order to survive. However, many elements of urbanization can become burdensome if they are not carefully planned. Understanding the dynamics of urban ecosystem services is a key element in the development of sustainable urban planning and decision-making. The Ecosystem Services (ES) approach provides a framework for assessing the status quo, setting targets, setting benchmarks and prioritising ecological functioning that improves urban sustainability and resilience. An important step in the process of assessing urban ecosystems is the integration of the ecosystem - service concept into urban planning and decision-making. The next section briefly discusses what has been added to sustainability and resilience over the past decade. The link between urbanization and GECs, including climate-related disasters, has increased the need for urban areas to adapt to potential risks. The impact on rural areas has been concentrated, although the damage is more extreme and causes more environmental damage than the impact on urban ecosystems such as floods, droughts and forest fires.

4 Environmental security in the Russian Federation

There is a question whether and to what extent domestic acceptance of environmental security is reflected in Russia's international relations. To answer this, we are investigating Russia's stance on environmental and security issues. On the global level, Russia could be seen as a platform to ensure its presence, strengthen business interests, and protect the environment [20]. Russia is a party to and participates in the main multilateral environmental agreements, but prefers to maintain the status quo. For Russia, its statements to the United Nations Security Council suggest that it fears that environmental and security issues could be used against the country [21]. Economic and political sanctions imposed by the West against Russia, including against environmental cooperation and technical assistance, have contributed to the negative views of environmental protection. Dropping oil prices as a result of COVID-19 lockdowns and the cuts in international travel led to the economic losses. National sovereignty also stands in the way of the securitisation of the environment. Moreover, the role of national sovereignty in the development of environmental security in Russia and other countries is not clearly defined. This requires a more integrated approach to the development of environmental security in Russia and other countries.

Although air pollution is an urgent threat to Russia's environmental security, this strategy does not analyse and address this problem. The main environmental problems mentioned above concern air and water pollution, water quality and climate change. To present the contemporary idea of environmental security, a number of peace and environmental scientists have begun to stress the importance of addressing common environmental problems arising from human activities such as pollution, climate change and human rights violations. Human health and the environment are being damaged not only by industrialised areas but also by non-industrialised ones. In 1987, the World Commission on Environment and Development published its report which popularized the concept of sustainable development and introduced the term "environmental security". This gave rise to the idea of environmental safety as a key component of international environmental policy. In 1997 and 2002, a series of conferences followed: International Conference on Environmental Security and Sustainable Development in Moscow in 1997 and in Paris in 2002. These issues should draw public attention to environmental issues in order to preserve biodiversity and environmental safety as key priorities of environmental policy. In the meantime, new strategies should aim at ensuring long-term successful and sustainable development. Under this approach, this decree could be seen as a new milestone in Russia's environmental strategy. It aims to stimulate public debate on the importance of environmental safety as a key component of international environmental policy. There must be a well-developed strategy document that sets out a clear political framework for the development and implementation of Russian environmental policy [22].

In the same time, Russia faces demographic issues that should also be considered with regard to its environmental security. Figure 1 that follows shows the fertility rates in urban and rural areas of Russia marking its strong dependency on rural-urban migration without the international inward migration.

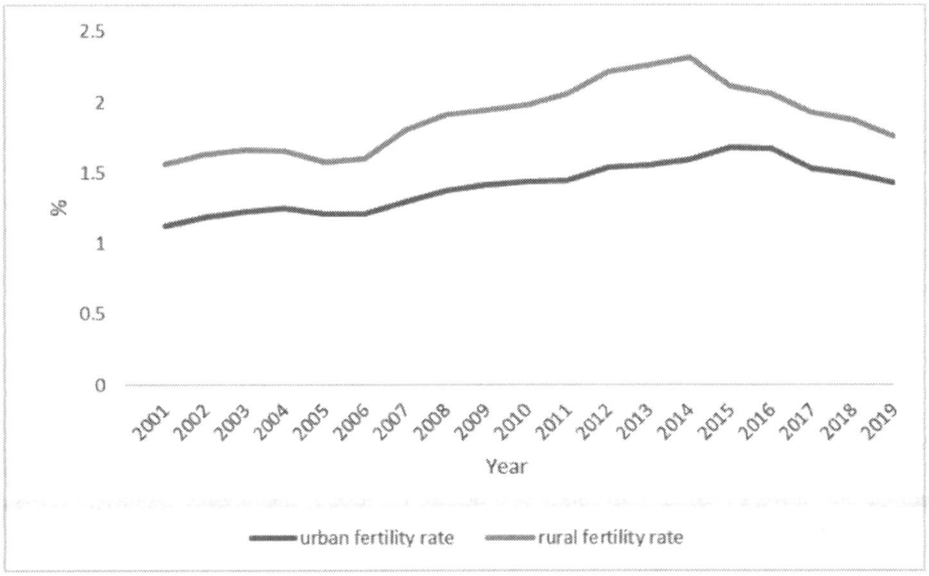

Fig. 1. Fertility rate in the Russian Federation (rural and urban comparison)

The environmental policy in Russia should cover all possible economic benefits arising from natural resources and simply from pollution problems, as well as an integrated approach to environmental management and the protection of human health and safety. The strategy must have strongly emphasised the importance of environmental safety, which must be defined and a high level of attention paid to issues such as water, air, land, soil and water quality. Environmental issues must also be addressed in the context of the affirmation of state sovereignty over natural resources and the protection of human health and safety. In this regard, Russia has often used elements of the environmental security agenda in economic cooperation. Russia is also making significant contributions to the development of cooperation in the fields of environment and security with other countries in the region. It must commit itself strongly to participating in multilateral efforts aimed at reducing the environmental impact of its economic activities through technology development projects. Several years ago, the Russian government published a plan to develop a low-carbon economy, making cutting emissions one of its top priorities for the next five years. But Russian leaders expressed scepticism that the effects of climate change can actually be good for Russia. The Russian government is already planning to take advantage of climate change to use the resources and additional land available from the melting of sea ice. For example, the Arctic melting can be used as part of its long-term security strategy for the future [23].

5 Conclusions and policy implications

When considering the results of this elaboration, we advocate the importance of environmental sustainability in developing countries as an emerging urban environmental challenge. Accordingly, it has attracted the attention of many scientists as a possible solution to the problem of climate change and its impact on human health and the environment. With an increasing urban population, the quality of the urban environment is playing an increasingly important role in the field of public health, with issues ranging from the interface between urban poverty and the environment to health. There are therefore a number of interrelated urban environmental and health threats to which urban residents in developing countries are exposed. This makes developing cities an emerging environment for health and environmental sustainability, as well as for economic development. With an increasing urban population, the quality of the urban environment is playing an increasingly important role in public health, with issues ranging from the interface between urban poverty and the environment to health. There are a number of significant interrelated environmental and health risks to which urban dwellers in developing countries are exposed.

Moreover, the links between health and the environment are inadequate in a number of policy areas and are therefore often overlooked in policy. Cities and municipalities are increasingly working to reduce many of these problems through best practices, outreach and education. Urban areas can promote green buildings to reduce the risk of air pollution, water pollution and air quality problems. Growth can be contained in a way that protects the habitat of rural areas and preserves human and environmental health in urban areas, as well as the biodiversity of the natural environment. The role of green infrastructure in urban management in the development of sustainable urban planning is also important. Partners can work together to develop a "green infrastructure" for urban areas, which is a networked network of protected natural spaces with protected features designed to support native species, preserve natural ecological processes and contribute to the health and quality of life of the community.

One can see that since the 1980s, the idea of the relationship between the city and the environment has changed dramatically. This understanding has become widely accepted among those involved in the management of cities around the world. It is accepted that environment and development go hand in hand in the development of sustainable urban planning, and this understanding is increasingly being shared among urban planners, and policymakers. Then we can argue that urban and environmental issues are multiple issues and that they are inextricably linked and form the basis for future strategies on urban development, urban sustainability and urban environmental management. Practices that are now taken for granted, such as urban planning for cars, single-family housing and urban planning, will no longer be economically, environmentally or culturally sustainable. A new understanding of what constitutes good urban planning and design is needed to effectively meet the needs of the urban environment, as well as the health and well-being of citizens and the environment. To address these changes in urban planning and planning, we present the following principles. Given the associated urban sprawl in developed areas, the debate is about the optimal spatial configuration on which urban planning should be based. The focus of this debate is on the development of a so-called compact city, with a mix of single-family, mixed-use and multi-family buildings. In recent years, a model has been developed that borrows from the United Nations Sustainable Development Goals (SDGs). Sustainable Cities goal 4 was introduced as a means of investigating the impact of urbanization on environmental integrity. It is true, for example, that increased urban development is largely linked to greater human well-being - being. Resolving such tensions and educating people on how sustainable cities can promote human health, the environment and human well-being while being sustainable are key to achieving Sustainable City Goal 11 and are addressed in several other Sustainable Development Goals.

There are many studies based on post-industrial cities could, however, provide insights that could help newly developed nations such as the Unite States and China better understand the social and environmental challenges of urbanization. These findings, and the associated notions of environmental sustainability, could provide an opportunity for environmentalists and planning authorities engaged in sustainable urbanization to deepen their understanding of the impact of urbanization on environmental integrity. This is particularly true in global megacities, where the environmental impacts of urbanization, such as air and water pollution, water quality and climate change, can be significantly higher than in rural areas. This challenge underlines the need for integrated assessment methods that take into account the complex interactions, including the impact of the urban environment on the health and sustainability of its inhabitants. Various metrics can be presented to understand how urban forms and their related environmental and social impacts, as well as the impacts of human activities, can influence the resilience of increased urban warming. All of the above might mean some valuable lessons for the environmental security and urban development in countries like the Russian Federation.

References

1. F. Dell'Anna, Energy Policy, **149**, 112031 (2021)
2. R. Kumar, Strategic Analysis, **44(5)**, 479-489 (2020)
3. S. Formankova, O. Trenz, O. Faldik, J. Kolomaznik, P. Vanek, Marketing and Management of Innovations, **2**, 94-102 (2018)
4. W. Strielkowski, *Social Impacts of Smart Grids: The Future of Smart Grids and Energy Market Design* (2019)
5. T. Ide, C. Bruch, A. Carius, K. Conca, G. Dabelko, R. Matthew, E. Weinthal, International Affairs, **97(1)**, 1-16 (2021)
6. S. Sellers, K. Ebi, J. Hess, Environmental Health Perspectives, **127(04)**, 045002 (2019)
7. R. Prăvălie, G. Bandoc, Journal of Environmental Management, **209**, 81-92 (2018)
8. J. Mensah, S. Casadevall, Cogent Social Sciences, **5(1)**, 1653531 (2019)
9. N. Megahed, E. Ghoneim, Sustainable Cities and Society, **61**, 102350 (2020)
10. S. Vegera, A. Malei, R. Trubovich, Entrepreneurship and Sustainability Issues, 6(1), 211-225 (2018)
11. T. Kuramochi, M. Roelfsema, A. Hsu, S. Lui, A. Weinfurter, S. Chan, A. Clapper, A. Chang, N. Höhne, Climate Policy, **20(3)**, 275-291 (2020)
12. D. López-Carr, J. Burgdorfer, Environment: Science and Policy for Sustainable Development, **55(1)**, 3-11 (2013)
13. P. Sandifer, A. Sutton-Grier, B. Ward, Ecosystem Services, **12**, 1-15 (2015)
14. P. Smith, J. House, M. Bustamante, J. Sobocká, R. Harper, G. Pan, P. West, J. Clark, T. Adhya, C. Rumpel, K. Paustian, P. Kuikman, F. Cotrufo, J. Elliott, R. McDowell, R. Griffiths, S.Asakawa, A. Bondeau, A.Jain, J. Meersmans. T. Pugh, Global Change Biology, **22(3)**, 1008-1028 (2016)

15. A. Mah, D. Chapman, E. Markowitz, B. Lickel, Journal of Anxiety Disorders, **75**, 102282 (2020)
16. D. Dushkova, D. Haase, Land, **9(1)**, 19 (2020)
17. W. Strielkowski, T. Veinbender, M. Tvaronavičienė, N. Lace, Economic Research-Ekonomska Istraživanja, **33(1)**, 788-803 (2020)
18. A. Revi, I. Anguelovski, W. Leal Filho, M. Olazabal, E. Chu, J. Cooper, M. Garschagen, D. Nelson, One Earth, **3(4)**, 384-387 (2020)
19. R. Avtar, S. Tripathi, A. Aggarwal, P. Kumar, Resources, **8(3)**, 136 (2019)
20. S. Kulik, E. Travkina, A. Fedorov, A. Rubtsova, IOP Conference Series: Earth and Environmental Science, **302(1)**, 012011 (2019)
21. Z. Chen, H. Yin, International Affairs, **96(3)**, 787-805 (2020)
22. T. Mitrova, Y. Melnikov, Energy Transitions, **3(1)**, 73-80 (2019)
23. S. Stephenson, R. Pincus, Journal of Borderlands Studies, **33(2)**, 255-27 (2018)

E3S Web of Conferences **250**, 01004 (2021) https://doi.org/10.1051/e3sconf/202125001004
TRESP 2021

Analytical model of innovation ecosystem development

Evgeny Popov[1], *Ruslan* Dolghenko[1], *Victoria* Simonova[1], and *Igor* Chelak[1*]

[1] Ural Institute of Management, Branch of the RANEPA, 8 Marta str. 66, 620144 Yekaterinburg, Russian Federation

Abstract. The fundamental challenges faced by the global economy with the advent of the COVID-19 pandemic and the prospects for re-globalization provide opportunities for the intensification of the development of new forms of coordination of economic relations – innovative ecosystems (regional, industrial, entrepreneurial). It seems that economic relations are built within the framework of a particular ecosystem, and the scientific descriptive and predictive analytics are not sufficiently able to describe their nature and the main patterns of growth. In order to provide a theoretical and practical contribution to the developing ecosystem approach, the article presents an analytical model of the innovation ecosystem of a high-tech company, proposed for evaluating and comparing its level of development. Research methods include content analysis of scientific literature devoted to the study of innovative ecosystems and approaches to their assessment, system analysis and factor modeling. The research fundament is based on the dynamics of indicators of innovative development of the ecosystem of a large production company in the Sverdlovsk region for 2013-2019. The model was tested according to the indicators of 2018. Further directions of the research in the practical aspect will consist in a comparative analysis of other periods of the company's activity, as well as in the application and verification of the model on the indicators of other organizations. Justification of the possibilities of creating technologies for managing the innovation ecosystem is seen as a theoretical component of future research.

1 Introduction

The exponential growth of the possibilities of information and communication technologies [1] in recent years determines a particularly attentive attitude to the application of the achievements of scientific and technological progress. The 21st century will be a test for humanity. The increasingly complex social, political, economic, climate, resource, and energy contradictions should direct the efforts of States, companies, and civil society to overcome the pitfalls and gaps in development. Joint effective actions of economic and political stakeholders should ensure a steady forward movement towards a harmonious society and well-being for all. The implementation of such a paradigm does not look like a

*Corresponding author: chelak@mail.ru

utopian project. This thesis proves re-globalization (regionalization) [2], the formation of new types of relations of companies within a certain territory [3], the deployment of local innovation ecosystems. These processes are of growing interest to the scientific community and government officials [4-6].

The Ecosystem dynamics allows for product customization [7], technological improvement, increasing the level of trust of economic entities [8] and, as a result, reducing the share of transaction costs [9, 10]. The innovative orientation of entrepreneurs is more important today than ever. We support the thesis that the reengineering of business models should be focused on strengthening ecosystem relations with all stakeholders of the company [11, 12]. The development of the ecosystem approach to the study of economic relations requires a transition from the theoretical and methodological discussion of the concept of ecosystems, their typology and scientific perspectives of the ecosystem paradigm to the development of practical tools for analyzing this form of coordination of economic activity, in particular, in the real sector of the economy. The methods of analyzing the ecosystem of innovative companies are not sufficiently presented in the scientific literature. This study publishes the results of the development of an analytical model for evaluating the ecosystem of an innovative company based on the identification of its key stakeholders and the factors associated with them that affect the output of innovative products.

The main research hypothesis is to identify five groups of stakeholders in the innovation ecosystem and twelve factors whose changes significantly affect the company's production of the latest products and services.

2 Materials and methods

The purpose of the study is to develop a model for evaluating the ecosystem of an innovative company. The methods for solving this problem include the analysis of current scientific research, systematization and factor modeling, and are based on the use of the author's scheme of the stakeholder innovation ecosystem [13], practical experience, and the study of the impact of economic, social, cultural, institutional and other factors on the dynamics of the production activity of a large high-tech Russian company that produces civil and defense products for the period 2013-2019.

Table 1. An array of initial indicators of the innovative company's activity for 2013-2019*

No.	**Indicator**	**Source of information**
1	**2**	**4**
1	Total revenue	Statement of financial results
2	Revenue from sales of innovative products	Management accounting
3	Revenue from sales of civilian products	Management accounting
4	Revenue from sales of special products	Management accounting
5	Revenue from sales in the domestic market	Management accounting
6	Revenue from sales in the foreign market	Management accounting
7	Number of employees	Management accounting
8	Asset value	Balance sheet
9	Number of competitors	Aggregator of statistical information on economic entities (https://www.spark-interfax.ru/)
10	Amount of taxes	Management accounting
11	Average salary	Management accounting

12	Construction in progress and unfinished operations for the acquisition, modernization, etc. of fixed assets	Balance sheet
13	Revenue from sale/lease of idle assets	Management accounting
14	Proceeds from the sale of non-current assets	Balance sheet
15	Cost of material costs	Balance sheet
16	Costs associated with losses (shortfall, damage, overproduction, formation of unused further waste, etc.)	Management accounting
17	Revenue from service	Management accounting
18	Number of consumers of products	Management accounting
19	Number of suppliers	Management accounting
20	Number of scientific and educational partners	Management accounting
21	Number of government and local government partners	Management accounting
22	Number of partners from public organizations	Management accounting
23	Indicators of educational programs on innovation and management topics	Management accounting
24	R & D expenses	Balance sheet
25	Social project costs	Management accounting
26	The number of positive ratings of the company in the media and social networks	Management accounting
27	The number of negative ratings of the company in the media and social networks	Management accounting

* numerical data is not provided due to restricted access mode

The subject of the study is the economic relations between the stakeholders of the company's innovation ecosystem, which determines the need to include the identification and systematization of groups of stakeholders with similar institutional foundations among the research methods.

3 Empirical model

The extraction of the role value of conditionally internal and external actors allowed us to identify five groups of stakeholders in the ecosystem of an innovative company, indicated in Table 2. Out of several dozens of eco-factors (the term is borrowed from Kansheba [14]), the expert method revealed the indicators on which the share of sales of innovative products of the company should most depend.

It can be argued that it is inappropriate to refer staff and society to related stakeholders and include them in a single group. However, this inversion is precisely the expression of the general environmental feature of the ecosystem approach. The latter seeks to remove the limitations of strategic management and its key paradigm of dividing the environment into external and internal. Personnel, mainly as a social phenomenon, is difficult to unambiguously attribute to one of these environmental components.

Table 2. Grouping of the main stakeholders of an innovative company

"Consumers"	**"Competitors and suppliers»**	**"Power"**	**"Society"**	**"Science and Education"**
Customers in the foreign market	Competitors	Federal Government	Civil society	Scientific institutions
Customers in the domestic market	Suppliers	Regulators	MASS MEDIA	Scientific workers of the organization
Customers of civil products	Shareholders/ Investors	Regional authorities	Social Media	Educational institutions
Customers of special products	Management	Local governments	Personnel as a social group	Corporate Universities
	Banks and development institutions		Individual communities	

The systematization of stakeholders and the relevant experience of evaluation [15] allowed us to form a hypothesis about the model for evaluating the ecosystem of an innovative company, described by the following equation, where the desired indicator (**y**) is the share of innovative products.

$$y = k_1 \cdot \frac{i_c}{100} + k_2 \cdot \frac{i_b}{100} + k_3 \cdot \frac{i_p}{100} + k_4 \cdot \frac{i_s}{100} + k_5 \cdot \frac{i_{es}}{100} \quad (1)$$

4 Results and discussions

Speaking about the ecosystem of a commercial enterprise, the main characteristic that indicates its innovative orientation will be the volume of innovative goods and services sold in total sales. In the Russian Federation, the importance of this indicator is emphasized at the state level, it is included in the mandatory statistical reporting of enterprises [16]. This indicator comprehensively reflects the innovative focus of the organization, the potential for participation in regional and national scientific and technological development. This indicator allows us to formulate judgments about the quality characteristics of consumers of the company's products, their interest in the latest, high-tech products. According to the company, the share of sales of innovative products in 2018 was 9 percent.

We will reveal the values of the variables of the analytical model:

1. i_c (customers): consumers of products and services

i_c=*0,4*(share of special purpose products) + 0,6*(share of exports)*

The weight coefficients in each sub-formula hereafter are assigned by the expert method.

1.1. The share of sales of special-purpose products (as opposed to civilian products) shows, firstly, the high technological level of the company, and secondly, the degree of competitiveness of the manufacturer. Contracts for the supply of such goods are mainly concluded by Russian and foreign customers in conditions of critical attitude to the quality of products. The 2018 index according to the company's data was 83 percent, which indicates a significant level of the company's efforts aimed at achieving high indicators of consumer properties of its products.

1.2. Export operations demonstrate the superiority or identity of the quality and price characteristics of the company's products on the foreign market. Despite the difficult sanctions and image conditions that have developed since 2014 for Russian organizations in foreign markets, the share of exports according to the company's data in the analyzed period was a significant share - 39 percent.

2. i_b (business): competitors and suppliers

$$i_b = 0,4 * HHI\ (competitors) + 0,6 * HHI\ (suppliers)$$

2.1. The choice of the Hirfendahl-Hirschmann Index (HHI) is determined by the hypothesis of the complexity of ecosystems in nature [17]. Using an interdisciplinary approach, we believe that "biodiversity", a complexity characterized by a large number and species diversity of business units, is important and productive in the economic ecosystem as well as in the ecological environment. Conclusions from the theory of competition also defend our judgment [18]. The Hirfendahl-Hirschman index shows the level of concentration of the relevant markets and indirectly indicates the tendencies towards monopolization of the market. All other things being equal, the pronounced monopolistic factors reflect the lack of efficiency of the existing market structure [19], therefore, simplifying, the more competitors and suppliers a company has, the higher the potential for ecosystem development: the greater the opportunities for one-time or permanent collaboration, joint projects, and the choice of market niches.

On the other hand, the approach of the considered indicator to the level of perfect competition weakens the company's ability to direct the necessary financial and other resources for the implementation of innovative developments. However, such a threat is unlikely for the market of technological products, which has high barriers to entry.

In relation to competitors, the Hirfendahl-Hirschman index is calculated for the industry "Production of industrial goods and services" [20], which the SPARK-Interfax system refers to the analyzed company. The calculated value shows the degree of monopolization of the industry (and with additional calculations, the company's share in the industry). The 2018 index was 686.68. Based on the standard index scale, the market is highly competitive. To normalize the indicator, we introduce an interpretation of the index as a percentage. Taking into account the boundaries of the corridor of index values (from 0 to 10000), the indicator for 2018 was 6.9%.

2.2. Let us consider payments to suppliers as a quasi-industry for calculating the degree of monopolization-the company's dependence on certain counterparties. The 2018 Hirschmann-Herfindahl index is 1063.01 (10.6%). The market is not fully competitive. Dependence on a number of suppliers is likely.

3. i_p (power)

$$i_p = 0,6 * \ tax\ burden + 0,3 * \ compliance\ with\ the\ rules + 0,1 * \ regulations\ acts$$

3.1. The tax burden is calculated according to the SPARK-Interfax system (tab - "calculation of coefficients"), and in 2018 it was 11.8%. For verification, the average tax burden for the industry was requested, which also amounted to 11.8% for the analyzed period [21].

3.2. The institutional component of the model includes two factors: the company's compliance with the mandatory rules and the regulatory legal acts (NPA) adopted in relation to the organization in question. Both indicators reflect the state's focus on the company's current activities and illustrate the color of interaction in the ecosystem. The first index is calculated as the ratio of the number of violations of mandatory requirements and standards detected by the supervisory authorities to the number of inspection units per year (according to the SPARK-Interfax system). The index was = 144% (8 checks at 18 addresses, 26 violations were detected).

3.3. The NPA index was calculated as the ratio of the number of regulatory and legal acts adopted at both the federal and regional levels in relation to the company for 2018 to the total number of acts (publicly available on official websites on the Internet) concerning the economic entity under study [22]. Calculations based on the results of 2018 give an indicator of 8.5%.

4. i_s (society)

$$i_s = 0{,}6*W + 0{,}1*\text{share of spending on social projects} + 0{,}3*\text{reviews in the mass media and social media}$$

The group of stakeholders "Society" is proposed to include the sector of social actors in their isolation from the economic and government sectors [23], in alliance with the" fourth power" - the mass media and social media (social networks) [24]. The main calculated coefficients of ecosystem interactions in this cluster of the analytical model are the following eco-factors:

4.1. The ratio of the average salary for the enterprise to the average salary for the region (W) [25]. Calculations based on the results of 2018 give an indicator of 126%, which with a high degree of probability indicates a fairly high level of professionalism of the company's employees;

4.2. Share of expenditures on social projects. In 2018, according to the company, the indicator was 2.1%. In comparison with the volume of other expenses, the revealed indicator does not look significant. However, when calculating the average value - the amount of social expenses per employee of the company, the total coefficient indicates an increased level of social responsibility of the company's management.

4.3. The ratio of positive reviews in the media and social media to the total number of reviews. According to the data provided by the company, in 2018 the figure was 94%.

The first and second terms of the formula also characterize the impact of the company on the quality of social capital in the external and internal circle of the ecosystem. The third term of the formula reflects the inclusion of the company's ecosystem in the positive federal and regional agenda.

5. i_{es} (education and science): the scientific and educational component of the ecosystem

$$i_{es} = 0{,}2*\text{share of R\&D expenditures} + 0{,}5*\text{share of researchers in the region} + 0{,}3*\text{share of partners}$$

5.1. The ratio of expenses for research and development (R&D) to the total amount of expenses demonstrates the intensity of innovation activity in the company. The indicator in 2018 was 4.25%;

5.2. The share of researchers in the region reflects the general "background" of the representation of science in the territory. For the Sverdlovsk region, this figure was 0.95% [26] of the total number of employees in 2018, which is significantly higher than in most other regions of Russia. This fact is explained primarily by the fact that the Ural Branch of the Russian Academy of Sciences is located in the Sverdlovsk region.

5.3. The ratio of the number of partners of the company from the total number of scientific and educational organizations in the region of presence allows us to identify the degree of development of communication with the considered segment of the regional innovation meta-ecosystem. The company's data for 2018 was 5.8%. Testing the model on the indicators of 2018 gives the following results:

i_c=0,4*83+0,6*39=56,6

i_b =0,4*6,9+0,6*10,6=9,1

i_p=0,6*11,8+0,3*144+0,1*8,5=51,13

i_s=0,6*126+0,1*2,1+0,3*94=104

i_{es}=0,2*4,25+0,5*0,95+0,3*5,8=3

$$y = k_1 \cdot 56.6/100 + k_2 \cdot 9{,}1/100 + k_3 \cdot 51.13/100 + k_4 \cdot 104/100 + k_5 \cdot 3/100$$

$$9 = k_1 \cdot 0{,}56 + k_2 \cdot 0{,}09 + k_3 \cdot 0{,}51 + k_4 \cdot 10{,}4 + k_5 \cdot 0{,}03$$

Key to the assessment of innovative ecosystems are stakeholder relationships, through which disparate organizations and other components of the socio-economic environment self-organize into ecosystems. At the same time, the energy that ensures the viability of the ecosystem is produced by the innovative orientation of the actors. Innovations are the essential basis of ecosystems in the social world, and if we are talking about a production, entrepreneurial ecosystem, a high - tech company-innovations here determine the survival, ensure the viability of the ecosystem, and competitiveness. By creating ecosystem effects, a permanent focus on the initiation, development, and dissemination of innovative ideas and products (R&D, new products and services, advanced technologies, continuous improvements in production, management, sales, service, communications, etc.) keeps the company, its employees, and stakeholders in good shape. Hence, the main objects of attention for orchestration in the ecosystem of the degree of innovation are seen as:

- the number and quality of internal and external impacts on innovation in its broadest sense;
- corporate and regional culture of support for new developments;
- external factors that particularly affect the innovation dynamics.

Giving priority to the (conditional) external environment, having developed a grouping of stakeholders and singling out the eco-factors of this segment of the innovation ecosystem, we have obtained an analytical model for assessing the eco-system, which takes into account:

a) the focus of consumers on access to the latest products, business ties, monopolistic influence of suppliers, competitors;

b) the complex impact of state structures that set a high bar through the institutional environment, national priorities (projects on import substitution, labor productivity, export expansion, introduction of digital technologies);

c) environmental social impact on the ecosystem;

d) the role of the scientific and educational environment, professional communities, and development institutions that are ready to provide impulses, resources, and expertise for the development of highly competitive production activities, establishing new contacts, and intensifying them.

The innovation ecosystem, which represents the unity of the socio-economic, cultural, and institutional environment and its constituent elements, by analogy with natural ecosystems, needs development resources. If the main resources (energy) of natural ecosystems are obtained through direct and indirect solar activity, then the energy impact in socio-economic ecosystems, which leads to their innovative evolution and expansion, is carried out at the expense of the key economic factor in the present increasingly complex world - entrepreneurial ability, without which neither labor, nor land, nor capital, nor information can be converted into meaningful goods.

The biggest innovations in the economy of recent decades clearly demonstrate how closely you need to work with other companies, organizations, and communities to create something much more complex, unique, and breakthrough. A single company will not be able to independently produce an iPhone, build and maintain a Large Hadron Collider, send an expedition to Mars, or develop high-tech (medical, optical, computer) equipment.

5 Conclusions

This article provides an example of a stakeholder analytical model, which is used to evaluate the level indicators of the innovation ecosystem of a high-tech company. The groups of stakeholders surrounding the company and embedded in its ecosystem are identified. Expert methods have been used to find eco-factors of the following components of the eco-system environment: client, business, government, public, scientific and educational. An example of the calculation of the model according to the data of 2018 of one of the largest high-tech companies in the Sverdlovsk region is presented. The next stage of research will be testing the model for a significant period of time, as well as at other enterprises.

The growth of computer capacity and the development of modern software should lead to the possibility of mathematical modeling and evaluation of the entire range of factors that affect ecosystems, both "local" and global, covering most socio-economic processes, up to the world economy as a super-ecosystem.

If one contemplates upon a more approximate stage of the evolution of the ecosystem approach, further study of ecosystem interactions requires careful attention to prioritization, ranking the importance of individual stakeholders. It seems that the entrepreneurial component is the main, meta-element that connects environmental relations to the ecosystem. Hence the importance of a detailed analysis of the control effects on the relations in the environment, management systems and technologies in the innovation ecosystem.

Acknowledgements

The article was supported by the Russian Foundation for Basic Research grant No. 20-010-00333.

References

1. L. Thomson, A. Kamalaldin, D. Sjödin, V. Parida, International Entrepreneurship and Management Journal, 1-28 (2021)
2. Z. Wang, Z. Sun, Journal of Chinese Political Science, 26(1), 69-87 (2020)
3. A.E. Plakhin, I.N. Tkachenko, M.V. Evseeva, Bulletin of the NGIEI, **8(111)**, 51-59 (2020)
4. L. A. Ramenskaya, The Manager, **4**, 16-28 (2020)
5. Ya.Yu. Radyukova, Yu.K. Arkhipova, V.Yu. Sutyagin, E.A. Kolesnichenko, Izvestiya Yugo-Zapadnogo gosudarstvennogo universiteta. Series: Economics. Sociology. Management, **9 (6),** 29-38 (2019)
6. Government of the Russian Federation, *http://www.garant.ru* (2020)

7. B. Hilton, B. Hajihashemi, C.M. Henderson, R.W. Palmatier, Industrial Marketing Management, **90,** 360-369 (2020)
8. A. Giovanini, P.F. Bittencourt, M. Uriona Maldonado, Revista Brasileira de Inovação, Campinas (SP), **19**, e020005-e020005 (2019)
9. J.H. Dyer, Strategic Management Journal, **18 (7)**, 535-556 (1997)
10. B.N. Panshin, Nauka i innovatsii, **3 (157)**, 17-20 (2016)
11. M. Talmar, B. Walrave, K.S.Podoynitsyna, J. Holmströmc, A.G.L. Rommea, Long Range Planning, **53 (4)**, 101850 (2020)
12. M.O. Soldak, Econ. promisl, **4 (88)**, 38–66 (2019)
13. E.V. Popov, V.L. Simonova, I.P. Chelak, Innovation, **6 (260)**, 46-53 (2020)
14. J.M.P. Kansheba, Small Enterprise Research, **27 (2)**, 110-124 (2020)
15. D. Nepelski, V. Van Roy, The Journal of Technology Transfer, 1-36 (2020)
16. Konsultant.ru, *http://www.consultant.ru/document/cons_doc_LAW_343020* (2021)
17. T.O. Kvålseth, Heliyon, **4 (10)**, e00846 (2018)
18. M. Porter, *Competition* (2000)
19. E. Carayannis, D. Campbell, International Journal of Social Ecology and Sustainable Development, **1(1)**, 41-69 (2012)
20. Spark, *https://www.spark-interfax.ru* (2021)
21. Nalog.ru, *https://www.nalog.ru/rn77/taxation/reference_work/conception_vnp* (2021)
22. Pravo, *http://pravo.gov.ru* (2021)
23. P. Oeij, S. Dhondt, S. Solley, A. Hill-Dixon, *Atlas of Social Innovation Future* (2018)
24. E.V. Popov, V.L. Simonova, O.V. Komarova, Bulletin of the UrFU, Series: Economics and Management, **18(2)**, 168-185 (2019)
25. Rosstat, *https://rosstat.gov.ru/labor_market_employment_salaries* (2021)
26. Rosstat, *https://rosstat.gov.ru/folder/14477* (2021)

 https://doi.org/10.1051/e3sconf/202125001005

Effects of climate change on environmental sustainability

Manuela Tvaronavičienė [1,2]*,

[1]Vilnius Gediminas Technical University (Vilnius Tech), Sauletekio 11, 10223, Vilnius, Lithuania
[2]General Jonas Zemaitis Military Academy of Lithuania, Silo 5A, 0322, Vilnius, Lithuania

Abstract. Adaptation strategies to the climate change include measures that can be taken to take account of the new climatic conditions. This paper aims at assessing the effects of climate change on environmental sustainability. This sustainability constitutes a major problem in many countries and regions around the world that experience industrial pollution, degradation of land as well as natural disasters caused by the global warming. The paper shows that adaptation strategies are often parallel strategies that can be integrated simultaneously with the management of natural resources. They can make resources more efficient and resilient to climate change. The paper shows that reducing the carbon footprint by more than 50 percent by 2030 and eliminating it by 2050 might be a viable solution how to tackle the climate change and support the environmental sustainability.

1 Introduction

In general terms, the strategies to mitigate the climate change can be narrowed down to two major areas: i) removing greenhouse gases (such as CO_2) from the atmosphere, and ii) preparing the societies to the effects of the climate change [1, 2].

Moreover, biodiversity should be seen as an important ally in dealing with climate change, rather than being portrayed as a victim of climate change [3, 4]. Protected areas with biodiversity that support the current and future climate can improve ecosystem services, including biodiversity, which is essential for mitigating climate change, such as food security, water quality and species protection [5, 6].

However, ecosystem-based solutions will be far from sufficient, and there is still a need to address the issue of fossil fuel emissions as a primary approach to mitigating climate change. Since the beginning of international climate efforts, they have focused primarily on reducing greenhouse gas emissions as a means of combating climate change, not on restoring biodiversity [7].

Where vulnerability is high, it can be extremely important to reduce risk in the short term even in the face of uncertainty, and where vulnerability to climate change is increasing, where it is not increasing to adapt. But climate change is much more than rising temperatures, and increasing climate variability will produce equal or greater impacts that need to be addressed. Many adaptation measures are part of an integrated mitigation - measures that reduce

* Corresponding author: manuela.tvaronaviciene@vilniustech.lt

greenhouse gas emissions. Measures to adapt forests to future conditions can help to maintain or increase their capacity to capture carbon [8, 9].

Energy-efficient and environmentally friendly technologies that reduce emissions from electricity generation and transport, build compactly and use renewable energy sources such as wind, solar, geothermal, biomass, hydro or biomass energy. Greenhouse gases (GHG) contribute primarily to global warming because they capture heat in the atmosphere. Of all the anthropogenic greenhouse gases emitted in 2004, 56 % came from carbon monoxide (CO), which is caused by the use of fossil fuels, while the second part came from CO_2 from deforestation (17 %) [10, 11].

This paper is structured as follows: Section 2 describes the links between the climate change and environmental sustainability. Section 3 focus on the anthropogenic activities and global warming. Section 4 focuses on the global climate change and the Environmental Vulnerability Index (EVI). Finally, Section 5 summarizes the findings and provides final conclusions and implications.

2 Climate change and environmental sustainability

The importance of the philosophy of sustainability that is being practised and implemented is becoming increasingly important for our communities as water resources, species and biodiversity decline. There is no doubt that urgent action is needed if we are to save the planet [12, 13].

Large temperature changes that have now became a reality on Earth will greatly accelerate other changes already underway in the climate system, such as changing precipitation patterns, increasing evaporation due to warming, and increasing temperature extremes [14]. It remains to be seen to what extent the extent of this development is reflected in elevated temperatures and extremes, but the fact that greenhouse gases influence aspects of the water cycle early on provides a strong argument for the role of greenhouse gas emissions in climate change [15]. Although GHG-induced warming was partially offset by a small amount of forced anthropogenic aerosol changes, temperatures will still rise over the next few decades. The models simulating changes in precipitation in the oceans are relatively robust, with moist regions becoming wetter and dry regions becoming drier, but not at the same rate. In addition, it appears that extreme rainfall levels in the oceans, such as in Greenland and Antarctica, are useful constraints on future changes, even if they apply to a relatively small part of the world's land area [16].

Evidence from high-resolution climate models suggests that the intensity of sub- daily extreme rainfall is likely to increase in the future, making the theoretically estimated decadence most likely in many regions. The representation of extreme precipitation has generally improved and confidence in model-based projections has increased. The range of 10-30 km, which is typically used in climate change studies, is still too small to explicitly represent sub- daily local heavy rainfall events [17]. More recently, weather forecasts with explicit convection have been used to predict extreme weather events such as floods and droughts. In the following, we propose that the use of sub daily extreme precipitation events as proxies for climate change risks will make it particularly easy to identify climate and climate change risks in the context of weather forecasting. To further underline the importance of heat extremes, we need to stress the fact that urban land has been estimated to be a proxy for the risk of climate change associated with extreme rainfall events. As described above, the direction and extent of the observed relative changes in urban areas correspond to the underlying measures of global temperature and precipitation changes in integrative quantitative models of total annual precipitation (mean annual total precipitation) [18].

Sustainability in responding to climate change should not be an overarching strategy that affects people's actions and responsibilities. The good news about climate change is that we know what is causing it and how to stop it, but we must not lose sight of other deeper environmental sustainability problems, which also require action. Climate science is one of the most important areas of scientific research in the world today, and we still have a lot to learn to fully understand this crucial problem [19]. Due to that many people rightly focused on the existential threat posed by climate change. Climate change will continue for decades to come, and no major measures will be taken to limit warming. The economic and political problems behind this are caused by the rapid growth of developing countries and the huge investments made worldwide in fossil fuels and other forms of fossil fuel use. In the future, not only our societies and ecosystems will commit to a much warmer world, but we will too. Ecosystem management and restoration play an important role and offer benefits, but only when used in conjunction with reducing fossil-fuel emissions can we mitigate the effects of climate change. We are truly in uncharted waters and we are contributing to these challenges to the best of our knowledge. As human civilization faces challenges of decarbonization, global greenhouse gas concentrations in the atmosphere are likely to stabilize, but carbon dioxide is rising, and the warming that will accompany it will occur much faster than we are experiencing today [20].

Even moderate changes in the climate are likely to have dramatic and damaging consequences, which will almost certainly affect every corner of the world. Around 8% of climate change is already taking place and can be directly linked to this global phenomenon. Global stability is being compromised, with less stability accompanied by higher pressure on governments, and the risk of civil unrest and political instability in countries such as China, India, and Brazil rising [21]. Urbanization and rural development affect many areas of biodiversity, including the increased risk of forest fires and habitat loss for wildlife. Changes in coastal land use also affect the distribution of wealth, as coastal countries provide a significant share of the world's natural resources, such as oil, gas, and coal.

The health effects of climate change include climate-related diseases such as asthma, allergies, respiratory diseases and cancer. Thence, the effects of human health and environmental change on human health and the environment need to be considered [22]. It provides a framework for quantifying the effects of changing air, land and ocean temperatures and reducing anthropogenic uncertainties in terms of health outcomes. All of these helps to make decisions at the local, national and global levels on health, environmental sustainability, health care and other issues. There are forms of environmental degradation about which we need to learn more and take measures to reduce them. We must face the inevitable effects of climate change in order to minimise the negative effects on human health and the environment, as well as on the health of the human population.

3 Anthropogenic activities and global warming

In the last 100 years, the average surface temperature of the earth has risen by more than 2 degrees Celsius. During this period, global surface temperatures, sea level rise and ocean acidification have increased in particular [23]. With regard to that, some pose the question of whether climate change is caused by natural events or human activities, also known as anthropogenic climate change [24]. Humans influence the climate and the temperature of the earth by burning fossil fuels, cutting down rainforests and keeping farm animals. Advocates argue that rising greenhouse gases in the atmosphere are a direct result of fossil fuel burning, and that this increase is causing significant and increasingly serious climate change, including sea-level rise, ocean acidification, and global warming. In fact, humans have been burning fossil fuels for more than 2,000 years. The gases in the Earth's atmosphere act like a glass greenhouse that captures the heat of the sun and prevents it from reaching space again. Over time, enormous amounts of greenhouse gases have been added to what is naturally occurring in our atmosphere, increasing the greenhouse effect and global warming. Nevertheless, earth warming follows and correlates with rising CO_2 emissions that can affect the climate, and correlates with the increase in global warming due to the burning of fossil fuels [25].

Generally, CO_2 is the greenhouse gas most commonly produced by human activities and responsible for human-induced global warming and other forms of climate change. Some scientists believe that human activities are the main cause of climate change. Nevertheless, the impact that man is causing is much greater in the current climate change cycle. For this reason, our current inaction on climate change and our inaction on global warming may be due not to human activity but to natural events. However, the current warming phase is occurring faster than past events, and the planet's climate is changing much faster than in past phases of climate change. One version of this is that the timescales differ from the Millennium, and that recent global warming is therefore only a phase in the upswing. The fact is that man has released large amounts of CO_2 by burning fossil fuels, and this is the main cause of recent global warming. It has become clear that humanity has caused most of the warming of the last century by releasing heat - the separation of gases commonly referred to as greenhouse gases into the force of modern life, such as carbon dioxide and methane [26].

However, the observed fingerprints of global warming exclude this argument, as natural cycles match these fingerprints. Natural circuits, for example, cannot explain why satellites measure less heat that escapes into space at the very wavelengths where CO_2 is absorbed. We have heard that the CO_2 emissions from human activities are low compared to those released into nature. The term "global warming" refers to the fact that the atmosphere is warming above the surface of the Earth. Global warming is a sign that the Earth has passed and will pass away in the future, and it is happening because greenhouse gas emissions are continuing to rise. In other words, the planet's climate will change even if atmospheric CO_2 and other greenhouse gas emissions continue to rise as a result of human activities. Climate change is happening, it is largely caused by human activities, and it will have many serious and potentially harmful effects in the coming decades. The scientific evidence paints a clear picture: climate change is occurring at least in the last few decades and possibly for many decades to come [27, 28]. This is because greenhouse gas emissions include carbon dioxide, the most important greenhouse gas, which has reached levels of concentration not seen on Earth for more than 400,000 years. These greenhouse gases act as a blanket, trapping the sun's heat at the Earth's surface and influencing the planet's climate system. The forces contributing to climate change include naturally occurring concentrations of greenhouse gases in the atmosphere and human activities. Greenhouse gases play a key role in keeping the planet warm enough to be inhabited, and greenhouse gas emissions from human activities (e.g. burning fossil fuels, oil and gas extraction, mining, agriculture) are a major cause of our rapidly changing climate. The records show that global warming, especially in recent decades, is occurring much faster than ever before and cannot be explained by natural causes alone. When we burn carbon-based materials, carbon dioxide is released by burning fossil fuels and absorbs heat and releases CO2 into the atmosphere.

4 Global climate change and the Environmental Vulnerability Index

The Global Environmental Vulnerability Index (EVI) is calculated and published by the global risk consultancy Maplecroft [29]. The new global ranking, which calculates the vulnerability of the world's largest and fastest-growing economies, including India, has found that they face significant risks to their populations, ecosystems, businesses, and environment. The Global Environment Vulnerabilities Index, a global assessment of national vulnerability to climate change, assesses 42 social, economic, and environmental factors to assess national vulnerabilities in three key areas [30]. Vulnerability mapping can include a combination of indicators selected for mapping and indicators that represent vulnerability within a given spatial unit. These indicators are integrated and the current assessment includes more than 1,000 indicators for each of the three key areas, as well as a number of other indicators. The physical component is represented by the number of hectares of vegetation and the presence of climate-related indicators - associated disasters and climate scenarios. The subsequent appearance of climate-sensitive diseases is also a physical component. This includes the increased burden of disease due to the frequency and severity of extreme weather events such as floods, droughts and forest fires. It brings together eight different populations that are disproportionately affected by the effects of climate change [31].

Mediterranean basin is a concrete example where the effects of climate change are quite obvious. There is an urgent need to identify and understand the factors that influence environmental vulnerability to climate change and to establish links between population trends and environmental health. The region is rich in natural resources and culture and a popular tourist destination with a population of more than 1.5 million and a high level of tourist activity. Low-altitude island communities are affected by rising sea levels, which are accompanied by a changing climate. At the basic level, economically weak communities are most likely to be ill - prepared for the effects of climate change because they lack the necessary financial resources. It is highly likely that they will be affected not only by the effects of rising sea levels, but also by dramatic changes in ecosystem services, including access to food [32].

Globally, climate change is affecting rural communities, which are heavily dependent on agriculture and natural resources for their livelihoods and access to food [33-36].

There are other non-climatic factors that also determine how much climate change will affect cities and countries. Measuring the future impacts of climate change can be a major challenge, as scientists' predictions of climate change cannot be completely accurate and many different factors come into play, such as rising temperatures, changes in precipitation and snow cover, and the effects of extreme weather events. It also seems necessary to weigh up what happens politically and socially in a region to see if a country can prepare for it. Many developing countries are also particularly vulnerable, because their livelihoods depend on natural resources [37, 38]. Overall, the EVI is a numerical index that reflects the status of the environmental vulnerability of a country or region. Since then, it has focused more on environmental threats from the misuse of the natural resources such as land, water and soil, and on human environmental management, which includes fossil fuel use, deforestation, land use change, human health and pollution. It was also decided that the vulnerability index should be easy to compile and based on indicators that are useful for comparing countries that reflect the relative vulnerability of countries. For example, the Mediterranean basin is connected to the Atlantic and the Mediterranean, as well as other regions [39]. It is a highly heterogeneous region where natural and anthropogenic activities interact in a complex way with climate variability [40]. Adaptivity is the ability of a species to cope and survive with changing conditions. For this reason, it is time to formulate a vulnerability index for all the regions of the world.

5 Conclusions

Overall, it becomes clear that climate change, in conjunction with socio-economic conditions, can increase vulnerability in both developed and developing countries. This can threaten the environmental sustainability and lead to many adverse effects to the nature and fauna. Thence, it is very important to join the efforts in both types of these countries and to take some action regardless of the level of income or life standards. All people on Earth are equally influenced by the climate change no matter which part of the planet they live in.

Broadly speaking, greenhouse gas emissions associated with transition risk fall under two main categories: mitigation activities and transition risks. Climate protection includes measures that can be taken to limit changes caused by human activities to the global climate. Research and science, which support climate protection, is trying to understand change itself and to improve the knowledge that underlies climate protection policy.

We must overcome climate change and address other critical challenges facing the planet. Given the scale of these challenges, which could take months or even years to resolve, the world cannot afford to pay attention to climate change or the broader issue of sustainability. This reality is accompanied by an increased risk of physical and economic dangers, and the only way to avoid them is to aggressively decarbonize the economy and everyday life.

References

1. D. Klenert, F. Funke, L. Mattauch, B. O'Callaghan, Environmental and Resource Economics, **76(4)**, 751-778 (2020)
2. O. Rueda, J. Mogollón, A. Tukker, L. Scherer, Global Environmental Change, 102238 (2021)
3. D. Brooks, E. Hoberg, W. Boeger, *The Stockholm paradigm: climate change and emerging disease* (2019)
4. L. Bevan, Energy Research & Social Science, **69**, 101580 (2020)
5. M. Gaglio, V. Aschonitis, L. Pieretti, L. Santos, E. Gissi, G. Castaldelli, F. Fano, Ecological Modelling, **403**, 23-34 (2019)
6. N. Seddon, A. Chausson, P. Berry, C. Girardin, A. Smith, B. Turner, Philosophical Transactions of the Royal Society B, **375(1794)**, 20190120 (2020)
7. H. Chenet, J. Ryan-Collins, F. Van Lerven, Ecological Economics, **183**, 106957 (2021)
8. A. Koutroulis, L. Papadimitriou, M. Grillakis, I. Tsanis, R. Warren, R. Betts. Global and Planetary Change, **175**, 52-63 (2019)
9. M. Morecroft, S. Duffield, M. Harley, J. Pearce-Higgins, N. Stevens, O. Watts, J. Whitaker, Science, **366(6471)**, eaaw9256 (2019)
10. N. Lee, S. Kim, J. Lee, Journal of CO_2 Utilization, **44**, 101414 (2021)
11. A. Shahsavari, M. Akbari, Renewable and Sustainable Energy Reviews, **90**, 275-291 (2018)
12. A. Wals, A. Benavot, European Journal of Education, **52(4)**, 404-413 (2017)
13. M. Smith, I. Smit, L. Swemmer, M. Mokhatla, S. Freitag, D. Roux, L. Dziba, Biological Conservation, **255**, 108985 (2021)
14. D. Koll, T. Cronin, Proceedings of the National Academy of Sciences, **115(41)**, 10293-10298 (2018)
15. M. Abdollahbeigi, Journal of Chemical Reviews, **2(4)**, 292-308 (2020)
16. N. Golledge, Wiley Interdisciplinary Reviews: Climate Change, **11(2)**, e634 (2020)
17. J. Schwaab, E. Davin, P. Bebi, A. Duguay-Tetzlaff, L. Waser, M. Haeni, R. Meier, Scientific Reports, **10(1)**, 1-9 (2020)
18. S. Scheuer, D. Haase, M. Volk, PLoS ONE, **12(12)**, e0189451 (2017)
19. D. Coen, *Climate in motion: Science, empire, and the problem of scale* (2019)
20. S. Brakman, H. Garretsen, A. Van Witteloostuijn, Social Sciences & Humanities Open, **2(1)**, 100034 (2020)
21. J. Warner, I. Boas, Environment and Planning C: Politics and Space, **37(8)**, 1471-1488 (2019)
22. J. Bell, C. Brown, K. Conlon, S. Herring, K. Kunkel, J. Lawrimore, G. Luber, C. Schrek, A. Smith, C. Uejio, Journal of the Air & Waste Management Association, **68(4)**, 265-287 (2018)
23. A. King, D. Karoly, Environmental Research Letters, **12(11)**, 114031 (2017)

24. G. Hoogendoorn, B. Sütterlin, M. Siegrist, Journal of Risk Research, **23(12)**, 1577-1589 (2020)
25. B. Ekwurzel, J. Boneham, M. Dalton, R. Heede, R. Mera, M. Allen, P. Frumhoff, Climatic Change, **144(4)**, 579-590 (2017)
26. D. Baldocchi, J. Penuelas, Global Change Biology, **25(4)**, 1191-1197 (2019)
27. S. Becken, B. Mackey, Journal of Air Transport Management, **63**, 71-83 (2017)
28. J. Lynch, M. Cain, D. Frame, R. Pierrehumbert, Frontiers in Sustainable Food Systems, **4,** 300 (2021)
29. Maplecroft, Climate Change Vulnerability Index, *https://www.maplecroft.com/risk-indices/climate-change-vulnerability-index* (2021)
30. H. Edmonds, J. Lovell, C. Lovell, Ecological Indicators, **117**, 106529 (2020)
31. C. Linares, J. Díaz, M. Negev, G. Martínez, R. Debono, S. Paz, Environmental Research, **182**, 109107 (2020)
32. N. Donthu, A. Gustafsson, Journal of Business Research, **117**, 284 (2020)
33. I. Elhassnaoui, Z. Moumen, M. Tvaronavičienė, M. Ouarani, M. Bendaoud, I. Serrari, I. Lahmidi, M.A.S. Wahba, A. Bouziane, D. Ouazar, M.D. Hasnaoui, Insights into Regional Development, **3(1)**, 80-103 (2021)
34. O. Eddelani, N.E. El Idrissi, S. Monni, Insights into Regional Development, **1(1)**, 6-18 (2019)
35. Z. Moumen, N.E.A. El Idrissi, M. Tvaronavičienė, A. Lahrach, Insights into Regional Development, **1(4)**, 301-317 (2019)
36. M.F. Faridi, M.M. Sulphey, Entrepreneurship and Sustainability Issues, **6(3)**, 1336-1345 (2019)
37. O. Wagdi, A. Hasaneen, Entrepreneurship and Sustainability Issues, **7(2)**, 962-976 (2019)
38. R. Akhmadeev, A. Redkin, N. Glubokova, O., Bykanova, L. Malakhova, A. Rogov, Entrepreneurship and Sustainability Issues, **7(2)**, 1149-1170 (2019)
39. T. Caloiero, S. Veltri, P. Caloiero, F. Frustaci, Water, **10(8)**, 1043 (2018)
40. C. Leduc, A. Pulido-Bosch, B. Remini, Hydrogeology Journal, **25(6)**, 1529-1547 (2017)

The role of science, technology and innovation in increasing the share of renewable energy worldwide

Anna Tarasova[1] *

[1] Department of Foreign Languages and Linguistics, Humanitarian Education Centre, Volga State University of Technology, Lenin square, 3, Yoshkar-Ola, 424000 Russian Federation

Abstract. A number of countries in the emerging markets are becoming leaders in manufacturing solar photovoltaic products. Thus, global companies are urged to compete in all stages of the supply chain for producing most of the installed solar modules around the world. Meanwhile, production costs are at record lows. The decisions made in the emerging markets will affect the course for the solar and wind industry and their role towards de-carbonization of global energy systems in the near future. Renewable energy is set of miscellaneous natural resources, which preserves non-renewable sources for further use in other sectors of the economy, and saves green energy for future eras. The way a country employs its potential in renewables will safeguard the country's energy security and the stability of electricity prices. Renewable energy is eco-friendly, as its operation does not produce any waste, or discharge pollutants into the atmosphere or water bodies. Our paper discusses the latest achievements of various scientific areas and industries, as well as it lists specific recommendations for national sustainable actions plans. These recommendations might be useful for policy-makers, working in adopting the transformation of their country's energy systems.

1 Introduction

In 2015 one hundred ninety-three states, including those from the member states of the Commonwealth of Independent States (CIS), agreed on the 2030 Agenda for Sustainable Development, which includes 17 Sustainable Development Goals (SDGs). The Sustainable Development Goals and their associated targets are global in nature and universally applicable, while taking into account different national contexts. The goals aim at achieving universal access to affordable, reliable, sustainable and modern energy for all [1]. This means supporting and implementing the UNECE flagship project "Pathways to Sustainable Energy" that relies on sustainable energy described in three pillars:

1. energy security, an even supply of energy for economic growth;

* Corresponding author: tarasova_a@bk.ru

2. energy for quality of life, providing energy available to all at all times;
3. energy and environment, limiting the impact of the energy system on climate, ecosystems and health.

International cooperation is a sine qua non for attracting investment in the development of renewable energy throughout the region and is one of the tools for expanding the use of renewable energy. One of the most important indicators of SDG is that by the year of 2030, strengthened international cooperation should facilitate access to research and technology in the field of clean energy, including renewable energy sources, energy efficiency and advanced clean technologies for the use of fossil fuels, and promoting investment in energy infrastructure and clean energy technologies [2,3]. Countries around the globe have already signed and ratified the Paris Climate Agreement, identified respective contributions and developed national action plans for their implementation. Most of the countries in the long-term plans for the development of the electric power industry provide for the large-scale development of renewable energy sources as one of the measures to limit greenhouse gas emissions. The expansion of the use of renewable energy sources became possible due to technical progress in this area, which made it possible to reduce significantly the cost of electricity production by wind and solar power plants of various types [4]. The electricity industry in many countries around the world is currently undergoing significant changes aimed at ensuring universal access to affordable, reliable, sustainable and modern energy for all. This goal can be achieved by the active integration of various traditional and renewable energy sources in a wide range of capacities from small-distributed generation facilities to large grid power plants, which entails the transformation of energy systems [5].

The ongoing technological changes are accompanied by the creation of an institutional framework that determines the regulatory, technological and economic rules for the reliable and efficient development and functioning of energy systems in new conditions. In other words, the process of creating an integral power management system, corresponding to the new structure of energy systems, is underway. Transformation of energy systems means an active process of creating political, market and regulatory conditions, as well as establishing practices for planning and operating energy systems that accelerate investment, innovation and the use of smart, efficient, reliable and environmentally friendly technologies [6,7]. Currently, the capacity of wind farms and solar photovoltaic power plants reaches hundreds and thousands of MW, and the energy sector is undergoing a fundamental shift towards a digitalized and decarbonized energy system. In this regard, an urgent task is to analyse the experience of large-scale development of renewable energy sources in a number of countries of the world, its impact, including negative, on traditional generation, on the electricity market and network infrastructure. Further step will be to develop specific recommendations on how to solve specific problems and obstacles in order to increase investment in renewable energy in each of the country involved, thus improving and optimizing cross-border regional energy cooperation [8].

2 Global trends in renewable energy deployment

Further development of cooperation among the countries in the electric power industry will be in the context of the transformation of energy systems, introduction of digital technologies, as well as expanded attention to environmental issues, energy efficiency, energy conservation and the development of renewable energy sources [9]. The implementation of the UN Sustainable Development Goals sets the following tasks for the Electric Power Councils in the countries-partners:

-to assist the implementation of international agreements ratified in terms of environmental protection and mitigation of climate change; energy efficiency and energy saving;

-to hold measures to create a database on the best available technologies in the field of ecology and energy efficiency, including for manufacturers of power equipment, taking into account the possibilities of import substitution;

-to encourage dissemination of information on best practices, exchange of experience, including training in the field of renewable energy and energy efficiency.

Countries worldwide commence using the tidal energy potential to the full. Currently, it is cost effective to use tides with a height of at least 4 meters. The height of the tides is highly dependent on the configuration of the coast [10]. In the inland seas, for example in the Black and Baltic seas, the tides are low. The greatest tides occur in deep inland bays, including river estuaries, hence, the highest known tidal heights (up to 18 m) are in the Bay of Fundy, Canada.

There are plots promising for the construction of tidal power stations in Russia, Great Britain, France, Norway, South Korea, China, Argentina, and the USA. Totally, the economically efficient potential for the use of tidal energy is currently estimated at 450 billion kWh per year; in the future, as tidal power plants improve, its value may significantly increase. Researchers around the world claim there are certain advantages of adopting tidal power plants [11]. These plants exploit renewable energy source, and do not generate waste. Moreover, they cause no flooding, occupying no land and minimally traumatizing fish and plankton that goes through the turbines.

Nevertheless, tidal power plants can be disadvantageous and highly scheduled, which means they are not stable in supplying electricity to the power system evenly, rather in several peaks during the day [12]. Hence, while implementing projects for the construction of specifically large tidal power stations with a capacity of several gigawatts, such a schedule can cause problems for the power system and require the construction of additional regulatory capacities, for example, in the form of a pumped storage power plant. Though the existing tidal power stations might have a certain effect on the local marine biocenosis, however, these changes do not lead to degradation of the biocenosis.

Since the Middle Ages people have used the tidal water farms; nevertheless, tidal power engineering has not received wide development and is still represented by several experimental power plants. The largest tidal power plant in the world is the French Rance with a capacity of 240 MW that was launched in 1966 [13,14]. Other examples of such plants in operation are the Annapolis in Canada with a capacity of 20 MW, Jiangxia in China with a capacity of 3.2 MW and Uldolmok in South Korea. South Korea is showing great interest in tidal energy, where the Sihwa station with a capacity of 254 MW is at the final stage of construction and the building of six more large tidal power stations is projected. Additionally, several more experimental small tidal power plants have been built in Norway and the UK.

One could ask the question as regarding to why the tidal energy has not been so widely used. The main reason is the lower economic attractiveness of such stations compared to traditional power plants. A classic tidal power station with high power requires the construction of an extended dam, the construction of a building in the sea area. Relatively small heads at high water flow rates necessitates the use of a large number of huge and expensive hydraulic units. Currently, a technology is being developed to reduce the cost of construction of tidal stations due to their construction from floating blocks and the use of orthogonal turbines [15].

Another example of renewable energy is solar energy, which has a colossal potential, far exceeding both current and future needs of humankind for electricity. If we cover only 0.7% of the globe with solar panels with a minimum efficiency of 10%, their power generation will exceed the generation of all power plants in the world. Hydropower, wind energy, and bioenergy also carry the energy of the Sun, which is spent on evaporating water, heating air masses and the process of photosynthesis of plants. Due to the tilt of the earth's axis and the associated change of seasons, the largest average annual flow of solar energy per unit area falls on the equatorial regions - Africa, Central and South America, Australia, India, Southeast Asia. In Russia, the most promising for the development of solar energy is the south of the European part of the country, especially the Krasnodar and Stavropol Territories, Rostov, Volgograd and Astrakhan Regions, Dagestan and Kalmykia [16].

Solar energy uses an almost inexhaustible renewable energy source; there are no emissions of pollutants into the environment during the production of electricity. Moreover, solar panels can be installed in almost any convenient location. Nonetheless, among the disadvantages of solar energy, we should mention strong dependence of production on the time of day, season, and weather conditions [17]. With a relatively small share of solar and other unregulated energy in the power system, about 10%), this feature does not cause significant problems. Additionally, a significant disadvantage is the high cost and, accordingly, the low economic efficiency of solar energy. Meanwhile, for large solar power stations, contamination of solar panels with dust, snow, and frost can become a problem.

Two methods of converting solar radiation energy into electricity that are being applied are photovoltaic and solar thermal. The photovoltaic principle is based on the generation of an electric current using solar cells, mainly silicon, collected in a panel. Panels of photovoltaic cells are easy to install, maintain and can be placed on any suitable area, such as roofs of houses, in this regard, it is solar energy based on photovoltaic cells that has become most widespread [18]. The solar thermal method is based on the concentration of sunlight using mirrors of various shapes. At the point where the sun's rays are concentrated, a heat carrier, such as water, is placed, which turns into steam and drives a turbine.

By the end of 2009, the total capacity of photovoltaic power plants alone in the world was about 23 GW, having increased 20 times in ten years. In 2010, at least 10 GW of capacity was commissioned at photovoltaic power plants. The largest photovoltaic plant in the world - Sarnia in Canada with a capacity of 80 MW was authorized [19]. About 70% of the capacity of solar power plants is concentrated in Europe, specifically in Germany, where the total capacity of solar power plants has exceeded 10 GW.

Solar energy is also developing in the USA and Spain. In the United States, back in 1991, the world's largest solar power plant with a capacity of 354 MW was launched at full capacity, and in 2010, construction began on a solar thermal power plant in California with a capacity of 1000 GW. In Spain, in 2009-2010, six large solar thermal power plants with a total capacity of 450 MW were accredited. However, there are still countries that do not recognize the potential of solar power stations and do not hasten to start including them in the energy systems [20].

The economic efficiency of solar thermal stations is highly dependent on their location, reaching a maximum in hot desert areas. Generally, the economic efficiency of solar energy in most cases is inferior to both traditional energy and other renewable energy sources. The payback of solar energy projects in developed countries is provided by various forms of government support and subsidies.

Apart from solar energy, global countries have commenced to use wind power plants. The United States are the leader in wind energy with 35 GW, followed by Germany with 26 GW and China with 25 GW.

Today, wind power stations provide about 2% of the world's electricity production. In a number of countries, these figures are significantly higher. Thus, in Denmark, wind farms provide about 20% of electricity generation, in Spain and Portugal - more than 10%. Wind power is developing dynamically. The total wind farm capacity in the United States increased 3.9 times, in France - 5.6 times, in China - 19.8 times. Some researchers project further development of the industry [21]. For example, by 2030 the European Union plans to receive about 15% of electricity from wind farms, and some of its member states would desire to get even more. Ireland plans to achieve by this time the generation of wind power stations at the level of 40% of the total electricity generation in the country.

The world's economically efficient wind capacity is 72,000 GW, which is well above current needs. The use of wind power stations is considered economically efficient when the average annual wind speed at an altitude of 50 m is more than 6.4 m/s. Most often, such conditions can occur in coastal areas. Nevertheless, wind power is underdeveloped in some areas. The reason for that can be that wind energy in terms of its economic efficiency is inferior to classical energy due to the long payback periods. The production of wind farms depends on weather conditions and is therefore variable [22]. Wind power stations cannot guarantee the supply of a clearly defined amount of electricity in a specific period. This feature of wind power stations, as well as a number of other power stations, that operate based on renewable energy sources, should be fixed in the rules for the functioning of the electricity market.

Most countries have various government support measures for wind energy; such measures are declared at the legislative level, but pragmatically they do not work due to the lack of necessary by-laws.

3 Technological drivers in renewable energy deployment

This section analyses the energy systems in the countries around the world. In particular, it examines generation and grids and analyses the state of renewable energy in existing systems, such as installed capacity of power plants and electricity; dynamics of the total installed capacity of power stations in 2000-2020; finally, it scrutinizes the importance of renewable energy systems and their potential in the world [23].

Hydroelectric power plants capacity differs from country to country. Thus, they provide about 20% of Russian and world electricity generation. However, in other countries the share of hydropower can be dramatically higher. For example, in Canada, closest to Russia in terms of natural conditions, hydroelectric power plants produce 58% of electricity, in Brazil - 86%, in Norway, which is known for its strict environmental legislation - 99%. The Table 1 below lists the top ten countries in the use of hydropower in 2019 [24].

Table 1. The top ten countries in the use of hydropower, GW

Country	HPP capacity, GW	HPP generation, billion kWh
China	196.79	585.2
Canada	88.974	369.5
Brazil	69.080	363.8
USA	79.511	250.6
Russia	44.426	176.0
Norway	27.528	140.5
India	33.600	115.6
Venezuela	13.26	86.8
Japan	27.229	69.2
Sweden	16.209	65.5

Currently, the global hydropower industry is experiencing a rebirth due to the active construction of hydropower stations in emerging countries such as China, Brazil, India, Pakistan, Iran and others. Only in China, it is drafted to double the capacity of hydroelectric power stations within the nearest ten years, this is in conditions when China is much ahead of other countries in terms of the installed capacity of renewable energy power stations [25].

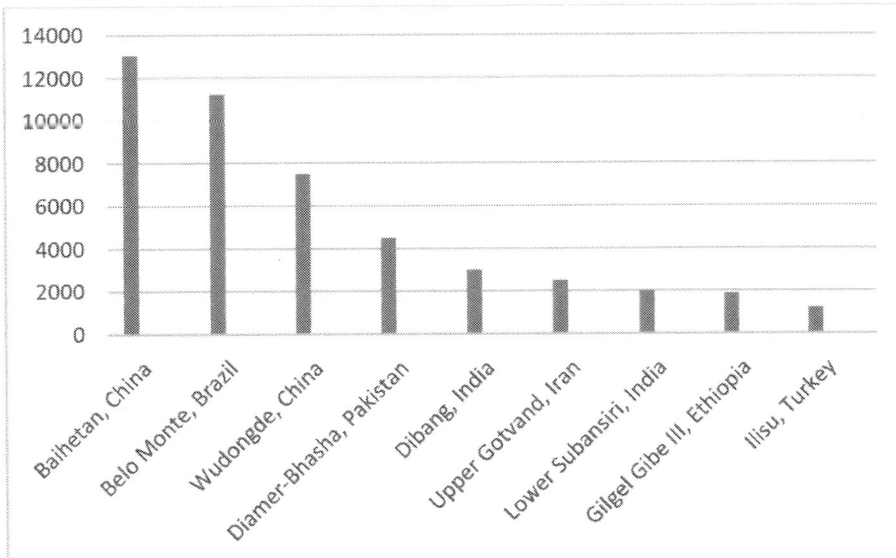

Fig. 1. Renewable energy power generated by world power stations (in bln kWh)

The energy sector is in transition and undergoing significant structural changes to ensure universal access to affordable, reliable, sustainable and modern energy for all. The key is transforming energy systems by integrating various conventional and renewable energy sources across a wide range of capacities. By creating policy, market and regulatory environments, countries can attract investment and accelerate innovation through smart grids, efficient, reliable and sustainable technologies. However, countries need to prepare to mitigate potential problems when trying to integrate renewable energy sources [26]. The future of the region's energy sector in accordance with various development scenarios looks different. Fossil fuel industries will be the hardest hit; however, they are essential for economic prosperity during the transition period.

It is designed that by 2050, half of the world energy will continue to be powered by fossil fuels under any economically viable scenario. In all regions electricity generation, district heating and the transport sector will depend on fossil fuels. Thus, investments should be spread over a wider range of low-emission technologies to ensure a rapid transition to sustainable energy [27]. The results can be used to develop specific recommendations for national sustainable energy action plans. In the process of transformation of energy systems in a specific country, the following actions might be helpful:

- determination of key course and target vision for the development of the electric power industry, taking into account the long-term perspective (for the period up to 2050);
- creation of an institutional framework, and in general - an integrated management system that defines regulatory, technological and economic rules for the optimal development and functioning of the electric power complex in the context of the transformation of energy systems in the country and in the world;
- creation of a joint program for the development of a technological and economic basis for building a future efficient and sustainable integrated energy system to achieve SDG 7.

For the development of renewable energy in the project beneficiary countries, the authors recommend taking the following measures:

https://doi.org/10.1051/e3sconf/202125001006

- integrate solar and wind energy into energy systems;
- create national institutions for future energy planning;
- improve methods for unifying relevant, reliable and timely statistics;
- develop legislative measures to support the integration of variable renewable energy sources into power systems;
- use international experience to harmonize national and international energy standards;
- use modern technologies for efficient production of solar and wind energy;
- implement measures to reduce greenhouse gas emissions in the energy sector;
- improve the conditions for the modernization of energy systems using generation based on solar and wind energy;
- develop guidelines for potential investors in renewable energy facilities;
- train national personnel to integrate solar and wind generation into energy systems.

Renewable energy technologies extensively rely on digital technologies, hence, digitalization of the energy system that is growing and becoming more akin, smart, anticipated and sustainable should become a key future research area. Certain advances of automation, electric and shared mobility are today influencing the transport infrastructure and electric vehicles. Incurring to the traverse nature of mobility the future will surely see the way smart power grids pair and merge with electricity such as solar and wind power and transport systems on a wide scale [28]. Potentially greater energy efficiency and optimized energy consumption will become benefits of such emergence. However, much will depend on the way consumers accept, politicians adopt the right measures and the technological progress that will guide akin, automated, electric and shared mobility.

4 Conclusions

Currently, the electricity industry in many countries of the world is undergoing significant changes, the goal of which is to ensure universal access to affordable, reliable, sustainable and modern energy sources for all – SDG 7. This goal is achieved through the integration of various conventional and renewable energy sources in a wide range of capacities from small distributed generation facilities to large grid power stations, which entails the transformation of power systems.

No doubts that certain factors determining the rapid transformation of energy systems in the world will either speed up or slow down the shift. The desire to improve the reliability and efficiency of energy systems, as well as the desire to enhance the availability of energy using innovative technologies may be sufficient conditions for the countries to start thinking of opting for renewable energy. Another good reason is striving to ensure a high level of environmental and climatic safety, as well as reducing the cost of technologies for the production and consumption of electricity, thus encouraging the electrification of the economy by expanding of digitalization and automation of energy systems.

However, the use of renewable energy sources in combination with increased energy efficiency is the main measures to achieve the specified climate solution; a successful transition will require due consideration of three key aspects: technical, economic and institutional [29]. Improving coordination between transmission and distribution network operators is a priority. Moreover, new actors, such as aggregators, should become members in governance. Many countries around the world create positive political, market and regulatory conditions, as well as adopt practices for planning and operating energy systems that accelerate investment, innovation and the use of smart, efficient, reliable and environmentally friendly technologies. Comprehensive planning is key to ensure that different types of generation sources, transmission and distribution systems, demand management systems, energy storage devices, and other systems operate in a coordinated manner. Innovations can further improve and reduce costs of renewable energy technologies. This can include material science in solar photovoltaic cells, the integration of electric vehicles into the power grid and enabling digital technologies into the energy systems.

References

1. K. Prandecki, Energy and Environmental Engineering, **2(4)**, 83 (2014)
2. I. Sotnyk, I. Shvets, L. Momotiuk, Y. Chortok, Marketing and Management of Innovations, **4**, 150-160 (2018)
3. W. Strielkowski, E. Lisin, E. Astachova, Entrepreneurship and Sustainability Issues, **4(4)**, 591-600 (2017)
4. I. Iddrisu, S.C. Bhattacharyya, Renewable and Sustainable Energy Reviews, **50**, 513 (2015)
5. T. Tolstykh, L. Gamidullaeva, N. Shmeleva, Journal of Open Innovation: Technology, Market, and Complexity, **6(4)**, 95 (2020)
6. T. Beliaeva, M. Ferasso, S. Kraus, E.J. Damke, International Journal of Entrepreneurial Behavior & Research, **26(2)**, 266 (2019)
7. D.-S. Oh, F. Phillips, S. Park, E. Lee, Technovation, **54**, 1 (2016)
8. M.S.S. Danish, M.E.L. Elsayed, M. Ahmadi, T. Senjyu, H. Karimy, H. Zaheb, Energy Reports, **6(2)**, 40 (2020)
9. M. Sakah, F.A. Diawuo, R. Katzenbach, S. Gyamfi, Renewable and Sustainable Energy Reviews, **79**, 544 (2017)
10. O, Zlyvko E. Lisin N. Rogalev G. Kurdiukova, International Economics Letters, 3(4), 124-138 (2014)
11. P. Woditsch, W. Koch, Solar Energy Materials and Solar Cells, **72(1-4)**, 11 (2002)
12. B.G. Gribov, K.V. Zinov'ev, Inorganic Materials, **39**, 653(2003)
13. X. Zou, L. Ji, J. Ge, Nat Commun, **10**, 5772 (2019).
14. N. Drouiche, P. Cuellar, F. Kerkar, S. Medjahed, T. Ouslimane, M.O. Hamou, Renewable and Sustainable Energy Reviews, **52**, 393 (2015)
15. S. Mittal, H. Dai, S. Fujimori, T. Masui, Applied Energy, **166**, 301 (2016)
16. Z.y. Zhao, S.-Y. Zhang, B. Hubbard, X. Yao, Renewable and Sustainable Energy Reviews, **21**, 229 (2013)
17. J.W. Busby, S. Shidore, Energy Research & Social Science, **23**, 60 (2017)
18. X. Yu, C. Cecati, T. Dillon, M. G. Simões, IEEE Industrial Electronics Magazine, **5(3)**, 49 (2011)
19. J. Xu, T. Liu, Energy Policy, **142**, 111538 (2020)
20. J. Wu, N.K.Tran, Sustainability, **10(9)**, 3067 (2018)
21. E. Lisin, W. Strielkowski, V. Chernova, A. Fomina, Energies, **11(12)**, 3284 (2018)
22. V. Stratigak, Water, **11(6)**, 1249 (2019)
23. M. van Staden, A. Marques, E. Villaseñor, Procedia, **57**, 840 (2014)
24. I.R. Chávez Urbiola, J.A. Bernal Martínez, J. Hernández Borja, C.E. Pérez García, R. Ramírez Bon, Y.V. Vorobiev, Energy Procedia, **57**, 24 (2014)
25. Y.A. Kaplan, Renewable and Sustainable Energy Reviews, **43**, 562 (2015)
26. D. Y.C. Leung, Y.Yang, Renewable and Sustainable Energy Reviews, **16(1)**, 1031 (2012)
27. R. Saidur, N.A. Rahim, M.R. Islam, K.H. Solangi, Renewable and Sustainable Energy Reviews, **15(5)**, 2423 (2011)

28. M. Shoaib, I. Siddiqui, S. Rehman, S. Khan, L.M. Alhems, Journal of Cleaner Production, **216**, 346(2019)
29. A. Mostafaeipour, M. Khayyami, A. Sedaghat, K. Mohammadi, S. Shamshirband, M.-A. Sehati, E. Gorakifard, International Journal of Hydrogen Energy, **41(15)**, 6200 (2016)

Analysis of some indicators by means of fuzzy logic on the example of Azerbaijani energy enterprises

Ulviyya Rzayeva[1]*, *Aida* Guliyeva[2], and *Narmin* Jafarova[3]

[1]Center of Digital Economy, Azerbaijan State University of Economics, 6, Istiglaliyyat str., AZ1001, Baku, Azerbaijan
[2]Department of Digital Technologies and Applied Informatics, Faculty of Digital Economy, Azerbaijan State University of Economics, 6, Istiglaliyyat str., AZ1001, Baku, Azerbaijan
[3] Moses Lab Marketing Agency, 156, Aynalı Plaza, Ahmed Rajabli str., Baku, Azerbaijan

Abstract. One of the most promising areas of scientific research in the field of analysis, forecasting and modelling of economic phenomena and processes is fuzzy logic. In the article with the use of the fuzzy logic apparatus (Fuzzy Logic toolkit of the MATLAB software package) a model for assessing the development of manufacturing indicators is developed. Key indicators for evaluating the activities of energy enterprises are analysed. The model is based on microeconomic indicators of the enterprises' efficiency and development. The presented indexes pass through the stage of fuzzification, at which each measure is assigned the membership function; then comes the decision stage; finally, the task ends by the stage of defuzzification, i.e., bringing the resulting fuzzy value to precise ones. The application of the Mamdani fuzzy inference algorithm allows getting an exact value of the consolidated indicator of enterprise development. The resulting value allows making operational management decisions in the energy enterprises in real-time.

1 Introduction

For the country's production potential, the efficiency of enterprises and their development act as the most important means of achieving national economic goals and interests. Azerbaijani manufacturing sector unites enterprises whose economic activity consists of the creation and processing of mostly petroleum products. This sector is of paramount importance in calculating GDP since its existence enables national economies to gain competitive advantages and enter the global trading environment. The article discusses how the effectiveness of this industry can be achieved by accelerating trade relations, as well as increasing the competitive advantages - with the help of technological advances, and improving the security of any type of assets - taking into account possible threats. The role of these indicators in assessing the development of manufacturing enterprises - trade, innovation and technology, and security - is the focus of this work. The resulting assessment

* Corresponding author: ulviyya.rzayeva@unec.edu.az

is necessary to identify strengths and weaknesses, but the lack of clear, quantitative values in the development concept can lead to incorrect results [1]. The fuzzy logic approach used, in which even ambiguity matters, has helped to obtain more adequate results.

Considering the role of consolidated indicators in the development of energy enterprises, it should be assumed that they perform the functions of providing information on the current level of growth and effectiveness of activities for a certain period to local and foreign investors. The role of consolidated information in the implementation of such functions will be determined by the needs of a particular manufacturer. Moreover, various groups of manufacturers are interested in different information: owners - in financial results and financial stability, investors and lenders - in creditworthiness, managers - in all aspects of the enterprise.

In the context of the formation of fundamentally new requirements for the informatization of society on the activities of the energy enterprises in Azerbaijan, the issues of development and implementation of an effective methodology for analysing consolidated quantitative information using new mathematical approaches (in particular, fuzzy) are relevant [2].

The article is structured as follows. The first section called methodology presents a set of goals, principles and the approaches used to create a logical organization of research. Section two reviews the related literature. The third section shows the scheme for constructing fuzzy rules, obtaining results with the use of MATLAB package and the built-in Fuzzy Logic Toolbox, as well as empirical results obtained by the described techniques. A discussion of the results and conclusion are featured in section four.

2 Materials and methods

Fuzzy logic, as its name indicates, is an alternative to classical logic, and allows us to quantify the uncertainty, variability, and inconsistency in the behaviour of economic systems. In the real world, especially in economics, there are many vague, inaccurate, uncertain, ambiguous knowledge. The main problem arises when these qualitative indicators need to be converted into quantitative ones. Such a tool for a more accurate reproduction of reality is the apparatus of fuzzy logic and fuzzy sets [3].

The article explores and analyses several indexes and their subindexes. Following them, we determine the input data and obtain the output variable using membership functions, which is the main tool of fuzzy logic. We used triangular fuzzy membership functions for both input and output data since they are more convenient for representing data in percent or proportions and their further processing [4]. The application of triangular functions leads to linear interpolation: possible phase transition construction based on triangular functions, gives an option of independent specification of the values of the reproduced function in the interpolation nodes. Therefore, this function type is used to determine appropriate scales and linguistic variables to the concrete valuation of both input and output variables [5].

When evaluating the relevant data on manufacturing enterprises, the fuzzy logic approach is more reliable, while a quantitative system does not give fuzzy concepts that are inherent for production economy. For the correct result, based on reliable indicators, a sufficient number of rules have been presented. The corresponding conclusion has been obtained by the MATLAB software. According to the study [6], using the Fuzzy Logic Toolbox GUI tools, a fuzzy output system can be created for any model.

The main goal of the work is to evaluate the activity of manufacturing enterprises in Azerbaijan based on consolidated indicators on trade, security, and innovation in energy enterprises. To solve this problem, a set of rules have been built that allow describing a set of fuzzy situations in the subject area and comparing them with each other.

As a result, rule tables have been built, which give us an opportunity to assess the current situation on the efficiency of the energy enterprises. The main advantage of this model is the establishment of fuzzy situations using linguistic variables. This makes it easy to understand the conditions under development and create a model for more complex situations. The Fuzzy Inference System (FIS) procedure is applied to quantify the qualitative levels of indicators, and based on the accepted criterion to find out the best alternatives for achieving our goal.

3 Literature review

Fuzzy logic is widely used in socio-economic researches: assessing the development of economic sectors, insurance, supporting decision-making, modelling the economies of regions and countries, forecasting, etc.

A review of the literature justifies the use of fuzzy logic to evaluate economic indicators, perform environmental, medical, engineering and other problems. Fuzzy logic methods allow the analysis of indicators of the object under uncertain conditions, for example, in cases where there is no any fully-fledged statistics, or qualitative data should be included among the studied indicators.

The MATLAB software package, containing the Simulink with Fuzzy Toolbox visual modelling tool, allows developing fuzzy models, evaluating their characteristics and graphically presenting the results.

Zlateva et al. in [7] presents a model of fuzzy inference for a comprehensive assessment of natural hazards in the southwestern region of Bulgaria, solving the multicriteria problem by the MATLAB software environment and the built-in Fuzzy Logic Toolbox package are used. An information system for disaster risk management is being developed, and the developed fuzzy system is presented as its integrated part.

The study [8] analyses the correlation between class attendance and student performance using fuzzy logic methods. The work is based on data from 330 students. Fourteen membership functions of fuzzy logic were used; implementation is carried out using the MATLAB Fuzzy Toolbox.

The aim of the study [9] is to develop an expert system using multiple FIS to predict daily trading performance. The authors argue that the developed model shows better results compared to the existing model with a similar goal.

The article [10] presents an approach that provides multi-purpose optimization of performance indicators of small and medium enterprises (SMEs). Value optimization was tested on a group of 197 SMEs; the results were obtained based on the fuzzy set approach.

Based on fuzzy rules the investigation [11] presents the development of a technical indicator for stock and currency markets when deciding on the securities' purchase and sale. As linguistic variables, the generalized volumes of the securities' purchase and sale, and their generalized prices are used.

The study [12] describes the experimental project of fuzzy controlling the fluid flow process. The simulation results clearly show that the intelligent fuzzy logic controller provides optimal control without increasing the fluid flow rate compared to the conventional PID (Proportional Integral Derivative) - controller.

One of the important areas of constructing an application with the fuzzy approach is the design of a fuzzy control system. In preparing a framework concept, an important step and a necessary guarantee is a system simulation. Article [13] describes methods of overcoming the working inconveniences that traditionally arise during simulations using fuzzy logic toolkits and the SILMULINK package.

The article [14] is devoted to simulation of fuzzy traffic control of traffic flow at a multi-line roundabout. The smart controller runs the waiting time and the queue length of vehicles in proportion to the length corresponding to the green and other phases. The management strategy provides for smooth traffic flow with minimal latency and queue length.

Zurek in [15] devotes his research in the medical field to the results of the implementation of a fuzzy inference system for the identification of four over-the-counter drugs.

The study [16] describes the automatic control of engine speed with switchable resistance using PID. The operation of the proportional-integral controller is simulated on a separate computer using the MATLAB/Simulink software. The fuzzy PID algorithm is compared with the conventional PID algorithm. The authors insist that using the proposed method, the engine reaches the reference speed in less time.

Thus, complex-structured systems operating under conditions of significant uncertainty, when applying the apparatus of the theory of fuzzy sets, give better results than when using known statistical or economic-mathematical models.

4 Empirical model

In this research paper, the main three microeconomic indicators of energy manufacturing enterprises were analysed to assess the efficiency level of the development of this kind of business. The actuality and importance of the indicators for both the national economy and the abovementioned industry type, particularly, are discussed as follows:

First, trade enables specialization and large-scale operation, provides access to technology and learning, offers increased competition, and increases the value of a country's natural resources. National trade is necessary to keep money in circulation within a country, which improves the economy and the living standards of its citizens. Likewise, trade between regions fosters equitable development, cultural, technical exchange and investments throughout the country. International trade, also known as foreign trade, is the one that encompasses the purchase and sale of goods and services between different countries and their markets. However, today we can no longer talk only about strictly business relationships. New technologies have globalized the planet, and cultural and political relations have also reached a world level. In this context, trade relations within and outside countries have gained significance in the energy manufacturing industry since it helps to use the benefits of value creation, gain competitive advantage and go globally easier and more beneficial than previous times.

Second, the possibility of competing products entering the energy market stimulates innovation activities and particularly encourages the invention of Innovations and New Technologies. When it comes to processes, which are an inseparable component of innovation in the energy manufacturing industry, companies are eager to invest in sectors where production technologies are changing rapidly. Judiciously designed competition policy in energetics contributes to innovation by facilitating the arrival of competing products, as well as by promoting the introduction of new technologies.

Moreover, referring to the functioning of processes of energy industry participants, safety and security can be defined as the ability of a system to operate while controlling the risks to people, property and environment at an acceptable level.

Undoubtedly, security is closely related to the economy. In particular, crime and violence can inhibit the production of goods and services, thus constituting a serious impediment to greater economic vigour. Insecurity affects countries both enterprise and macroeconomic level. It dampens the ability to empower performance, as well as, negatively affects reputation, since reputation directly affects the number of foreign investments. It can be sure that it has a great emphasis on economic dynamism. Therefore, a favourable environment of tranquillity and public peace in industries is justified by the potential to increase sustained economic growth.

4.1 Trade

Throughout time, the main goal of enterprises was to maximize profits, while the trade acts as the most important and necessary condition without which this goal cannot be achieved. The paper presents the following subindexes of energy trade:

- Energy export in the trade performance, since the indicator of trade in manufacturing enterprises, is more dependent on fluctuations in the value of direct or indirect exports [17];
- Direct energy export in the trade performance, as the motivational mechanism for the development of production enterprises, is based on the comparative advantages that companies can obtain by direct export of manufactured products [18];
- The energy sales, since the prospects for direct exports specifically depend on the target sales markets, without applying to market channels [19];
- Production dependence on imports. This indicator is used to analyse the dependence of production enterprises on the import of intermediate goods for energy manufacturing [20];
- The share of total inputs used in the production which comes from the domestic origin resources; if the economy and enterprises can be self-sufficient for further manufacturing processes, it can gain competitive advantages in nearly all situations [21].

The first step of the model construction is the fuzzification of our input variables. As their membership function, we use a triangular function. For the first term _Trade_, we took six input variables according to the number of our subindexes:

Input 1. The proportion of energy enterprises engaged in export directly or indirectly (Minimum 10% of their sales);

Input 2. The proportion of energy enterprises engaged in export directly (Minimum 10% of their sales);

Input 3. Share of total sales engaged in export directly;

Input 4. The proportion of enterprises, which use foreign material inputs or supplies;

Input 5. The degree of total inputs that imported or has a foreign origin (%);

Input 6. The degree of total inputs produced completely in the domestic market (%).

Each input variable has five triangular membership functions. A linguistic expression for a fuzzy set of input variables is shown in Table 1:

Table 1. Fuzzy set of input variable of the term _Trade_

Range	Interval	Linguistic Variable
0 to 25	(0, 0, 25)	Poor
0 to 50	(0, 25, 50)	Fair
25 to 75	(25, 50, 75)	Average
50 to 100	(50, 75, 100)	Good
75 to 100	(75, 100, 100)	Excellent

The membership function of our output variable is a triangular. We will present it not in tabular but in graphical form (Figure 1), taken from the MATLAB editor:

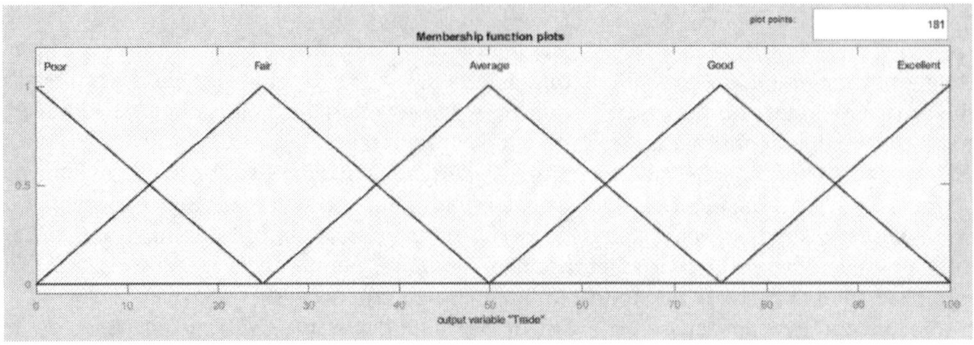

Fig. 1. The membership function of the output variable of the term _Trade_

The next step in our modelling is the inference procedure according to the IF-THEN rules. Our combination of rules is as follows:

1) If (Proportion of enterprises engaged in export directly or indirectly is Poor) then (Trade is Poor);

2) If (Proportion of enterprises engaged in export directly or indirectly is Fair) then (Trade is Fair);

3) If (Proportion of enterprises engaged in export directly or indirectly is Average) then (Trade is Average);

4) If (Proportion of enterprises engaged in export directly or indirectly is Good) then (Trade is Good);

5) If (Proportion of enterprises engaged in export directly or indirectly is Excellent) then (Trade is Excellent);

6) If (Proportion of enterprises engaged in export directly is Poor) then (Trade is Poor);

7) If (Proportion of enterprises engaged in export directly is Fair) then (Trade is Fair);

8) If (Proportion of enterprises engaged in export directly is Average) then (Trade is Average);

9) If (Proportion of enterprises engaged in export directly is Good) then (Trade is Good);

10) If (Proportion of enterprises engaged in export directly is Excellent) then (Trade is Excellent);

11) If (Share of total sales engaged in export directly is Poor) then (Trade is Poor);

12) If (Share of total sales engaged in export directly is Fair) then (Trade is Fair);

13) If (Share of total sales engaged in export directly is Average) then (Trade is Average);

14) If (Share of total sales engaged in export directly is Good) then (Trade is Good);

15) If (Share of total sales engaged in export directly is Excellent) then (Trade is Excellent);

16) If (Proportion of enterprises engaged in export directly is Poor) and (Proportion of enterprises, which use foreign material inputs or supplies is Poor) then (Trade is Poor);

17) If (Proportion of enterprises engaged in export directly is Fair) and (Proportion of enterprises, which use foreign material inputs or supplies is Fair) then (Trade is Fair);

18) If (Proportion of enterprises engaged in export directly is Average) and (Proportion of enterprises, which use foreign material inputs or supplies is Average) then (Trade is Average);

19) If (Proportion of enterprises engaged in export directly is Good) and (Proportion of enterprises, which use foreign material inputs or supplies is Good) then (Trade is Good);

20) If (Proportion of enterprises engaged in export directly is Excellent) and (Proportion of enterprises, which use foreign material inputs or supplies is Excellent) then (Trade is Excellent)

21) If (Share of total sales engaged in export directly is Poor) and (The degree of total inputs that imported or has foreign origin is Poor) then (Trade is Poor);

22) If (Share of total sales engaged in export directly is Fair) and (The degree of total inputs that imported or has foreign origin is Fair) then (Trade is Fair);

23) If (Share of total sales engaged in export directly is Average) and (The degree of total inputs that imported or has foreign origin is Average) then (Trade is Average);

24) If (Share of total sales engaged in export directly is Good) and (The degree of total inputs that imported or has foreign origin is Good) then (Trade is Good);

25) If (Share of total sales engaged in export directly is Excellent) and (The degree of total inputs that imported or has foreign origin is Excellent) then (Trade is Excellent);

26) If (Share of total sales engaged in export directly is Poor) and (The degree of total inputs produced completely in the domestic market is Poor) then (Trade is Poor);

27) If (Share of total sales engaged in export directly is Fair) and (The degree of total inputs produced completely in the domestic market is Fair) then (Trade is Fair);

28) If (Share of total sales engaged in export directly is Average) and (The degree of total inputs produced completely in the domestic market is Average) then (Trade is Average);

29) If (Share of total sales engaged in export directly is Good) and (The degree of total inputs produced completely in the domestic market is Good) then (Trade is Good);

30) If (Share of total sales engaged in export directly is Excellent) and (The degree of total inputs produced completely in the domestic market is Excellent) then (Trade is Excellent);

Our output function will be determined according to the minimax principle [22]. The results of our calculations, according to which the output variable is equal to 15.8 (shows the level of "poor") are presented in the Figure 2:

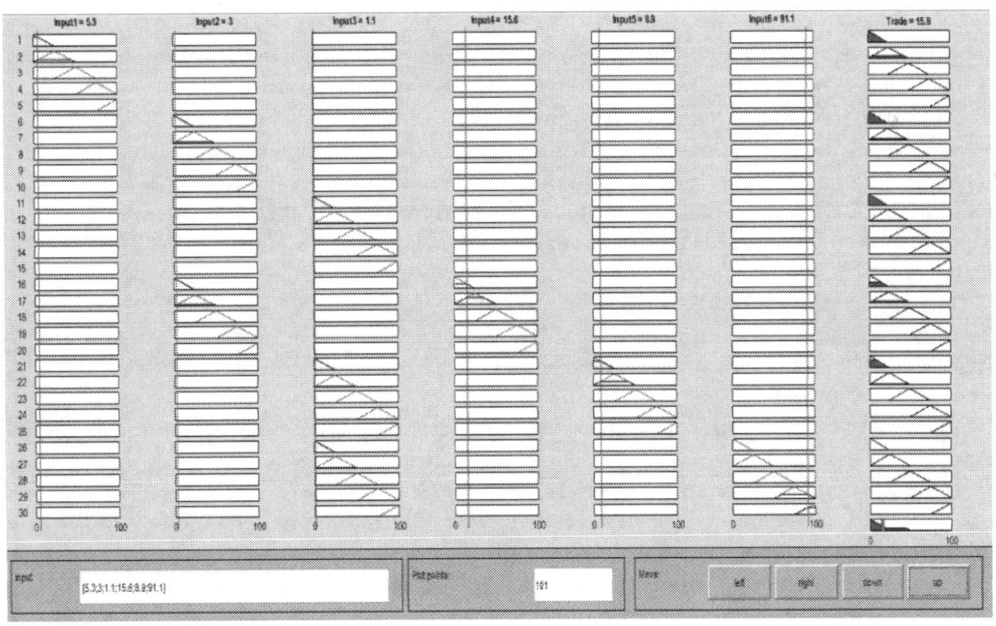

Fig. 2. The result calculations for term _Trade_

4.2 Innovation & technology

Over the past decades, interest in innovation among scientists and economists has increased significantly, which is natural, since innovation has become one of the main irreplaceable components of economic growth. Innovations are a prerequisite for the development of production, improving the quality and increasing the variety of products and services, and both in developing and developed market economies, they are the driving force of competition; thanks to them they can use modern technology and organization of production, to ensure the success and effectiveness of the enterprises. Therefore, the following subindexes are included in the paper:

- Licensed and imported technology holders from foreign companies. This subindex indicates on the importance in solving the strategic problems of energy enterprises entering the international market, increasing competitiveness and economic efficiency as well as presents how licensing promotes innovation [23].
- Enterprises, which has their website platform. As a rule, the establishment of websites for commercial organizations is significant. Its specific functions-image, information, advertising serves to increase its commercial function, which is considered to be positively related to the improvements of the enterprise and its technological environment [24].
- Energy enterprises, which launched a new product or service, as an introduction of a new product is a basis for business development and innovation management [25].
- Innovation methods in enterprises. Firms put a process innovation method into operation because they ensure a qualitative increase in the efficiency of processes or production of goods that are in demand by the market and has a great impact also on accelerating economies of scale [26].
- Enterprises that invest on Research and Development, whereas R&D is the most important task of the "knowledge economy". It also helps to increase the pace of development of the energy manufacturing sector of the economy, the transition from uncontrolled economic growth to sustainable development [27].
- Holder firms of international quality certificates. Whereas special attention needed to pay to the analysis of problems of international standardization and technical regulation of national production, this kind of measures are necessary for the implementation and further improvement of the process of normative and technical regulation of the international economy [28].

The first step of the model construction is putting our input variables into the Fuzzy Logic Designer. We need to use a triangular function for their membership functions. For our output variable of the term _Innovation and Technology_, we took 7 input variables due to the number of our subindexes:

Input 1. The proportion of enterprises, which imported technologies from foreign companies;

Input 2. The proportion of enterprises, which has their Web site platform;

Input 3. The proportion of enterprises, which use business e-mails to communicate with customers and suppliers;

Input 4. The proportion of enterprises, which launched a new product or service;

Input 5. The proportion of enterprises, which put a process innovation method into operation;

Input 6. The proportion of enterprises, which invests on Research and Development (R&D);

Input 7. The proportion of enterprises, which holds internationally-recognized certificates for quality.

The next step in our modelling is the inference procedure according to the IF-THEN rules. Our combination of rules is as follows:

1) If (The proportion of enterprises, which imported technologies from foreign companies is Poor) and (The proportion of enterprises, which launched a new product or service is Poor) and (The proportion of enterprises, which put a process innovation method into operation is Poor) then (Innovation & Technology is Poor);

2) If (The proportion of enterprises, which imported technologies from foreign companies is Fair) and (The proportion of enterprises, which launched a new product or service is Fair) and (The proportion of enterprises, which put a process innovation method into operation is Fair) then (Innovation & Technology is Fair);

3) If (The proportion of enterprises, which imported technologies from foreign companies is Average) and (The proportion of enterprises, which launched a new product or service is Average) and (The proportion of enterprises, which put a process innovation method into operation is Average) then (Innovation & Technology is Average);

4) If (The proportion of enterprises, which imported technologies from foreign companies is Good) and (The proportion of enterprises, which launched a new product or service is Good) and (The proportion of enterprises, which put a process innovation method into operation is Good) then (Innovation & Technology is Good)

5) If (The proportion of enterprises, which imported technologies from foreign companies is Excellent) and (The proportion of enterprises, which launched a new product or service is Excellent) and (The proportion of enterprises, which put a process innovation method into operation is Excellent) then (Innovation & Technology is Excellent);

6) If (The proportion of enterprises, which has their Web site platform is Poor) and (The proportion of enterprises, which use business e-mails to communicate with customers and suppliers is Poor) then (Innovation & Technology is Poor);

7) If (The proportion of enterprises, which has their Web site platform is Fair) and (The proportion of enterprises, which use business e-mails to communicate with customers and suppliers is Fair) then (Innovation & Technology is Fair)

8) If (The proportion of enterprises, which has their Web site platform is Average) and (The proportion of enterprises, which use business e-mails to communicate with customers and suppliers is Average) then (Innovation & Technology is Average);

9) If (The proportion of enterprises which has their Web site platform is Good) and (The proportion of enterprises, which use business e-mails to communicate with customers and suppliers is Good) then (Innovation & Technology is Good);

10) If (The proportion of enterprises which has their Web site platform is Excellent) and (The proportion of enterprises, which use business e-mails to communicate with customers and suppliers is Excellent) then (Innovation & Technology is Excellent);

11) If (The proportion of enterprises, which launched a new product or service is Poor) and (The proportion of enterprises, which put a process innovation method into operation is Poor) then (Innovation & Technology is Poor);

12) If (The proportion of enterprises, which launched a new product or service is Fair) and (The proportion of enterprises, which put a process innovation method into operation is Fair) then (Innovation & Technology is Fair);

13) If (The proportion of enterprises, which launched a new product or service is Average) and (The proportion of enterprises, which put a process innovation method into operation is Average) then (Innovation & Technology is Average);

14) If (The proportion of enterprises, which launched a new product or service is Good) and (The proportion of enterprises, which put a process innovation method into operation is Good) then (Innovation & Technology is Good);

15) If (The proportion of enterprises, which launched a new product or service is Excellent) and (The proportion of enterprises, which put a process innovation method into operation is Excellent) then (Innovation & Technology is Excellent);

16) If (The proportion of enterprises, which invests on R&D is Poor) and (The proportion of enterprises, which holds internationally-recognized certificates for quality is Poor) then (Innovation & Technology is Poor);

17) If (The proportion of enterprises, which invests on R&D is Fair) and (The proportion of enterprises, which holds internationally-recognized certificates for quality is Fair) then (Innovation & Technology is Fair);

18) If (The proportion of enterprises, which invests on R&D is Average) and (The proportion of enterprises, which holds internationally-recognized certificates for quality is Average) then (Innovation & Technology is Average);

19) If (The proportion of enterprises, which invests on R&D is Good) and (The proportion of enterprises, which holds internationally-recognized certificates for quality is Good) then (Innovation & Technology is Good);

20) If (The proportion of enterprises, which invests on R&D is Excellent) and (The proportion of enterprises, which holds internationally-recognized certificates for quality is Excellent) then (Innovation & Technology is Excellent);

Our output function will be determined according to the minimax principle. The results of our calculations, according to which the output variable is equal to 29 (shows the level of "fair") are presented in the Figure 3:

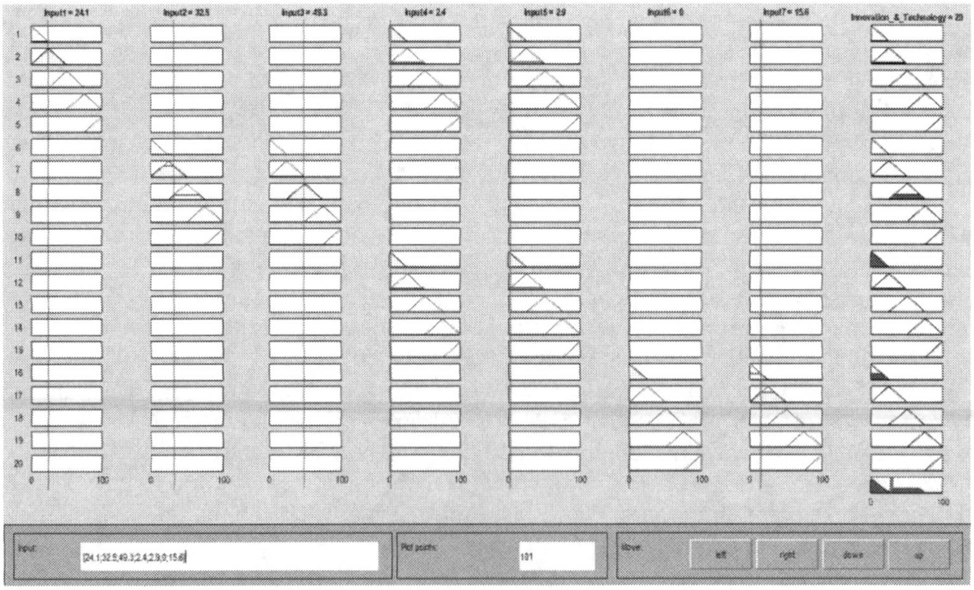

Fig. 3. The result calculations for term _Innovation and Technologies_

4.3 Security

The external and internal environment of any enterprise always bears its unique features, opportunities, and threats related to the profile of activity, size of production, number of staff, geographical and regional features and other factors. An auxiliary function, designed to protect the enterprise from internal and external threats, is the security function. This function should be built on the principle of "a threat to security is primary for measures to protect the enterprise from threats" [29]. This principle exists to ensure that the enterprise's measures to protect it from real and potential threats that are dangerous and vital for the economic security of the businesses. Following other functions of the businesses, the security system must be adequate to threats and correspond to the financial capabilities of the enterprise.

- The proportion of enterprises making payments for security or protection issues. An analysis is made of the experience of market participants who take measures to determine necessary actions to minimize threats and participants, establish the reasons that prompted them to terminate their activities or simply take preventive or corrective measures against any kind of violence [30];
- Average security costs if the enterprises make payments for security or protection (estimated as the percentage of annual sales). A significant problem for the domestic enterprises is increasing the effectiveness of financial costs. The elimination of ineffective costs is needed to improve the process of managing innovation and production activities. Therefore, average security costs are worth to be assessed as an additional expense for enterprises [31];
- Average losses or any damages because of burglary (estimated as the percentage of annual sales). This subindex is needed to the accurate calculation of losses from theft and shortage which should begin with an acquaintance with the procedure for determining the amount of losses from theft, shortage, destruction of tangible or intangible assets, for instance, intellectual property as well as any cyber-attack cases [32];
- The proportion of total product value, which was lost when delivering to the domestic markets. There are many reasons: severe frosts, transport accidents, fires, theft, etc. However, theft as the reason for the spoilage or disappearance of the goods is the most difficult one to mitigate the risks completely for avoid additional costs and delay in shipping which can result in loss of customer value [33];

The first step of the model construction is putting our input variables into Fuzzy Logic Designer for fuzzification. We need to use a triangular function for their membership functions. For our output _Security_ we took 6 input variables due to the number of our subindexes:

Input 1. The proportion of enterprises making payments for security or protection issues;

Input 2. If the enterprises make payments for security or protection, average security costs (estimated as the percentage of annual sales);

Input 3. The proportion of enterprises has losses or any damages because of burglary;

Input 4. If the enterprises experience losses, average losses or any damages because of burglary (estimated as the percentage of annual sales);

Input 5. The proportion of total product value, which was lost when delivered to domestic markets;

Input 6. The proportion of enterprises, which describes violence, theft, and other disorders as the main limitation.

The next step in our modelling is the inference procedure according to the IF-THEN rules. Our combination of rules is as follows:

1) If (The proportion of enterprises has losses or any damages because of burglary is Poor) and (The enterprises experience losses, average losses or any damages because of burglary is Poor) then (Security is Excellent);

2) If (The proportion of enterprises has losses or any damages because of burglary is Fair) and (The enterprises experience losses, average losses or any damages because of burglary is Fair) then (Security is Good);

3) If (The proportion of enterprises has losses or any damages because of burglary is Average) and (The enterprises experience losses, average losses or any damages because of burglary is Average) then (Security is Average);

4) If (The proportion of enterprises has losses or any damages because of burglary is Good) and (The enterprises experience losses, average losses or any damages because of burglary is Good) then (Security is Fair);

5) If (The proportion of enterprises has losses or any damages because of burglary is Excellent) and (the enterprises experience losses, average losses or any damages because of burglary is Excellent) then (Security is Poor);

6) If (The proportion of enterprises making payments for security or protection issues is Poor) and (The enterprises experience losses, average losses or any damages because of burglary is Poor) then (Security is Excellent);

7) If (The proportion of enterprises making payments for security or protection issues is Fair) and (The enterprises experience losses, average losses or any damages because of burglary is Fair) then (Security is Good);

8) If (The proportion of enterprises making payments for security or protection issues is Average) and (The enterprises experience losses, average losses or any damages because of burglary is Average) then (Security is Average);

9) If (The proportion of enterprises making payments for security or protection issues is Good) and (The enterprises experience losses, average losses or any damages because of burglary is Good) then (Security is Fair);

10) If (The proportion of enterprises making payments for security or protection issues is Excellent) and (The enterprises experience losses, average losses or any damages because of burglary is Excellent) then (Security is Poor);

11) If (The proportion of enterprises, which describes violence, theft, and other disorders as the main limitation is Poor) then (Security is Excellent);

12) If (The proportion of enterprises, which describes violence, theft, and other disorders as the main limitation is Fair) then (Security is Good);

13) If (The proportion of enterprises, which describes violence, theft, and other disorders as the main limitation is Average) then (Security is Average);

14) If (The proportion of enterprises, which describes violence, theft, and other disorders as the main limitation is Good) then (Security is Fair);

15) If (The proportion of enterprises, which describes violence, theft, and other disorders as the main limitation is Excellent) then (Security is Poor);

16) If (The enterprises make payments for security or protection, average security costs are Poor) and (The enterprises experience losses, average losses or any damages because of burglary is Poor) then (Security is Excellent);

17) If (The enterprises make payments for security or protection, average security costs are Fair) and (The enterprises experience losses, average losses or any damages because of burglary is Fair) then (Security is Good);

18) If (The enterprises make payments for security or protection, average security costs are Average) and (The enterprises experience losses, average losses or any damages because of burglary is Average) then (Security is Average);

19) If (The enterprises make payments for security or protection, average security costs are Good) and (The enterprises experience losses, average losses or any damages because of burglary is Good) then (Security is Fair);

20) If (The enterprises make payments for security or protection, average security costs are Excellent) and (The enterprises experience losses, average losses or any damages because of burglary is Excellent) then (Security is Poor);

21) If (The proportion of enterprises making payments for security or protection issues are Poor) and (Proportion of total product value, which was lost when delivered to the domestic markets, is Poor) then Security is Excellent);

22) If (The proportion of enterprises making payments for security or protection issues are Fair) and (Proportion of total product value which was lost when delivered to the domestic markets is Fair) then Security is Good);

23) If (The proportion of enterprises making payments for security or protection issues are Average) and (Proportion of total product value, which was lost when delivered to the domestic markets, is Average) then Security is Average);

24) If (The proportion of enterprises making payments for security or protection issues are Good) and (Proportion of total product value, which was lost when delivered to the domestic markets is Good) then Security is Fair);

25) If (The proportion of enterprises making payments for security or protection issues are Excellent) and (Proportion of total product value, which was lost when delivered to the domestic markets is Excellent) then Security is Poor);

26) If (The proportion of enterprises making payments for security or protection issues are Poor) and (The proportion of enterprises, which describes violence, theft and other disorders as a main limitation is Poor) then (Security is Excellent);

27) (The proportion of enterprises making payments for security or protection issues are Fair) and (The proportion of enterprises, which describes violence, theft and other disorders as a main limitation is Fair) then (Security is Good);

28) (The proportion of enterprises making payments for security or protection issues are Average) and (The proportion of enterprises, which describes violence, theft and other disorders as a main limitation is Average) then (Security is Average);

29) (The proportion of enterprises making payments for security or protection issues are Good) and (The proportion of enterprises, which describes violence, theft and other disorders as a main limitation is Good) then (Security is Fair);

30) (The proportion of enterprises making payments for security or protection issues are Excellent) and (The proportion of enterprises, which describes violence, theft and other disorders as a main limitation is Excellent) then (Security is Poor).

Our output function will be also determined according to the minimax principle. The results of our calculations, according to which the output variable is equal to 87.7 (shows the level of "excellent") are presented in Figure 4:

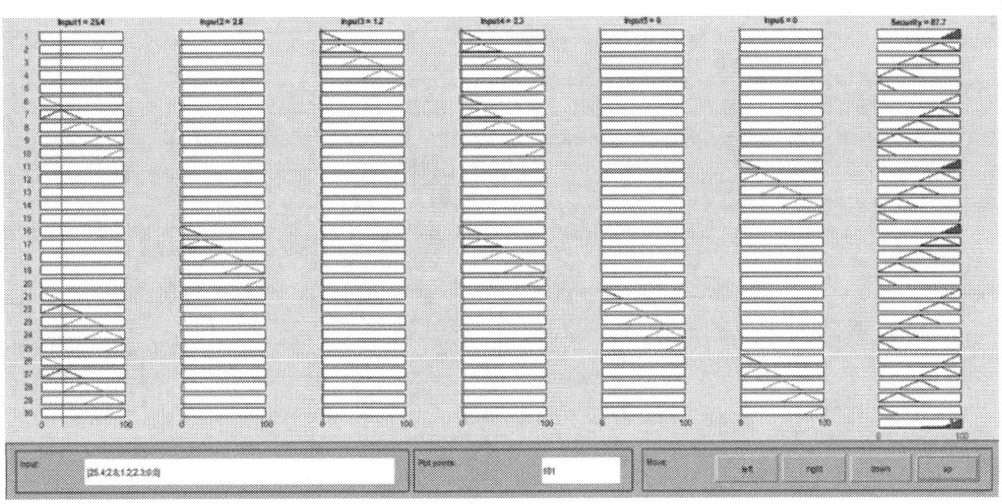

Fig. 4. The result calculations for term _Security_

5 Conclusions

Assessing the development of energy manufacturing enterprises in terms of trade, innovation and technology, and security makes it possible to identify the main problem areas of these sub-sectors. The results obtained allow us to state that, firstly, national trade has "weak" competitiveness, because enterprises have little interest in direct or indirect export. Secondly, innovation and technology based on the results of the study are presented as "fair" due to the lack of investment in national R&D and a very small number of firms that are holders of internationally recognized quality certificates. However, in the security sector, an "excellent" result was obtained in Azerbaijan, since it is not considered as an obstacle to both production and logistics operations. Finally, all three indicators for the efficiency of energy manufacturing enterprises include important sub-indices for assessment, and a fuzzy tool allows identifying the results, combining qualitative and quantitative estimates.

The rationale for the application of the fuzzy sets' methods and the theory of fuzzy logic for assessing the effectiveness of the manufacturing enterprises' activities is given. World indicators with their sub-indices that impede the development of national manufacturing enterprises are analysed. The paper proposes a fuzzy production model and implements the process of fuzzy modelling of the rules sets by using the specialized Fuzzy Logic Toolbox of software package MATLAB. Fuzzy inference is realized on the base of the Mamdani algorithm. The sequence of actions when designing a fuzzy model for evaluating various indicators is determined. Input and output variables at the verbal level are defined, the sets of fuzzy rules are established for subsequent decision-making. The features of inference are considered and an analysis of the results of the study, confirming the feasibility of using fuzzy methods for solving estimation problems, is shown.

In this paper, three important aspects of manufacturing enterprises were evaluated. In further studies, this approach can be used in various sectors of the economy to reliably determine the expected results. The once-developed process of evaluating indicators and choosing the right methodology will help determine the strengths and weaknesses of the subject area in formulating basic assumptions and evaluating important criteria.

References

1. F.M. Martin, T. Ellen, *A Practical Approach* (1994)
2. A. Meier, W. Pedrycz, E. Portmann, *Fuzzy Management Methods* (2013)
3. L.A. Zadeh, Fuzzy Sets, Fuzzy Logic, and Fuzzy Systems, **394** (1996)
4. W. Pedrycz, Fuzzy Sets and Systems, **64(1)**, 21 (1994)
5. D. Shepherd, F. Shi, IFAC Proceedings Volumes, **31(16)**, 435 (1998)
6. T. Ramya, A.C. Kannan, R.S. Balasenthil, B.A. Bagirathi, Advanced Materials Research, **984** (2014)
7. P. Zlateva, L. Pashova, K. Stoyanov, D. Velev, 2nd International Conference on Education and Management Technology, **13** (2011)
8. S.Z. Iqbal, International Journal of Innovative Science and Research Technology, **3(8)**, 440 (2018)
9. M. Rubell, G. Lincy, J.C. Jessy, Expert Systems with Applications, **44**, 13 (2016)
10. M. Stefanovic, S. Nestic, A. Djordjevic, D. Djurovic, I. Macuzic, D. Tadic, M. Gacic, Journal of Engineering Manufacture, **231(1)**, 15 (2015)
11. M. Shahjalal, S. Abeda, K.M. Nirmal, K. Khodadad, International Journal of Applied Economics and Finance, **6**, 53 (2012)
12. Ch.Sh. Besta, A.K. Kastala, P.R. Ginuga, R.K. Vadeghar, IFAC Proceedings Volumes, **46(32)**, 349 (2013)
13. X.Q. Xu, Applied Mechanics and Materials, **494**, 1306 (2014)

14. A. Che Soh, L. Guan Rhung, H. Sarkan, International Journal on Computer Science and Engineering, **2(4)**, 924 (2010)
15. E.E. Zurek, D.J. Romero, H.D. Blanco, C.A. Gutiérrez. Rev. Univ. Medellín, **12(22)**, 109 (2013)
16. A. Uysal, S. Gokay, E. Soylu, T. Soylu, S. Çaşka, Measurement and Control, **52(7-8)**, 1137 (2019)
17. G. Buturac, D. Mikulić, P. Palić, Economic Research-Ekonomska Istraživanja, **32(1)**, 101 (2019)
18. M. Funke, R. Ruhwedel, Weltwirtschaftliches Archiv, **138**, 97 (2002)
19. K. Karagöz, Procedia Economics and Finance, **38**, 446 (2016)
20. Y. Simachev, M. Kuzyk, N. Zudin, Foresight And STI Governance, **10(4)**, 25 (2016)
21. A. Moeuf, S. Lamouri, R. Pellerin, S. Tamayo-Giraldo, E. Tobon-Valencia, R. Eburdy, International Journal of Production Research, **58(5)**, 1384 (2020)
22. M.R. Casals, M.A. Gil1, P. Gil, European Journal of Operational Research, **27(3)**, 371 (1986)
23. N. Kumar, Economic and Political Weekly, 20(35), **M103** (2020)
24. Y. Zhao, S. von Delft, A.M. Thomas, T. Buck. Long Range Planning (2019)
25. Y. Lee, G.C. O'Connor, JAMS, **31**, 241 (2003)
26. M. Daksa, M. Yismaw, S. Lemessa, J Innov Entrep, **7(6)**, (2018)
27. H. Wang, P. Liang, H. Li, R. Yang, Procedia Computer Science, **91**, 122 (2016)
28. K. Blindab, A. Mangelsdorf, J. Pohlisch, International Journal of Production Economics, **198**, 50 (2018)
29. A. Bisong, S.M. Rahman. International Journal of Network Security & Its Applications, **3(1)**, 30 (2011)
30. J. Jang-Jaccard, S. Nepal, Journal of Computer and System Sciences, **80(5)**, 973 (2014)
31. S. Romanosky, Journal of Cybersecurity, **2(2)**, 121 (2016)
32. F. Jalilian, M. Vejdani, M. M. Alavijeh, F. Mostafavi, Avicenna J Neuro Psych Physio, **1(2)**, e24260 (2014)
33. C. Vignali. British Food Journal, **103(2)**, 97 (2001)

Energy tourism in Serbia: online representation of the hydro energy tourism

Darko *Dimitrovski*[1] *

[1] University of Kragujevac, Faculty of Hotel Management and Tourism in Vrnjačka Banja, Vojvođanska 5a, Vrnjačka Banja, 35240, Serbia

Abstract: Energy tourism is an emerging type of tourism that received increasing attention in the recent period. Hydro plants were part of the energy tourism offer that has been attracting the tourist throughout the time. Thus, revealing the online representation of their textual and visual content is of considerable interest for countries that aim to enhance this specific tourism activity. Serbia has an unfavourable structure of exploited energy sources with a significant share of renewable source, in contrast to considerable hydro energy potential, both for energy creation and tourism usage. By acknowledging the current state of the tourism use of hydro plants, understanding its attraction and experience value, and future activities that could derive from it is an initial step in formulating tourism management and marketing efforts.

1 Introduction

Nowadays, tourists are driven by unique and novel experiences. Thus, 'energy regions' [1] have become a novel tourism product that has the capacity to attract the considerable interest of the contemporary tourist. Population inhabited outside energy landscapes is not fully aware of the aesthetic, educational, experiential, spatial and environmental aspects of 'energy regions' in the tourism context [2], so visitation of these regions should generate substantial interest. Energy-tourism encircles various tourist initiatives: educational trails, exhibitions, guided tours to energy camps [3]. Finally, 'energy tourism has considerable societal relevance [3], especially regarding clean energy solutions, climate change and sustainability in general.

Energy tourism can be regarded as a segment of special interest tourism [4] and/or industrial tourism segment [5]. Energy tourism „involves visits by tourists to former, retired, or regenerated sites, as well as to still operational energy sites where some facilities, services, or activities have been provided specifically for tourists' use" [6, p. 1398]. According to Frantál & Urbánková [6] energy–tourism nexus conceptualization, the following three perspectives were highlighted: energy as a driver of tourism, energy as a constraint of tourism and energy as a tourist attraction. Jiricka et al. [3] proposed two approaches towards the creation of the energy tourism product. The first one is expert-based energy-tourism, while the second relies on energy experience tourism.

* Corresponding author: darko.dimitrovski@kg.ac.rs

Some previous studies [7,8,9] were focused on the nexus between energy and tourism, emphasizing the relationship between tourism turnover and energy consumption. Conversely, some recent studies [3,8,10] have initiated the energy tourism discussion that nurtures the value of energy tourism attractions and their visitation's experiential dimension. Thus, the proposed study follows the attraction/experience-based trend and investigates Serbia's underinvestigated energy tourism product.

The shift within energy tourism research was also made from traditional energy sources towards exploiting renewable energies in a tourism context [11]. First attempts at examining renewable energies as a niche tourism product have been made by Jiricka et al. [3]. Frantál & Urbánková [6] study of energy tourism has acknowledged both renewable and traditional energy sources as energy tourism examples. However, hydro energy and plants as the first known electric energy source have unfortunately not received well-deserved attention. Thus, the proposed study will further examine the phenomena by depicting the value and experience embedded in a hydro-plant facility and its tourism visitation.

Previous studies [3,6] have implemented demand and supply evaluation of the specific energy tourism sites, however, there is still limited understanding of their online representation, especially in the context of the (hydro) energy tourism supply. Thus, it was interesting to systematize and analyze the current tourism offer regarding Serbia's hydro energy, especially how this offer was textually and visually presented to the potential visitors.

Thus, the paper aims to identify and analyze the online content related to hydro energy tourism attractions and accompanying experiences in Serbia. In this way, the existing Serbia hydro energy tourism offer was systemized, leading to the enhancement of its visitations in the future. The contribution of the study was both theoretical and practical. Theoretically, energy tourism as a niche topic benefits the novel approaches, while practically it raises the awareness of its existence and importance of renewable sources for achieving sustainability.

2 Materials and methods

Serbia has a long history of hydro energy use. For example, hydro plant "Pod Gradom" was the first hydro plant in Serbia and, among the first ones in Europe, initially set in motion in 1900, solely four years after launching the Niagara Falls power plant. "Pod Gradom" was built following Nikola Tesla's patents and inventions introduced by his close friend and collaborator professor Đorđe Stanojevic. Electrification of Serbia using hydro energy continued with the following hydro-power plants, such as Vucje built in 1903, Sveta Petka in 1908, Gamzigrad in 1909, Moravica in 1911, Temac in 1940, etc.

Serbia's current energy policy is based predominantly on traditional energy sources (see Table 1). Thus, giving proper attention to its hydro energy potential and its valorisation as a tourism resource would strengthen an initiative of restructuring the current energy policy towards renewable energy sources, with particular attention given to hydro energy.

Table 1. Electricity generated in Serbia (2018)

Coal	22,546 GWh	65.29%
Hydro	11,329 GWh	32.81%
Wind	124 GWh	0.36%
Solar	10 GWh	0.03%
Gas	399 GWh	1.15%
Biomass	120 GWh	0.35%
Total	34,525 GWh	100

A two-step content analysis approach was employed to propose a comprehensive understanding of Serbia hydro energy tourism's online representation. The first step includes mapping a pool of sampled Serbia hydro energy tourist attractions (hydro-plants) and accompanying experiences and showing their visibility on Google. The second step encircles analysing online content of previously singled attractions and their accompanying experiences. In the first step, to identify Serbia's hydro energy tourism potential and chose the appropriate keywords for Google search, an examination of hydro energy tourism attractions (larger and smaller hydro plants) was deployed. Thus, to identify larger hydro plants names, the online energy portal was used [12], while small hydro-plants names were chosen based on the map proposed on elektroenergetika website [13]. As a final Google query, names of identified larger and smaller hydro plants were combined with the keyword "hydro plant energy tourism".

An index called "visibility ratio" [14] was calculated by dividing the number of results presented by the total number of Web pages indexed by Google containing the hydro-plant name. Only the first ten webpages retrieved through Google search based on the hydro energy tourism search were further analyzed regarding its content. The websites were analyzed to identify a direct resemblance to energy tourism, following the identified websites' content. A combined approach focus on both attractions [6] and accompanying experiences [3] was deployed.

3 Results and discussions

Initially, Serbia energy tourism and a more focused search term Serbia hydro plant energy tourism were examined within the Google trends search (both globally and on a national level), however, it has not generated enough data to implement analytics. Thus, a manual search was implemented in line with the methodology proposed by Xiang et al. [14].

Overall, 45,735,430 Web pages are related to 36 hydro-plants (11 large and 25 smaller) as hydro energy tourist attractions. However, only a very small fraction of indexed Web pages are shown as search results (mean=696 Web pages) accessible to a user (490.163), resulting in an overall visibility ratio of 0.010%. Considering that most engine users only view the first search result page, the actual visibility ratio is much lower. Djerdap hydro-plant, as an internationally renowned and largest hydro plant on the Danube in Europe, have a high visibility ratio of 0.113. Interestingly, some of the hydro plants that were small and not of such importance had indicated visibility ratios that are higher in comparison to Djerdap hydro plant, for example, it has been noted for Sicevo (0.705), Vucje (0.102) and Temac (0.264) small hydro plants (see Table 2).

Table 2. Google results for the Serbia hydro plants

Destination/site	Google results		
	Total Indexed	**Results presented**	**Visibility Ratio**
Large hydro-plants			
1. Djerdap	266.000	30.200	**0.113**
2. Pirot	5.680.000	28.400	0.005
3. Vrla	7.560.000	32.300	0.004
4. Bajina Basta	668.000	15.200	0.023
5. Uvac	536.000	28.700	0.053
6. Kokin Brod	157.000	3.480	0.022
7. Bistrica	7.140.000	33.000	0.005
8. Potpec	61.400	29.400	**0.479**
9. Zvornik	5.420.000	39.200	0.007
10. Ovcar Banja	78.200	863	0.011
11. Medjuvrsje	38.400	8	0.001
Total	**27.605.000**	**240.751**	**0.066**
Small hydro-plants			
12. Radaljska Banja	69.000	2.870	0.041

13. Vrelo	4.030.000	38.000	0.009
14. Pod Gradom	2.430.000	25.200	0.010
15. Turica	157.000	8	< 0.001
16. Kratovska reka	4.730	9	< 0.001
17. Seljasnica	240.000	8	< 0.001
18. Moravica	484.000	29.300	0.061
19. Raska	7.610.000	27.200	0.003
20. Sokolovacka	608.000	8	< 0.001
21. Gamzigrad	166.000	31.500	0.190
22. Jelasnica	317.000	9	< 0.001
23. Sicevo	37.700	26.600	**0.705**
24. Sveta Petka	1.610.000	13.000	0.008
24. Temac	254.000	25.900	**0.102**
25. Vucje	113.000	29.800	**0.264**
Total	**18.130.430**	**249.412**	**0.270**

Source: Author's research

Within the second phase of the research, each of the hydro tourism attraction and websites on its first search page has been individually examined regarding the online content. Based on the gathered data of large hydro plants presented in the previous table, the following conclusions should be made:

Only one sizeable hydro plant (Djerdap) (Figure 1) provide some reference to an organized tour of its facilities, however, only as a part of the tour package provided by a travel agency and not specifying details of what kind of experience they offer.

Departure to Đerdap HE plant and a tour of one of the largest HE plants in Europe [15]

Fig. 1. Djerdap (Iron Gate) hydro plant
Source: http://www.tos.org.rs/node/990

Moreover, considerable attention was also given to the artificial lakes that have been formed as a result of the existence of the investigated hydro plants (Bajina Basta, Potpec, Uvac, Bistrica, Zvornik and Kokin Brod). For example, Trip Advisor's textual content was directed to the lake-related activities and opportunity to experience them as a tour-package activity, however, limited to the lake itself, ignoring the context of the hydro plant experience:

Where there are hydro power plants, there are also lakes, and on the, of course, beaches. On the developed beach in Perućac, during the summer, five to six thousand people come every day to cool down. Next to the lake there is a hotel with bungalows, open plane, and a restaurant. From there you have a beautiful view over the lake and the 90 meter high and 461 meter long dam. Reservations for the hotel in Perućac can be made through the agency "EPS turs" [16].

The Perucac artificial lake has been extensively visualized, however, solely one photo of the Bajina Basta hydro plant was proposed.

Finally, some interest was given to hydro plants' investments (Bistrica HP, Zvornik HP, Ovcar Banja and Medjuvrsje) (see Table 3).

Table 3. Large hydro-plants content analysis

Attraction	Website address	Textual content	Visual content (Number of images)
1. Djerdap	https://serbianadventures.com/en/adventure/view/197/djerdap-gorge-and-rajac-wine-cellars	√	/
2. Pirot	/	/	/
3. Vrla	/	/	/
4. Bajina Basta (Perucac)	https://www.tripadvisor.com/Attraction_Review-g1028708-d10365086-Reviews-Perucac_Lake Bajina_Basta_Central_Serbia.html https://www.nacionalnarevija.com/en/tekstovi/Br%2021/Osvetljavanje%20-%20Drinsko-limske%20elektrane.html	√	**1**
5. Uvac	http://www.serbia.com/visit-serbia/natural-beauties/unique-nature/uvac-a-natures-masterpiece/ https://www.republiktours.com/explore/uvac-lake/ https://balkangreenenergynews.com/eps-gets-approval-in-serbia-for-gornja-drina-hydropower-project-in-bih/	√	
6. Kokin Brod			
7. Bistrica			
8. Potpec	https://www.intellinews.com/vast-tide-of-floating-waste-threatens-balkan-hydropower-plants 199558/?source=serbia/	√	/
9. Zvornik	https://turizamrs.org/en/zvornicko-jezero/ https://renewablesnow.com/news/serbias-eps-to-complete-overhaul-of-zvornik-hpp-by-end-2019-599334/	√	/
10. Ovcar Banja	https://www.ekapija.com/en/news/188866/revitalization-of-hydroelectric-power-plants-on-the-drina-river-finally-started-first\	√	/
11.Medjuvrsje	/	/	/

Based on the gathered data of smaller hydro plants presented in table 3, the following conclusions should be made:

Most of the figures were identified within the group of smaller hydro plants. For example, the ratio between larger and smaller visual (photographic) representations of hydro plants was 1:27. The smaller hydro plants' photo content was directed to their architecture and equipment (mainly designed based on Tesla's principles in power plant operation and electricity transmission implemented in practice by Djordje Stanojevic as a close friend of Nikola Tesla).

In contrast to larger hydro plants that have been interpreted within a broader tourism context, the smaller hydro plants were generally tourism-focused and widely discussed as tourism attractions within the analyzed website content. Several of the smaller hydro plants (Pod Gradom, Sveta Petka and Vucje) were attracted the interest of ERIH (European Route of Industrial Heritage) as they act as anchor points under the patronage of ERIH (Figure 2).

Fig. 2. Hydro plant Pod Gradom (left) and Vucje (right)
Source: https://www.erih.net/i-want-to-go-there/site/pod-gradom-hydroelectric-power-plant; https://www.erih.net/i-want-to-go-there/site/vucje-hydroelectric-power-plant

Cultural route 'Tesla Ways' has included most of the smaller hydro plants (Vrelo, Turica, Pod Gradom, Moravica, Sicevo, Sveta Petka, Temac and Gamzigrad) as locations along the route. In 2019 'Tesla Ways' has become the first official cultural tour of the Council of Europe dedicated to a scientist. Involving them as a part of the route has increased their tourist potential and positioned them in the broader international context (see Table 4).

Table 4. Small hydro-plants content analysis

Attraction	Website address	Textual content	Visual content (Number of images)
12. Radaljska Banja	/	/	/
13. Vrelo	http://www.hotelitara.mod.gov.rs/en/day-trips/reka-vrelo-i-stecci-u-peruccu	√	/
14. Pod Gradom	https://www.erih.net/i-want-to-go-there/site/pod-gradom-hydroelectric-power-plant http://e2012.drustvotermicara.com/english/news/conference-excursion http://nmuzice.org.rs/locations-of-the-national-museum-uzice/hidroelectric-power-plant/?lang=en	√	11
15. Turica	/	/	/
16. Kratovska reka	/	/	/
17. Seljasnica	/	/	/
18. Moravica	https://ivatourism.org/en/tourist-offer/what-to-see/cultural-monuments.htm https://serbianadventures.com/en/adventure/destinations/218/Ivanjica http://tonp.rs/water/?lang=en	√	1
19. Raska	/	/	/
20. Sokolovacka	/	/	/
21. Gamzigrad	https://serbianadventures.com/en/adventure/view/377/nikola-tesla-s-route-industrial-heritage-of-nis-area-three-day-tour https://serbianadventures.com/en/adventure/view/130/magical-eastern-serbia-2-days http://www.istnews.com/ebanje34.htm	√	2
22. Jelasnica	/	/	/
23. Sicevo	/	/	/
24. Sveta Petka	http://visitniskabanja.org/en/hidroelektrana-sveta-petka/ https://www.erih.de/da-will-ich-hin/site/sveta-petka-hydroelectric-power-plant	√	4

24. Temac	https://www.masina.rs/eng/private-hydropower-plants-destroying-natural-resourses/	√	1
25. Vucje	https://www.atlasobscura.com/places/vucje-hydroelectric-plan http://kmi.vtsns.edu.rs/KMI_2020/radovi/5-KMI_Razno/KMI_razno-5.pdf https://naturetraveloffice.com/en/team-building/avantura-porecje-vucje/ https://www.erih.net/i-want-to-go-there/site/vucje-hydroelectric-power-plant	√	8

4 Conclusions

Based on the Google web pages analysis, Serbia's hydro energy tourism attraction's overall visibility is relatively low, especially since most of the identified and analysed web pages (within the analysed first ten) is not related to the keyword search. Moreover, a large number of searches was limited only to one search result page.

The textual and visual online representation of the Serbian hydro energy tourism offer was grounded on a combined expert and experience-based energy tourism approach [3]. An expert-based approach could be referred to hydro plant "Pod Gradom" visitation as a part of the social programme within the International conference on power plants. Analysis of the online content has recognized an initial effort to create hydro plant experiences, both as a 'stand-alone' offer or tour packages (especially those within the ERIH route and Tesla Ways). Unfortunately, this offer has a limited correlation with the 'traditional tourism offers', thus, additional effort should be invested in complementing artificial lake activities formed by larger hydro plants with the hydro plant experience itself. Thus, this issue deserves additional attention, especially considering future tourism management and marketing efforts.

Visual content follows abundant textual material identified within the small hydro plants' content analysis. The particular interest was given to the smaller hydro plants' architecture and technology in line with their importance of industrial heritage guardians acknowledged by ERIH and the Tesla Ways route. The majority of analyzed small hydro plants examples were driven by the value and uniqueness of its technical facilities, which are opened to the public via guided tours. Thus, the analyzed offer is predominately driven by its industrial heritage and educational purposes that derive from it. It is in line with the Kolblmüller and Bärntaler [17] argument that knowledge transfer (technological facts and benefits discussed during guided tours or seminars) was an essential element of energy-tourism. To some degree (primarily for small hydro plants), environmental education and raising environmental awareness were pointed out, which could direct awareness towards renewable energies, especially considering Serbia's energy structure and its comparison to EU countries. Unfortunately, other tourist experiences that derive from the hydro plant visitation are neglected.

Finally, looking from a perspective of the scale of the facilities, surprising results were obtained. One could expect that larger hydro plants will dominate the offer, however, except Djerdap HE, other larger hydro plants were not considered as tourist attractions per se, yet only in the context of their dams and accompanying artificial lakes and experiences that derive from it. Thus, it could be concluded that based on the current online representation, their value was recognized only as a part of broader natural scenery. Moreover, analysis of Djerdap hydro plant energy tourism web pages, regardless of its continental importance, has limited reference to tourism, especially regarding opportunities of its visitation and proposed experiences. The main reason behind that acquired findings is that these facilities are considered essential energy suppliers in public property, so their visitation is strictly forbidden as a strategic national importance resource.

The study was also faced with several limitations. Analysis of the Google search results reflect the specific point in time, and thus it is challenging to make generalizable conclusions based on these findings. The study was also faced with the limited material written in English, the issue expected for the initial phase of the energy-tourism development and its perception on a national scale. Thus, future research endeavours should be directed in evaluating the existing online offer on the Serbian language, as it will provide a more realistic insight into the investigated phenomena.

References

1. P. Spath, H. Rohracher, Research Policy, **39(4)**, 449–458 (2010)
2. M. Pasqualetti, Geographical Review, **90**, 381–394 (2000)
3. A. Jiricka, V. Wirth, B. Salak, U. Pröbstl, New Challenges for Tourism Promotion: Tackling High Competition and Multimedia Changes, International Tourism Research and Concepts, 85-103 (2012)
4. B. Trauer, Tourism Management, **27(2)**, 183-200 (2006)
5. A. Otgaar, Tourism Management Perspectives, **4**, 86-91 (2012)
6. B. Frantál, R. Urbánková, Current Issues in Tourism, **20(13)**, 1395-1412 (2017)
7. C. Işik, E. Doğan, S. Ongan, Economies, **5(4)**, 40 (2017).
8. M.K. Nižić, Z.Š. Grdić, A. Hustić, Academica Turistica-Tourism and Innovation Journal, **9(2)**, (2017).
9. A. Sharif, S. Saha, N. Campbell, A. Sinha, D.M. Ibrahiem, Current Issues in Tourism, **23(23)**, 2930-2949 (2020)
10. A. Jiricka, B. Salak, R. Eder, A. Arnberger, U. Pröbstl, In: Brebbia, C.A., Pineda, F.D., Sustainable Tourism IV (2010)
11. Chen J.S./ Sloan P./Legrand W.: Sustainability in the Hospitality Industry: Principles of Sustainable Operations von Butterworth Heineman. (2009)
12. Energetski portal, *https://www.energetskiportal.rs/predstavljamo-hidroelektrane-u-srbiji/* (2014)
13. Elektroenergetika, *https://www.elektroenergetika.info/he-sr.htm* (2021)
14. Z. Xiang, D.R. Fesenmaier, M.Y. Hyun, K. Wober, *http://scholarworks.umass.edu/cgi/viewcontent.cgi?article=1374&context=ttra* (2016)
15. Serbian Adventures, *https://www.elektroenergetika.info/he-sr.htm* (2021)
16. Tripadvisor. Perucac lake, *https://www.tripadvisor.com/Attraction_Review-g1028708-d10365086-Reviews-Perucac_Lake Bajina_Basta_Central_Serbia.htm* (2021)
17. M. Koblmüller, J. Bärntaler, Presentation on the conference: 'Energieregionen – Energie für Regionen', Bruck/Leitha (2009).

E3S Web of Conferences **250**, 02003 (2021) https://doi.org/10.1051/e3sconf/202125002003
TRESP 2021

Optimization and diversification of natural gas supply in Ukraine

Vadim Tkachuk[1]*, *Andriy* Skrypnyk [2], *Viktoriia* Baidala[3], *Nataliia* Klymenko[2], *Yuriy* Namiasenko[2]

[1]Department of Labour Economics and Social Development, Faculty of Economics, National University of Life and Environmental Sciences of Ukraine, Heroiv Oborony 11, 03041, Kyiv, Ukraine
[2] Department of Economic Cybernetics, National University of Life and Environmental Sciences of Ukraine, Heroiv Oborony 16, 03041, Kyiv, Ukraine
[3]Department of Economic Theory, National University of Life and Environmental sciences of Ukraine, 03041, Kyiv, Ukraine

Abstract. This article considers different variants of natural gas supply in Ukraine which are based on the main tendencies of natural gas regional markets and some important factors, such as: possibility to use natural gas storage system, different changes in government regulation and innovations in the energy generation and supply sphere. The purpose of this work is to substantiate the current gas supply strategy in Ukraine on the basis of quantitative methods. The main results of the article are: Numerical estimation of the positive effect, gained from the using of natural gas storage system; Numerical estimation of the positive effect from the diversification of the natural gas supply in Ukraine in terms of decreasing the potential role of Gazprom monopoly. The population response evaluation on the gas price rising. This article provides an effective tool for assessing the possible consequences of decisions of state bodies when moving away from the existing monopoly of gas supply system and the macroeconomic effect of different gas supply options. The quantitative assessment of the gradual convergence of regional markets and the creation of a global natural gas market is proposed

1 Introduction

Nowadays it became quite clear that very optimistic forecasts about reaching 20% of the renewable energy mark in the Ukraine total energy consumption by 2020. If we consider three main components of renewable energy: bioenergy, solar and wind energy then primary hopes in "Energy strategy of Ukraine to 2035" have been laid on bioenergy. The reason for such kind of forecasts was an overestimation of bioenergy raw materials volumes which are formed during agrarian production. But this way requires [1] billions of investments in creating an appropriate infrastructure (technology, qualified staff, transport) [2]. Despite of the increasing interest to renewable energy, the natural gas still remains the main source of heating and electricity production even in the European countries [3,4,5].

* Corresponding author: vadtkachuk@hotmail.com

Apart from this, a significant negative impact on Ukrainian bioenergy development has the cost of efficient bio energy generators, the lowest price of which approximately equals 1 million USD [6], due to the greater financial availability of solar energy, the pace of its development turned out to be significantly higher [7] This means that only big agrarian companies can afford purchasing of them. Also tariffs of renewable energy have a significant impact on choosing a type of energy generation [8].

As a result, the rate of renewable energy growth in Ukraine is extremely low. The state of Ukrainian traditional energy is rather critical [8,9]. Nuclear generation provides more than 50% of total consumed energy but it consists primarily of reactors which have age of 30 years (30 years is an upper time limit of operating period). After modernization any reactor can safely work for only 10 years. It should be noticed that despite the conducted modernization of nuclear energy equipment with aging the probability of incidents is increasing [10], [11].

The solution of this problem is also rising up the level of traditional energy generation efficiency (which also provides the reduction of environmental pollution) [12,13].

Considering all above mentioned, the natural gas supply in Ukraine still is on the top of energy development agenda including the diversification of gas supply to Ukraine.

This work shows that natural gas supply diversification leads to reducing overall costs, approximately on 5-10%. The significant disadvantage of this work was not taking into account existing underground gas storage facilities in Ukraine, the total volume of which equals 30 billion of cubic meters [9, 14]. At this time all storage facilities are used only on 40% of their entire capacity. Significant seasonal price fluctuations which exist on different regional gas markets [15] suggest that using of natural gas storage system can have enormous economic effects. Despite the fact that Black Sea is closed for liquid natural gas (LNG) carriers it would be also useful to estimate the possible economic benefits of natural gas supply from one of the biggest global LNG supplier Qatar [16] to Black Sea ports of Ukraine. The great interest of researchers causes the question of global natural gas market formation.

Researchers' views on the problem of global natural gas formation differ significantly. Most of them insist that there is some stable tendency toward global gas market formation due to the development of innovative technologies for extraction and transportation of natural gas. Looking back to 2009, the first hypotheses about globalization of natural gas markets to the level of intercontinental trade and significant integration of the market prices were put forward [17,18]. A hypothesis about possible natural gas markets integration due to increasing volumes of LNG trade across the Atlantic Ocean basin was proposed. LNG is also considered as one of the main factors which have a great impact on natural gas price formation [19].

There were also some attempts to model the European natural gas markets in form of a two-stage game [20]. By using nonlinear optimization, the main result of the research: biggest supplier of natural gas in Europe – Russia, continues to lose its market share, while the share of liquefied natural gas import from different countries continues to growth.

The basic model of global natural gas market carries out optimization of the future productive, transport and storage capacities [21]. Besides the pipeline transportation of natural gas, the potential transportation of LNG was also taken into account. The given model allows additionally to model strategic behaviour of different players in the market (for example, the behaviour of producers).

It was proved that European natural gas market is characterized by oligopolistic supply structure which consists mainly from external suppliers herewith the capacity of transport infrastructure (pipelines and storage facilities) has a key role during determination of trade flows.

The results of modelling show a large inverse influence of the distance factor on the import volumes of pipelined and liquefied natural gas [22]. At the same time, importers on the liquefied natural gas market are characterized by large capacity of storage facilities, what can't be said about the European natural gas market which highly dependents on the pipeline supply system.

2 Materials and methods

Natural gas has always been one of the main sources of primary energy consumption in Ukraine. At this time, it takes the second place after nuclear energy. Although the growth of energy efficiency is extremely slow some positive vibes can be observed and it has impact on reducing power consumption. The same impact on total power consumption reducing has the transition of Crimea and some parts of Donetsk and Luhansk regions under control of Russia. This effect is also exacerbated by the local economic crisis of the recent years (Fig.1).

In the dynamics of gas consumption (which consists of two components: consumption by the population and industry) during global economic crisis 2008-2009 a change of ranks appeared: consumption by the population has started steadily exceed consumption by industry.

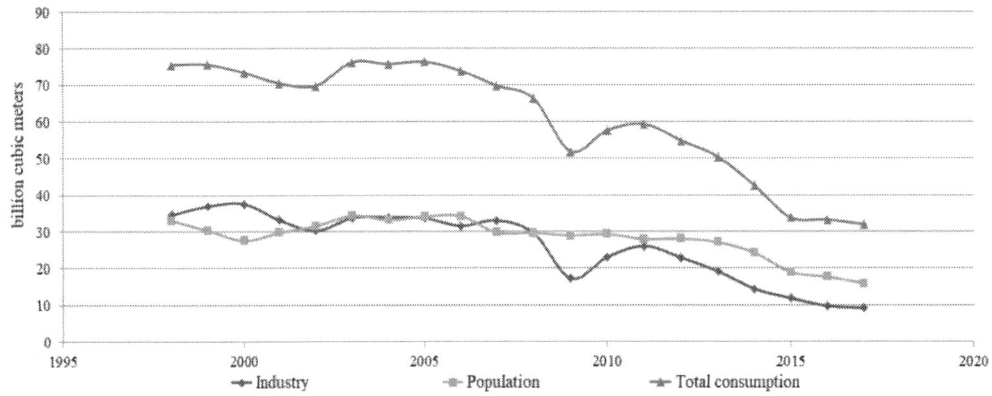

Fig.1. Natural gas consumption by industry and population in Ukraine, 1998-2017

At this it becomes quite hard to forecast whether this tendency toward natural gas consumption decreasing will be taking place in the future or not. In our opinion, there are some signs of natural gas consumption stabilizing at the estimated level of 30-32 billion cubic meters. But newly announced intent to raise the natural gas price by government in 2020-2021 may lead to significant reduction of gas consumption by the population.

It is known that Ukraine has at its disposal many gas storage facilities of considerable volumes, which all together can store almost the equivalent of an average annual consumption. Therefore, it is a question of large interest how the including of gas storage factor in the optimization problem will affect the total natural gas supply expenditure. Taking into consideration climatic conditions of Ukraine it should be noted that gas storage facilities in general work as consumption regions during summer season and as supply regions during winter season.

Using of gas storage facilities is also stipulated by the existing difference in spot and futures prices during winter and summer periods. The influence of seasonality and the seasonal variation on a price level was estimated by a two-factor model [14]. This model used the weekly price observations of the Henry Hub natural gas contracts.

Using available data with regard to seasonal price variation on Henry Hub it became possible to measure the price variability per 1 thousand cubic meters (common measuring of natural gas volumes in Eastern Europe region). Natural logarithm for the price of 1 (one million British Thermal Units) MMBtu during summer period equals 1.57 and during winter period equals 1.68. This means that summer price of 1 MMBtu equals 4.8 USD for the Henry

Hub market and winter price is respectively 5.4 USD. In terms of cubic meters, we get the summer price of 170 USD and winter price of 194 USD per 1000 cubic meters. Thus, the average seasonal difference of prices for the last several decades on Henry Hub market approximately equalled 24 USD. This also means that by the end of the winter season the price is reduced by 12,4%, and during autumn season it's being increased by 14.1% [14].

Let's consider the geographic location of gas storage facilities in Ukraine. To simplify the calculation of the optimization model, all gas storage facilities (total capacity of 31 billion cubic meters) were grouped according to their geographical location: i) West gas storage (capacity of 20.9 billion cubic meters); ii) Southwest gas storage (capacity of 4.5 billion cubic meters); iii) North gas storage (capacity of 3.1 billion cubic meters); iv) East gas storage (capacity of 2.5 billion cubic meters).

Further the total capacity of each gas storage group will be related to the main consumption regions: i) West gas storage (Lviv, capacity of 25.4 billion cubic meters); ii) North gas storage (Kyiv, capacity of 3.1 billion cubic meters); ii) East gas storage (Kharkiv, capacity of 2.5 billion cubic meters).

In order to solve the optimization model it is necessary to have some data about the cost of natural gas storage. Let us formulate the optimization model of natural gas supply which include the gas storage system and price difference in summer and winter periods.

At first denote: j - index of natural gas supply region, j = 1,…., m (external and internal sources all together); i - index of natural gas consumption region, i = 1,…., n. D_i - volume of the natural gas which was consumed in i region ($D = \sum_{i=1}^{n} D_i$ - total consumption); d_{ij} - distance from j supplier to i consumer (measured in 100 kilometres); $\rho_{ji};\mu_{ji}$ - matrices of natural gas supply volumes from j supplier to i consumer during summer and winter seasons respectively (measured in 1000 cubic meters); p_0 - transportation cost of natural gas (measured in USD for 1000 cubic meters per 100 km.); Pr – natural gas price during summer season (in USD for 1000 cubic meters); f*Pr – natural gas price during winter season (f>1), Sl - capacity of l gas storage facility (l= 1,...,k); $\Delta -$ natural gas storage cost (measured in USD for 1000 cubic meters during 6 months; $V_j -$ maximum volume of natural gas which can be provided by the j supplier.

During summer season gas storage facilities are considered as ordinary consumers and their consumption volumes are added to the total consumption but during heating season the same gas storage facilities can be already considered as suppliers and the stored natural gas can be added to other suppliers.

The supply of gas in the summer and the heating season are considered separately.

The objective function which has to be minimized consists of expenditure on purchasing, storage and natural gas transportation:

$$W(\rho,\mu) = C_d(\rho,\mu) + C_T(\rho,\mu) \Rightarrow \min \tag{1}$$

The cost of purchasing and storage can be stated as:

$$C_d(\rho,\mu) = \Pr\sum_{i=1}^{n}\sum_{j=1}^{m}\rho_{ji} + f\cdot\Pr\sum_{i=1}^{n}\sum_{j=1}^{m}\mu_{ji} + (\Pr+\Delta)\sum_{j=m+1}^{m+k}\sum_{j=1}^{m}\mu_{ji} \tag{2}$$

The last equation determines the cost of natural gas which was supplied directly to consumer's regions during the summer and heating seasons and the cost of natural gas stored in gas storage facilities for the next sixth months (the same level of price as it had been in summer, but also adding to it the cost of storage).

Transportation cost:

$$C_T(\rho,\mu) = p_0\sum_{i=1}^{n+k}\sum_{j=1}^{m}\rho_{ji}\cdot d_{ji} + p_0\sum_{i=1}^{n}\sum_{j=1}^{m+k}\mu_{ji}d_{ji} \tag{3}$$

The first part of the equation (3) determines the transportation cost of the natural gas during summer period and the second part determines the same transportation cost but only for heating season.

There is block of constraints which is necessary for regional volumes of consumption and maximum storage capacity of gas storage facilities:

$$\sum_{j=1}^{m}(\rho_{ji}+\mu_{ji}) + \sum_{j=m+1}^{m+k}\mu_{ji} \geq D_i; i = 1,2,...,n$$

$$\sum_{j=1}^{m}\rho_{j,m+l} \leq S_l; l = 1,2,...,k$$

$$\sum_{i=1}^{n}\mu_{li} \leq \sum_{j=1}^{m}\rho_{jl}; l = 1,2,...,k \tag{4.1- 4.4}$$

$$\sum_{i=1}^{n+k}\rho_{ji} + \sum_{i=1}^{n}\mu_{ji} \leq V_j; j = 1,2,...m$$

The first constraint (4.1) means that supply from primary suppliers and storage facilities have to satisfy the consumption volumes of each region. The second constraint (4.2) means that storage facilities cannot hold volume of natural gas which exceeds their maximum capacity. The third constraint (4.3) means that natural gas supply from storage facilities doesn't have to exceed the volume of injected natural gas. The last constraint (4.4) means that each supplier can't exceed its own export limits.

Besides this there is a need to include a natural constraint concerning consumption ratio during summer and heating periods. At the first phase it was considered that summer consumption equals 40% of total annual one and heating period consumption has the other 60%. The full list of natural gas suppliers on the west border of Ukraine is given in table 1.

At first let's consider merely the effect of geographic diversification of gas supply. To eliminate the price factor let's give each of the suppliers the same price. In this formulation of the optimization, it's advisable to add Russia as one of the suppliers. There are two possible solution of the model:

Monopolized gas supply from Russia, at the price of 220 USD for 1000 cubic meters; Objective function = 14.807 billion USD; Purchasing cost 8.474 billion USD; Transportation cost = 6.333 billion USD.

Gas supply only from EU countries (table 1), at the price of 220 USD for 1000 cubic meters; Objective function = 8.102 billion USD; Purchasing cost 8,474 billion USD; Transportation cost = 2.762 billion USD.

These calculations show the positive effect of reducing Transportation cost from natural gas supply diversification. The result of natural gas supply optimization in prices of 2016 including storage facilities is presented in table 1. The upper part of the table corresponds to the gas supply during the summer period, the lower part represents gas supply during heating period.

Table 1. Matrix of natural gas supply in2016 prices o including storage facilities, billion cubic meters*

	Donetsk	Kyiv	Lviv	Chernivtsi	Kharkiv	Cherkasy	West, s	North, s	South, s
Drozdovichi		2.93	2.22						
Uzhhorod									
Beregove				1.32		1.30	5.26		
Chernivtsi							4.00		
Poltava	0.60				0.63	1.14		4.11	1.52
Kharkiv					3.07				
Drozdovichi									
Uzhhorod									
Beregove									
Chernivtsi									
Poltava									
Kharkiv					**4.93**				
West, s		**3.95**	**3.32**	**1.99**					
North, s		**0.45**				**3.66**			
South, s	**0.90**				**0.62**				

*(natural gas supply during winter season is denoted by bold numbers, groups of gas storage facilities denoted by letter "s")

Objective function = 8.71 billion USD. Purchasing cost = 6.01 billion USD. Transportation cost = 2.70 billion USD. According to the optimization model 14.9 billion cubic meters of natural gas have to be pumped into the storage facilities. The West group of storage facilities has to be filled in with 9.26 billion cubic meters, the North group with 4.11 billion cubic meters and East group with 1.52 billion cubic meters. It should be noted that during winter season almost all gas supply was performed from storage facilities except Kharkiv region where natural gas was taken from the domestic producer.

Let's compare the result of optimization model in prices of 2016 which include storage facilities with the result of the same optimization but without including storage facilities (table 2). Prices in both models were taken at the same level.

Objective function in case of not using storage facilities equalled 11.71 billion USD; the purchasing cost equalled 9.54 billion USD and transportation cost was 2.17 billion USD. In this case the price difference was taken at the level of 14% (f=1.14). Comparing the obtained objective functions of both optimization models prove the advantage of gas storage using.

Table 2. Matrix of natural gas supply in prices of 2016 without including storage facilities, billion cubic meters*

	Donetsk	Kyiv	Lviv	Chernivtsi	Kharkiv	Cherkasy
Drozdovichi		2.93	2.22			
Uzhhorod						
Beregove				1.32		
Chernivtsi						
Poltava	0.60				1.25	2.44
Kharkiv					2.45	

Drozdovichi			**3.32**			
Uzhhorod						
Beregove		**4.40**				**0.16**
Chernivtsi				**1.99**		**0.69**
Poltava	**0.90**					**2.81**
Kharkiv					**5.55**	

*(natural gas supply during winter season is denoted by bold numbers)

A similar comparison was made for the gas supply in prices of 2017. Since the average weighted price of 2017 (table 3) was significantly lower than the price of 2016 - $\overline{\Pr}_{2016} = 269USD/1000м^3; \overline{\Pr}_{2017} = 196USD/1000м^3$, both objective functions of 2017 had lower values then in 2016.

Table 3. Import of natural gas to Ukraine from different countries in 2017

Country	Import, billion cubic meters	Import cost, million USD	Price USD/ 1 тис. м3
Germany	3.700	847.8	229.1
Poland	1.500	343.2	228.8
France	1.020	237.9	233.2
United Kingdom	0.542	127.3	234.9
Hungary	0.768	178.9	232.9
Czech Republic	0.252	55.6	220.6
Slovakia	0.222	55.3	249.1
Luxembourg	0.179	40.7	227.4
Italy	0.067	15.9	237.3
Austria	0.048	11.2	233.3
Domestic producer	20.500	-	179.6

Including storage facilities into optimization model has significant benefits. The value of objective function in case of using storage facilities is less than in case of its absence (7.6 and 9.2 billion USD respectively). The main factor which defines the cost difference is that purchasing cost of natural gas in case of using storage facilities is less by 1.5 billion USD although the transportation cost is 0.5 billion USD more. Just like in case of 2016, 14.9 billion cubic meters were pumped into the gas storage facilities, but the volumes of the pumped in natural gas differ for different groups of storage facilities. The next question arises: "How in real world volumes of purchased natural gas will differ from optimized values?"

Volume of available natural gas in storage facilities at the beginning and end of winter season allows us to estimate the volume of natural gas which was pumped out from the storage facilities of Ukraine (fig. 2). The average level of natural gas consumption at the time interval of 2006-2016 was 12.7 billion cubic meters, the maximum consumption during these years was 16.9 billion cubic meters and the minimum consumption was 8.5 billion cubic meters. The volume of not used natural gas in storage facilities after winter season in average equals 9.7 billion cubic meters but during the last years its level decreased to 6 billion cubic meters, which at present time can satisfy national consumption for approximately two months. Significant natural gas residue leads to efficiency decreasing of storage facilities using however in case of gas shortages during heater season possible losses will be essentially higher. The possibility of such situation was shown by winter season of 2017-2018 when the rapid decrease of stored gas led to restrict of consumption.

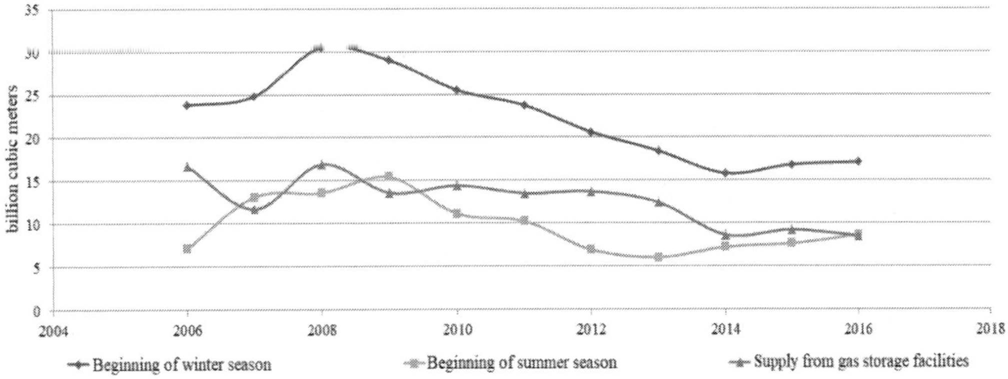

Fig. 2. Stored natural gas in storage facilities, 2000-2017

In any case, practically a double decrease of pumped natural gas in storage facilities, which had been taking place during recent years, indicates the incomplete use of existing opportunities to receive dividends from availability of storage facilities. It should be noted that calculated volume of used natural gas from gas storage facilities during heating season equals 14.9 billion cubic meters and is approaching the average level of gas supply from storage facilities during winter season (12.7 billion cubic meters) on the time range of 2006-2016. Optimal volumes of consumed natural gas were obtained under assumptions that consumption of natural gas during winter season equals 60% of annual level and winter prices are on 14% higher. Certainly, there may be other ratios of natural gas consumption and prices which lead to different optimal volumes of stored natural gas (Table 4).

Table 4. Different optimal solutions of the optimization model which include gas storage facilities, calculated on the data of 2016*

	5%	10%	15%	20%
50%	1) 7.60 2) 5.03 3) 12.66	1) 7.63 2) 5.06 3) 12.66	1) 7.67 2) 5.10 3) 12.66	1) 7.70 2) 5.13 3) 12.66
55%	1) 7.57 2) 4.98 3) 13.77	1) 7.61 2) 5.02 3) 13.77	1) 7.65 2) 5.06 3) 13.77	1) 7.69 2) 5.10 3) 13.77
60%	1) 7.54 2) 4.94 3) 14.89	1) 7.59 2) 4.99 3) 14.89	1) 7.63 2) 5.03 3) 14.89	1) 7.68 2) 5.07 3) 14.89
65%	1) 7.52 2) 4.90 3) 16.00	1) 7.57 2) 4.95 3) 16.00	1) 7.62 2) 5.00 3) 16.00	1) 7.67 2) 5.04 3) 16.00
75%	1) 7.48 2) 4.89 3) 18.23	1) 7.54 2) 4.95 3) 18.23	1) 7.60 2) 5.01 3) 18.23	1) 7.66 2) 5.07 3) 18.23

Source: own calculation

Horizontal line represents percentage of price rising during winter season. Vertical line represents the share of natural gas which is being consumed during winter season. Main elements of the table:

1) Objective function, billion USD;

2) Cost of purchased natural gas, billion USD;
3) Volume of used natural gas from gas storage facilities, billion cubic meters
Available gas storage facilities have a significant effect on the reduction of the objective function (cost of natural gas supply). It reduces the impact of rising prices during winter season.

But even at 20% of price rising and 75% of consumption rising during the winter season the maximum loading of gas storage facilities (18.2 billion cubic meters) doesn't exceed 60% of their total capacity. Access to gas storage facilities is so called damper of the excessive variability of the main financial indicators: transportation cost and cost of purchasing. In order to prove this statement, the estimations of natural gas price variance and variation coefficients, cost of purchased natural gas and volumes of its consumption from storage facilities for different combinations of price rising and volumes of consumed gas during winter season were made (table 5).

Table 5. Descriptive statistics of various optimal solutions for optimization model which include storage facilities

	Objective function, billion USD	Cost of purchased natural gas, billion USD	Volume of natural gas in storage facilities, billion cubic meters
Average value	7.610	5.016	15.110
Variation	0.004	0.004	3.866
Standard deviation	0.060	0.066	1.966
Coefficient of variation, %	0.800	1.300	13.000

Based on the obtained data it can be said that the largest variability is observed for optimal volumes of pumped in natural gas. Other parameters of objective function and purchasing cost stay almost unchanged: maximum value of variation coefficient for purchasing cost is 1.3%, while for optimal volume of stored natural gas it exists at the level of 13%.

In recent years, the significant reduction of natural gas consumption in Ukraine has been taking place, as it was mentioned this happens due to political and macroeconomic factors.

To the end of 2018 IMF recommends to raise the price of natural gas for population ones. That's why it's necessary to estimate volumes of consumed natural gas after price rising in order to calculate the optimal value of gas storage facilities loading. For accomplishing this estimation, the time series (1998-2017) of natural gas prices and volumes of its consumption by population were used.

3 Conclusions

Overall, in recent years, there has been a tendency toward decreasing of the natural gas prices difference on various regional markets. The variance of the average weighted global natural gas price can be considered as one of the main indicators of global natural gas market formation. The variance has significantly decreased during recent years, which also improves the possibilities of gas supply diversification for each importer of natural gas.

Quantitative experiments of solving the optimization model of natural gas supply in Ukraine with the condition that all suppliers have the same price confirmed the efficiency of the natural gas supply diversification, which have to be continued in the future by the efforts of Ukrainian diplomacy. These experiments revealed a significant impact of the natural gas storages availability on minimizing total cost of purchased natural gas. Due to the growth of pumped into gas storage facilities volumes of natural gas it became possible to maintain practically a stable level of gas supply cost (even under condition of significant rising of natural gas price and consumption during winter season).

With the help of econometric methods of time series analysis, an estimation of the elasticity of natural gas consumption by the population according to the price level was made (demand turned out to be inelastic according to the price level; the elasticity value equals E = -0.24). On the basis of the calculated elasticity the estimation of the total consumption was also obtained. Including Odessa hub (LNG from Qatar) to the model as one of the supplies significantly reduces the cost of gas supply. Therefore, effective external economic steps toward achieving transit agreement with Turkey have to be made. Besides this agreement opens up prospects of natural gas re-export to Eastern European countries from Ukrainian gas storage facilities

References

1. T. Shindina, J.Streimikis, Y.Sukhareva, Ł. Nawrot, Economics and Sociology, **11(2),** 334-344 (2018)
2. A.Skripnik, *Energy sector of Ukrainian economy from the perspective of social prosperity* (2017)
3. M. Tvaronavičienė, D. Prakapienė, K. Garškaitė-Milvydienė, R. Prakapas, Ł. Nawrot, Economics and Sociology, **11(1)** (2018)
4. A. Frank, P. Osmundsen, M. Sandsmark, Energy Journal **27(2)** (2006)
5. S. Amir, N. Das, O. Langhelle, Jo. Roy, M. Assadi, Energy Science & Engineering, **7(4),** 1075–1094 (2019)
6. A. Skrypnyk, N.Klymenko, M. Talavyria, A. Goray, Y.Namiasenko, International Journal of Energy Sector Management,**14 (2),** 468-481 (2019)
7. D. Zherlitsyn, A. Skrypnyk, N. Rogoza, S.Saiapin, T.Kudin, Estudios de Economia Aplicada, 2021, **38(4),** 3994 (2021)
8. A. Skrypnyk, Y. Namiasenko, O. Sabishchenko, International Journal of innovative technologies in economy, **1 (13),** 120-128 (2018)
9. Gas Balances in Ukrainian Underground Storages, https://naftogaz-europe.com/article/en/englstorage (2018)
10. J.Haverkamp, *Lifetime extension of ageing nuclear power plants: Entering a new era of risk,* (2014)
11. O. Dyachuk, M. Chepelev, R. Podolets, G. Trypolska, *Ukraine's transition to renewable energy by 2050* (2017)
12. V.Kadiyevskyy, N.Klymenko, Actual Problems of Economics, **152(2),** 313-320 (2014)
13. A.Skrypnyk, O. Zhemoyda, N.Klymenko, L.Galaieva, T. Koval, Journal of Ecological Engineering, **22(3),** 275-288 (2021).
14. A. Kaminskyi, M. Nehrey, M., International, Journal of Industrial Engineering and Production Research, **31(4),** 579–586 (2020)
15. B.Fladmark, G.Grimstad, *Seasonality in Natural Gas* (2013)
16. World Energy Outlook 2020, https://www.iea.org/reports/world-energy-outlook-2020# (2020)
17. K. Rosendahl, Special Issue of The Energy Journal, 57-69, (2009)
18. C. Corinne, A. Creti, B., Energy Policy, **36(11),** 4235–4246 (2008)
19. A. Neumann, Special Issue of The Energy Journal, 187-199 (2009)
20. F.Holz, C.Kemfert, *A Strategic Model of European Natural Gas Supply (GASMOD)* (2016)
21. H. Hecking., T.Panke, EWI Working Paper,**12(06),** 5-11 (2012)
22. S. Catalan, *Determinants of demand for natural gas and liquefied natural gas* (2016)

E3S Web of Conferences **250**, 03001 (2021) https://doi.org/10.1051/e3sconf/202125003001
TRESP 2021

Renewable energy technologies on the path towards decentralized low-carbon energy systems

Natalya Danilina[1*], *Irina* Reznikova[1]

[1] Institute of Engineering and Environmental Security, Togliatti State University, Belorusskaya str. 14, 445020 Togliatti, Russian Federation

Abstract. Renewable energy technologies (RET) that emerged as a result of the shift towards the renewable energy sources (RES) which aims at setting the path towards decentralized low-carbon energy systems intended for tackling global warming are becoming key elements of the smart grids of the future. Our paper applies the economic, social and technological model of the renewable energy platforms to the energy markets of the 21st century. The paper analyses the growing importance of the individual players (prosumers) on the energy market, especially when it comes to the renewable energy generation and trading. It shows that modern advanced information and communication technologies enabled the energy prosumers to trade their energy and information in two-way flows. All of these might be important for the transition towards sustainable economy and green technology.

1 Introduction

The switch to renewable energy is made possible by technological advances and a societal push towards sustainability, but it is not without its challenges. The gradual transition from the fossil energy to the renewable energy sources means that oil, gas, and coal will remain the dominant energy source for at least the next few decades [1, 2]. This means that the increase in energy demand for fossil fuels is likely to peak in a generation or more, allowing traditional and established energy companies to continue to thrive. The global energy sector does not face the prospect of painful changes in the short term, but the Paris Agreement's climate targets are absent from the transition scenario. However, in many countries (like the United States), renewable energy and CO_2 emissions, which may have peaked last year, are unlikely to meet the goal of a Paris climate agreement [3, 4].

Wind and solar power technologies have achieved the wide popularity among all renewable energy sources because they are the most attractive economic alternatives and because government subsidies that support their fleet ramp-up are being phased out [5, 6]. The growth of renewables is displacing coal, and wind and solar power have already begun to influence the development of new coal-fired power plants. Other energy sectors such as

* Corresponding author: danilina@tltsu.ru

natural gas, oil and coal will remain relatively stable. As a result, renewable energy is facing the same challenges as other energy sectors such as natural gas, oil, and coal, but with different challenges [7, 8].

Technology can play a key role in addressing the challenges of long-term energy storage and regulating the ups and downs of renewable energy supply. Renewable energy technologies such as wind, solar, hydropower, geothermal, biomass, biofuels, wind turbines, hydropower and solar photovoltaics offer the opportunity to accelerate the transition to renewable energy, while increasing synthetic fuel production and reducing fossil fuel emissions in sectors ranging from steel to food production, chemical industry and fertilisers [9].

Moreover, technological improvements and economies of scale have made renewable energy cost-effective - and competitive with fossil fuels, even without subsidies. This has created the conditions for the production of clean energy to be promoted by low-carbon energy sources such as wind, solar and geothermal energy. New solar plants are approaching cost parity for fossil-fuel power generation, and wind is already competitive which makes the renewable energy more than viable.

2 Smart grids and energy platforms

A smart grid is an electricity grid that uses digital technologies to create a more efficient, reliable and cost-effective power distribution system [10]. Nowadays, smart grid approach is implemented by load organizations, but getting the technologies that are most suitable and convenient can be a demanding task. They have several advanced features that ensure reliable network organization. The current electricity distribution network is under pressure due to the increasing energy consumption and rapidly increasing demand for energy using renewables and some novel approaches to producing and transferring energy [11]. Table 1 that follows show some examples of the renewable energy platforms operating as a part of smart grid in several European Union countries as well as the United Kingdom and United States.

Table 1. Examples of the renewable energy platforms

Platform	Country	Novelty	Pricing
Enyway www.enyway.com	Germany	Intermediary between renewable energy producers and customers	Traditional contract with a combination of kWh-prices dependent on the suppliers
SonnenCommunity https://sonnenbatterie.de	Germany	Stable power supply utilizing batteries and provision of surplus power pool	Sales of batteries to prosumers
Vandebron www.vandebron.nl	Netherlands	Offering green energy from local suppliers ("know your clean energy supplier")	Monthly subscription fee for providers and consumers
Piclo www.openutility.com/piclo	United Kingdom	Government-supported information service that helps to match the preferences of renewable energy providers, energy consumers and distribution use of service	Supported by the UK government and venture capital

Mosaic www.solarmosaic.com	United States	Peer-to-peer solar power platform with the possibility for everyone to join and become a local solar contractor	Flat monthly payment for prosumers

All around the world, stakeholders and policymakers want to simplify the control of a smart grid from the perspective of companies and homeowners with its prosumer solution. With tailored software, prosumers can help create demand programs, manage, automate, and automate their use, and improve overall reliability of power supply, making them more efficient and reliable. Such systems are not only going provides uninterrupted power, but also to work at lower cost, reducing the carbon footprint [12, 13].

With regard to this shift, it is predictable that traditional network companies, which face limited customer exodus, will try to become leaner by using new digital technologies to improve operational efficiency. Some companies claim to be able to create a self-renewing network using its smart grid technology and smart contracts. Operators must become system stabilisers in markets where the level of decentralised generation is moderate to high and where consumers maintain their grid connectivity by selling surplus electricity into the grid. Energy aggregation platforms bring decentralized energy sources to a single platform, allowing clusters of generators to operate like a large power plant. Investing in smart technologies would enable to dynamically balance demand and supply in real time. Managing demand from intermittent renewable energy sources and demand for intermittent renewable energy from all sources can be achieved. A typical example of such a platform is a virtual power plant. Such a plant allows renewable energy sources to be integrated and avoids the usual pitfalls of scalability and predictability [14, 15].

Virtual power plants enable the system to be optimised on the basis of regional - specific grid requirements and can supply economically produced electricity. The platform can store excess electricity and, if required, supply it, balancing the grid. Energy companies develop renewable energy site and grid modelling software that delivers faster and more accurate results than traditional solutions. The software was described as a utility - out-of-the-box solution that provides a solution for renewable and non-renewable energy systems. Sustainable energy strategy aims for an energy and internet network that includes smart grids, smart power plants and smart energy storage.

These approaches are being implemented in many countries but the real showcase would be Germany that is now at the forefront of implementing and using the green energy technologies and renewable energy. The transformation of the German energy supply system is one of the greatest current and future challenges for Germany [16, 17]. This is because the sector is becoming more flexible in managing the demand and supply of electricity. The so-called "smart grid" of the future will be able to better match fluctuating renewable energy sources with demand in the future and vice versa. The smart grid is designed as a modernised electricity grid that facilitates mutual digital communication between suppliers and consumers through an intelligent measurement and monitoring system that receives and transmits information. This will enable 'smart grids' to control the flow of energy from renewable energy sources such as wind, solar and hydropower, and to use AI to collect measurement data and provide consumer information. This allows business processes for energy retailers, energy suppliers and network operators to be automated.

3 Citizens participating on the energy transition

In order to achieve such transitions, it is essential that the public is involved constructively, one needs to analyse this commitment through the concept of energy dialogue and energy citizenship. Citizenship in the energy field is created through policy making and planning, with potential measures marked by the need to address issues such as climate change, energy security, renewable energy and the future of the economy [18]. In a large part of the focus groups and interviews, many participants discussed how dependent they are on fossil fuels, as opposed to others who prefer renewable energy technologies. This should include the energy dialogue which covers a range of issues such as energy security, climate change, energy efficiency and the role of renewable energy in the economy [19, 20].

From an instrumental, moral and ethical perspective, it is important to involve the entire socio-economic spectrum in the transition to low-carbon technologies. On the instrumental side, this would help to avert or mitigate the profound social adjustment that will inevitably lead to social unrest, and reduce resistance to proposed transitional solutions. If decentralised energy is to play a major role in the transition to a low-carbon future, the implementation community must be supported by a more effective approach to public engagement. The results presented show that this was initially unsuccessful in several respects. Specifically, it is about the shortcomings that occur before renewable energy systems are installed, about the lack of support for residents and about whether we can effectively solve the problems that arise after implementation. It is clear that there are concerns, but they are ineffective and bad practices that do not take into account the subtleties and needs of any community.

In the same time, renewable energy solution involving regular prosumers might be very advantageous and profitable, since the profit that the traditional energy companies are generating is removed from the scheme. One can see the difference using the example from the renewable energy pricing by Vandebron.nl, a company from the Netherlands [21] (see Figure 1 below).

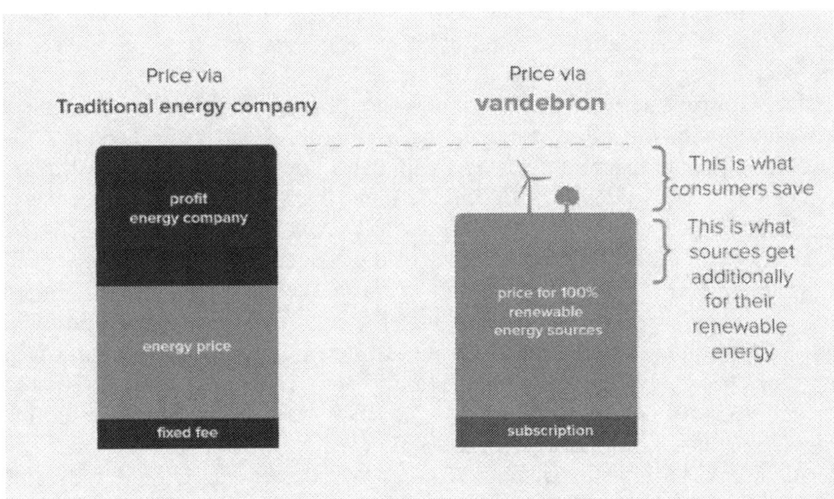

Fig. 1. Advantages of the renewable energy for the consumers on the example of vandebron.nl

This approach, while laudable, implies real challenges in terms of public participation. We have an approach that says that there is an urgent need to tackle new renewable energy projects, which need to be more transparent and transparent than traditional fossil fuel electricity generation.

Increased use of renewable energy sources such as wind, solar and geothermal will help ensure that each country has secure energy supplies and reduce greenhouse gas emissions to slow climate change and boost investment, new jobs and business. Many local planning authorities are legally required to set up advisory councils and are responsible for planning, building and installing new and installed renewable energy capacity. Planning plays a role in selecting sites where local environmental impacts are acceptable. Local planning authority is typically a statutory advisory body with a considerable annual budget for planning and planning services. It is clear that the contribution that decentralised local energy solutions can make to the development of renewable energy and the decarbonisation of the energy system in the context of the national renewable energy plans can have long-range impacts on the development of the low-carbonized future renewable energy systems [22]. While new business models challenge traditional large energy providers, we say we are pursuing similar innovations. As businesses and cities move forward, current market designs and existing regulations are not yet ripe for more decentralized energy models. There are obstacles that mean that there is no level playing field, but decentralised projects can be implemented because they are an inherently cost-effective form of infrastructure. The operating costs of a renewable energy project are much lower than they would be if there were no direct fuel costs associated with fossil fuel production such as coal, oil, and gas. A small, low-cost renewable power plant in a small district could be a great advantage for neighbouring counties as a source of low-cost electricity.

4 Digital platforms and the Internet of Energy

The state narrative links the development of the electricity sector to modernity, energy security, and industrialization and normalizes the temporal need for coal by providing an incentive to strengthen fossil-fuel-based infrastructure when newly planned coal-fired power plants go online. This state narrative, coupled with the declared mitigation of climate change through state-directed direct management of the energy sector, is justified by the need to achieve the “green” agenda. The tension between these seemingly contradictory goals seems to have been resolved by portraying it as a conflict between the state and the private sector over the future of electricity generation and consumption.

Digital industrial platforms are crucial because they capture data from machines, make it accessible for monitoring and controlling applications, enable third parties to develop data-based applications and connect different users and application developers. These platforms will be crucial for energy service providers to provide consumers with data services and to use new technologies such as electric vehicle charging to pave the way for maximising the use of renewable energy and decarbonising energy. Digital industrial platforms can also be used to support farmers by combining location data with data from the Internet of Things (IoT) and field sensors, allowing seeds and optimal water use to be adapted to specific fields to ensure efficient productivity (as well as the Internet of Energy (IoE) which is a part of the IoT and includes all electrical devices, sensors and smart meters connected to the Internet or linked within the smart grid) [23, 24]. In this way, the ongoing COVID-19 pandemic has accelerated the development of a digital industrial platform in which the utilities industry plays a central role in providing sustainable services to consumers. Digital industrial platforms can ultimately be equipped with the appropriate business models. The industry that enables regulators has taken risks and adapted to new standards by using new technologies and new business models such as smart meters, smart grids and smart grid systems. Some IoT platforms are touted as those that promote secure and reliable power supplies, with the industry trying to question itself and explore the possibilities. Regulators will play a key role in creating the platform that will enable us to remove these barriers. The platform is also remarkably popular with entrepreneurs, investors and private companies. There are also a number of examples of how to harness the power of community networks to develop business. In the next phase, known as "cloud computing", the IT companies’ unique software architecture, which enables faster and more efficient deployment of technologies that are easily scalable across the company, is designed to accelerate the evolution into a platform-based company [25]. To shape the future, rather than waiting for it, our innovation centres are trying to harness innovation by supporting start-ups in technologies with significant potential. Digital platforms are increasingly regulated, with Japan the latest country to pass new laws to regulate their operation. They work with local investors, scientists, governments and others to find solutions to today's most pressing problems. Governments are also empowered to assess compliance, publish the results of their assessments and take the necessary measures to enforce compliance, including the issuing of fines for infringements. Thence, the role of digital platforms in the future of energy supply in Europe and the United States is very important and the impact of their regulations can be crucial for the development of new technologies such as renewable energies and energy storage. In addition, there is a question of whether the future of distribution system operators lies in the platform business or in the ownership of basic assets. One has to look at the role of digital platforms and the potential impact of the regulatory framework for digital technologies. In a year of continuous change, the energy companies' commitment to improving customer service in the regulated electricity and gas business has remained constant. The strength of the strategy lies in the combination of our many strengths as we continue to create the most efficient, efficient and competitive distribution network operators. This might help to overcome the challenges and strengthen energy customers now and in the future. To make smart networks smarter and more resilient and keep pace with growth, investments in new renewable energy technologies such as smart grids, smart electricity poles and meters, smart meters and new infrastructure, should be generous and ongoing. The investment in people, systems and digital platforms might also enable energy companies to streamline the installation process to conveniently meet service agreements and enable meter installations to scale consistently and reliably.

5 Conclusions and policy implications

In the conclusion, we can say that people need to take control of their own energy future and invest in a decentralized network that lowers prices, improves energy security, reduces carbon emissions, and makes their communities more prosperous and resilient. Decentralised energy is not yet a commonly understood concept, but it broadly refers to energy produced outside the main grid, including micro-renewable heating and cooling systems. It can cover any energy and the use of a combined heat and power plant will fuel competition as the number of energy suppliers increases. This allows energy-consuming communities to be integrated into a single energy system, such as a power plant or heating and cooling system. If consumers also become both energy producers and prosumers, the system would lead to a decline in electricity generation when old power plants are obsolete. Consumers could then sell electricity back into the grid, offering a way to supply a national grid with backup electricity.

The growing importance of the prosumers on the energy market, especially when it comes to the renewable energy generation and trading is becoming a reality thanks to the advanced information and communication technologies which enable these prosumers to trade their energy and information in two-way flows similar to the principles of the sharing economy. People were ready to accept Airbnb and Uber, so they might be also fine with the peer-to-peer energy sharing networks. All of these might be important for the transition towards sustainable economy and green technology.

In addition, it becomes apparent that the existing forms of social engagement do not generate the speed and scale of change required for the transition from the fossil-based economy to the ones based on the renewable energy sources. The main reason might be the lack of public participation in the development and implementation of new energy technologies. For example, houses, wind turbines, solar power plants and wind farms, as well as solar thermal power plants. A more connected approach will give policymakers the evidence they need to properly address the concerns of different citizens, recognize that low-carbon policies often disappear under the radar, and capitalize on the synergies of otherwise isolated public engagement.

References

1. D. Jie, X. Xu, F. Guo, Energy, **220**, 119644 (2021)
2. I. Sotnyk, I. Shvets, L. Momotiuk, Y. Chortok, Marketing and Management of Innovations, **4**, 150-160 (2018)
3. M.N. Dudin, E.E Frolova, O.V. Protopopova, A.A. Mamedov, S.V. Odintsov, Entrepreneurship and Sustainability Issues, **6(4)**, 1704-1713 (2019)
4. W. Strielkowski, E. Lisin, I. Gryshova, Romanian Journal of European Affairs, **16(4)**, 68-77 (2016)
5. P. Joskow, Oxford Review of Economic Policy, **35(2)**, 291-331 (2019)
6. F. Van Triel, T. Lipman, Energies, **13(20)**, 5277 (2020)
7. G. Chebotareva, W. Strielkowski, D. Streimikiene, Journal of Cleaner Production, **269**, 122110 (2020)
8. C. Magazzino, M. Mele, N. Schneider, Energy, **219**, 119586 (2021)
9. M. Javed, T. Ma, J. Jurasz, M. Amin, Renewable Energy, **148**, 176-192 (2020)
10. F. Al-Turjman, M. Abujubbeh, Future Generation Computer Systems, **96**, 579-590 (2019)
11. M. Buth, A. Wieczorek, G. Verbong, Energy Research & Social Science, **53**, 194-205 (2019)
12. M. Lovati, X. Zhang, P. Huang, C. Olsmats, L. Maturi, Buildings, **10(8)**, 138 (2020)
13. S. Sharda, M. Singh, K. Sharma, Sustainable Cities and Society, **65**, 102517 (2020)
14. P. Dey, D. Das, A. Latif, S. Hussain, T. Ustun, Sustainability, **12(17)**, 6979 (2020)
15. L. Ju, Q. Tan, H. Lin, S. Mei, N. Li, Y. Lu, Y. Wang, Energy, **196,** 117078 (2020)
16. M. Sait, U. Chigbu, I. Hamiduddin, W. De Vries, Resources, **8(1)**, 7 (2019)
17. K. Hansen, B. Mathiesen, I. Skov, Renewable and Sustainable Energy Reviews, **102**, 1-13 (2019)
18. D. Brown, S. Hall, M. Davis, Energy Research & Social Science, **66**, 101475 (2020)
19. I. Cabelkova, W. Strielkowski, I. Firsova, M. Korovushkina, Energies, **13(7)**, 1742 (2020)
20. R. Ostfeld, D. Reiner, Energy Policy, **140**, 111332 (2020)
21. C. Park, T. Yong, Energy Procedia, **128**, 3-9 (2017)
22. C. Nolden, J. Barnes, J. Nicholls, Renewable and Sustainable Energy Reviews, **122**, 109722 (2020)
23. M. Ayaz, M. Ammad-Uddin, Z. Sharif, A. Mansour, E. Aggoune, IEEE Access, **7**, 129551-129583 (2019)
24. S. Marzal, R. González-Medina, R. Salas-Puente, G. Garcerá, E. Figueres, IEEE Internet of Things Journal, **6(4)**, 7241-7252 (2019)
25. N. Hossein Motlagh, M. Mohammadrezaei, J. Hunt, B. Zakeri, Energies, **13(2)**, 494 (2020)

Renewable energy: a bibliometric analysis

Anna Rosokhata[1]*, *Mariia* Minchenko [1], *Liliya* Khomenko [1], *Olena* Chygryn [1]

[1]Department of Marketing, Academic and Scientific Institute of Business, Economics and Management, Sumy State University, 2, Rymskogo-Korsakova st., 40007 Sumy, Ukraine

Abstract. The aims of this paper are to identify existing research on renewable energy; identify the most influential publications, authors, organizations to understand the research areas. The paper is centered on the bibliometric analysis based on the Web of Science database for the key phrase' renewable energy' in the article title until 2020. We analyzed the publications about renewable energy by years, by research areas, by geography, by research organizations and research sponsors, by journals, by citations of journals, authors, publications, co-occurrence by keywords. Analysis of the most cited publications and authors, analysis of co-occurrence by keywords was performed using VOSviewer. We performed hierarchical cluster analysis, and clusters were selected using VOSviewer. We found 17805 scientific publications on renewable energy published in scientific journals (51.7% of publications were published during 2016-2020). The most popular research areas are energy fuels, engineering, science technology, environmental sciences, ecology, and business economics. The majority of papers was published by the scientists from the United States, China, and India. In addition, we identified six main research clusters. They are related to an optimizer, renewable energy, biomass, co2 emissions, model, desalination. *Conclusions*: This analysis confirmed the author's hypothesis about the definition of new scientific horizons of renewable energy research. Our results can help scientists interested in renewable energy looking for research funding and research project risks based on renewable energy.

1 Introduction

All over the world, the study of renewable energy draws interest of various scientific schools, economics, and especially stakeholders, who can benefit from the use of renewable energy technologies [1–7]. Resource constraints, sustainable energy, biomass, co2 emissions make it possible to establish the hypothesis of the transfer of scientific interests from energy fuels and environmental sciences ecology to other areas. The definition of the role and place of renewable energy in business economics, mechanical engineering, science technology and compute science is gaining more and more popularity [8–17].

In addition, the essential study of modern scientific schools on renewable energy has a number of patterns, confirming that it is possible to obtain sound theses to determine the

* Corresponding author: a.rosokhata@kmm.sumdu.edu.ua

features of optimization of financial, intellectual and labor resources in the development of renewable energy research in modern space [18–23].

Modern energy systems are undergoing a revolutionary change. Significant influence on these changes is lobbied by various organizations of world importance and economically developed countries [24–26]. We are currently observing:

- adjusting the operation of more flexible architectures of energy systems;
- development of intelligent networks and their interaction with energy storage and use technologies;
- scaling the market of active consumers, for whom the origin of energy is important;
- reorganization of the electricity market and development of cloud technology technologies, energy net.

But not in all countries you can see support for the development of renewable energy. Although these areas of research and implementation at the industrial level can be actively promoted in the digital environment [27–29]. In this paper, we aim to examine the place of renewable energy in various areas of research [30–34], as well as to identify the relationship between renewable energy and economic aspects. The main objectives of this study are to substantiate the following assumptions:

1. Renewable energy research vectors are moving towards business economics. The economic benefits of implementation and the financial risks that businesses may incur during the implementation of renewable energy projects are becoming an increasingly important issue.
2. The issue of renewable energy is interesting to absolutely all countries and the geographical location of scientific schools should be ignored. It is more appropriate to emphasize the study of research sponsorship, ie those stakeholders who may later become the owners of these studies results.
3. There is a subjective statement that renewable energy is more related to CO2 emissions and their system of optimizing storage and using. Modern researchers are already paying more attention to such issues as desalination, hydrogen energy, power electronics, biomass.

Therefore, our research focuses on the peculiarities of the development of research in the field of renewable energy, and determining the patterns of this issue.

The main direction of this paper is bibliometric analysis of the scientific literature on renewable energy. It allows to form a holistic view of existing research on selected topics, to identify the most influential publications, authors and organizations, to analyze the directions of existing and future research.

The aim of the work is to analyze existing research of renewable energy, to identify the most influential publications, authors, organizations patterns of development research areas.

This study analyzes the impact of journals, authors and articles, analyzes research areas, research geography, renewable energy organizations and sponsors such studies. In the end of the article the analysis of co-occurance by keywords and recommendations for possible further research is proposed.

2 Materials and methods

At the beginning of the analysis, the methods used for bibliometric analysis [35–38] were analyzed. As a result, the analysis application of the most cited publications and authors, analysis of co-occurrence by keywords using the program VOSviewer was selected. To do this, we first selected all scientific publications on renewable energy based on the Web of Science database for the keyword phrase "renewable energy" in the article title until 2020 inclusive. 17805 scientific publications were found.

Based on 17,805 publications found in the Web of Science database, a sample of 2,000 publications was formed. All publications were sorted by the number of citations from the largest to the smallest, the first 2000 publications were included in the sample as the most relevant (probability 95%, confidence interval 2.06%).

It was analysed of several sets of publications on renewable energy by years (1974-2020), by research areas, by geography, by organizations that conduct research and sponsor research, by publications, by citations of publications, authors, publications, co-occurrence by keywords. For each indicators it were identified the ten most popular fields of research, countries, organizations, publications, authors, publications, which were displayed using graphical and tabular methods.

It was performed the hierarchical cluster analysis and it were selected clusters using VOSviewer based on keywords. Keywords with more than 45 repetitions excluding geographical names were selected, a sample of 2000 publications from the Web of Science database (95% probability, 2.06% confidence interval).

3 Results and discussions

Carrying out bibliometric analysis and searching for scientific publications on issues related to the energy system is the key to the formation of a system removing communication and organizational barriers to stakeholders' interaction. In order to properly construct these approaches, 17805 scientific publications (articles, reviews, book sections, abstracts, etc.) published during 1974–2020 were found in the Web of Science database, primarily for the analysis of renewable energy publications. The number of publications by year is presented in Fig. 1.

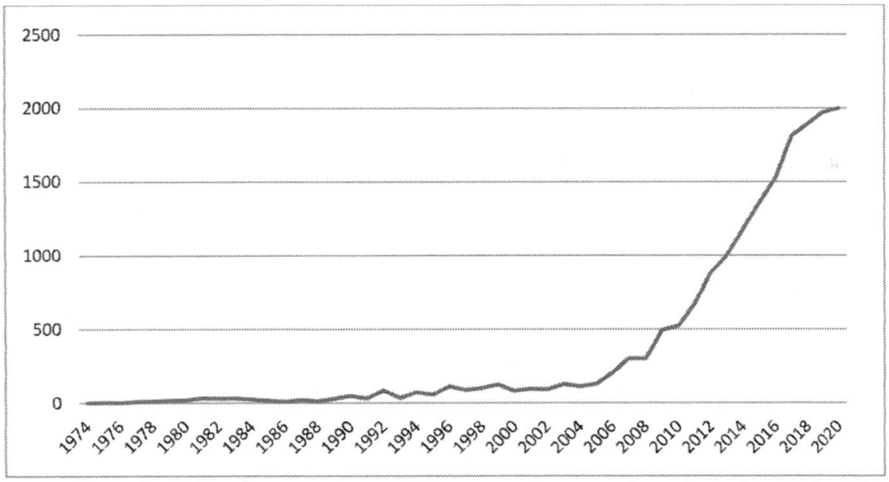

Fig. 1. Number of publications on renewable energy by years based on the Web of Science database (created by the authors)

As can be seen from Fig. 1, 51.7% of publications were published during 2016-2020, which indicates a relatively recent increase in research in this area.

Research on renewable energy is being studied by scientists in various fields of application. The ten most popular areas of research are presented in Fig. 2.

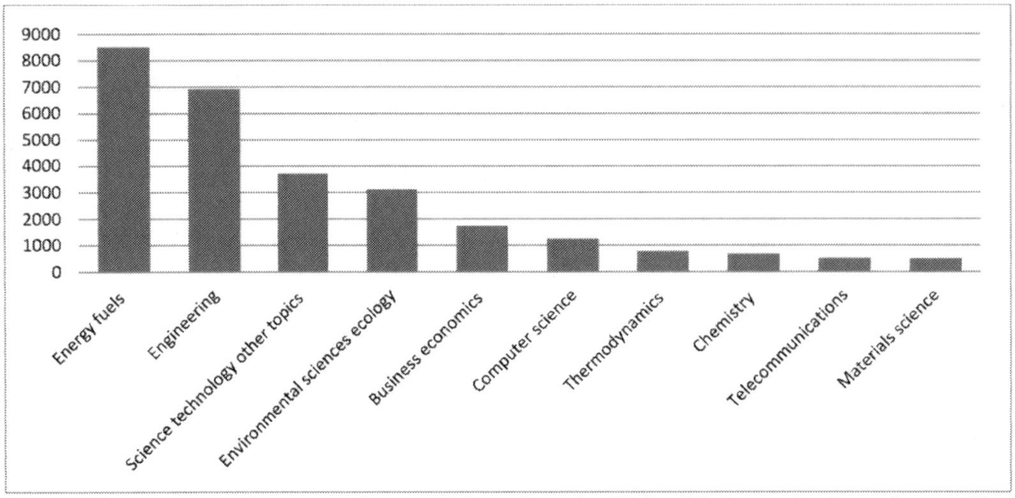

Fig. 2. The most popular areas of research based on the Web of Science database (created by the authors)

As can be seen from Fig. 2, most research concerns Energy fuels (47.8%), Engineering (38.9%), Science technology other topics (20.9%), Environmental sciences ecology (17.5%) and Business economics (9.8%).

Researchers from around the world deal with renewable energy issues. In fig. 3 presents the countries whose scientists have written the most scientific works.

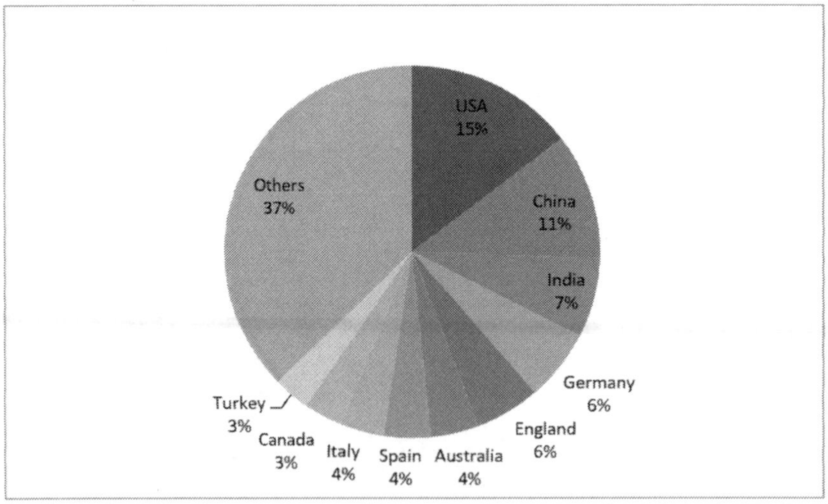

Fig. 3. Number of publications on renewable energy by country based on the Web of Science database (generated by the authors)

As can be seen from Fig. 3, most papers have been published by researchers from the United States (2606 papers), China (1951 papers) and India (1223 papers). At the same time, in some regions, including Ukraine, this topic is insufficiently studied (59 works).

More than 500 organizations conduct research on selected topics. The ten organizations that conduct the most research on selected topics are presented in Fig.4.

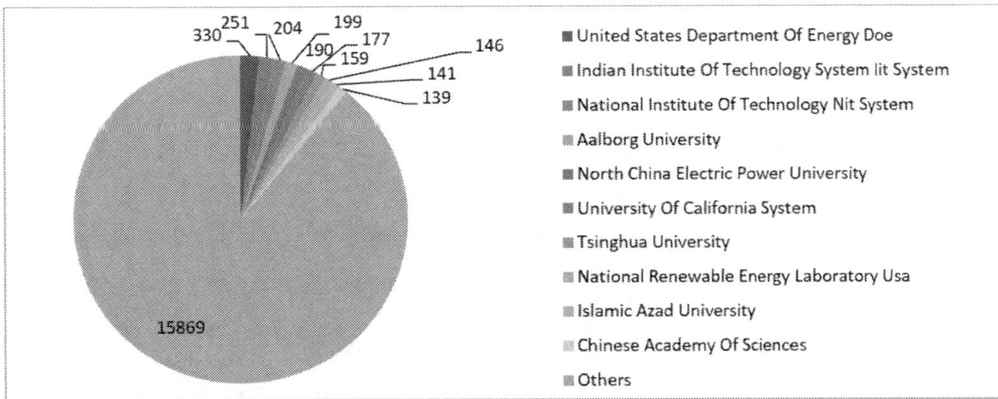

Fig. 4. The organizations that conduct the most research on the basis of the Web of Science database (formed by the authors)

Acording Fig. 4, the most renewable energy issues are dealt with by the United States Department of Energy Doe (1.9%), Indian Institute of Technology System Iit System (1.4%), National Institute of Technology Nit System (1.1%), Aalborg University (1.1%), North China Electric Power University (1.0%).

The research is sponsored by more than 500 organizations. In fig. 5 presents the ten organizations that most often invest in the development of research of renewable energy.

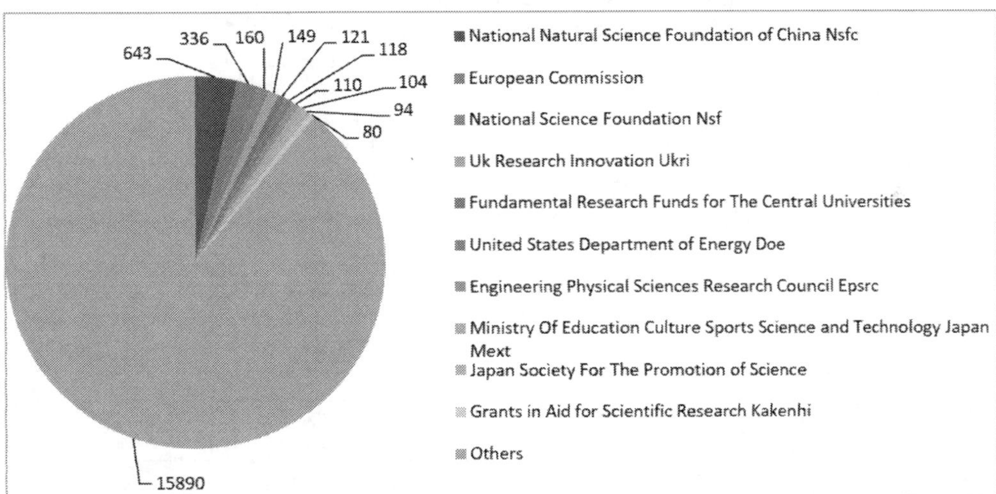

Fig. 5. Organizations that sponsor renewable energy research based on the Web of Science database (created by the authors)

The most frequent sponsors of the research are the National Natural Science Foundation of China Nsfc (3.6%), the European Commission (2.1%), the National Science Foundation Nsf (0.9%) and Uk Research Innovation Ukri (0.8%). (Figure 5).

Renewable energy publications were published in more than 100 publications, with almost a quarter of all publications were published in ten journals (Figure 6).

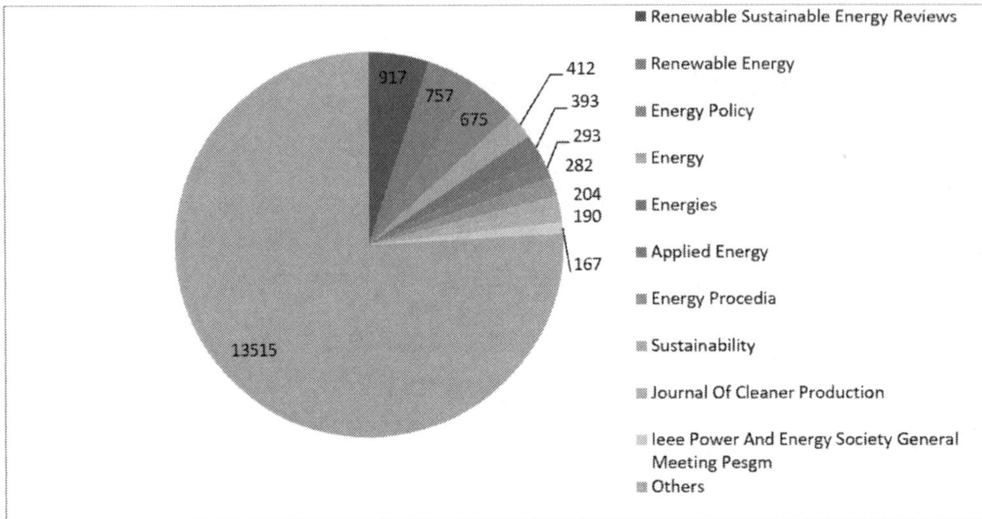

Fig. 6. Journals with the largest number of publications by country based on the Web of Science database (created by the authors)

The most popular magazines are "Renewable Sustainable Energy Reviews" (5.2%), "Renewable Energy" (4.3%), "Energy Policy" (3.8%), "Energy" (2.3%), "Energies". (2.2%). It can indicate the influence of these publications on selected topics.

To determine the influence of journals, we analyzed citations. The most cited publications with at least five publications are presented in Table 1.

Table 1. The most cited publications on renewable energy based on a sample from the Web of Science database (created by the authors)

Journals	Number of citations	Number of publications
Renewable Sustainable Energy Reviews	43720	443
Energy Policy	22471	227
Renewable Energy	13317	180
Energy	9041	116
Applied Energy	8602	100
Ieee Transaction on Industrial Electronic	4412	20
Energy Economics	4119	41
Energy Conversion and Management	3503	47
International Journal of Hydrogen Energy	2720	26
Ieee Transaction on Power System	2401	19

Thus, both lists includes only five publications: Renewable Sustainable Energy Reviews, Energy Policy, Renewable Energy, Energy and Applied Energy. It indicates their impact on the subject.

In order to determine the most influential scientists in the chosen field, we identified the most cited authors. The most cited authors with at least five publications are presented in Table 2.

Table 2. The most cited authors on renewable energy based on a sample from the Web of Science database (generated by the authors)

Authors	Number of citations	Number of publications
Lund H.	5898	23
Mathiesoen B.V.	2825	8
Connolly D.	1992	6
Apergis N.	1808	7
Payne J.E.	1755	7
Ozturk I.	1434	10
Walker G.	1310	7
Devine-Wright P.	1300	8
Saini R.P.	1295	12
Blaabjerg F.	1145	6

Each authors has at least six publications, and Lund H. 23 publications that provided the largest number of citations (Table 2). It is indicated the influence of the authors' data on the topic of renewable energy.

Table 3. The most cited articles on renewable energy based on the Web of Science database (created by the authors)

Authors	Title article	Journal	Year of publication	Number of citations
Carrasco J.M. et al. [39]	Power-electronic systems for the grid integration of renewable energy sources: A survey	Ieee Transactions on Industrial Electronics	2006	2237
Panwar N. L. et al. [40]	Role of renewable energy sources in environmental protection: A review	Renewable & Sustainable Energy Reviews	2011	1233
Kempton W & Tomic J. [41]	Vehicle-to-grid power implementation: From stabilizing the grid to supporting large-scale renewable energy	Journal of Power Sources	2005	1092
Wuestenhagen R., Wolsink M. & Buerer M.J. [42]	Social acceptance of renewable energy innovation: An introduction to the concept	Energy Policy	2007	1082
Turner J.A. [43]	A realizable renewable energy future	Science	1999	1012
Dincer I. [44]	Renewable energy and sustainable development: a crucial review	Renewable & Sustainable Energy Reviews	2000	902
Banos R. et al. [45]	Optimization methods applied to renewable and sustainable energy: A review	Renewable & Sustainable Energy Reviews	2011	869
Connolly D. et al. [46]	A review of computer tools for analysing the integration of renewable energy into various energy systems	Applied Energy	2010	857

Ellabban O., Abu-Rub H. & Blaabjerg F. [47]	Renewable energy resources: Current status, future prospects and their enabling technology	Renewable & Sustainable Energy Reviews	2014	818
Barton J.P. & Infield D.G. [48]	Energy storage and its use with intermittent renewable energy	Ieee Transactions on Energy Conversion	2014	743

Table 3 shows that all articles were published before 2014, with the oldest article published in 1999.

To understand the areas of research, a co-occurrence analysis was performed by keywords. Figure 7 presents the results of hierarchical cluster analysis with the number of repetitions from 45 except for geographical names in the sample. The sample includes the 2000 most cited publications based on the Web of Science database (95% probability, 2.06% confidence interval).

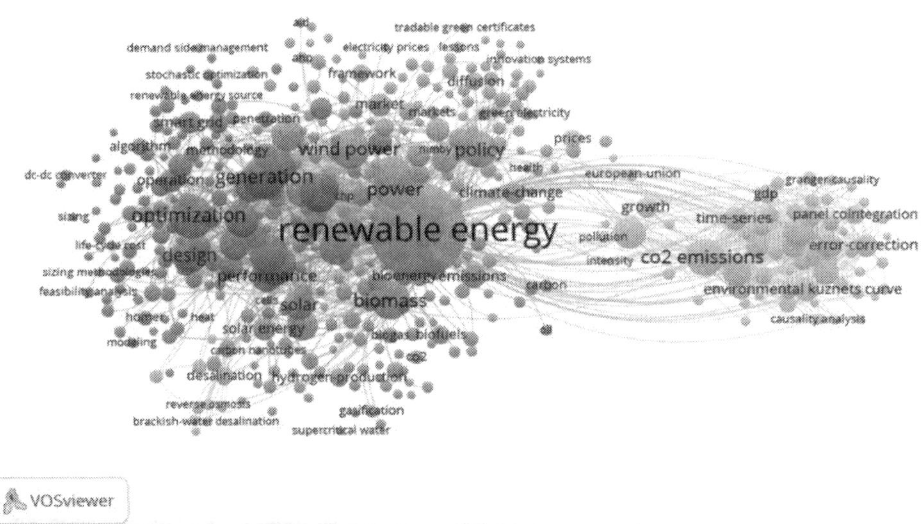

Fig 7. Analysis of co-occurrence by keywords (Web of Science database, sample of 2000 publications, more than 45 keyword duplications excluding geographical names).

Fig. 7 shows that there are six main clusters:

Cluster 1 (123 items): optimizator (Links: 213, Total links straight: 689, occurrences: 116)

Cluster 2 (106 items): renewable energy (Links: 414, Total links straight: 2656, occurrences: 549)

Cluster 3 (102 items): biomas (Links: 200, Total links straight: 529, occurrences: 93)

Cluster 4 (65 items): co2 emissions (Links: 145, Total links straight: 733, occurrences: 86)

Cluster 5 (47 items): model (Links: 167, Total links straight: 353, occurrences: 63)

Cluster 6 (15 items): desalinetion (Links: 70, Total links straight: 145, occurrences: 22).

4 Conclusions

In general, this paper considers the historical development of renewable energy research in different areas: temporal, geographical, authorial, sponsorship of scientific fields, implementation of scientific journals. This approach has identified a few patterns that can justify further research.

The chosen time lag of 45 years showed the relevance of this topic in the last 10 years and allowed to divide the time period into 2 stages. There was a constant interest in the topic (up to 200 publications per year) until 2006. There was intense revitalization of scientists to renewable energy in 2006-2020. Based on this, we can conclude about the relevance of this topic and the fact that in the future it is necessary to more meaningfully study the period 2006-2020 with detailing and make the additional divisions into stages in this time period.

In this work, we were able to confirm the study vector of renewable energy. Indeed, almost 2/3 of all work belongs to energy fuels and engineering. But in the research of this issue, concentrated in the active period of 2006-2020, we saw a trend of increasing research renewable energy in business economics, mechanical engineering, science technology and computer science.

An interesting result of the study concerns the researchers affiliation to the country. We have suggested that the topic of renewable energy is most interesting for economically developed countries that can afford to invest in this area. But the analysis revealed that most of the work has been published by researchers from the United States, China and India. None of the G7 countries, except the United States, is actively involved in the scientific development of renewable energy. It turns out that our assumption about the non-neglect of geographical localization must also be taken into account. Therefore, it is necessary to pay attention to the work of the following research centers: United States Department of Energy Doe, Indian Institute of Technology System Iit System, National Institute of Technology Nit System, Aalborg University, North China Electric Power University.

But the issues of research sponsorship has shown the role of the European Union and the United Kingdom in the study of renewable energy, where they influence research through the European Commission and Uk Research Innovation Ukri.

A rather pleasant fact is that the world community is interested in renewable energy issues and actively presents its own results in those journals that are authoritative. And in these 5 magazines more than 80% of all publications of explored authors are published. Authors citation indexes indicate their publications relevance and the renewable energy influencing in general.

Conducting co-occurrence research on keywords allowed us to identify six main clusters related to optimization, renewable energy, biomass, CO2 emissions, model, desalination. This analysis confirmed the author's assumptions about the definition of new scientific horizons of renewable energy research.

The results can be of value to those scientists who are interested in the topic of renewable energy, especially the prospects for the development of the financial investment rationale, risk research of projects based on renewable energy.

References

1. A. El Amri, R. Boutti, S. Oulfarsi, F. Rodhain, B. Bouzahir. Financial Markets, Institutions and Risks, **4(4),** 31-44 (2020)
2. A. El Amri, R. Boutti, F. Rodhain. Financial Markets, Institutions and Risks, **4(2)**, 52-64 (2020)
3. M. Smith. A SocioEconomic Challenges, **2(1),** 21-34 (2018)
4. D. Bhowmik. SocioEconomic Challenges, **3(2),** 15-30 (2019)

5. O. Lyulyov, T., Pimonenko, A. Kwilinski, H. Dzwigol, M. Dzwigol-Barosz, V. Pavlyk, P. Barosz. Energies, **14(2),** 373 (2021)
6. D.T.A. Marcel. SocioEconomic Challenges, **3(2),** 63-69 (2019)
7. I. Didenko, K. Volik, T. Vasylieva, S. Lyeonov, N. Antoniuk. E3S Web of Conferences, 202 (2020)
8. Y. Samusevych, A. Vysochyna, T. Vasylieva, S. Lyeonov, S. Pokhylko, E3S Web of Conferences, 234 (2021)
9. S. Furmaniak, P.A. Gauden, A. Patrykiejew, R. Miśkiewicz, P. Kowalczyk, Scientific Reports, 15407 (2018)
10. He, Shuquan. Business Ethics and Leadership, **3(4),** 43-48 (2019)
11. V. Pavlyk, SocioEconomic Challenges, 4(1), 122-128 (2020)
12. V. Panchenko, Yu. Harust, Ya. Us, O. Korobets, V. Pavlyk, Marketing and Management of Innovations, **1,** 256-264 (2020)
13. Ya. Us, T. Pimonenko, T. Tambovceva, J-P Segers, Health Economics and Management Review, **1(1),** 48-59 (2020)
14. R. Miskiewicz, Energies, **13(22),** 6106 (2020)
15. R. Miskiewicz, Polityka Energetyczna, **21(2)**, 49-62 (2018)
16. Y. Kharazishvili, A. Kwilinski, O. Grishnova, H. Dzwigol, Sustainability, **12(21**), 8953 (2020)
17. H.Dzwigoł, M. Dzwigoł–Barosz, Z. Zhyvko, R. Miskiewicz, H. Pushak, Journal of Security and Sustainability Issues, 8(3), 307-317 (2019)
18. O. Y. Halan, Marketing and Management of Innovations, 1, 196-206 (2018)
19. I. Sotnyk, I. Shvets, L. Momotiuk, Y. Chortok, Marketing and Management of Innovations, **4,** 150-160 (2018)
20. R. Vanickova, Marketing and Management of Innovations, **2,** 56-67 (2020)
21. H.Dkhili, L. B. Dhiab, Marketing and Management of Innovations, **4**, 252-268 (2019)
22. V. Pavlyk, Financial Markets, Institutions and Risks, **4(1),** 117-123 (2020)
23. L. Starchenko, S. Lyeonov, T. Vasylieva, T. Pimonenko, O. Lyulyov, E3S Web of Conferences, 234, 00015 (2021)
24. O. Chygryn, A. Rosokhata, O. Rybina, N. Stoyanets, E3S Web of Conferences, 234, 00004 (2021)
25. I. Vakulenko, L. Saher, L. Syhyda, S. Kolosok, A. Yevdokymova, E3S Web of Conferences, 234, 00020 (2021)
26. Ye. Ziabina, T. Pimonenko, L. Starchenko, L. SocioEconomic Challenges, **4(4),** 160-174 (2020)
27. H. S. Andrade, G. Loureiro, Business Ethics and Leadership, 4(2), 86-95
28. A. Rosokhata, O. Rybina, A. Derykolenko, V. Makerska, Research in World Economy, **11(4),** 42–52 (2020)
29. H. Dzwigol, M. Dzwigol-Barosz, Academy of Strategic Management Journal, **19(5**), 1-7, (2020)
30. Y. Bilan, T. Vasylieva, S. Lyeonov, I. Tiutiunyk. Entrepreneurial Business and Economics Review, **7(2**), 27-43 (2019)
31. O. Chygryn, O. Lyulyov, T. Pimonenko, S. Mlaabdal. Engineering Management in Production and Services, **12(4**), 92-104 (2020)
32. O. Lyulyov, T. Pimonenko, N. Stoyanets, N. Letunovska. Research in World Economy, 10(4), 97-105 (2019)
33. M.P. Bhandari. Business Ethics and Leadership, **3(1),** 101-108 (2019)
34. K. Djalilov, S. Lyeonov, A. Buriak, Risk Governance and Control: Financial Markets and Institutions, 5(4CONT1), 178-187 (2015)
35. L. Saher, L. Syhyda, O. Korobets, T. Berezianko, E3S Web of Conferences, 234, 00011 (2021)

36. Y. Bilan, T. Pimonenko, L. Starchenko, E3S Web of Conferences, **159**, 04037 (2020)
37. L. Khomenko, L. Saher, J. Polcyn, Health Economics and Management Review, **1(1),** 20-36 (2020).
38. A. Zolkover, V,Terziev, Business Ethics and Leadership, **4(3)**, 107-118 (2020)
39. J. M. Carrasco, L.G Franquelo, J.T. Bialasiewicz, E. Galvan, R. Portillo, M.M. Prats, J.I. Leon, N. Moreno-Alfonso, IEEE Transactions on Industrial Electronics. **4(53)**, 1002-1016 (2006)
40. N. L. Panwar, S.C Kaushik, S. Kothari, Renewable & Sustainable Energy Reviews. **3(15),** 1513-1524 (2011)
41. W. Kempton, I. Tomic, Journal of Power Sources. **1(144)**, 280-294 (2005)
42. R. Wuestenhagen, M. Wolsink, M. J. Buerer. Energy Policy. **5(35)**, 2683-2691 (2007)
43. J.A. Turner, Science. **5428(285)**, 687-689 (1999)
44. I. Dincer I. Renewable & Sustainable Energy Reviews. **2(4),** 157-175 (2011)
45. R. Banos, F. Manzano-Agugliaro, F. Montoya, C. Gil, A. Alcayde, J. Gome, Renewable & Sustainable Energy Reviews. **4(15)**, 1753-1766 (2011)
46. D. Connolly, Applied Energy, **4(87)**, 1059-1082 (2010)
47. O. Ellabban, H. Abu-Rub, Renewable & Sustainable Energy Reviews. **39**, 748-764 (2014)
48. J.P. Barton, IEEE Transactions on Energy Conversion, **2(19)**, 441-448 (2014)

Bionic organization as a stage of production enterprise development in a digital transformation process

Vitalii Cherepanov[1]*, *Evgeny Popov*[1], *Victoria Simonova*[1]

[1] Ural Institute of Management, Branch of the RANEPA, 8 Marta str. 66, 620144 Yekaterinburg, Russian Federation

Abstract. This article embraces a concept of bionic organization as a form of production enterprise which main attribute is an integration of digital technologies in the management processes and the production lines from the product idea to its delivery to a consumer. Digital technologies reinforce human competencies and abilities of analogous machines, playing a role of independent component of executive and production system. At the same time, digital technologies become full-fledged members of the organization, enriching it with bionic treats.

An idea of full human substation in production management with digital technologies is far from being practicable, but a tendency towards this change is quite clear. The main purpose of this this article is to discover main features of bionic organization as an intermediate evolution stage on the way of full human substitution with machines and digital technologies. Additionally, the article focuses on the interconnections between bionic organization, digital transformation, digital maturity and management system. This analysis was made for a production enterprise set upon a way to deep digital transformation.

1 Introduction

Nowadays, the modern management process is focused not only on the staff management, but also on its reinforcement with technological solutions and on inclusion of machines and artificial intelligence as independent agents into the production process and executive system of production enterprise.

The emergence of new communication tools such as messengers, videoconferencing services, group interaction tools and chatbots makes modern interactions in organizations instant, actually uninterrupted and deprived of hierarchical constraints. This new communication pattern demands changes in management. Furthermore, this form of communications is possible not only between humans, but also between humans and machines and even between systems and machines.

On the way towards full human substitution with machines and digital technologies in production there are a lot things to be done and determined from the management methods, legislation and competency development point of view. But it is already clear that full-

* Corresponding author: v.cherepanov@outlook.com

autonomous production enterprises will appear sooner or later. These production peers will not be organizations as we used to reckon those ones, but they will demand specific approach in management, strategy development and engineering role of human. Finally, market will always demand compliance of proposed product to human necessities, those constantly change. As it is well-known that only "alive" systems are able to adapt to the environmental conditions and circumstances, changing its inner organization or structure [1]. However, this adoption is a goal of a modern organization on the way of Industry 4.0 development, as it suggested by Acatech [2]. The main purpose of transition towards Industry 4.0, contrary to many opinions, is not fully blown automation of production processes, but attainment of agility, readiness to changes and mobility to get necessary competitive advantages in a struggle for consumers.

Strengthening of normal biologic treats or performance with electronic or electromechanical devices is bionic [3]. Is it possible to apply this concept to an organization? The main goal of this article is to discover main features of bionic organization as an intermediate stage of organization evolution and to find a place of this stage in entire digital transformation process.

On this way we took into account that concept of "bionics" was introduced by Jack E. Steele, as a merger of "biology" and "electronics". Nowadays, this direction of science is focused on biologic methods and systems features application towards engineering tasks and modern technologies [4].

As we discovered earlier, a number of researches have already made some steps towards concept of bionic organization. This concept is actively developing by Boston Consulting Group in a cycle of articles and conferences.

This article contains a research of main features of bionic organizations and authors vision of a role of bionic organization as a stage of organizational evolutions in digital transformation, pointed to full-autonomous production enterprise as a finite form of this transformation. A relevance of the research is proved by increase interest towards artificial intelligence, increase of production efficiency and decrease of human physical interconnections in order to reduce the rate of spread of viral infections, as a main output of COVID-19 pandemic, connected with human labor [5, 6].

2 Materials and methods

There were two methods used in this research: i) discovering of scientific, popular scientific and information publications connected with a topic of the research; and ii) modelling of changes the essence of main features of production enterprise when its key agents of managements systems and production are replaces with machines and digital technologies.

In this research, the authors lean on their own outputs on digital technologies. They consider digital technologies as technologies those support actions and interactions with reduction of cognitive limits of humans, overcoming time and space, equally between humans, devices, machines and systems as equal agents with certain role defined by creator of digital technology or platform.

In addition, the authors draw from their own typology of digital maturity levels of an organization, those include the following stages of digital transformation (see Table 1 that follows).

In analysis process, we discovered a plethora of publications on the topic of the research and on related topics. Google, analytical system Teqviser, online library Wiley, Elsevier ScienceDirect, Springer Open, Informs, arXiv of Cornell University, insights of McKinsey, Boston Consulting Group, PriceWaterhouse Coopers and Bain&Co. were used to assess an to classify them [7].

Table 1. Levels of digital maturity (for production enterprise)

№	Level of digital maturity (towards digital technologies usage)	The essence
1.	Lack of	Fully manual labor or usage of analogous machines
2.	Existence	Usage of digital technologies in social life in communications, but not in production process of management system
3.	Application	Usage of digital technologies to particular tasks and problems
4.	Using	Composition of processes and interactions based on digital technologies
5.	Substitution	Gradual substitution of management functions of humans to digital technologies
6.	Autonomy	Transition towards a stage, when digital technologies create new processes of interactions, means of labor or technologies and manage a production process. Enterprise operates without humans

Note: Authors' typology of digital maturity levels is getting ready to be publish in an individual article in Russian

The main search criteria were the following (see Table 2).

Table 2. Keywords used in literature review

Direction of the "bionic organization"	Related topics
Bionic company, bionic organization, bionic plant, autonomous production, bionic management, bionic (in economics and management)	Digital transformation, digitalization, digital technologies, artificial intelligence, patterns of organizational structures, stages of organizational development

When modelling the essence of changes of production enterprises in digital transformation process, there were distinguished production machines, information systems, means and methods of interactions, technologies, processes and roles as components of management and production system. The conducted research was also based on published researches on bionic organizations and related fields.

3 Prevailing views on bionic organizations

One of the first mentions of bionic organization was published in Clay Carr's cycle of researches, published from 1989 to 1990. In the first research in 1989, Carr distinguished three features of bionic organization [8]:

1. Key agents of activities include not only humans, but computer systems.
2. Any agent of activity or employee has an access towards massive storage of an information and knowledge of organization.

3. All organization members including computer systems can interact with each other using common methods.

Actually, Carr described modern organization with digital assistants, chatbots, artificial intelligence and messengers for communications. Considering such an organization as a bionic, he emphasized that computer systems become full-fledged members of processes. In the early 2000s there were mentions [9] of bionic organization, using technologies in its processes. After these researches there were no detailed mentions of bionic organization in scientific researches and publications on economics, managements and related fields. Further development of bionic organization concept took place in 2018-2019 in articles of PriceWaterhouse Coopers [10] and Boston Consulting Group [11]. The main point of those publications is that technological solutions and especially digital technologies such as artificial intelligence and big data, allow to enhance human competencies and broaden capabilities of organization [12].

Other modern researches note [13], that technologies and approaches to knowledge management could be developed if those points are supported by big data and artificial intelligence. This confirms Clay Carrs's withdrawals about availability of knowledge in bionic organization as its main feature, because this enhanced knowledge can widen human competencies.

Apart from working with data, in modern organizations there are a lot of digital assistants [14] which play humans roles in matters of support, requests handling and so on. But this role should be regulated and defined, because digital assistant cannot just substitute human. The matter of responsibility for the function delegated to a digital assistant becomes critical, because digital technologies are not able to be responsive, but humans do not already perform a function.

A similar output researches offer [15] in a matter of artificial intelligence and its integration in modern management systems. This integration should be regulated, in order to avoid humans fears and deep concerns those could arise because some parts of power and control go to machines and digital technologies.

At the same time, digital technologies development move workplaces away from productions facilities and creates significant trend towards coworking forms of interactions and workplaces structuring [16]. This change also demands correct interaction rules and approaches to work with production objects.

Social interaction is still in high demand, but it lost a necessity of direct physical contact in knowledge transfer processes, especially because of the COVID-19 pandemic. Social networks play the increasingly important role. Those networks become something like courts and platforms in international collaboration of corporations [17] which play a role of agents in social networks.

Finally, modern business tends to deeper integration of software in its processes those begins to drive business [18]. Dependence of the management and productions systems on software and information and communication technologies is different and defined by industry, character of activity and ability to digitize products and services, but a common tendency towards software key role is obvious.

Modern organizations have a lot of described features, but is it correct to consider those organizations bionic? In what points are bionic organization connected with digital transformation and is it correct to view both concepts mutually in context of digital technologies? This research allows us to make the following outputs in return to these questions.

4 Vision of bionic organization as a stage of organization's evolution in digital transformation

First of all, it should be mentioned that the published researches on the topic in question does not answer the questions about decision-making process for humans and digital agents in modern business. Who does define these decisions and make a structure and routines of organizational processes?

Actually, if an organization uses digital technologies to make its production more efficient, but decisions are made by humans, that means that digital technologies do not influence on the organization. It also means that the organization is still the same in its essence and DNA and has no bionic features. This kind of organizations, where all the main decisions and competencies are fixed to humans, should be considered as "biological".

The finite goal of a digital transformation process of production enterprise is a level of digital maturity, where invested funds and efforts bring an expected effect and organization is still competitive, profitable and ready to produce products and services demanded by target consumers. This process is usually connected with the goal to attain the highest level of digital maturity in every process in this organization.

The finite stage of digital transformations is a full-autonomous organization, that works without a role, but with a task of a human. This organizational stage has main specific: organizations stop being an organization and becomes a technical system. This stage is specific, because organization loses its main features and components: functions, staff, competencies, relationships, human, behavioral and social capital and in some extent even knowledge. Finally, it can be concluded that at this stage organization becomes autonomous (technical) system, but not an organization. This system is under controls of a human or a group of humans, who compose a kind of a new group with specific type, but they are not organization in itself, because all of its components were substituted by the digital technologies and machines. It is evident that because of its complexity this new technical autonomous system could not be run by this group completely. This system demands quite a bunch of deep competencies and knowledge to operate and take care of it. This system will demand skills of external engineering companies, architects and experts in system development and management. The group of people running the system will be focused only on its management and definitions of tasks for it.

Digital transformation up to the level of autonomy is impossible without going through certain stages of organizational evolutions from digital transformation point of view. We reckon that one of these stages is a stage of bionic organization. Actually, this is a stage of transforming a biological organization, where decisions are made by humans, to an autonomous (technical) system, where decisions are made by digital technologies and humans become a kind of owner or even specifier.

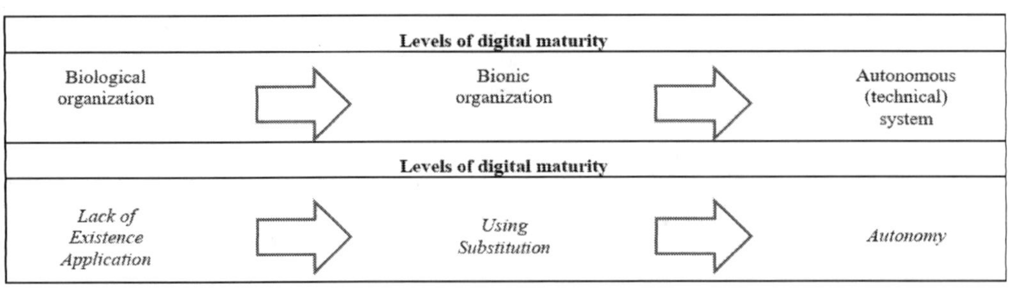

Fig. 1. Stages of organizational evolution in digital transformation process

Considering the approach of defining an organizational type based on decisions owner, it should be remarked that an organization could be depicted as a bionic if only managerial and production solutions are made by digital technologies (e.g. artificial intelligence (AI)) at least partly. Thereby, the main stages of organizational evolution include the following (see Figure 1).

Social dynamics and struggle for power and control in a bionic organization also get changes on this stage. A lot of decisions move from humans to digital technology components and organization faces a new issue, who will be responsible for these decisions.

A human is not limited in a space of one's decisions. This space is limited with abilities, competencies, moral convictions or values. Every machine or technology has a limited space of decisions defined by its designer, creator or specifier. This machine or technology can also be limited by a complex of supported algorithms in its decision space. This means that a human defined a space of decisions is responsible for certain decisions made by a machine or digital technology that is run with such a space of decisions.

Machine or digital technology is stronger than a human in tasks with certain or in general finite space of decision variants or actions. However modern machines produce art objects with high value [19], decision space of these machines is finally limited by developer's imagination and competencies. At the same time human imagination is almost infinite, except intrinsic borders those a human made for oneself personally.

Discovering an engagement of digital technologies in decision making process in bionic organization, we should remark that managerial tasks on these process steps comes down to defining a space of decisions with certain alternatives or limits with detailed parameters of such decisions [20]. For example, it is necessary to define temperature limitations to change, amounts to be paid automatically and so on.

Further digital technologies development, as it was mentioned before, will lead to complete human exclusion from the production process, but a key aspect of that is a disappearance of relationships in organization. When relationships disappear from the executive system, executive system moves to tasks set-up and constraining space of decisions. This key change will create autonomous (technical) system those main features of organization fade. This system also loses a main feature of biological organization – an ability to change and adopt to environment conditions and, finally, to market conditions. At this stage of organizational evolution connections between "human" and "machine" become definite: a human defines a space of decisions for digital technologies and is responsible for system's adoption to the environment and digital solutions support all system's processes in defined limits and provide it with demanded level of agility and desired efficiency.

This content of digital transformation concept makes a clear evidence of a further relevance of human in production facilities and especially in social frameworks. Organization needs an adoption to the environment it works in and to the market it is aimed at even at the stage of autonomous (technical) system. This adoption is an alien feature of technical systems, hence human have to play a role of an architect of the adoption to market and environment for this system in future, as a biological organism.

5 Conclusions

Overall, the main results of the research can be summarized as follows: First of all, it becomes apparent that bionic organization is an intermediate stage of organizational evolution in a digital transformation process between biological organization towards autonomous (technical) system.

Second, an organization could be considered as bionic if it has transformed to a digital maturity level "using" or "substitution" from the digital technologies point of view. These levels are among ones defined by authors of the article in their own concept of digital maturity.

Third, functions transfer from a human to digital technologies and solutions, including artificial intelligence, doesn't redeem managers from a responsibility for decisions made by these technological solutions. On the contrary, making a way from bionic organization towards autonomous system, create new managerial task to define a space of decisions and limits for digital technologies in production and executive system.

Fourth, a concept of bionic organization allows defining stages of organizational evolution from the digital transformation point of view. This path starts with biological organization and finalizes with autonomous (technical) system.

Finally, a demand of compliance to extrinsic terms, the environment and of organizational redesign form new human role in disruptive digital transformation in economics and society.

Further elaboration of these points might be focused on shaping up a deep concept of organizational transformation on every stage of its evolution from digital technologies point of view and on defining every single role of organization's members who should be played on the way of efficient digital transformation at its every stage.

References

1. D. Provan, D. Woods, S. Dekker, A. Rae, Reliability Engineering & System Safety, **195**, 106740 (2020)
2. Acatech, *https://en.acatech.de/publication/industrie-4-0-maturity-index-update-2020* (2020)
3. Merriam-Webster Unabridged Dictionary, *https://www.merriam-webster.com/dictionary/bionic* (2021)
4. DaytonDailyNews, *http://www.daytondailynews.com* (2009)
5. Digitaltrends.ru, *https://digitaltrends.rt.ru/reports/multiple_sources_rating* (2021)
6. McKinsey, *https://www.mckinsey.com/featured-insights/future-of-work/the-future-of-work-after-covid-19#* (2021)
7. Teqviser, *https://teqviser.ru* (2021)
8. C. Carr, Performance Improvement, 28(9), 44-46 (1989)
9. G. A. Cohen, J. S. Parkinson, Perspectives on Business Innovation 8, 79-84 (2002)
10. M. Everson, J. *Sviokla, https://www.strategy-business.com/article/The-Bionic-Company?gko=41603* (2018)
11. R. Hutchinson, L. Aré, J. Rose, A. Bailey, *https://www.bcg.com/en-ru/publications/2019/bionic-company* (2019)
12. S. Singh, A. El-Kassar, Journal of Cleaner Production, **213**, 1264-1273 (2019)
13. W. Strielkowski, *Social Impacts of Smart Grids: The Future of Smart Grids and Energy Market Design* (2019)
14. R. Rawassizadeh, T. Sen, S.J. Kim, C. Meurisch, H. Keshavarz, M. Mühlhäuser, M. Pazzani, Transactions on Pervasive Computing and Interaction, 1, 163–174 (2019)
15. A. Trunk, H. Birkel, E. Hartmann, Business Research, 13, 875–919 (2020)
16. R.B. Bouncken, S. Kraus, J.F. Martínez-Pérez, International Entrepreneurship and Management Journal, 16, 1465–1481 (2020)
17. I.R.P. Cuypers, G. Ertug, J. Cantwell, A. Zaheer, M. Kilduff, Journal of International Business Studies, 51, 714–736 (2020)
18. R. Alt, J.M. Leimeister, T. Priemuth, S. Sachse, N. Urbach, N. Wunderlich, Business & Information Systems Engineering, 62, 609–621 (2020)
19. BBC, *https://www.bbc.com/russian/news-45984669* (2019)
20. B. Gavurova, R. Bacik, R. Fedorko, M. Rigelsky, Marketing and Management of Innovations, **2**, 186-200 (2018)

Economic analysis of alternative energy development in Nagorno-Karabakh and surrounding regions

Aida Guliyeva[1]*, *Parvin* Mammadzada[1,2], *Turan* Mamedov[3]

[1]"SABAH" center, Azerbaijan State University of Economics, UNEC, Istiqlaliyyat str. 6, Baku, Azerbaijan.
[2]Department of World Electric Power Industry, International Institute of Energy Policy and Diplomacy at Moscow State University of International Relations (MGIMO University) under MFA, 76 Prospekt Vernadskogo, Moscow, Russian Federation
[3]Department of Economics of Oil and Gas Industry, State University of Oil and Gas "Gubkin University" (Gubkin University), 65 Leninsky Prospekt, Moscow, Russian Federation

Abstract. This paper studies the main opportunities for stimulating investments in the alternative energy development through implementation of state plans for creation and support of favourable socio-economic conditions in Nagorno-Karabakh and 7 surrounding regions. Due to the longevity of projects on renewable energy sources, as well as the traditional financing of infrastructure projects, it is extremely important that political and economic interests were protected by law. At the present time there are active project work on development and optimisation of energy sector, and therefore understanding of basic principles is quite diverse in public and private sectors. Main conclusions and recommendations are developed and presented for economic development of renewable energy sources (RES) in Nagorno-Karabakh and surrounding regions of Azerbaijan.

1 Introduction

The ever-increasing demand for energy, limited hydrocarbons and environmental challenges via traditional energy production have enforced the need of RES use. The declining level of provision of the global economy with gas and oil reserves has marked the trend of a new energy reality. In these conditions, the interest of industrially developing consumers in renewable energy sources (RES) has been formed, that is, solving the energy problem in the face of an ever-increasing demand for energy resources has become an agenda and an urgent message in the scientific world.

In order to support RES development, the government approved the "State Program on the Use of Alternative and Renewable Energy Sources in the Republic of Azerbaijan" by Decree No. 462 dated October 21, 2004, for stimulating RES development and preparation of short-, medium- and long-term strategic action plans [1]. In April 2016, official Azerbaijan approved the Paris Agreement on Climate Change, in which RES implementation and CO2

* Corresponding author: guliyevaaida@ymail.com

reduction were identified as priority areas of activity [2]. Renewable energy targets include increasing their share by 2020 to 20% in electricity generation and to 9.7% in total energy consumption and reducing losses in gas distribution systems [3]. At the same time, it is planned to commission power plants based on renewable energy sources with a total capacity of 2 thousand MW. By 2030, the target is to reduce carbon dioxide emissions by 35% compared to the 1990 baseline [2,3]. At present, RES covers 16.8% of total electricity production [2,4]. In 2017, alternative energy covered 2.0 billion kWh of electricity and 3789.2 Gcal of thermal energy, which saved 504.5 million cubic meters of natural gas in the same year and prevented the emission of 898.0 thousand tons of carbon dioxide. gas (CO2) into the atmosphere [4].

According to forecasts, by 2030 RES will provide 35-50% of global electricity production and 19-25% of total energy consumption. Renewables will face a bigger challenge in penetrating the automotive fuel market, providing only 7% of the world's fuel [5]. The growth rates of consumption of primary energy resources will decrease to 1.6% per year by 2030; energy consumption per capita will grow by 0.7% per year [6].

The aim of the paper is to cover theoretical basis for realizing methodological and practical recommendations for RES implementation due to expansion of international economic cooperation in the electric power industry of Azerbaijan. To achieve this goal the following tasks were set:

- Determine the criteria, methods of macroeconomic factors of international cooperation;
- Analyze the modernization and innovative development of the Azerbaijani economy;
- Determine the ratio of reducing dependence on petroleum and increasing role of RES in energy;
- Formulate practical recommendations for creating a regional model of the electric power transmission system.

2 Materials and methods

The nature of research is mixed, using exploratory and inductive types of research. Subsequently, a qualitative methodology was used to gain insight into alternative energy and design techniques. The following structure is proposed for the methodology: at the first stage, various scientific articles were used to determine the structure and scheme for the introduction and use of alternative energy. At the second stage, information obtained through data collection through interviews, observations and studies of advanced countries in the field of alternative energy was used. The methodological basis of the paper will be:

1) local and international regulations of RES implementation;
2) the provisions of the doctrine of environmental and energy law;
3) a set of theoretical and practical methods for solving macroeconomic problems associated with the transition to innovative development;
4) methods and tools for stimulating renewable energy projects.

Considering global electricity issues in the work, the information and empirical base was compiled by the annual reports of the SAARES, IRENA, IEA, World Bank, MoE of Azerbaijan, UN, REN21, UNCTAD, as well as the Electricity Directive of EP, Regulation on Access to Electricity Networks and legislative acts of Azerbaijan.

Azerbaijan, an exporter of net energy, meets energy demands through the domestic extraction of hydrocarbon resources, specifically natural gas and oil. Also shared with neighbor countries (Russia, Kazakhstan, Turkmenistan, and Iran), Caspian Sea reservoirs proved oil resources were estimated at 7 billion barrels in 2018, comparable to the North Sea's capacity decades ago [7].

Table 1 reveals that Azerbaijan's internal energy consumption is five times less than its export. Consolidation of the reduced wood utilization as a raw material for energy, with the volatile nature of hydropower production, had led to a decline in primary renewable energy production. Individually, the reduction in wood use is attributed to the installation of vapor heaters and the advancement of the region heating system [8].

Table 1. Energy balance in selected years, 2010–2017

Indicators (Tonne of oil equivalent)	**2010**	**2014**	**2015**	**2016**	**2017**
Primary products, including:	68 254.6	61 132.0	61 084.2	59 977.6	57 036.0
Raw oil	52 312.5	43 295.9	42 835.5	42 240.9	39 810.4
Natural gas	15 555.6	17 565.1	17 947.3	17 463.5	16 967.6
Renewable energy and wastes	386.5	271.0	301.4	273.5	258.0
Net imports of energy products	- 54 300.2	- 45 869.5	- 45 747.4	- 44 693.0	- 40 789.6
General power supply	12 566.5	15 085.5	15 569.4	15 393.5	15 471.9
Transformation processes	-2 692.7	-3 491.3	-3 819.7	-3 731.0	-4 020.6
Energy for own consumption	931.1	1 188.2	1 167.0	1 133.0	1 147.7
Total final energy consumption	6 710.6	8 241.7	8 304.3	8 644.9	8 210.0

According to the Ministry of Ecology and Natural Resources of Azerbaijan (MENR), environmental safety is of paramount importance, the policy is primarily aimed at reducing human influence on the nature and coordinating environmental protection. Taking into account the demands, the other policy direction is the efficient use of natural resources, including RES implementation. Assessing the country's needs in global environmental problems, developing solutions for them and making decisions through cooperation with international organizations is the third aspect of Azerbaijan's environmental policy [9].

The State Program on the Use of Alternative and Renewable Energy Sources in the Republic of Azerbaijan was approved by the Presidential Decree No. 462 of October 21, 2004 and became a new milestone in Azerbaijan's energy policy [1]. The goal of the State Program was to promote the generation of electricity from renewable and environmentally friendly sources and more efficient use of hydrocarbon energy sources.

Given that RES is a relatively new and unconventional sector in Azerbaijan, the country has nevertheless achieved significant success. First, these are legal problems for both the production and use of renewable energy sources. RES investments in Azerbaijan are mainly carried out by the government and international organizations, and the draft law is still under discussion. Secondly, these are technical problems, primarily related to technology transfer. Renewable energy technologies are expensive investments, and given that Azerbaijan imports them from other countries, it is necessary to simplify technical procedures. For a long time, the country's electricity has been dominated by oil and natural gas, which have a significantly established nature of technology and experience, which constitutes serious competition for renewable energy sources. The third problem is low awareness. There is a lack of support from international organizations and local authorities to raise public awareness of RES in Azerbaijan [10].

Azerbaijan is completely electrified, meaning the electrification level reaches 100%. The country's total installed capacity is over 7,141 MW, with 60% derived from thermal power (TPP), specifically, 15 thermal and 14 hydroelectric power plants. TPP is the largest power plant in the country with capacity over 2,400 MW (oil-based) [4,9]. Also, factories are constructed, taking into account the requirements of industrial design for efficiency and reliability [4]. The latest TPPs are advanced and operate on the use of gas turbine steam cycle technologies [4]. Considering installed capacity of renewable sources, hydro energy is leading (1,134 MW), followed by wind (66 MW), biomass (38 MW) and solar PV (33 MW). Table 2 presents data on electricity generation by source in 2019 [4,9].

Table 2. Electricity production by RES source (GWh)

Energy source	**2010**	**2011**	**2012**	**2013**	**2014**	**2015**	**2016**	**2017**
Thermal	15 262.7	17 618.0	21 167.1	21 729.6	23 249.3	22 859.9	22 761.0	22 344.9
Hydro	3 446.3	2 675.8	1 821.0	1 489.1	1 299.7	1 637.5	1 959.3	746.4
Wind	0.5	-	-	0.8	2.3	4.6	22.8	22.1
Solar	-	-	-	0.8	2,9	4.6	35.3	37.2
Biomass	-	-	-	134.1	173.5	181.8	174.5	170.3
Total	18 709.5	20 293.8	22 988.1	23 354.4	24 727.7	24 688.4	24 952.9	24 320.9

According to the Ministry of Energy, over 4,000 MV of solar and up to 500 MV of wind energy potentials and 25% of national hydro resources, or over 2.56 bln m3 of water per year formed in Nagorno-Karabakh and surrounding areas (Figures 1 and 2).

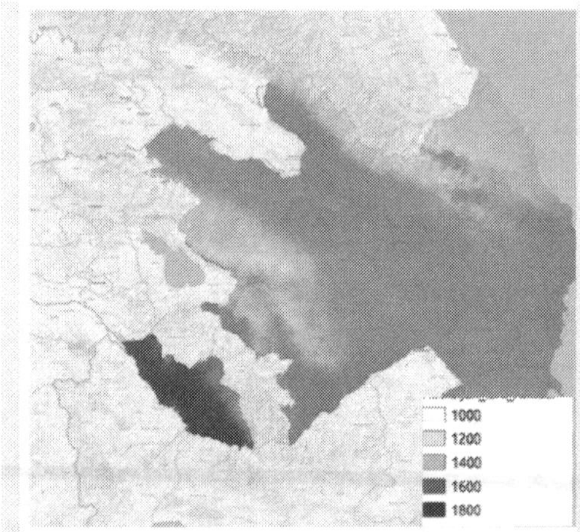

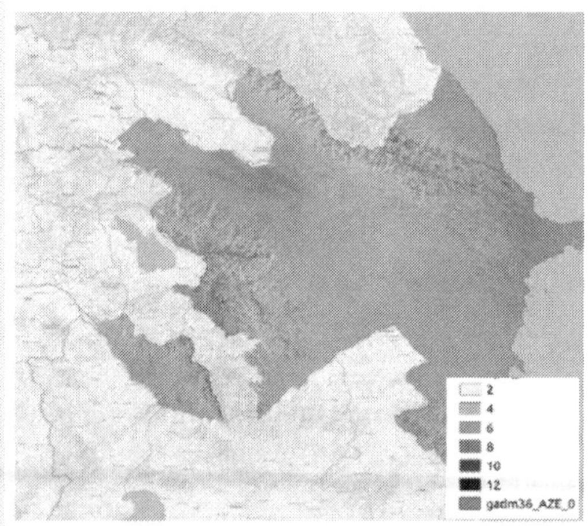

Fig. 1. Annual global horizontal solar emission (kWh/m2) **Fig. 2.** Wind speed at 100 meters (m/s)

Improving secure energy supply, creating decentralized power grids, reducing the share of power outages and diversifying energy sources provide and accelerate access to energy for residents of Azerbaijan and remote areas in the liberated territories. In a country dominated by oil and gas, alternative energy will provide a catalyst for new job opportunities, technological innovation, and new opportunities for economic creation in Nagorno-Karabakh and surrounding regions [12].

3 Results and discussion

The main objectives of the development of alternative energy should be to reduce risk, stimulate deployment and reduce losses. The facts testify to the fact that a significant share of breakthrough innovations comes from new firms that challenge existing business models. Thus, state actions to eliminate barriers to entry foreign companies have positive impact on stimulation of low-carbon energy technologies. Electricity markets in the European Union, as well as in other regions of the world, undergo significant transformations that lead to growth in terms of electricity safety. We are witnessing a rapid growth of renewable, often decentralized electricity generation, which is due to the goals and policies in the field of decarbonization and a sharp reduction in the cost of energy use. Meanwhile, there is reduction of traditional, centralized and more flexible generating capabilities. Another factor that affects the change in the landscape of electricity markets is the availability of innovative low-carbon solutions to balance the demand and supply. Figure 3 shows an illustration of possible changes in the landscape power system for the Azerbaijani power supply system based on renewable energy sources [13].

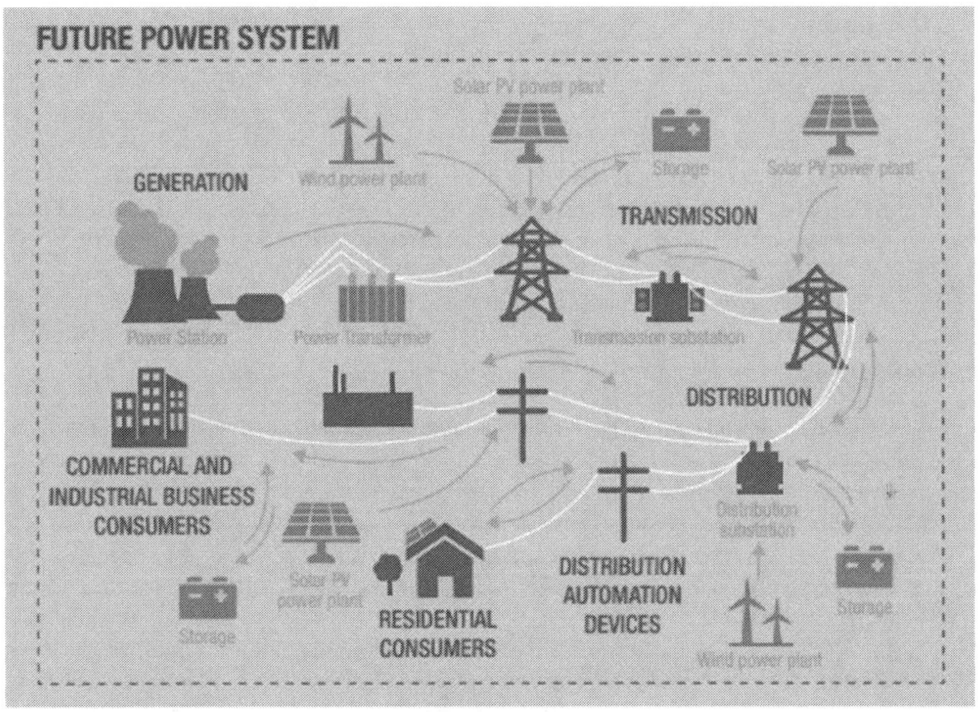

Fig. 3: Future low carbon decentralized power system

The basic principle of government as a market maker is to exert competitive pressure between buyers and sellers. The goal is to harness the power of the markets to achieve broader policy goals. Well-designed market processes can avoid the constraints of competition from more traditional command-and-control approaches. The market mechanism, in which the government plays a role in the design, control and application of the market, must provide the necessary incentives for the effective provision of goods or services while achieving the same policy objectives [14,15].

4 Conclusions

Azerbaijan has made colossal steps in the energy sector over the past decade; electricity generation's capacity surpasses 7,100 MW and expands by approximately 40%. The primary reason for improving electricity generation capacity is close attention to combining an innovative generating capacity over 2.3 GW from 2006 to 2015. Azerbaijan actively partners with several countries, domestic and foreign organizations, and financial institutions in RES in-depth development [4,9].

The sustainability of Azerbaijan's power system additionally depends on the advancement of renewable energy sources. For instance, the country can improve the security of supply through enhanced utilization of batteries consolidated with distributed renewable power systems. In addition, renewable energies provide a technical solution for the rapid deployment of production capacity due to a relatively shorter predicted adaptation time than traditional sources, taking into account the projected growth in national energy needs.

Government issued Presidential Decree No. 1138, on Strategic Roadmap to improve utilities (electricity and heat, water and gas supplies), which consists strategic vision to 2020, long-term vision to 2025 and target vision to post-2025. Over the next 5-10 years, to increase generating capacity by 1000 MW, investments will be made in the supply of sufficient quantities electricity in addition to the planned investment [16].

Economic Energy Efficiency and Renewable Energy Initiatives that affect the demand for energy services tend to change the patterns of energy consumption by business and residential consumers by reducing the amount of energy required for a given level of production or service. Initiatives to increase demand for energy efficiency in the field of alternative energy led to direct costs and savings. Supply-side policies generally support RES development, hybrid heating and clean energy transfer. The direct impact of supply-side initiatives arises from the costs supported by the initiative, as well as the energy savings and potential savings in energy costs through fuel swaps [17,18].

Concluding remarks on the economic efficiency of using RES implementation in government programs are listed below:

- Azerbaijan should obtain foreign financial and technical assistance in production of eco technologies;
- Sustainable development should be applied in all sectors of the economy, taking into account the problem of climate change;
- Short and long-term sustainable energy development programs in Azerbaijan should provide an opportunity for foreign energy investors to work closely with national energy companies;
- International finance to support investments in climate change adaptation should be integrated into the country's renewable energy policies.

In a conclusion, it should be noted that greenhouse gases accumulate and can remain in the atmosphere for decades or centuries, affecting the global climate system for a long time. Because of this, measures such as energy efficiency and renewable energy that immediately prevent or reduce greenhouse gas emissions can create long-term and positive benefits for the atmosphere and human health.

References

1. Presidential Decree "On the state program for the use of alternative and renewable energy sources in the Republic of Azerbaijan" (2004)
2. IRENA, *https://www.irena.org/publications/2017/Jan/Planning-for-the-renewable-future-Long-term-modelling-and-tools-to-expand-variable-renewable-power* (2017)
3. Directive 2009/72/EC of the EP and of EC concerning common rules for the internal market in electricity (2009)

4. N. Vidadili, E. Suleymanov, C. Bulut, C. Mahmudlu, Renewable and Sustainable Energy Reviews, **80**, 1153-1161 (2017)
5. D. Gielen, F. Boshell, D. Saygin, M. Bazilian, N. Wagner, R. Gorini, Energy Strategy Reviews, **24**, 38-50 (2019)
6. U. Deichmann, A. Reuter, S. Vollmer, F. Zhang, *https://papers.ssrn.com/sol3/papers.cfm?abstract_id=3113873* (2018)
7. OPEC, *https://www.opec.org* (2018)
8. State Statistical Committee, Energy of Azerbaijan: Statistical Yearbook, **5**, 23, (2018)
9. IRENA, *https://www.irena.org/-/media/Files/IRENA/Agency/Publication/2019/Dec/IRENA_RRA_Azerbaijan_2019.PDF* (2019)
10. IEA, *https://www.iea.org/reports/world-energy-outlook-2017* (2017)
11. IRENA, *https://www.irena.org/-/media/Files/IRENA/Agency/Publication/2019/Feb/IRENA_Innovation_ToU_tariffs_2019.pdf?la=en&hash=36658ADA8AA98677888DB2C184D1EE6A048C7470* (2019)
12. AERA, *http://regulator.gov.az/en/news/127* (2020)
13. World Energy Council, *https://www.worldenergy.org/publications/entry/world-energy-scenarios-2019-exploring-innovation-pathways-to-2040* (2019)
14. IEA, *https://www.iea.org/reports/energy-policies-beyond-iea-countries-eastern-europe-caucasus-and-central-asia-2015* (2015)
15. M. Petitet, D. Finon, T. Janssen, Energy Policy, **103,** 30-46 (2017)
16. Presidential Decree, On a strategic roadmap for the development of utilities (electricity and heat, water supply and gas supply) in the Republic of Azerbaijan No. 1138 (2016)
17. IRENA, *https://www.irena.org/publications/2015/Jun/Renewable-Energy-Target-Setting* (2015)
18. IRENA, *https://www.irena.org/publications/2015/Jun/Renewable-Energy-Auctions-A-Guide-to-Design* (2015)

Emerging clean energy markets are drivers for solar and wind industries

Marina Faminskaya[1]*, *Elena* Potekhina[1]

[1] Faculty of Information Technology, Russian State Social University, 4 Wilhelm Pik str., Moscow, 129226, Russian Federation

Abstract. Emerging markets are rapidly developing and outperforming developed markets in terms of onshore wind and solar power. Moreover, emerging markets are now getting drivers for the enhancement of global renewable energy as they are seen as territories with the greatest potential for expansion of wind and solar capacities. This study aims to investigate the casual links between financial investments by gas and oil giants and long-term vision around clean energy. We analysed various reports on procurement tools used by corporations, purchasing renewable energy. This gave us an idea about the availability and frequent usage of different instruments employed around world. The empirical results show that state support and policy development both have positive impact on state and private investments in the long run. The influence of energy development is more evident in emerging clean energy markets. The findings of the this study suggest that as renewable energy targets price and performance parity with traditional sources around the world, and is getting able to improve grid efficiency thus strengthening its competitive edge with new technologies, the hurdles to its adoption are disappearing.

1 Introduction

The solar and wind industries and markets initially emerged in developed countries, which we define as the 33 high-income OECD member countries, though over the time, these industries have shifted to emerging markets, i.e. all less developed countries. Emerging markets outpaced developed markets in onshore wind power growth in 2013, and later in 2016 in photovoltaic growth. These areas accounted for 63% of all new global investments in renewable energy sources in 2017, which widened the gap with developed countries to record levels. Currently, the total capacity of such facilities exceeds the indicator for the developed world [1,2]. Emerging markets declined the cost of renewable energy, which allowed them to bypass developed countries in terms of deployment of renewable energy sources, reduce the volume of carbon intensive developments and innovate in such a way that those initiatives are useful both for emerging and developed markets.

As the world leader in renewable energy growth, China is a driving force for other emerging markets. China has recorded the largest increase in the number of solar and wind installations and the total installed capacity for the last 5 years [3,4]. China is the only market

* Corresponding author: faminskaya@mail.ru

https://doi.org/10.1051/e3sconf/202125003005

with a capacity of over 100 gigawatt of both of these types of energy. The country accounted for more than half of all new solar installations, as well as two-thirds of solar energy production from photovoltaic panels. Eight of the world's top ten solar energy suppliers are based in China, and the three largest Chinese wind energy companies collectively have the largest share of the wind energy market [5,6]. Moreover, China is the only developing country to rank among the top 10 recipients of net cross-border investment and the top 10 investors, with China being the only emerging market country in the latter category [7]. China set a record for net cross-border investment since 2015, and up to the first half of 2017, the country has invested $ 2.23 billion in wind and solar energy in 11 other developing countries thus attracting domestic investment in these industries from 13 countries for a total of US $ 1.34 billion [8].

Even excluding China, emerging markets are driving the development of renewable energy globally and have the greatest potential for growth, though emerging markets are not ahead of developed markets in terms of the volume of annually commissioned wind and solar capacities [9,10]. However, China's share in terms of added capacity in emerging markets has declined since 2016, both for wind and solar. Elsewhere, emerging markets outperform developed markets even without China's performance. Auction prices for solar and wind energy broke recent records in Mexico and the UAE in 2017 [11,12]. These countries also recorded the world's lowest demand price for wind and solar energy. With the help of energy auctions, India was able to become the most competitive renewable energy market in the world as new players entered the competition. India and Turkey doubled their solar capacity in 2017, with India recently raising its already high bar on renewables. The country plans to increase its renewable capacity to 227 GW by 2022 [13]. Over the past two years, all new facilities have been in emerging markets. Countries with the highest rates of RES investment in GDP (including the Marshall Islands, Rwanda, Solomon Islands, Guinea-Bissau and Serbia) are also emerging markets. Finally, a huge opportunity for the development of renewable energy is given by the African territories, lying to the south of the Sahara - the largest market in the world not yet covered with electricity. For population groups living in regions with a low population density and highly isolated from power grids, the best solution would be solar installations for households, which allow paying for consumed electricity only [14].

According to the International Energy Agency, in the next two decades, most people without access to electricity will be getting it via distributed photovoltaic solar power plants and micro-grids. Developed countries have already benefited from market and product developments that originally emerged in developing countries. As an example, auctions for the sale of renewable energy supplies are a trend driven primarily by emerging markets, resulting in a sharp drop in renewable energy prices worldwide [15]. Some solar and wind energy products targeted at emerging markets are now being introduced to developed markets because of a "reverse innovation" process. Thus, micro-power grids designed to electrify areas not served by traditional power grids have found application in remote mines for developed countries. Consequently, corporations are getting more powerful in facilitating technology transfer between developed and developing countries that are adopting renewable energy sources.

2 Materials and methods

The data used for this paper were taken from the reports on various procurement tools employed by corporations, purchasing renewable energy in new ways, involving participants from an increasing number of industries. As far as corporations are concerned about procurement quality, Power Purchase Agreements (PPAs) are becoming the preferred instrument [16]. Such PPEs provide an optimal opportunity to obtain additional capacity, while being available for the most part only to large companies. Meanwhile, there has been a trend to aggregate companies to enhance access to this tool for smaller players as well. The largest companies set goals for the introduction of RES into their supply chains, and therefore, encourage that smaller players purchase RES.

Currently, corporations produce or purchase renewable energy, hence, a number of procurement instruments can be used at different scales. One of the most widely used tools employed by 57 countries is the certificate. The certificate certifies that a company has met mandatory requirements or voluntary targets in the field of renewable energy. The certificate does not reflect the full cost of using RES and does not always provide the possibility of obtaining additional benefits in the future. Another tool is green tariffs that is available in 39 countries (mainly European), but they are the least used and transparent instrument of all. In 2017 corporations in 10 countries signed agreements to purchase a record 5.4 GW of renewable energy. This record had been broken by a wide margin, with 7.2 GW of renewable energy purchased in 28 markets worldwide by the end of July 2018 [17]. PPEs offer more value-added benefits and greater cost savings compared to certificates and green tariffs, as well as compared to the cost of electricity from traditional sources. However, for smaller companies, this tool is less available. All instruments mentioned above are available in North America and most European countries. These developed countries continue to be the leading markets for corporate electricity purchases. Situation in other countries (such as Russia) might be different [18].

However, companies in other sectors are also increasing their purchases of renewable energy, and emerging markets are helping to strengthen this trend. Emerging markets such as Brazil, India and Mexico are also offering a full range of renewable energy tools and are showing an increase in renewable energy purchases from international and local companies.

3 Results and discussions

Many oil and gas companies have already accumulated significant experience in implementing projects in mature markets in Europe and the United States.

The RES market is developing unevenly around the world and has varying degrees of maturity in developed and developing countries. Various regions have their own RES growth drivers. The strategy for the participation of oil and gas companies in renewable energy projects largely depends on the regions in which they are represented (see Figure 1). As a rule, the countries that are leaders in the development of renewable energy sources are states that are limited in their own resources of fossil fuels.

The development of renewable energy for them is a way to ensure energy security and compliance with the requirements of the Kyoto Protocol, which limits the level of emissions when using solid fuel sources. In many countries, the development of renewable energy sources becomes part of a long-term state strategy; therefore, measures to support the development of renewable energy sources are adopted at the state level, which contributes to the growth of attractiveness (including profitability) and the active implementation of renewable energy projects by international companies.

Renewable energy projects are becoming part of the development strategy of the largest oil and gas companies, which is supported by the volume of their investments.

Independent and state-owned European oil and gas companies retain their historically established global leadership in the development of renewable energy sources. Table 1 that follows below provides an overview of the global investments into renewable energy by a selected group of countries.

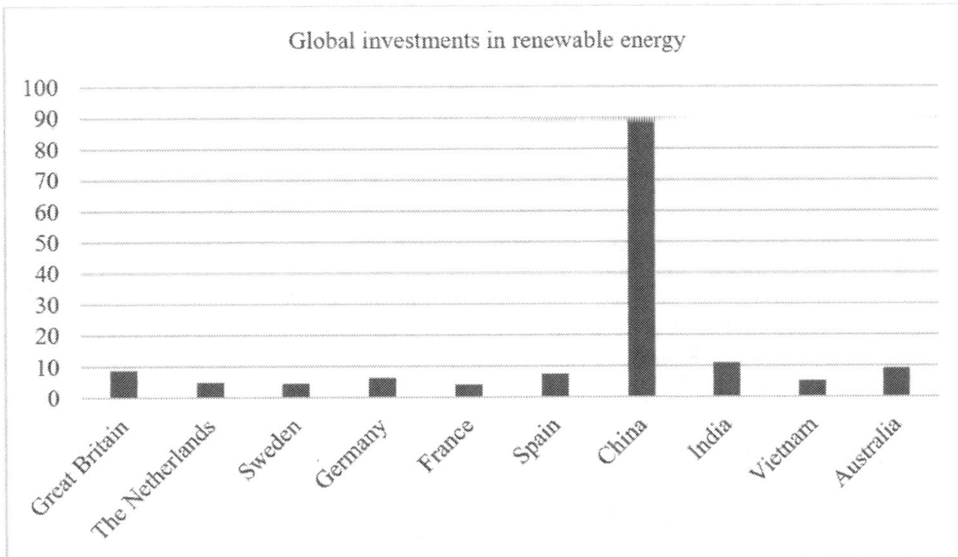

Fig. 1. Global investments in renewable energy (in billions of U.S. dollars) [19]

Companies such as Shell, Total, ENI, Equinor and BP are the largest investors in renewable energy projects. 3-5% of the total investments of these companies were on renewable energy sources and related projects as part of the energy transition strategy in 2018 as an average.

Table 1. Renewable energy investments of major gas and oil companies [20]

Oil and gas company	Planning horizon	Investments in RES (billion US dollars)	Share in total company investments (%)
Shell	2021-2025	2-4	8
ENI	2018-2020	1.1	3.6
Total	2019-2023	1.5-2	10-15
Equinor	2020-2025	0.7-1.5	n/a
Saudi Aramco	Up till 2030	30-50	1.7

In the near future, large national companies from oil and gas producing countries may compete with international giants, who are shaping a long-term vision around clean energy. The state-owned company Sonatrach in Algeria plans to increase its solar activities. Asian companies Pertamina (Indonesia) and Sinopec (China) claimed they are ready to continue the development of renewable energy sources. Oman national company PDO said that up to 20% of its revenues in 10 years will be coming from new businesses such as solar, wind and wave energy.

Oil and gas companies have not yet proposed a single business model in the field of renewable energy; each company forms a strategy that reflects its long-term vision and the current state of its core business. By the early 2030s, Shell intends to become a leader in the world of power generation, including the implementation of renewable energy projects. By integrating the entire value chain from generation to distribution and offering B2B and B2C products, the company plans to maximize its return on investment. Total, on the contrary, adheres to the asset-light model by gradually reducing its share in renewable energy projects from the development stage to commissioning (up to 50%). During the operational phase, the company remains the main operator.

4 Conclusions

The development of solar and wind energy has recently reached new frontiers. Renewable energy is no longer just a popular, but a preferred source of energy supply. As alternative energy sources achieve price and performance parity with traditional sources around the world, demonstrate the ability to improve grid efficiency and strengthen their competitiveness with new technologies, the barriers to their adoption are gradually disappearing. Solar and wind energy are already some of the cheapest energy sources in the world and have significant potential for further development, as a lot of trends have not yet been fully implemented.

The cost of renewable energy is declining, and the successful integration of renewables is well under way, supported by the development of technologies that provide even greater efficiency and empower companies. Meanwhile, the demand for renewable energy sources is steadily growing. Today, solar and wind energy can almost completely satisfy three key consumer needs: stability of energy supply, affordable cost and environmental friendliness. In markets leading the renewable energy sector, such as Denmark, supranational, national and local interests are aligned with these requirements. In other countries, such as the United States and Australia, where national governments are reluctant to participate in global carbon reduction initiatives, cities, communities and corporations have taken on the most important roles in these endeavors. They are stepping up their efforts to fill the vacuum, and therefore the demand for renewable energy continues to grow.

Finally, emerging markets, which will see the most noticeable increases in electricity demand with further development in electrification, are pursuing a leading position in the solar and wind energy market. This means that the case for renewable energy is stronger than ever before.

References

1. IRENA (2018), Renewable capacity statistics 2018, International Renewable Energy Agency (IRENA), Abu Dhabi
2. E. Lisin, W. Strielkowski, V. Chernova, A. Fomina, Energies, **11(12)**, 3284 (2018)
3. Peidong, Y. Yanli, Shi jin, Z. Yonghong, W. Lisheng, L. Xinrong, Renewable and Sustainable Energy Reviews, **13(2)**, 439 (2009)
4. W. Liu, H. Lund, B. Vad Mathiesen, X. Zhang, Applied Energy, **88(2)**, 518 (2011)
5. F. Dinçer, Renewable and Sustainable Energy Reviews, **15(1)**, 713 (2011)
6. T.M. Razykov, C.S. Ferekides, D. Morel, E. Stefanakos, H.S. Ullal, H.M. Upadhyaya, Solar Energy, **85(8)**, 1580 (2011)
7. S. Gopal, J. Pitts, Z. Li, K.P. Gallagher, J.G. Baldwin, W.N. Kring, Energies, **11(10)**, 2804 (2018)
8. G. Ragosa, P. Warren, Journal of Cleaner Production, **235**, 854 (2019)
9. A. Elia, M. Kamidelivand, F. Rogan, B. Ó Gallachóir, Renewable and Sustainable Energy Reviews, **138**, 110488 (2021)
10. R. Vanickova, Marketing and Management of Innovations, **2**, 56-67 (2020)
11. A.J. Al-Yasiria, M.A. Alib , R.S. Alic , H.N. Bekheet, International Journal of Innovation, Creativity and Change, **11(3)**, 376 (2020)
12. N. Martinez, N. Komendantova, Energy Policy, **145**, 111744 (2020)
13. Baris Memduh Eren, Nigar Taspinar, Korhan K. Gokmenoglu, Science of The Total Environment, **663**, 189 (2019)
14. A.K. Aliyu, B. Modu, C.W. Tan, Renewable and Sustainable Energy Reviews, **81(2)**, 2502 (2018)

15. A. K. Nasr, M. K. Kashan, A. Maleki, N. Jafari, H. Hashemi, Entrepreneurship and Sustainability Issues, **7(3)**, 2526-2541 (2020)
16. B. Tranberg, R. T. Hansen, L. Catania, Energy Economics, **85**, 104567 (2020)
17. B. Usher, *Renewable Energy: a primer for the twenty-first century* (2019)
18. A. Kuzmin, E. Volkova, A. Fomina, International Journal of Energy Economics and Policy, **9(1)**, *130–136 (2019)*
19. *Statista, https://www.statista.com/statistics/186818/north-american-investment-in-sustainable-energy-since-2004* (2021)
20. D. Bogdanov, A. Gulagi, M. Fasihi, C. Breyer, Applied Energy, **283**, 116273 (2021)

The heterogeneous effect of democracy, economic and political globalisation on renewable energy

Oleksiy Lyulyov[1], *Tetyana* Pimonenko[1 *], *Alexey* Kwilinski[2], and *Yana* Us[1]

[1]Department of Marketing, Academic and Research Institute of Business, Economics and Management, Sumy State University, Rymskogo-Korsakova st. 2, 40007 Sumy, Ukraine
[2]The London Academy of Science and Business, Baker St. 120, W1U 6TU, London, England

Abstract. This paper summarises the arguments and counterarguments within the scientific discussion on international cooperation's role in combatting climate change and its impacts. The primary purpose of the research is to determine renewable energy development reliance on democracy and globalisation levels. The objects for analysis are Ukraine and countries with different democracy regimes: full democracy (Finland, Denmark, Spain), flawed democracy (Poland, Slovakia, Hungary, and the Czech Republic) and hybrid democracy (Ukraine, Turkey and Montenegro). To gain the research goal, the authors examined data on the share of renewable energy, GDP per capita, labour force and gross fixed capital formation from 2012 to 2019. The data was retrieved from the Eurostat database, World Data Bank, KOF Swiss Economic Institute and the Economist Intelligence Unit. The following methods and tests were used: Levin, Lin, and Chu test; Augmented Dickey-Fuller Fisher and Phillips-Perron Fisher unit root test; Im, Pesaran, Shin's panel unit root tests. The authors used the Pedroni test to cointegration among variables. The Fully Modified OLS and Dynamic OLS panel cointegration techniques were applied to evaluate a statistically significant longer-term relationship between variables. The findings confirmed that for countries with the hybrid regime, the changes in political and economic globalisation provoked the rapid growth of renewable energy compare with countries from full and flawed democracy.

1 Introduction

Nowadays, international climate change commitments are on the agenda. The pointed question is to mitigate and adapt to adverse climate changes while decreasing human health and environmental risks and providing global economic, social, and energy preparedness. The urgent action to combat climate change and its impacts need international cooperation. It stands to mention that the climate consensus is expressed by the several signed international agreements such as the United Nations Framework Convention on Climate Change, Kyoto Protocol to the UN Framework Convention on Climate Change, Paris climate agreement, etc.

* Corresponding author: tetyana_pimonenko@econ.sumdu.edu.ua

In December 2019, the EU Council presented the European Green Deal (EGD) to achieve a carbon-free economy with net-zero greenhouse gas emissions until 2050. Green energy transition drives the energy-efficient potential, renewable energy sources and green transport, circular economy development, carbon capture and storage, smart network services, the extension of bioenergy etc.

EGD opens the window of opportunities for global energy collaboration to boost economic growth. It should provide a positive energy balance, gain energy independence, create fair competition in the energy market, increase the share of energy generations from renewable sources, develop a proportional energy tariffs policy, etc. To be an active actor of global combating climate change, different countries have accepted the Paris Agreement Goals in reliance upon national strategies and priorities. However, many countries are falling behind in the case of their National Energy and Climate Plans.

Nowadays, the pandemic challenge opened new opportunities to renovate the economics sustainably. Green energy transformation is considered to be a vector toward energy independence. Energy efficiency and renewable energy sources are the determinants attributes of economic growth. Thus, energy-efficient advance allows a considerable decline in the different energy generation for GDP growth and social welfare. It triggers all-electric economy development by renewable energy share gains and reduces fossil fuel consumption. Thus, advance in energy efficiency and renewable energy are global priority orientations.

It worth noting that GDP energy intensity is on a downtrend. However, it is still high, especially for Ukraine. Figure 1 shows that, in 2019, the global total energy consumption per unit of GDP decreased by 26.2% compared to 2000, while in Ukraine – by 56.3% [2]. However, in 2019, Ukrainian GDP energy intensity was by 2.11 times compared to the global level. In turn, in Poland (one of the biggest trade partners of Ukraine), the level of GDP energy intensity was lower by 2.7 times in 2019.

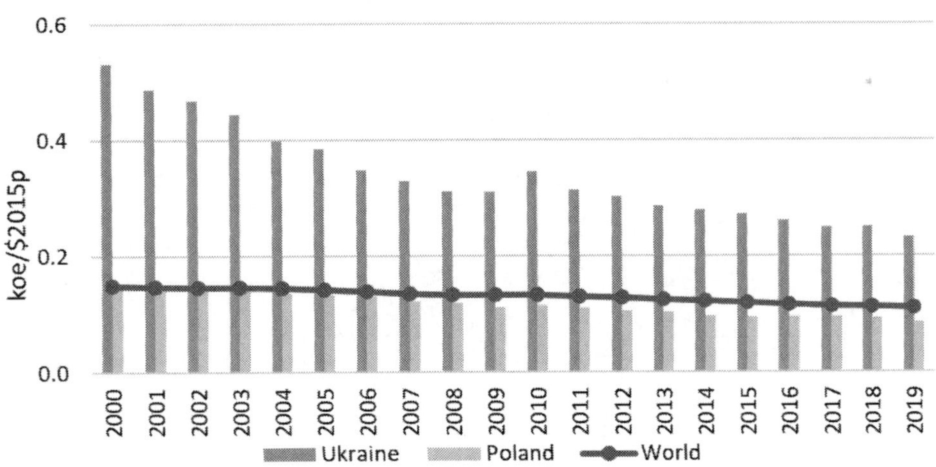

Fig. 1. Energy intensity (total energy consumption per unit of GDP), 2020-2019
*Source: compiled by the authors based on the data [2].

The analysis of statistical data showed that Ukrainian total energy consumption per unit of GDP is stratospheric. Therefore, the green energy transition requires massive investments and expenditures in energy and consumer sectors to increase energy efficiency by implementing new technologies in production, transportation, and energy consumption. Notably, renewable energy sources are considered the most powerful instrument in decarbonising the national and global economies.

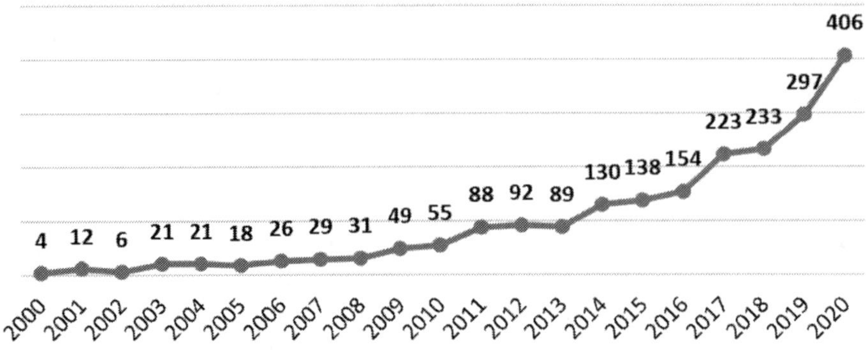

Fig. 2. Publication activity o "renewable energy" and "economic growth", 2000-2020
Source: developed by the authors based on [1].

Many scientific discussions were made to unveil the impact of economic, social and political spheres on developing energy-efficiency and promoting renewable energy sources. The findings proved the snowballing growth of scientific interest in investigating renewable energy development's economic influence (Fig.2). In could be assumed that in 2014, the number of publications increased by 31% and stated to rapid growth when the Intergovernmental Panel on Climate Change presented the Fifth Assessment Report on the anthropogenic impact on climate change in 2013.

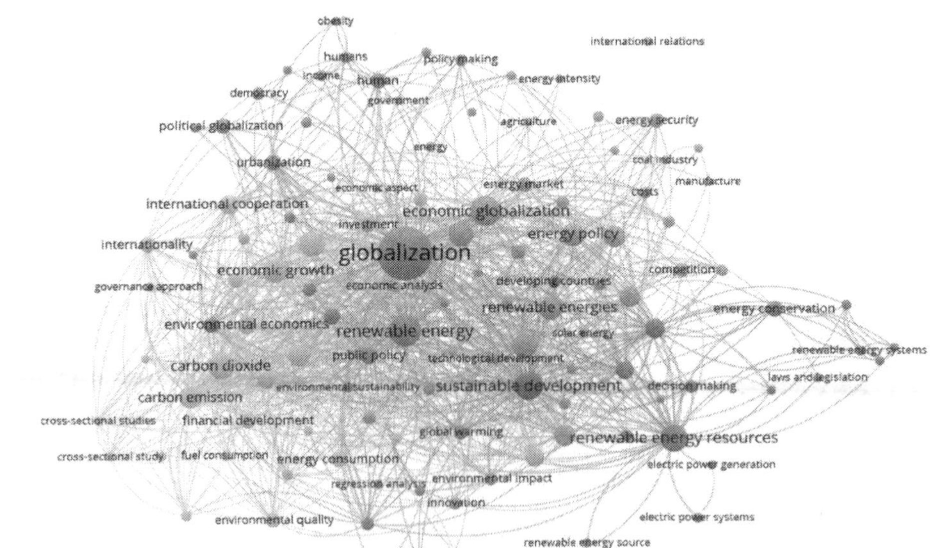

Fig. 3. The visualisation of co-occurrence network map on "renewable energy" and "economic growth", 2000-2020
Source: developed by the authors based on [1].

To find the main directions in publication activity, this study applied the visualisation of similarities method described in the several scientific papers [3; 4; 5; 6]. The practical realisation was provided with the VOSviewer software tool. The study sample consisted of

2000 of the most cited documents published in the high-reliable scientific journals indexed by the Scopus database from 2000 to 2020.

The findings proved the strong relationship between economic growth and renewable energy. Figure 3 shows the strong link strength between the term "renewable energy" and "economic growth". This couple of terms co-occurred 147 times in the investigated documents published from 2000 to 2020. The total link strength of the "renewable energy" is higher – 833, while "globalisation" – 719.

Therefore, the network analysis of keyword co-occurrence demonstrates three big clusters which indicate the research direction of analysed documents. The first biggest cluster (blue) consists of 42 terms which demonstrates the scientists' interest in developing the globalisation process. The second cluster (green) shows that scientists devoted a lot of attention to economic growth and economic globalisation in renewable energy consumption (link strength is 79). In turn, the third (purple) and forth (red) clusters allowed concluding that the issues of renewable energy were investigated under economic globalisation (link strength is 39), climate change (link strength is 35), in prospects of political globalisation (link strength is 35), energy efficiency (link strength is 27) and others.

The above analysis showed that research directions such as looking for best energy practices, new methods, innovative technologies, green financing, economic mechanisms etc., in mitigating and adopting against adverse climate are prioritised on the international level to develop renewable energy carbon-free economy. The obtained results indicated that the scientific community has significant progress in investigating the role of economic and political globalisation in developing and extending renewable energy [7–11].

In the studies [13–15] the findings confirmed the linking between economic growth, renewable energy and greenhouse gas emission. The authors noted that green investments boost economic development by renewable energy while decreasing the adverse greenhouse gas emissions [16; 17]. The influence of energy efficiency aspects on economic welfare was considered in the studies [1826].

However, the obtained results showed the scarcity of publications devoted to investigating the impact of determinants of democracy and globalisation processes on renewable energy development [27–33]. In turn, Fig. 4 allowed noticing that growth of democracy (DI) and globalisation level (KOF) provoke energy-efficient development).

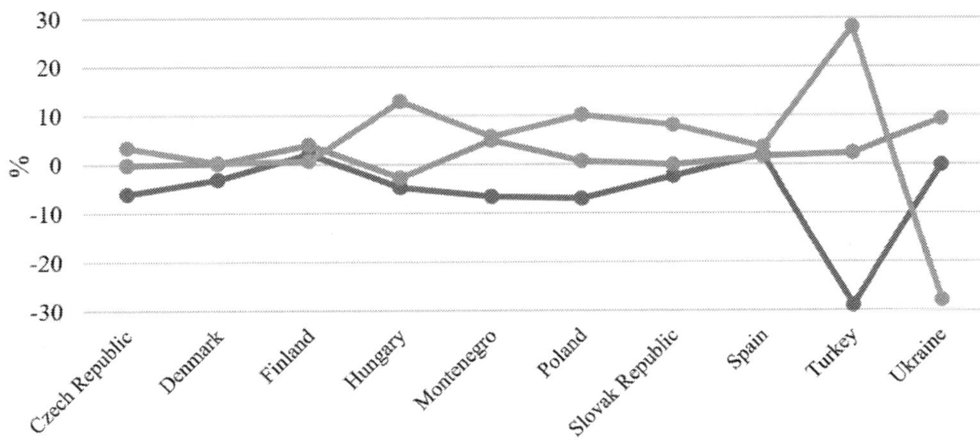

Fig. 4. Changes oEEEE, DI and KOF in 2019 compared to 2012
*Source: developed by the authors based on the data [34; 35; 36].

Besides, this research direction is relatively new. The study [38] analyses the impact of democracy and renewable energy on the volume of carbon emission. The authors emphasised that the trade openness, population, economic growth and foreign direct investment forced behind the carbon emissions. However, economic growth could decrease the volume of carbon emission if democracy was accounted for. In the paper [39] the author proposed instruments to accelerate energy transition under political influence. The author indicated that collaboration between government and business in the energy sector promotes economic decarbonisation by increasing the renewable-driven buyers and increasing the number of government energy policy supporters by the corporate renewable buyers.

The paper [40] provides the analysis of policy choices impact on renewable energy development. The authors concluded that the EU membership, federalist political system and the state of the existing energy supply system were the main drivers in renewable energy development. The findings showed that supporting renewable policy in electricity production did not depend on fossil and nuclear energy shares in the economy's national energy supply and CO2 intensity. In the study [41] the scientists analysed the determinants of adopting climate-relevant policies based on the examples of developing and emerging countries. The authors concluded that the probability of adopting renewable energy policies is higher for more developed counties. Besides, EU membership facilitates renewable policy adoption, while natural endowments for producing renewable energy had less influence on governments to adopt renewable energy policies.

Given the above results, this study investigates the impact of economic and political globalisation on renewable energy for EU countries with different democracy regimes and Ukraine and Turkey as the potential EU candidates.

2 Materials and methods

The core hypothesis of the paper was:

H1: The type of political regime, the level of economic and political globalisation influences renewable energy.

For the analysis, the study used the Cobba-Duglas function framework, which explains that total production relates to labour and capital inputs [42–45]. In the papers [46; 47] the authors used the modified function (extending the traditional model by the renewable energy and others explanatory variables) to explain the relationships among renewable energy, carbon emissions and gross domestic product. The findings [48; 49] confirmed that globalisation allowed declining the energy demand by spreading green innovations.

Considering it, the study used the modified function (1) where economic and political globalisations were chosen as explanatory variables:

$$RE = f(K; L; GDP;\ EG; PG) \qquad (1)$$

where RE – the share of renewable energy in the final energy consumption; EG – the level of economic globalisation; PG – the level of political globalisation; K – the gross fixed capital formation; L – total labour force (people from ages 15 and older who supply labour); GDP – gross domestic products.

The study sample was generated by EU countries with different political regimes involving Ukraine and Turkey (potential EU members). The Economist Intelligence Unit was a source countries' Democracy Indexes for dividing them be three groups (Table 1): Full democracy (Finland, Denmark, Spain), Flawed democracy (Poland, Slovakia, Hungary and the Czech Republic) and Hybrid regime (Montenegro, Turkey and Ukraine).

Table 1. Countries by regime types (Democracy Index 2020)

Country	Regime type	Country's score 2020	Description
Finland	Full democracy	9.2	Political culture respects and reinforces civil liberties and fundamental political freedom.
Denmark		9.15	
Spain		8.12	
Czech Republic	Flawed democracy	7.67	Elections are fair and accessible, and fundamental civil liberties are honoured. However, political culture is underdeveloped.
Slovakia		6.97	
Poland		6.85	
Hungary		6.56	
Montenegro	Hybrid regime	5.77	Government pressure on political opposition, electoral frauds, corruption, etc.
Ukraine		5.81	
Turkey		4.48	

*Source: compiled by the authors based on the data [34].

The data used in this study were obtained from the Eurostat database (share of renewable energy in final energy consumption), KOF Swiss Economic Institute (KOF Globalization Index and its dimensions), World Data Bank (GDP, Labour force and Gross fixed capital formation) for 2012-2019 years.

Table 2. Variables, denotation, and their meaning for analysis

Variables	Denotation	Meaning	Source
Economic Globalisation	EG	The economic dimension demonstrates international trade and business activity state, trade flows, international investment, international trading constraints and taxes, etc.	KOF Swiss Economic Institute
Political Globalisation	PG	Political dimension characterises country membership in international organisations, ratification of international multilateral agreements, the number of embassies and other foreign delegation in the country, etc.	
Renewable energy	RE	The share of renewable energy in the final energy consumption	Eurostat
Gross Domestic Product	GDP	The sum of value added (differences between producers' gross output and the value of intermediate goods and services consumed in production) from all producers.	World Data Bank
Capital	K	gross fixed capital formation in US$	
Labour	L	total labour force (people from ages 15 and older who supply labour for the production of goods and services	

*Source: compiled by the authors based on [35; 36; 37].

The descriptive statistic of the data presented in Table 3. All data were in logarithm before calculation. The study used the software EViews for the analysis.

Table 3. The summary of the descriptive statistic of the selected variables

	RE	GDP	K	L	PG	EG
Mean	1,21	4,23	10,54	6,68	1,94	1,90
Median	1,18	4,25	10,73	6,67	1,96	1,91
Maximum	1,64	4,80	11,44	7,37	1,99	1,96
Minimum	0,30	3,33	8,91	5,40	1,77	1,75
Std. Dev.	0,34	0,38	0,63	0,58	0,05	0,04
Skewness	-0,83	-0,53	-1,27	-0,71	-2,11	-1,27
Kurtosis	3,57	2,81	4,31	3,06	6,48	4,69
Jarque-Bera	9,32	3,51	24,47	6,14	89,77	27,90
Probability	0,01	0,17	0,00	0,05	0,00	0,00
Sum	86,82	304,84	759,12	480,82	139,56	136,74
Sum Sq. Dev.	8,04	10,15	28,27	23,70	0,20	0,11

*Source: compiled by the authors.

The findings of correlation analysis allowed concluding that political and economic globalisation had the highest correlation. It justified using two modified models of function (1) with separately involving political and economic globalisation.

3 Empirical model

Considering the results mentioned above of analysis, under the investigation, the function (1) presented as two panel cointegration equations:

Model 1: $$RE_{it} = \delta + \alpha K_{it} + \beta L_{it} + \gamma GDP_{it} + \delta EG_{it} + \varepsilon_{it} \quad (2)$$

Model 2: $$RE_{it} = \delta + \alpha K_{it} + \beta L_{it} + \gamma GDP_{it} + \delta PG_{it} + \varepsilon_{it} \quad (3)$$

where α, β, γ, δ – regression's parameters which evaluate and explain the elastic of output relate to share of the renewable energy in the final energy consumption, economic globalisation, political globalisation, labour, capital; ε – the error term; i=1,..., N; t=1,..., T.

At the first stage, the study checked the stationarity of the data using the panel unit root tests (the null hypothesis – the selected variables were non-stationary): Levin, Lin, and Chu test (LLC); Augmented Dickey-Fuller Fisher (ADF) and Phillips-Perron Fisher) unit root test; Im, Pesaran, Shin's panel unit root tests (IPS). The next step – check the cointegration between variables using the Pedroni residual cointegration test.

If cointegration exists, at the last stage, the statistically significant longer-term relationship between variables will be checked using the Fully Modified OLS (FMOLS) and Dynamic OLS (DOLS) panel cointegration techniques.

4 Results and discussions

The first stage is checking the stationarity of the variables using the panel unit root tests. The findings confirmed the stationarity of all variables at the first level.

Table 4. The findings on stationarity analysis of the variables

Tests	Statistic Parameters	Variables					
		At level					
		RE	GDP	L	K	EG	PG

Levin, Lin & Chu	Statistics	-1,67	-6,17	-16,13	-5,26	-4,17	-0,32
	Probability	0,05	0,00	0,00	0,00	0,00	0,37
Im, Pesaran and Shin W-stat	Statistics	0,95	-1,02	-1,86	-0,91	-2,33	0,37
	Probability	0,83	0,15	0,03	0,18	0,00	0,65
ADF-Fisher Chi-square	Statistics	9,67	26,11	31,08	26,63	35,67	12,89
	Probability	0,94	0,097	0,03	0,09	0,00	0,79
PP-Fisher Chi-square	Statistics	17,84	8,21	20,94	5,48	47,24	20,19
	Probability	0,47	0,97	0,28	0,99	0,00	0,32
Tests	**Statistic Parameters**	**at 1st difference**					
Levin, Lin & Chu	Statistics	-4,39	-4,95	-3,23	-5,43	-3,04	-6,53
	Probability	0,00	0,00	0,00	0,00	0,00	0,00
Im, Pesaran and Shin W-stat	Statistics	-1,16	-0,21	-0,64	-0,56	1,66	-0,02
	Probability	0,00	0,00	0,00	0,00	0,00	0,00
ADF-Fisher Chi-square	Statistics	27,02	17,08	22,33	21,13	29,16	18,59
	Probability	0,00	0,00	0,00	0,00	0,00	0,00
PP-Fisher Chi-square	Statistics	69,65	19,21	43,01	19,51	91,22	40,84
	Probability	0,00	0,00	0,00	0,00	0,00	0,00

Source: developed by the author.

The results of stationarity analysis confirmed that only EG was stationary at the level in all tests. Besides, all variables had become stationarity at the 1st level. The findings allowed rejecting the null hypothesis of non-stationarity at 1% significance. It allowed testing the cointegration among variables using the Pedroni panel cointegration test.

Table 5. The results of cointegration tests among selected variables for model 1 with economic globalisation

Test	Within-dimension				Test	Between-dimension	
	Stat.	Prob.	Stat.	Prob.		Stat.	Prob.
			weighted				
panel v-statistic	0,05	0,48	-0,93	0,83	group rho-statistic	4,10	1,00
panel rho-statistic	2,69	1,00	2,56	0,99	group PP-statistic	-6,71	0,00*
panel PP-statistic	-3,50	0,00*	-4,83	0,00*	group ADF-statistic	-5,38	0,01*
panel ADF-statistic	-2,89	0,00*	-4,79	0,02*			

Note: * represents significance at the 1% level.
Source: developed by the author.

Table 6. The results of cointegration tests among selected variables for model 2 with political globalisation

Test	Within-dimension				Test	Between-dimension	
	Stat.	Prob.	Stat.	Prob.		Stat.	Prob.
			weighted				
panel v-statistic	-2,56	0,99	-2,32	0,99	group rho-statistic	3,9	1,0
panel rho-statistic	2,31	0,99	2,29	0,99	group PP-statistic	-5,6	0,0*
panel PP-statistic	-1,57	0,00*	-2,79	0,00*	group ADF-statistic	-2,7	0,0*
panel ADF-statistic	-1,45	0,00*	-2,39	0,01*			

Note: * represents significance at the 1% level.
Source: developed by the author.

The findings (Table 5 and 6) confirmed the statistical significance at 1% level for six out of eleven test probabilities. Thus, the hypothesis of non-cointegration among selected variables could be rejected. It allowed applicating of the Fully Modified Ordinary Least Square (FMOLS) and Dynamic Ordinary Least Square (DOLS) panel cointegration techniques for checking the long-run relationship among selected variables. The finding of FMOLS and DOLS presents in Table 7.

Table 7. The findings of a long-run relationship among selected variables (FMOLS and DOLS panel cointegration techniques)

Tests	Stat. Param.	Model 1 with political globalisation											
		Dependent Variables RE											
		Full democracy				*Flawed democracy*				*Hybrid regime*			
		Independent Variables											
		L	K	PG	GDP	L	K	PG	GDP	L	K	PG	GDP
F	Coeff.	0,79	1,2	-1,11	1,35	1,68	0,62	0,26	-0,25	0,82	0,07	3,63	0,18
	Prob.	0,46	0,00*	0,26	0,00*	0,38	0,09**	0,04**	0,60	0,00*	0,65	0,00*	0,38
D	Coeff.	0,22	1,28	0,84	1,44	0,58	0,55	0,43	-0,39	0,87	0,12	3,62	-0,22
	Prob.	0,05**	0,00*	0,03**	0,01*	0,01*	0,02**	0,02**	0,44	0,03**	0,75	0,01*	0,68
Tests	Stat. Param.	Model 2 with economic globalisation											
		L	K	EG	GDP	L	K	EG	GDP	L	K	EG	GDP
F	Coeff.	1,03	1,13	0,13	1,36	2,37	0,74	0,33	-0,33	0,18	0,24	0,54	0,46
	Prob.	0,45	0,01*	0,00*	0,01*	0,18	0,01**	0,05**	0,21	0,02**	0,01*	0,03**	0,04**
D	Coeff.	0,39	1,19	0,12	1,38	2,28	0,72	0,31	-0,40	0,20	-0,68	0,42	1,51
	Prob.	0,82	0,02**	0,06**	0,03**	0,22	0,03**	0,04**	0,22	0,76	0,36	0,00*	0,08**

Note: * and ** represents significance at the 1% and 5% levels; F – FMOLS techniques; D – DOLS techniques

Source: developed by the author

The results of long-run relationship analysis confirmed that for the countries from full democracy the group an increase of 1% in gross domestic product and capital (in both FMOLS and DOLS panel cointegration techniques), labour and political globalisation (DOLS panel cointegration techniques) lead to increasing renewable energy by 1.35, 1.2, 0.84 and 1.28, respectively. In the model with economic globalisation, the same indicators gross domestic product, capital and political globalisation had a statistically significant impact on renewable energy at 1% and 5%.

In the model with political globalisation for the countries with flawed democracy, all variables excluding GDP (DOLS) positively impacted growing renewable energy. Thus, increasing labour, capital and political globalisation allowed increasing the renewable energy by 0.58, 0.55, 0.43, respectively. In the model with economic globalisation only two indicators (labour and economic globalisation) were statistically significant at 1% and 5%.

However, in model 1 for countries with the hybrid regime, an increase of 1% in political globalisation and labour (in both FMOLS and DOLS panel cointegration techniques) allowed increasing the renewable energy 0.82 and 3.63 (FMOLS), 0.87 and 3.62 (DOLS), respectively and in the model 2 with FMOLS panel cointegration techniques, increasing of all variables by 1% lead to increasing of renewable energy.

The findings allowed concluding that for countries with the hybrid regime, the changes in political and economic globalisation provided the rapid growth of renewable energy compared with countries from full and flawed democracy.

5 Conclusions

Given the traditional economic model, economic growth requires more resource while increasing the adverse environmental impact. The carbon-neutral model of economic development promotes renewable energy demand. However, the aforementioned contradicts with principles of sustainable development. Therefore, this paper provided a bibliometric analysis of publication activity to determine the main research directions on the linkage between renewable energy and globalisation process in the countries. Using the VOSviewer software tool allowed visualising the network map of keywords' co-occurrences of the high reliable documents indexed by the Scopus database from 2000 to 2020. The obtained results proved the strong link strength between renewable energy, political and economic globalisations.

In turn, there were visualised three significant clusters that indicate the scientists' interest in 1) developing energy policy toward carbon-neutrality; 2) economic development under the influence of energy consumption, CO2 emission and energy consumption; 3) considering the relationship between renewable energy development, political and economic globalisation, Herewith, a comprehend analysis of identified research directions showed that exploring the best energy practices, new methods, innovative technologies, green financing, economic mechanisms etc., in mitigating and adopting against adverse climate were prioritised on the international level to develop renewable energy and carbon-free economy.

Thus, the findings confirmed that in countries with different levels of democracy and political regimes, the changes in the core economic parameters (labour, capital and gross domestic product), economic and political globalisations lead to extending renewable energy with different amplitude. Similar findings were obtained in the papers [39–41]. Thus, the increasing of political globalisation by 1% provoke the growth of renewable energy by 0.84 for full democracy countries (DOLS), 0.43 for flawed democracy (DOLS), 3.62 for the hybrid regime (DOLS). Thus, the countries with the hybrid regime should focus on implementing the mechanism to strengthen political stability and provide political globalisation. Besides, the growth of economic globalisation by 1% lead to increasing of renewable energy by 0.12 (FMOLS) and 0.13 (DOLS) for full democracy countries, 0.33 (FMOLS) and 0.31 (DOLS) for flawed democracy, 0.42 (FMOLS) and 0.54 (DOLS) for the hybrid regime. The increasing of trade and financial openness allowed improving economic globalisation. It allowed increasing the speed of extending and penetration of renewable energy. The increasing of political and economic globalisation allowed attracting additional green investments and innovation for spreading renewable energy.

Acknowledgements

This research was funded by the grant from the National Research Foundation of Ukraine "Stochastic modelling of road map for harmonising national and European standards for energy market regulation in the transition to a circular and carbon-free economy" 0120U104807); the Ministry of Education and Science of Ukraine (0121U100468, 0120U102002 and 0121U100469).

References

1. Scopus database, *https://www.scopus.com* (2021)
2. Enerdata. Global Energy Statistical Yearbook 2021, *https://yearbook.enerdata.net/total-energy/world-energy-intensity-gdp-data.html* (2021)
3. L. Saher, L. Syhyda, O. Korobets, T. Berezianko. E3S Web of Conferences, **234**, 00011 (2021)
4. Ya. Us, S. Bilan, T. Pimonenko, R. Seliga, G. Ostasz. Proceedings of the 35th IBIMA Conference: 1-2 April 2020, Seville, Spain (2020)
5. A. Zolkover, V. Terziev. Business Ethics and Leadership, **4(3)**, 107-118 (2020).
6. H. Eddassi. SocioEconomic Challenges, **4(2)**, 67-77 (2020)

7. T. Vasylieva, S. Lieonov, I. Makarenko, N. Sirkovska. Marketing and Innovation Management, **4,** 350-357 (2017)
8. M. AUgbaka, A. Awujola, T. Shcherbyna. SocioEconomic Challenges, **3(4),** 5-12 (2019)
9. T. Pimonenko, J. Us, D. Leus, S. Fedyna. Bulletin of Sumy State University. Economy Ser, **2**, 57-67 (2017)
10. T. Vasilyeva, S., Bilan, K, Bagmet, R. Seliga. Economics and Sociology, **13(1),** 271-294 (2020)
11. Y. Kharazishvili, A. Kwilinski, O. Grishnova, H. Dzwigol. Sustainability, **12(21),** 8953 (2020)
12. H. Dzwigol, M. Dzwigol-Barosz. Academy of Strategic Management Journal, **19(5**), 1-7 (2020)
13. R. Miskiewicz. Polityka Energetyczna, **21(2),** 49-62 (2018)
14. K. Djalilov, S. Lyeonov, A. Buriak. Financial Markets and Institutions, 5(4CONT1), 178-187 (2015)
15. A. Kasych, M. Vochozka. Marketing and Management of Innovations, **2**, 298-305 (2017)
16. O. Chygryn, A. Rosokhata, O. Rybina, N. Stoyanets. E3S Web of Conferences, **234,** 00004 (2021)
17. Y. Samusevych, A. Vysochyna, T. Vasylieva, S. Lyeonov, S. Pokhylko. E3S Web of Conferences, **234**, 00012 (2021)
18. H. Dzwigoł, M. Dzwigoł–Barosz, Z. Zhyvko, R. Miskiewicz, H. Pushak. Journal of Security and Sustainability Issues, **8(3),** 307-317 (2019)
19. O. Chygryn, Y. Bilan, A. Kwilinski. Marketing and Management of Innovations, **3**, 358-370 (2020)
20. T. Pimonenko. Economics and Region. **4(71)**, 5-15 (2018)
21. D. T. Am. Financial Markets, Institutions and Risks, **3(2),** 69-78 (2019)
22. A. El Amri, R. Boutti, S. Oulfarsi, F. Rodhain, B. Bouzahir. Financial Markets, Institutions and Risks, **4(4),** 31-44 (2020)
23. He. Shuquan Business Ethics and Leadership, **3(4),** 43-48 (2019)
24. E. Jafarzadeh, He. Shuquan. Business Ethics and Leadership, **3(2)**, 53-62 (2019)
25. H.D. Kaya. SocioEconomic Challenges, 3(3), 5-12 (2019)
26. S. Mahapatra, A. Pandey, B. Narayanan. SocioEconomic Challenges, **4(2),** 23-30 (2020)
27. M.A. Khan, A. Kishwar. Financial Markets, Institutions and Risks, **4(2),** 108-114 (2020)
28. A. El Amri, R. Boutti, F. Rodhain. Financial Markets, Institutions and Risks, **4(2**), 52-64 (2020)
29. T. Kurbatova, R. Sidortsov, I. Sotnyk, I., O. Telizhenko, T. Skibina, R. Hynek. Problems and Perspectives in Management, **17(3),** 464 (2019)
30. V. Panchenko, Yu. Harust, Ya Us, O. Korobets, V. Pavlyk. Marketing and Management of Innovations, **1,** 256-264 (2020)
31. V. Bozhkova, O. Ptashchenko, L. Saher, L. Syhyda. Marketing and Management of Innovations, **1,** 73-82 (2018)
32. A. Bonamigo, D. Mendes. Business Ethics and Leadership, **3(4),** 66-73. (2019)
33. N. Letunovska, O. Dalechin, K. Bieliaieva. Practical aspects of business planning in the system of investment project implementation. Marketing and Management of Innovations, **3**, 226-235 (2017)
34. O. Chygryn. Centre for Studies in European Integration Working Papers Series, **(6)**, 6-13 (2017)
35. Y. Bilan, T. Vasylieva, S. Lyeonov, I. Tiutiunyk, I. Entrepreneurial Business and Economics Review, **7(2),** 27-43 (2019)
36. T. Vasylieva, V. Kasyanenko. Actual Problems of Economics, **144(6),** 50-59 (2013)
37. M. Kordos. Marketing and Management of Innovations, **3,** 341-353 (2019)

38. The Economist Intelligence Unit, *https://www.eiu.com/n/campaigns/democracy-index-2020* (2021)
39. Eurostat, *https://ec.europa.eu/eurostat/databrowser/view/nrg_ind_eff/default/table?lang=en* (2021)
40. KOF Globalisation Index, *https://kof.ethz.ch/en/forecasts-and-indicators/indicators/kof-globalisation-index.html* (2021)
41. World Data Bank, *https://data.worldbank.org* (2021)
42. S. Adams. Journal of Cleaner Production, 118245 (2019)
43. Z. Tzankova. Energy Research and Social Science, **67**, 101504 (2020)
44. L. Schaffer, T. Bernauer. Energy Policy, **68,** 15-27 (2014)
45. M. Stadelmann, P. Castro. Global Environmental Change, **29**, 413-423 (2014)
46. M. E. BildiriciMEE. Biomass Bioenergy, **50,** 19–24 (2013)
47. E. Dogan, B. Turkekul. Environ. Sci. Pollut. Res **23**, 1203–1213 (2016)
48. C.T. Tugcu. Int. J. Energy Econ. Policy **3**, 307 (2013).
49. C. T. Tugcu, I. Ozturk, A. Aslan. Energy Econ, **34**, 1942–1950 (2012)
50. K. Saidi, A. Omri. Environmental research, **186**, 109567 (2020).
51. D. H. Vo, A. T. Vo, C. M. Ho, H. M. Nguyen. Renewable Energy, **161**, 278-292 (2020)
52. J. Baek, Y. Cho, W. W. Koo. Ecol. Econ. **68 (8–9),** 2255–2264 (2009)
53. G. Gozgor, M. K. Mahalik, E. Demir, H. Padhan. Energy Policy, **139**, 111365 (2020)

Natural and economic effects of renewable energy sources in the developing countries: a case of Azerbaijan

Zhala Rzayeva[1] *Aysel* Guliyeva[2*], and *Arzu* Miriyeva[1]

[1] SABAH CENTER, Azerbaijan State University of Economics (UNEC), 6, Istiglaliyyat str., AZ1001, Baku, Azerbaijan
[2]International Economy department, Azerbaijan State University of Economics (UNEC), 6, Istiglaliyyat str., AZ1001, Baku, Azerbaijan

Abstract. This paper focuses in the natural and economic effects of renewable energy sources in the developing countries using a case of Azerbaijan. Traditional fuel supplies dependent on gas, oil, and coal are highly useful for the development of a country's economy. However, certain detrimental effects of these commodities on the environment have limited us to use these resources under clear cut-off points. As a result, we were able to turn our intuition into renewable energy sources. By using renewable energy sources, social, natural, and financial challenges can be ignored because these sources are considered harmless to the atmosphere, as there are little or virtually no fumes discharged, and toxic gases such as carbon dioxide, carbon monoxide, and sulfur dioxide. Renewable electricity would be a major advantage for power generation in the near future so we will reuse these assets to generate useful electricity. The paper discusses these and other issues with the implementation of renewable energy and assess the potentials of Azerbaijan in adapting this type of energy in the future.

1. Introduction

Alternative (renewable) energy will quickly become a significant contributor to climate change and these resources will be reused and produced usable energy repeatedly [1]. The three types of energy supplies that are commonly used are fossil, renewable, and nuclear [2]. Hydroelectric power, solar energy, wind energy, biomass, ocean oil, biofuel, geothermal fuels, and other green energy sources account for approximately 20 percent of overall global energy output.

On account of the expanded interest for energy because of the world's quickly developing populace, petroleum derivatives like coal, gas, and oil will be utilized to fulfill the need, bringing about impractical circumstances and various issues like petroleum product consumption, ecological and geological clashes, nursery impact, a dangerous atmospheric deviation, and change in the renewable energy is called clean energy since it is harmless to the ecosystem and discharges fewer gases; it is likewise advantageous to society across all

* Corresponding author: aysel.guliyeva10@gmail.com

angles, including monetary, social, and natural [3]. Today, around 1.6 billion individuals do not have the access to electric power, and about 1.1 billion do not have access to drinking water [4]. Environmentally friendly power supplies can meet the entirety of the world's energy needs while likewise securing the climate and guaranteeing energy proficiency. Alongside the various advantages of these devices, a few disadvantages exist, for example, creation difference because of occasional changes, which is an ordinary event for wind and hydroelectric force plants; thus, exceptional plan and care are required, which are met by equipment and programming on account of advances in PC innovation. Table 1 groups the major environmentally friendly power sources and their different applications; it is anticipated that sustainable power will be perhaps the most basic sources later on; Table 2 gauges the world's sustainable power sources situation by 2040.

Table 1. The most popular renewable energy sources and their usage

Energy resources	The choice for energy transfer and consumption
Hydropower	production of electricity
Biomass	Production of heat and fuel, pyrolysis, gasification, and digestion
Geothermal	Hydrothermal, hot rock, and urban heating
Solar	Solar lighting systems, solar dryers, and solar ovens
Direct solar	Photovoltaics, solar power generation, and heating systems
Wind	Wind turbines, wind generators, and windmills
Wave	a variety of styles
Tidal	Tidal drain, barrage

There are several challenges in the twenty-first century, particularly in developing countries, such as meeting the demands of thousands of humans who still lack access to easy, sustainable energy supplies and engaging in the international context of sustainable, reduced energy generation.

Fortunately, the aim of lowering greenhouse gas emissions may be compatible with other energy-related priorities, such as the production of sustainable indigenous resources and the elimination of local forms of pollution. If they also lead to other social and economic growth goals, renewable energy strategies are more likely to be successful. To optimize positive synergies where they occur, policymakers should evaluate strategies to get the most optimal solutions [5].

Since it is relatively sunny and windy, as well as a large hydro, biomass, and geothermal deposits, Azerbaijan has a huge untapped renewable energy capacity. The government is working on legislation to encourage the use of clean energy sources [6]. Auctions and tenders are proposed as funding structures in the draft clean energy legislation. It also includes the development of other legal papers, such as a PPA draft and a connection agreement. Also, guidelines for auctions and rules for the implementation of net-metering/net-billing systems are being drafted.

Table 2. A global renewable energy scenario in place by 2040

2001	**2010**	**2020**	**2030**	**2040**	**2050**
Total consumption (million tons equivalent)	10,038	10,549	11,425	12,352	13,310
Biomass	1080	1313	1791	2483	3271
Large hydro	22.7	266	309	341	358
Geothermal	43.7	86	186	333	493
Small hydro	9.5	19	49	106	189

Wind	4.7	44	266	542	688
Solar thermal	4.1	15	66	244	480
Photovoltaic	0.1	2	24	221	784
Solar thermal electricity	0.1	0.4	3	16	68
Marine (tidal/wave/ocean)	0.05	0.1	0.4	3	20
Total RES	1365,5	1745,5	2964,4	4289	6351
Renewable energy contribution source (%)	13.6	16.6	23.6	34.7	47.7

As a result, the ongoing discussion on energy security must be expanded to include the potentially disastrous effects of high energy prices in a hydrocarbon-based global energy system, as well as climate change. Extreme weather conditions will restrict access to energy supplies, and climate rises will rapidly alter the pattern of global energy markets, resulting in new and often negative interdependence relationships. Developing countries need more reliable and sustainable energy supplies, as well as new technical solutions that allow for more efficient use of conventional and renewable energy sources. In the nations of the South, more predictable help for environmentally agreeable advancement dependent on renewables is required. The equivalent can be said for worldwide advancement help bundles that incorporate innovation sharing and environmentally friendly power.

We all should save electricity, increase energy conservation, and prioritize renewable energy sources. On a global scale, we should encourage renewable energies, as well as energy conservation and efficiency technologies. Meeting these problems is in both developing and developed countries' best interests. Renewable energy must be promoted by the European Union and its member countries. They have the political, analytical, and technical skills, as well as the financial capital, to not only direct the global discussion on renewable energy but also to push global action. Renewable energy is the best solution to energy poverty because it reduces CO_2 emissions, slows the global warming trend, and provides a viable model for long-term energy strategies [7]. It is also a business opportunity, not only a requirement.

2 Policies on renewable energy in developing countries

The main impediments to renewable energy adoption in developed countries derive from a lack of technical know-how to build appropriate power systems. The main concern for adopting technology in developed countries is their feasibility and appropriateness, as well as their ease of use. The second is an educated and convinced public to use this energy, and the third is a barrier that removes effort in political and large economic concerns to the strong promotion of the above implementations so that policymakers can carry out this agenda. Apart from the above factors, societal changes in behavior are a major influence on society's ability to respond to these developments, and these will be more difficult to solve if these problems are ignored. Despite some issues, renewable energy in developed countries, especially in Asia, has shown how to inspire efficient renewable energy installation to meet their rising energy needs. These commitments are centered around limited scope environmentally friendly power sources including little hydro, biomass, and geothermal energy.

Table 2 shows the general yield of sustainable power for power generators, just as the portion of limit conveyed in created nations. As per this table, the net inexhaustible force limit in created nations rose by 88.5 percent somewhere in the range of 2003 and 2009, expanding from 58.1 GW in 2003 to 109.5 GW in 2009. Wind, sunlight based photovoltaic, and biomass power are the three principal supporters of sustainable power item advancement from 2003 to 2009, with wind power representing 1,333.0 percent (from 3 GW in 2003 to 40

GW in 2009), sun based photovoltaic representing 25.0 percent, and biomass power representing 33.3 percent. Sustainable electric energy projects were the harbingers of generously expanding clean energy creation volumes to kill ozone-depleting substance emanations from 1993 to 2007.

Table 3. RES by type (2003 and 2009)

RES energy by type	2003 in GW		2009 in GW	
	All countries	Developing countries	All countries	Developing countries
Small hydropower	56	33	60	40
Wind power	40	3	159	40
Biomass power	35	18	54	24
Geothermal power	9	4	11	5
Solar photovoltaic power	1.1	0	0.7	0
Solar thermal power	0.4	0	21	0
Ocean power	n.a	m.a	0.3	0
Total renewable power capacity	141.5	58.1	306	109.5
Total hydropower (including the Large Hydro Power)	872	399	980	580

Source: REN 21 [8]

Developing countries have established small power producers (SPP) systems to support renewable energy production during those years. They began with the addition of new solar small power projects (SPP) to the nation's energy source, which was funded by private power projects and shared risk with the government. Ferrey [9] conducted an analysis to test the small power supply program (SPP) in five Asian countries with advanced systems (e.g. Thailand, Indonesia, India, Sri Lanka, and Vietnam). He discovered that by an open mechanism involving consumer, producer, and lender trust, additional incentives representing the fuel variety and environmental benefits are used to assist higher-cost renewable energy and smaller SPP.

3 Renewable energy resources in the Republic of Azerbaijan

Azerbaijan is a country with huge sustainable power creation potential. The nation has a plentiful breeze and sunlight-based assets, just as promising biomass, geothermal, and hydropower potential. The public authority has effectively defined an objective of adding 420 MW of environmentally friendly power limit by 2020 to understand this chance. The nation has continued to send projects utilizing designing, acquisition, and development (EPC) contracts, guided by this objective. Notwithstanding, in contrast with the size of the country's accessible assets and long-haul objectives, the commonsense arrangement has been negligible [10, 11].

Past monetary broadening, a higher portion of an environmentally friendly power in the energy blend will have a few benefits. Sustainable power can make new openings in a world that have generally been overwhelmed by oil and gas, by taking into consideration mechanical advancement and opening up new business sectors for the monetary worth turn of events and GDP development. Besides, expanded environmentally friendly power arrangement, joined with proceeded with enhancements in energy proficiency, could lessen homegrown oil and gas utilization, opening up freedoms to create extra income through sends out and decrease homegrown sponsorships.

Currently, Azerbaijan is a net exporter of energy. Oil, gaseous petrol, and power are traded from the area. Azerbaijan's energy needs are met principally through homegrown creation, which is right now reliant on the extraction of the country's hydrocarbon saves, specifically oil and petroleum gas. The Caspian Basin has massive oil holds, which Azerbaijan imports to Russia, Kazakhstan, Turkmenistan, and Iran are which are assessed to be seven billion barrels in 2018 [12], which is similar to the North Sea saves from many years prior. Azerbaijan trades multiple times its all-out conclusive energy utilization, as demonstrated in Table 4. The decrease in essential sustainability and waste creation is because of fluctuating hydropower age and diminished utilization of wood as a feedstock for energy creation. Upgrades in region warming and the development of gas-terminated water radiators likewise decreased the utilization of wood.

Table 4. Energy balance of Azerbaijan in 2010–2017 (ktoe)

	2010	**2014**	**2015**	**2016**	**2017**
Primary production, including:	68 254.6	61 132.0	61 084.2	59 977.6	57 036.0
Crude oil (with NGL)	52 312.5	43 295.9	42 835.5	42 240.9	39 810.4
Natural gas	15 555.6	17 565.1	17 947.3	17 463.5	16 967.6
Renewables and waste	386.5	271	301.4	273.5	258
Net imports of all energy products	- 54 300.2	- 45 869.5	- 45 747.4	- 44 693.0	- 40 789.6
Total energy supply	12 566.5	15 085.5	15 569.4	15 393.5	15 471.9
Transformation processes	-2 692.7	-3 491.3	-3 819.7	-3 731.0	-4 020.6
Energy industry own-use	931.1	1 188.2	1 167.0	1 133.0	1 147.7
Total final energy consumption	6 710.6	8 241.7	8 304.3	8 644.9	8 210.0

Source: State Statistical Committee of Azerbaijan [13]

In its Strategic Road Map on National Economic Perspectives of Azerbaijan, the country's administration underlines the significance of differentiating the economy and decreasing dependence on the oil and gas areas. In this specific circumstance, expanding the portion of sustainable power in the energy blend can give a few benefits, including expanded government assistance and occupation creation, energy solidness, financial enhancement, and GDP development [14, 15].

Sustainable power, by offering new roads for specialized advancement and new businesses for monetary worth improvement in a world customarily overwhelmed by oil and gas, will catalyze new position open doors in a nation generally overwhelmed by oil and gas. The country's broad involvement with the oil and gas area gives a strong establishment to the appropriation of environmentally friendly power advances. For instance, aptitude acquired from the seaward oil and gas framework has been utilized to build up the seaward wind industry in Norway, while mastery acquired from seaward oil and gas foundation has been utilized to build up the seaward wind industry in Canada.

The State Oil Company of Azerbaijan Republic (SOCAR) – the nation's entirely state-possessed public oil and gas organization – is a central participant in the country's energy area. SOCAR has effectively introduced wind and sun-based PV in the Tagiyev oil and gas area and is investigating sustainable power alternatives for providing the ability to seaward offices while keeping up its dependence on conventional fuel sources.

In addition, expanded sustainable power organization, joined with proceeded with upgrades in energy effectiveness, could lessen homegrown oil and gas utilization, giving a chance to produce extra income through sends out and decrease homegrown sponsorships. Sustainable power is additionally the most mainstream low-carbon elective for Azerbaijan's environmental objectives. In its Nationally Determined Contribution (NDC) under the Paris Agreement, the nation swore to diminish GHG emanations by 35% by 2030 contrasted with the 1990 base year. To meet this objective, the NDC as of now accentuates the utilization of option and sustainable power sources.

The advancement of environmentally friendly power could likewise add to energy framework flexibility in Azerbaijan. For instance, appropriated sustainable power frameworks, joined with expanded utilization of battery stockpiling, can improve the security of supply. Given the extended expansion in public energy interest, sustainable power likewise gives a specialized answer for the requirement for quick force age limit organization, attributable to generally more limited lead times for project development when contrasted with traditional sources. The public authority ought to figure a drawn-out technique to raise public mindfulness and limit concerning sustainable power, focusing on state organizations, neighborhood organizations, and inhabitants. Accreditations might be fused into preparing educational programs, with an accentuation on improving neighborhood administration and support abilities to improve open positions. Specific limit working in sustainable power strategy and guideline, just as sustainable power project evaluation strategies, may uphold government offices. Courses laying out the possibilities for setting up sustainable power organizations, just as subtleties on accessible government subsidizing plans and accomplices, could be given to the private area. Affirmation and preparing in an assortment of sustainable power applications and related advancements for both force and non-power applications can likewise be valuable.

A greater public understanding of the effects of green energy and how they relate to Azerbaijan's overall economic condition may be beneficial. A series of information events could be used to organize public awareness, especially in areas with a high potential for renewable energy growth. Also, various other formats, such as social media promotions, contests, and challenges for students in schools, higher education institutions, and colleges, as well as municipal activities in specific towns and cities, may be designed to increase awareness.

4 Conclusions

Overall, the customary characteristic assets like oil, gas, and coal are extremely fundamental for the improvement in economies of a world. A country like Azerbaijan is for the most part dependent on conventional fuel sources, notwithstanding perceiving the negative effects on wellbeing and the environment, like the nursery impact, worldwide environmental change impact, etc. Azerbaijan is blessed with both environmentally friendly power sources, for example, the hydro, wind, and geothermal, and it is additionally an ideal area for the sun-based force age. Nonetheless, the greatest difficulties in producing power from clean energy assets are subsidizing and governmental issues. For the gathering of elective fuel sources, all contemplations like ozone harming substance outflows, asset supply, land prerequisites, water use, financial impacts, and force age cost are thought of. Wind power age is thought to have minimal measure of water use, a minimal measure of relative ozone harming substance outflows, and the best friendly impacts. It is perceived as quite possibly the most feasible environmentally friendly power fuel source, behind just hydropower, sunlight-based force energy, and geothermal energy. Biomass is believed to be ideal for limited scope production in light of the fact that it saves a lot of carbon. The appropriate utilization of environmentally friendly power energy frameworks will bring about neighborhood occupations, improved wellbeing, work openings, work development, client inclination, expectation for everyday comforts change, social control creation, pay creation, segment changes, social security creation, and local area advancement. Alongside the upsides of efficient power energy assets, they are hard to fabricate and are helpless against nearby ecological and climate conditions. In contrast with different projects, their determining, execution, and arrangement require additionally thought and experience.

References

1. R. Bora, R. Wang, F. You, ACS Sustainable Chemistry & Engineering, **8(43)**, 16350-16363 (2020)
2. I. Čábelková, W. Strielkowski, D. Streimikiene, F. Cavallaro, J. Streimikis, Technological Forecasting and Social Change, **163**, 120477 (2021)
3. S. Kolosok, T. Pimonenko, A. Yevdokymova, N. Hajiyev, M. Palienko, L. Prasol, Marketing and Management of Innovations, **4**, 50-60 (2020)
4. O. Durodola, V. Nabunya, M. Kironde, C. Nevo, J. Bwambale, World Water Policy, **6(2)**, 176-201 (2020)
5. S. Kalyugina, W. Strielkowski, L. Ushvitsky, E. Astachova, Journal of Security & Sustainability Issues, **5(2)**, 297-304 (2015)
6. N. Vidadili, E. Suleymanov, C. Bulut, C. Mahmudlu, Renewable and Sustainable Energy Reviews, **80**, 1153-1161 (2017)
7. E. Lisin, D. Shuvalova, I. Volkova, W. Strielkowski, Sustainability, 10(4), 1111 (2018)
8. REN21, *https://www.ren21.net/reports/global-status-report* (2020)
9. S. Ferrey, *http://regulationbodyofknowledge.org* (2004)
10. A. Ciarreta, E. Ahmadov, Economics, **6(2)**, 75-95 (2019)
11. U. Aydine, *https://www.econstor.eu/handle/10419/222759* (2019)
12. A. Ibrayeva, *http://www.zbw.eu/econis-archiv/bitstream/11159/2112/1/1028133111.pdf* (2018)
13. State Statistical Committee of Azerbaijan, *https://www.stat.gov.az/source/balance_fuel/?lang=en* (2020)
14. R. Pomfret, *The Central Asian economies in the twenty-first century: Paving a New Silk Road* (2019)
15. S. Humbatova, N. Hajiyev, International Journal of Energy Economics and Policy, **11(1)**, 341 (2021)

Impact of alternative energy sources on development of neo-industrial foreign trade specialization

Elena Andreeva[1*], *Alla* Golovina[1], and *Artem* Ratner[2]

[1]Ural State University of Economics, 8 Marta str. 62/45, 620144 Yekaterinburg, Russian Federation
[2]Institute of Economics of the Ural Branch of Russian academy of sciences, Moskovskaya str. 29, 620014 Ekaterinburg, Russian Federation

Abstract. This paper identifies the structure of the energy balance in the world and in individual countries was analyzed, trends and factors in the development of alternative types of energy carriers. The impact of their development on the volume and structure of international energy trade is substantiated. Its development prospects are presented for developed countries that are leaders in the development of alternative energy sources, and for developing countries for which the price factor is more important. Prospects of development of Russian energy exports in conditions of increasing importance of alternative types of energy carriers in the world economy and international trade are considered.

1 Introduction

Energy is very important in the life of mankind, for this reason it occupies a significant place in the structure of international trade. Energy export plays an important role for highly energy-rich countries; for the countries being not much provided with energy resources the import plays an important role. Therefore, Russia ranks 6th (6.2%) in the world oil reserves, 1st in the world natural gas reserves (19.1%) and 2nd in the world coal reserves (15.2%) (see Figures 1-3 that follow).

Moreover, in terms of gas, the countries following from the 6th position in terms of security have a resource of 4 ½ or more times less than Russia. As regards the coal, the countries following from the 6th position on security have a resource 4 and more times less than Russia.

Therefore, for Russia, as for a country being highly provided with energy resources, and as for a major exporter of oil and gas, despite the importance of diversification process, the energy export continues to be of high importance for filling the national budget and for creating the job places. In this regard, the hypothesis is proposed, that the development of alternative energy in the world and in the leading countries – importers of Russian energy resources have an impact on international energy carriers trade in general and Russian energy exports in particular. The purpose of the analysis is to investigate this hypothesis for confirmation or non-confirmation.

* Corresponding author: elenandr@mail.ru

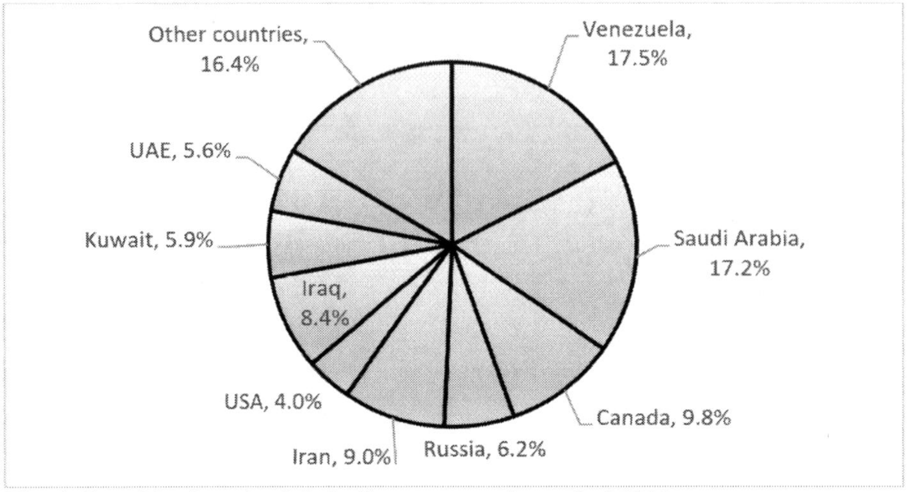

Fig. 1. Russia's share in global oil reserves at the end of 2019
Source: British Petroleum (2020)

The literature emphasizes that many countries, including those being provided with energy resources, attach great importance to the issue of depletion of the mineral resource complex. This depletion refers to the most important environmental indicators of subsoil use [1, p. 555]. It also appears to reflect the extent of pollution from these resources use (in the case of fuel resources: the scale of carbon dioxide emissions, of oil spills).

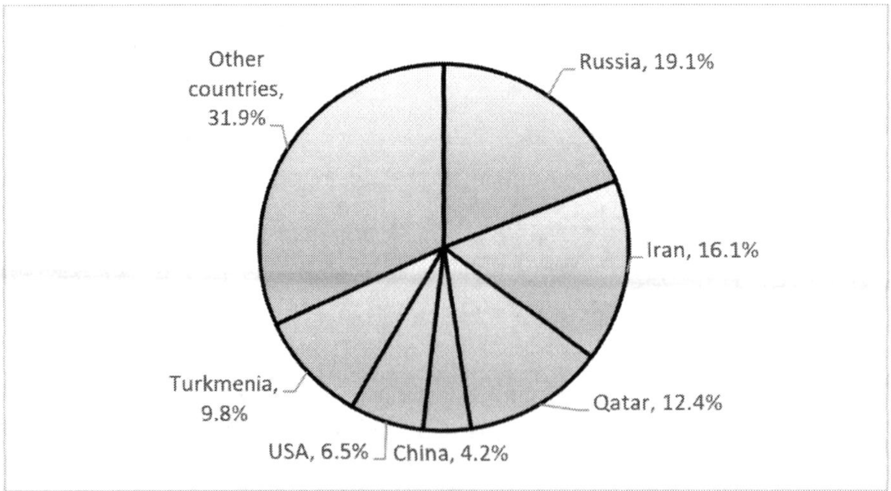

Fig. 2. Russia's share in global gas reserves at the end of 2019
British Petroleum (2020)

Other indicators being identified include the rate of waste generation during the subsoil resources' extraction, during the use and disposal of these wastes [2, p. 862].

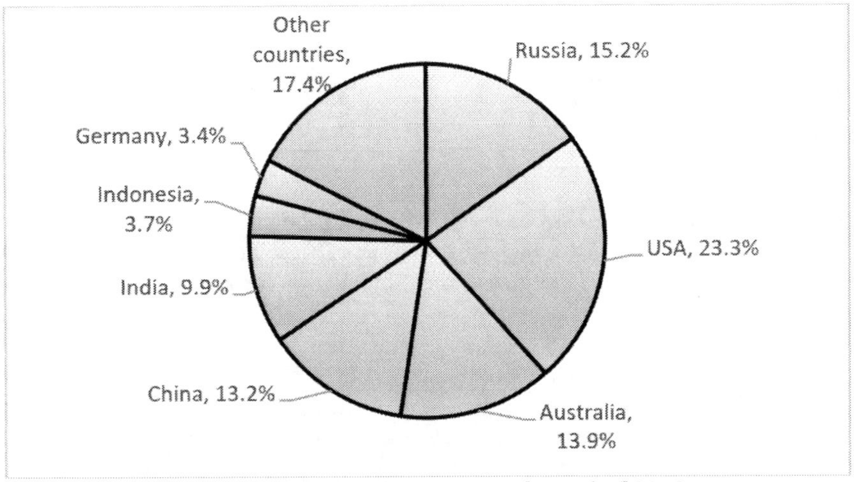

Fig. 3. Russia's share in global coal reserves at the end of 2019
Source: British Petroleum (2020)

Even the countries being provided with energy resources are set by search of technologies of energy saving and of low-carbon (i.e. alternative) power engineering [3, p. 23-24]. Energy conservation and the development of renewable energy sources are attractive even to foreign investors (as shown by studies on the example of almost two dozen of Central and South American countries) [4, p. 16]. Moreover, for Russia itself, the transition to a low-carbon development trajectory is recognized as relevant, which is associated with the requirements of international agreements [5, p. 672].

It is noted in the literature that even China, where coal accounts for 3/5 of primary energy consumption, faces the problem of excessive coal mining capacity. On the basis of the analysis in research literature, the advantages of the variant of removal from the market of inefficient capacities are noted – against the variant of even reduction of all capacities [6, p. 1073, 1090-1091]. At that, the decrease in international demand for coal of a country (as on an energy carrier with the largest emissions of carbon dioxide) leads to a decrease in its exports, which leads to the closure of coal mines and, ultimately, to a decrease in the supply of coal for the national power engineering of this country [7, p. 11-12].

And the literature investigates the issues of country behavior in these conditions in the international market. In particular, an example is being explored of how a national economy is working on the problem of reducing the efficiency of coal industry and reducing foreign demand for coal due to climate policy. In particular, the example of Poland is taken: previously it was a major exporter of coal, and became a net importer [8, p. 263-264, 267].

As for the Russian economy, it is noted that in the conditions of global power transition to not carbon sources, the introduction of energy's economy is of great importance for it [9, p. 53-55, 62]. In order to generate export revenues in the case of a global energy transition, the falling part of raw export can be offset by the development of high-tech export. For this are important innovations, investments into them, the formation of a pool of specialists in the relevant industries, cooperation of large companies with small ones, including startups [10, p. 585]. In this context, it is emphasized that the comparative advantages of the Russian economy lie not only in the provision by energy resources, but also, for example, in developed technologies of nuclear energy mechanical engineering, which are attractive for developing countries. Therefore, taking into account the energy deficit in South Africa, Russia and other countries can enter its market with these technologies [11].

But energy carriers however continue to be a critical component of Russian export. Thus, the question remains be relevant of what prospects Russian energy carriers can have in the global hydrocarbon market, taking into account the development of alternative energy.

2 Materials and methods

It would be useful to analyze the structure of the energy balance in the world and in individual countries, to identify trends in the development of alternative energy carriers, as well as the volume and structure of international trade by energy carriers and the prospects for its development.

The development of alternative energy carriers is determined primarily by an array of factors. The development of the climate policy aimed at reducing the carbon dioxide emissions in the atmosphere in order to reduce the greenhouse effect, makes it more profitable to use the energy sources that produce energy with minimal carbon dioxide emissions. According to forecasts, by 2040, coal and oil will account for 50 to 42% of the global energy balance (depending on whether a conservative variant of energy consumption development will be implemented or an energy transition variant), gas – for 25-26%, while alternative sources – for up 25 to 32%. And this – despite the fact that 2015 coal and oil gave 59%, gas – 22%, and alternative sources – 19% (Figure 4).

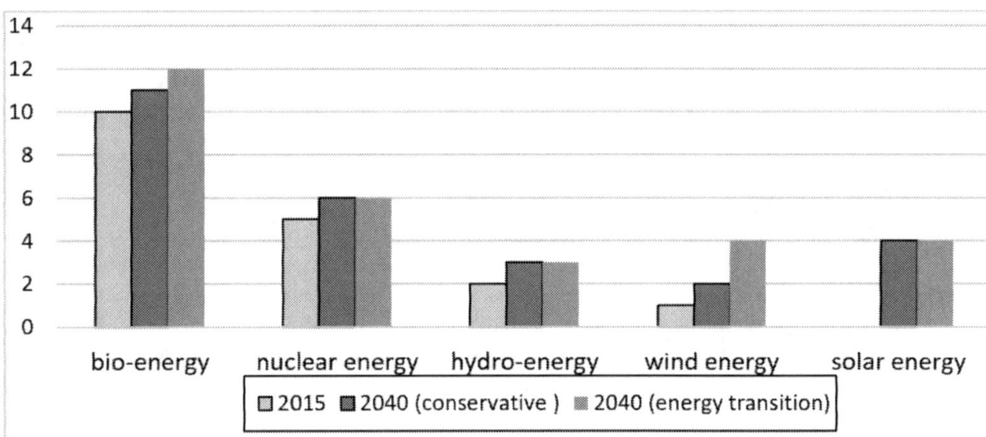

Fig. 4. Share of non-fuel energy sources in the global energy balance, %
Source: British Petroleum (2020)

Thence, for example, in Germany 2019 renewable sources provide 46% of the country's electricity. The use of energy resource, which gives the largest emissions of carbon dioxide, the coal, will be legally limited. By 2038, and if possible, by 2035, it is planned to stop all coal-fired power plants.

Economic stimulation's tools are also used. So, a fee for carbon dioxide emissions for industrial facilities was introduced: 2019 it amounted to 25 Euro/t, which made the gas electric generation more profitable than the coal one, due to the smaller specific size of emissions by gas power engineering. For the use of cars, this fee will be the same to 2021, and 2025 – 55 Euro/t, 2026 – 55-65 Euro/t. The transition of housing heating to renewable energy sources will be stimulated, including through hybrid heating systems. Since 2026, diesel boilers should not be installed or updated in new houses.

As a result, in 2019 the share of coal in electricity generation in Germany decreased: of the brown – by 22%, of the stone coal – by 33%.

An important factor for the development of alternative energy carriers are safety considerations (within the framework of environmental policy), also being related to nuclear energy, especially after the disasters of Chernobyl (1986) and Fukushima (2011), taking into account the risk of man-made disasters and the issue of storage of spent nuclear fuel. In this regard, Germany also plans to stop the production of nuclear energy: about 2/3 of the number of nuclear power plants has already been stopped, and by 2022 it is planned to stop the remaining ones.

3 Prospects of development of demand for energy resources in the world

In general, the forecasted consumption of primary energy worldwide will increase by 2040 on 17-27% [12, p. 26]. The development of this segment in international trade is not even. If developed economies predominantly reduce energy consumption, then developing economies increase. In addition, as already noted, the commodity structure of this increase will be redistributed towards greener types of energy resources. This is illustrated by an example of the transport sector: the share of electricity in its energy balance grows rapidly (from 1% in 2015 to 11-32% in 2040, depending on the implementation of a conservative variant or energy transition variant), the share of gas in the case of an energy transition grows (up to 6%) (Table 1).

Table 1. Energy balance of transport sector, %

Power source	2015	2040	
		Conservative variant	Energy transition variant
Oil products	93	82	59
Electric power	1	11	32
Gas	4	4	6
Synthetic liquid fuel	3	3	3

Source: British Petroleum (2020)

At that, the share of coal in global energy consumption though is projected to decrease, but will continue to remain significant: it is expected by dynamics from 28% (2015) to 19% (2040) – and this if there is an energy transition to renewable energy sources (and in the case of a conservative scenario, the share will become 23%) (Figure 4). The peak of coal consumption in the European Union and the Commonwealth of independent states was passed until 1990, but in Africa and the Middle East it is not expected to come until 2040 (and in a conservative scenario – also in the developing countries of Asia and South America) [12, p. 28-29]. Developing Asian countries are significantly ahead by the absolute volumes of coal use in electricity generation. Even in the case of an energy transition, their coal use will amount to about 6 thousand Terawatt-hour in 2040, or about 1/3 of all energy sources in these countries. While in Europe these volumes will not exceed hundred – a couple of hundreds Terawatt-hour (that is a couple of percent) [12, p. 33]. Coal consumption growth by 2040 is expected in India (by ½ million tons), in developing countries of Asia (except China); consumption will remain comparable to current levels in China and developed Asia. China and India will remain the largest consumers of coal [12, p. 62, 66].

This is due to the fact that in developing economies (of Asia and Africa), due to economic coercion, the more important factor is not the environmental friendliness of energy resource, but its cost, and here coal wins.

Regarding the geographical reorientation of international trade in energy carriers, it can be noted that Russia, starting since increase of protectionism and international sanctions imposed in the 2010s, has to analyze the possibility of reorienting the export flows from developed to developing countries. One of the Russian important trading partners is China. In the context of coal export, it may be noted that in December 2020 China stopped the import of Australian coal (its scale is evidenced by the fact that it occupies the 2nd position in Australian coal exports). This case shows that theoretically there are opportunities to find demand for Russian coal, in the case of the refusal of a number of developed countries, including Germany, from coal. However, the transport opportunities are in question here: the question is of the presence of a reserve for loading the railway leading on east.

Of mineral non-renewable fuels, only gas is projected to increase its share in global energy consumption by 2040: from 22 to 25-27% (Figure 4). Including in Europe – on about 2-4 tens of billion cubic meters; in developing Asian countries (except China and India) – on 2 hundred; in China – on about 4 hundred billion cubic meters. In the countries actively pursuing the environmental policy, to the gas is given priority in electricity generation. So, in Europe, the number of terminals for receiving the liquefied natural gas (LNG) was increased.

At that, the gas production in Europe will decrease on a hundred billion cubic meters by 2040, while in Russia – will increase on 2 hundred billion cubic meters. But a comparable increase is expected also in the United States and in the rest of North America, in China, the Middle East, and Africa. At that, it is expected that unconventional (i.e. new) technologies will be increasingly used in gas production: biogas production, coal gasification, production of methane of coal beds, shale gas production, development of new fields, gas transfer to liquefied state for transportation [12, p. 56-57]. In these conditions of technological competition, Russia has to improve the production of natural gas and the logistics of its delivery, including its production in hard-to-reach areas and in harsh climatic conditions, liquefaction and delivery in such conditions. So, in 2017, gas production and LNG production were launched in the Arctic region – in Yamal. Gas liquefaction will be carried out using the latest technologies (modules for the construction of the plant were produced in China, Indonesia, Belgium), as well as transportation to foreign consumers (ice-class methane tankers were built in the Republic of Korea).

4 Conclusions

Thus, as the analysis has shown, the proposed hypothesis was confirmed: the development of the use of alternative (new) types of energy sources, conscious, environmentally conditioned, will affect international trade in energy resources, including Russian energy export. Demand for traditional fuels – coal, oil and gas – will remain high, but there will be a redistribution towards gas in the structure of energy consumption, on the background of a deliberate decrease in coal and nuclear energy consumption by several countries. Reducing the demand for traditional energy sources will lead to increased competition in international markets of their sale, which will force exporting countries to offer more competitive solutions (in terms of price, of convenience and speed of delivery), that will be possible for the exporter if new technological solutions are developed and implemented that reduce the cost of production, primary processing and logistics (delivery, storage) of resources. At that, the competition for markets of sale will occur in the context of the gradual depletion of the energy fields usual by availability level. New technological solutions will be also required by the development of more hardly accessible fields and by the organization of transportation of goods from hardly accessible areas.

Thus, the development of alternative energy will gradually reduce the demand for some Russian energy energetic commodities (coal; also oil), reorient Russian supply geographically to developing countries and form a request for the development of technicality, for the use of new technologies in the preparation of Russian fuel and energy export (for example, technologies for the development of hardly accessible gas fields, liquefaction and transportation of liquefied natural gas, including in harsh (Arctic) climatic conditions).

Acknowledgements

The article was prepared by financial support of RFBR (Russian Foundation for Basic Research), project number 20-010-00806-A.

References

1. V.G. Loginov, M.N. Ignateva, V.V. Yurak, Economy of Region, **16(2)**, 547 (2020)
2. I.G. Polyanskaya, V.V. Yurak, Economy of Region, **14(3)**, 851 (2018)
3. T.F. Agbanike, C. Nwani, U.I. Uwazie et al, Latin American Economic Review, **28**, 8 (2019)
4. M. Ben Jebli, S. Ben Youssef, N. Apergis, Latin American Economic Review, **28**, 2 (2019)
5. A.Y. Kolpakov, Studies on Russian Economic Development, **29(6)**, 665 (2018)
6. W. Wu, B. Lin, Computational Economics, **55(4)**, 1073 (2020)
7. M.C. Rademeyer, Mineral Economics (2021)
8. W. Suwala, Mineral Economics, **31(1-2)**, 263 (2018)
9. A.A. Makarov, Studies on Russian Economic Development, **31(1)**, 52 (2020)
10. E.L. Andreeva, E.V. Sapir, D.A. Karkh, I.A. Karachev, Economy of region, **15(2)**, 576 (2019)
11. H. Dkhili, Marketing and Management of Innovations, **3**, 333-344 (2018)
12. Forecast of power engineering development in the world and in Russia, INEI RAN, "Skolkovo", *https://energy.skolkovo.ru/* (2019)

Sustainability of economic systems and business enterprises

Svetlana Kalyugina[1]*, Natalia Snegireva[2], *Zaur* Shalashaa[3], *Polina* Shmygaleva[4]

[1] North-Caucasus Federal University, Pushkin str.1, Stavropol 355009, Russian Federation
[2] North-Caucasus Social institute, Goleneva str. 59A, Stavropol, 355000, Russian Federation
[3] Institute of Economics and Law, Academy of Sciences of Abkhazia, Sukhum, Dzizdzaria str. 40, Abkhazia
[4] Institute of Economics and Law, Stavropol Branch of RANEPA, Lermontova str. 189, Stavropol, 355002, Russian Federation

Abstract. Our paper analyses the issues of sustainability of economic systems and business enterprise at micro and meso levels. Dynamics are analysed in a global bio-geophysical subsystem linked to the macroeconomic system, the environment and the human economy. The paper proposes a conceptual framework that looks into the compromise between economic growth and sustainability and concludes that both can be achieved if relevant efforts are spent and funds are allocated. Moreover, it analyses the risks and assesses economic and financial stability using the Altman and Altman-Sabato models. In addition, the research addresses the role of the macroeconomic system, the environment and the human economy in the sustainable development of ecosystem functions and its impacts on business enterprises. Our results might become the guidelines for the governments seeking to combine the development of their economic systems with the principles of sustainable development.

1 Introduction

In general terms, sustainability can be considered profit-making by combining profits with sustainability values by using a combination of elements of environmental, social, economic and economic development in the microeconomy [1, 2]. These include the use of renewable energy, energy efficiency, water saving, waste management, environmental protection and waste management. Many developing countries can skip the ruthless industrialization model and integrate directly into a clean, sustainable development model that focuses on the increased use of renewable energy to supply their growing needs [3]. Sustainable development is a broad term that includes all forms of growth to meet the needs of the current generation without endangering future generations. The Unite Nations defines sustainability as "*a form of the economic and social development in the micro-level economic system*" [4].

* Corresponding author: skaliugina@ncfu.ru

The essence of economics is to formulate the welfare - the welfare of the people - by maximizing the social welfare function (SWF) [5]. Hence the link between economics and sustainable development. The principles for sustainable development are intertwined with sound economic principles presenting a suitable and solid guide for economic actors. In another words, the economic activities that create jobs, income - earning opportunities for jobs and economic growth create opportunities to achieve the best results for the sovereign wealth funds. This is the aim of economic operators, who focus on maximising profits not only for themselves but for society as a whole. Since the economy is the source of human activity, effective governance is crucial to achieving sustainable development [6]. It is imperative that governments undertake the necessary work to provide basic services and coordinate the activities of economic actors in order to ensure that strategies for achieving sustainable development are coordinated. Responsible governments will ensure proper coordination of economic activities and coordination between government, private and public sectors [7].

In fact, direct and indirect cooperation in combination with open innovations (e.g. open source and open access) exhibits a variety of dynamic interactions in open innovations [8, 9]. The role of open cooperation in the sustainable development of ecosystem functions is very important here. Open innovation offers an opportunity for cooperation between the macroeconomic system, the environment and the human economy. None of these anthropocentric frameworks takes into account the dynamics of ecological systems, but all analyse the system and its processes, whether or not they are relevant to humans or the environment.

2 Indicators for assessing economic and financial stability

Economic and financial stability take into account business forecasts and assesses the risk to the financial system of economic and financial stability [10, 11]. In addition to assessing risks to households and businesses, the macroeconomic forecasts and stress tests are conducted. Stress tests have become an important part of the process of risk assessment in the banking sector and can shed light on how exposed financial institutions are to macroeconomic changes [12, 13]. These models explain the actual losses quite well and also have a good result - samples of real estate, but not as good as the bank's own data. It turns out that these changes have a positive impact on the growth of economic growth and financial stability.

However, one can note that growth in selected economics is only positively determined by the ratio of turnover to market efficiency, due to the presence of FDP [14]. In addition, one can also note that the ratio of traded shares to total assets in the banking sector and the total number of bank loans is both positively and significantly influenced by economic growth, regardless of whether indicators for the development of the banking sector are selected [15]. In this context, it should be noted that the number of scientific papers and econometric studies examining the relationship between economic growth and financial stability and the development of the banking sector has increased significantly over the last two decades.

Nevertheless, the main conclusion that comes from the above is that there is a positive correlation between the ratio of sales to market efficiency and growth in selected economies. Balancing financial stability, innovation and economic growth and other economic indicators in economic development. Although known sources for the relationship between financial stability and economic growth, including the International Monetary Fund (IMF), the World Bank and the European Central Bank, new sources have emerged in recent years. Financial stability represents a lack of instability, but this approach is based on the assumption that the basis for financial sustainability is the ability of financial markets to reach their full potential without the need for central bank intervention. This approach to understanding "financial sustainability" emphasises the role of the financial sector in economic growth and stability, and requires an understanding of its role in the economy as a whole and its relationship to economic and financial growth [16, 17]. For example, in China the Bank of China has its main objectives aimed at ensuring price stability at home, managing the exchange rate and promoting economic growth. The bank recently promised to maintain its prudent monetary policy by fine-tuning its policies - while respecting its long-term inflation target of 2.5% [18]. One needs to monitor the conditions and developments of the financial system as part of the economic environment, and monitoring of financial stability receives regular attention. After the financial crisis, the Federal Reserve in the United States developed a framework for systematically tracking risk. Stricter macroprudential measures can tighten financial conditions, increasing the likelihood that monetary policy responses will be needed. There must be a clear and consistent assessment of the financial and economic stability of the United States economy [19]. When making monetary policy decisions, we must be aware of the risks to the economy, the financial markets and the global economy as a whole. Regular attention is important, as we have learned from the financial crisis, because vulnerabilities can build up over time, even in periods when conditions appear favourable. The idea behind measuring financial stress is to prevent it from culminating in systemic risk. The indicators are then combined to create an index that uses the appropriate statistical techniques to reflect its evolution in the economy [20].

Since different market variables behave differently and due to different market variables, a single composite index can be used and expected to better understand the prevailing stress. Improving supervision and financial supervision requires a better understanding of financial stress and its impact on financial stability. Central banks can measure financial burdens by creating a Financial and Stress Index (FSI) to monitor the functioning and resilience of the financial system. The IMF has developed the Financial Stability Index (FSI), the world's first and most comprehensive financial and economic stress measurement. These indicators of financial sustainability include one that characterises the financial stability of a country's financial system and its financial institutions. These indicators include aggregated information on the level of risk in financial markets, as well as indicators that characterise the quality of the markets in which financial institutions operate and the level of liquidity in those markets [21, 22].

In order to define the integral indicator for assessing economic and financial stability the Z-model devised by Altman [23] can be used. The model can be presented in the following way:

$$Y_0=0.012X_1 +0.014X_2 +0.033X_3 +0.006X_4 + 0.999X_5 \quad (1)$$

where X_5 - sales profit/A.

There is also an improved version of the Altman model that is called the Altman-Sabato model [24]. This model has high accuracy but does not take into account the specifics of particular industries. The model can be presented in the following form:

$$p_z=1/(1 + e - 4.28 - 0.18X_3+0.01X_1-0.08X_2-0.02X_3-0.19X_4) \quad (2)$$

where p_z – possibility of bankruptcy according to Altman-Sabato model;
X_1 = short-term liabilities/assets;
X_2 = net profit/A;
X_3 = monetary funds/A;

X_4 = *EBIT*/ receivable interests.

In the research literature, the following modification of the model are also used for calculations of the same relationship:

$$ROE=ROS * total\ assets\ turnover * leverage\ ratio \quad (3)$$

where ROS – return on sales.

$$ROE = t * r * ROS * total\ assets\ turnover * leverage\ ratio \quad (4)$$

where:

t - tax burden ratio,
r - percent burden ratio.

Hence, equations (3) and (4) determine the profitability of sales based on the net profit which allows comparing disparate values such as net profit and turnover.

3 Value added and net profit in sustainable development

Company managers need to consider equity to adjust interest-rate relationships between owners and operators, adjust taxes - correct the cost of capital, and improve investment levels. Economic value creation and net profit in sustainable development both embed the analysis of the economic value of equity and capital. Growth refers to a company's ability to pursue sustainable development and is an important indicator for measuring the company's future performance based on the estimated rates of revenue growth in this paper. Growth means that companies of scale are growing, market competitiveness is increasing and profitability of companies is increasing. A general definition of business growth is the potential value that a company continuously achieves by exploiting or exploiting unused resources. In addition, managers in low-growth companies with little development potential tend to be overinvested, while high-growth companies with high development potential (e.g. large companies) are abundant, and that lower-growth companies have little or no development opportunities. The terms used to describe the economic justification of a proposed investment can be understood as a benefit that is exactly in line with the cost of the proposed project.

Similarly, the benefits that arise at the time of cost collection will also bring the same benefits that will continue to accrue in the future. In this sense, ERR is the result that not only the cost of living in the partner country is taken into account, but also the benefits for the country as a whole. A crucial basic concept is that sustainable business models create and use ecological and social surpluses, by which we mean sustainability investments that can have secondary value through higher returns. This will allow cereal producers to achieve net CO_2 offsets from their production by reducing emissions. Agricultural banks can include net greenhouse gas sequestration in the terms of payment of the offsets, thereby promoting the sustainability of farms and enabling the bank to accumulate new assets from available carbon offsets. They can also offer farmers preferential financing for the introduction of renewable agriculture or include carbon credits in their payments. A fundamental prerequisite for sustainability is the transformation of an industry's business model (business models) to reflect value for customers and society by accepting a premium for achieving performance levels and achieving certain performance goals.

Consumers and businesses are also offering businesses behavioural skills as new assets in business, and as we have seen, they are paid premiums for this. A major challenge for the research community is to develop more appropriate business models that capture and reflect this added value by providing sustainable buildings that are operated while reducing resource consumption and maintaining adequate utilities for residents. However, this has not been a priority for the academic community and is indeed not even a priority for many in the business community. To limit managers' investment behaviour and improve investor efficiency, a sensible valuation and index system must be put in place. Performance indicators must be established to determine whether evaluation outcomes can directly change managers' workflows and behaviour. It must also enable fine-tuning and problem-solving, and examine how values can actually be validated in practice, and how performance data can be shared with stakeholders to receive feedback and feed-forward. The EVA is an evaluation index proposed in the 1980s that measures a company's value - its ability to create. This is a method of assessing a company's performance and financial results, based on the value of its invested capital and its ability to create and its financial performance. BOC's ROE and REVA are the lowest on this list, lagging behind other commercial banks. This is acceptable if one takes into account the level of equity and other factors, but you can withdraw it. Analysing for certain reasons, the state-controlled stock exchange trading bank generated low net operating income after tax and paid a large amount of capital, resulting in a low RVA. Overinvestment in state-owned enterprises is more severe compared to private holding companies, but improvements in the external governance environment can prevent overinvestment to some extent. They find that cash dividends can effectively constrain investment behaviour and have a good inhibitory effect. Economic value added is an important factor for the development of sustainable development in China and other countries, especially in rural areas.

4 Sustainability of enterprise economy

Sustainable development presents itself as an alternative to such problems by focusing on finding alternatives that meet current and existing needs without compromising on meeting the needs of future generations. Entrepreneurship has proven to be an effective wealth creation mechanism and is seen as a driver of change, innovation and economic growth. Sustainable entrepreneurship builds on the foundations of entrepreneurship and extends it to the environmental and social concerns that are taken into account by existing companies. Sustainable entrepreneurship thus becomes relevant and focuses on preserving nature, supporting the life of communities, and creating services and products that reflect individual well-being - being, the economy, and society. Entrepreneurship is also seen as a viable alternative when faced with environmental, social and economic challenges such as climate change and poverty, as it pursues objectives that go beyond economic benefits. Market opportunities include addressing sustainability concerns by providing new products and services that reduce energy and resource consumption. Sustainable businesses are recognizing the growing scientific evidence and growing evidence that growing social inequalities are destroying the economy, the environment, and global society. They recognize these changes by understanding that the consumer public is increasingly concerned about the challenges of a sustainable world. This module is intended to provide a framework for developing a business plan for the transition to a sustainable business model with a focus on energy efficiency, sustainability and environmental sustainability. The current challenges facing companies worldwide are presented in the context of the United Nations Global Compact [25].

The economic sustainability of companies is to act responsibly and to make a profit without causing undue harm to the world. This means recognizing that even very small businesses are part of a larger global economy. It is important to consider how your business practices affect others ecologically, culturally and socially. Regenerative capitalism addresses precisely the need to restructure the economy and to risk the opportunities and risks that the absence of such a change might entail in the short term. Economic sustainability can refer to how an economy functions to protect the social and environmental elements of its business environment and its employees. When small businesses make ethical choices for growth, their owners' benefit, and we look at the benefits that this can bring both globally and at the corporate level. In this article, we want to examine our idea of economic sustainability, ensure that it is interconnected, and consider it as more holistic. The second definition adds environmental and social factors to the mix to sustain economic production, such as water, energy, food, and energy efficiency. Another important economic objective in this context is the Sustainable Development Goal No. 9 (SDG9), a part of the SDGs developed by the United Nations which relates to innovation and infrastructure in industry.

5 Sustainability of economic systems

Thence, sustainable development requires improving the quality of life for all by increasing the number of jobs, education, health care and social services for all, reducing poverty by at least 75%, and reducing the use of fossil fuels and fossil fuels - intensive industries [26, 27]. Conservation and the environment: To preserve our ecological heritage and natural resources for future generations and to preserve natural habitats, economically viable solutions need to be developed to reduce resource consumption, stop pollution and preserve the natural habitat. Today's interconnected global economies require an integrated approach to promoting responsible, long-term growth, to ensure that no nation or community is left behind.

People worldwide need access to a wide range of goods and services, from food and energy to health care and education. These include education, health, housing, transport, energy, food, water and sanitation, and health and safety. All these pieces are part of the puzzle of a sustainable society, because they are fundamental parts of everyday life. Sustainable development is a way of thinking that thinks about the development of economic and social systems, not just economic systems. It represents the need to find ways of achieving economically, socially and ecologically sustainable development.

Sustainable capitalism is fundamentally different from environmental protection because it adheres to the principles of social justice, human rights, and the rule of law. It internalizes direct social and environmental costs and benefits, and does not rely on free markets to distribute environmental and social capital. Many people doubt whether a moral and just society can ever be limited and guided by the principles of social justice, human rights and the rule of law.

Although no other economic system has found that it can compete with capitalism in terms of the allocation of resources that are rightfully owned by individuals and economies, capitalism remains open to discussion. Even if sustainable solutions do not yield gains, they will succeed in terms of social justice, human rights, and the rule of law. It is now clear that economies must be designed to reflect the priorities and values associated with maintaining habitable conditions on the planet.

While the way we create value in today's economy depends on a systemic rejection of environmental priorities, design and industry must prioritize the needs of the environment and the economic system. Diversion of the economic design sector towards economic and service structures and systems that support socially distributive and ecologically regenerative design. Achieving a sustainable future will require a set of policies and institutions that keep the economy within the limits set by nature. In this paper, we argue that the acceptance of natural resource constraints that limit the long-term growth of material production and consumption is unnecessary to protect and sustain the biophysical systems that underpin the sustainable development of the global economy and the environment as a whole.

More significantly, an optimal tax that balances the private benefits and social costs of consumption would support the creation of an economy that strikes an appropriate balance between society's needs and natural resources. In a growth-oriented society, this offers an alternative to critics who warn that the cost of achieving environmental sustainability would jeopardize the economy. This is an important concept that needs to be communicated across various channels and mass media outlets in order to convince the members of the general public that sustainable development is economically viable and does not need to compromise the economic growth and prosperity.

6 Conclusions and implications

In this research, we discussed the economic and environmental challenges that people and the planet face and presented solutions to integrate the needs of the planet, people and profit. This is the largest corporate citizenship initiative in the world, addressing the sustainability of businesses and their impact on the environment and human well-being. Environmental and eco-entrepreneurship appears to be able to solve environmental problems by creating economic value. Sustainable value, then, means creating a sustainable business model and its impact on the environment and human well-being - being. Environmental sustainability and the associated corporate strategies that are understood can be incorporated into the creation of sustainable value for companies.

One can see that sustainability therefore consists of three pillars: the economy, society and the environment. This is due to the fact that economics has a number of well-developed areas, such as social psychology, economics and system economics. Economics, in combination with other social and behavioural sciences, is important for understanding the trade-offs that accompany the development of a sustainable economy for all life on Earth and for the long-term sustainability of the planet. The combination of economics and earth system science is crucial to understanding how it might be possible to change human behaviour in order to achieve sustainable development.

Overall, it appears that the application of economic principles and empirical knowledge should be crucial to meet the challenges of sustainable development, given the limited resources of the earth. The economic and earth system sciences, as well as other social and behavioural sciences, have a long and complex history of the humanities. This approach is strongly bottom-up, suggesting that a sustainable future will only be achieved if the biophysical and social conditions needed to support economic activity and human prosperity are maintained for this and the next generation.

References

1. I. Nesterova, Journal of Cleaner Production, **262**, 121382 (2020)
2. T. Vasylieva, S. Lieonov, I. Makarenko, N. Sirkovska, N. (2017). Marketing and Management of Innovations, **4**, 350-357
3. A. Rogalev, I. Komarov, V. Kindra, O. Zlyvko, Entrepreneurship and Sustainability Issues, 6(1), 429-445 (2018)
4. United Nations, *https://www.un.org/sustainabledevelopment/economic-growth*
5. M. Fotros, E. Torkamani, Advances in Mathematical Finance and Applications, **4(2)**, 89-101 (2019)
6. J. Mensah, S. Casadevall, Cogent Social Sciences, **5(1)**, 1653531 (2019)
7. M. D. Jensen, Public Administration, **95(1)**, 249–268 (2017)
8. J. Yun, X. Zhao, K. Park, L. Shi, Sustainability, **12(11)**, 4379 (2020)

9. J. Yun, X. Zhao, K. Jung, T. Yigitcanlar, Sustainability, **12(12)**, 5076 (2020)
10. I. Cabelkova, W. Strielkowski, Society and Economy, **35(4)**, 513-529 (2013)
11. W. Allen, G. Wood, Journal of Financial Stability, **2(2)**, 152-172 (2006)
12. C. Sahin, J. de Haan, E. Neretina, Journal of Banking & Finance, **117**, 105843 (2020)
13. P. Gai, S. Kapadia, Oxford Review of Economic Policy, **35(4)**, 586-613 (2019)
14. S. Adeusi, O. Dada, O. Adeosun, International Journal of Academic Research in Business and Social Sciences **8(9)**, 1119-1132 (2018)
15. E. Nizam, A. Ng, G. Dewandaru, R. Nagayev, M. Nkoba, Journal of Multinational Financial Management, **49**, 35-53 (2019)
16. C. Paun, R. Musetescu, V. Topan, D. Danuletiu, Sustainability, **11(6)**, 1713 (2019)
17. S. Kalyugina, W. Strielkowski, L. Ushvitsky, E. Astachova, Journal of Security & Sustainability Issues, **5(2)**, 297-304 (2015)
18. R. Vasudevan, Review of Keynesian Economics, **9(1)**, 83-108 (2021)
19. P. Ireland, Journal of Applied Corporate Finance, **31(4)**, 68-76 (2019)
20. A. Neveu, Journal of Economic Interaction and Coordination, **13(2)**, 241-281 (2018)
21. M. Power, Review of International Political Economy, **12(4)**, 577-599 (2005)
22. D. Delpini, S. Battiston, G. Caldarelli, M. Riccaboni, PLoS ONE, **14(5)**, e0217141 (2019)
23. E. Altman, Journal of Investment Consulting, **19(1)**, 15-22 (2019)
24. E. Altman, M. Iwanicz-Drozdowska, E. Laitinen, A. Suvas, Applied Economics, **52(37)**, 4092-4111 (2020)
25. G. Orzes, A. Moretto, M. Ebrahimpour, M. Sartor, M. Moro, M. Rossi, Journal of Cleaner Production, **177**, 633-654 (2018)
26. A. Kasych, M. Vochozka, Marketing and Management of Innovations, **2**, 298-305 (2017)
27. A. Omri, F. Belaid, Journal of Environmental Management, **278**, 111483 (2021)

https://doi.org/10.1051/e3sconf/202125004002

University 4.0 within the context of the sustainable development of higher education

Larisa Gorina[1]*, *Ekaterina Polyakova* [1]

[1] Institute of Engineering and Environmental Security, Togliatti State University, Belorusskaya str. 14, 445020 Togliatti, Russian Federation

Abstract. The process of development of universities from the places where medieval elites met to conduct alliances and partnerships to the modern biotechnological and digital institutions of higher education, research and business took several centuries and went through four main stages. This paper describes the role and the place of the University 4.0 within the context of the sustainable development of higher education in the 21st century. We show that University 4.0 is designed to promote the technology industry through partnerships that support research, marketing, workforce development, as well as entrepreneurship. In keeping with the cooperative nature efforts directed and driven by the policy-makers and relevant stakeholders need to be dedicated to promoting modern universities as hub for the development of intelligent, connected technologies and services that would lead to the sustainable development of higher education that would reflect the challenges of our time.

1 Introduction

Modern universities face many challenges which creates a necessity for them to establish themselves as local and global leaders. In order to achieve that goal, they should build relationships with both internal and external stakeholders [1,2]. In order to build successful relationships, university managers must have a strong understanding of the stakeholders in their institutions and their interests. Today, the leadership skills in the universities can be learned through mentoring and self-development programs offered by organizations specializing in the professional development of university leaders [3, 4]. Moreover, these skills can include understanding mission, vision, goals, values, and morality, transferring behavioural models, and influencing campus culture. The relationship of trust between university stakeholders as well as other interest groups such as lecturers, staff and students are particularly influenced by the perception of the university management [5]. Research on trust-based and integrated approaches and models has shown, key variables that determine successful relationships are influenced in particular by leadership and leadership skills, as well as a strong understanding of the university's mission and goals [6].

Understanding the role of leadership can be particularly important in the context of any university's mission, mission and specific goals. The change in the academy has also changed the ability of university presidents to lead and steer the institution into the future. One can

* Corresponding author: gorina@tltsu.ru

rank the six skills presidents must possess when they take office, including vision, vision, and vision for the university, values, ethics, morality, leadership, and behavioural models. For leaders at the top of an institution, these forces can collide in many ways, from a lack of resources and resources to a lack of leadership, or even a combination of both. One can distinguish the University 4.0 management competencies for higher education, which provide university leaders with skills related to managing the resources, resources and resources of a university and management competence [7, 8].

Management skills at universities include relevant management competencies such as planning, budgeting, planning and financing, communication skills and business and leadership skills. Interpersonal skills that are critical to university managers include interpersonal skills such as communication, management skills, and other leadership skills. Relevant stakeholders need to understand what qualities boards should prioritize in interviews and hiring a new president or chancellor.

The American Council on Education (ACE) published a list of the five most important areas occupied by presidents at the time, including reporting to 65 presidents, leading senior-level teams, governing-board relationships, and leadership skills. The perception of university presidents describes duties and responsibilities, which include leadership skills such as communication, management skills, interpersonal skills and organizational ability. The American Association of Colleges and Universities (ACU), a national association of more than 1,000 colleges and universities, describes the perception of a college president, including his role as president and the role of the board of trustees [9, 10]. These might include basic materials, often dealt with in master's and student programs, outlining the basic skills and abilities needed to prepare successfully for the world of work.

Ultimately, there are also individual areas of competence, each of which is a key element within the university's workforce development strategy and the key to successful workplace preparation. One should be aware of her or his abilities so that they have a better understanding of how their lecturer or teacher, class and tasks help them to build the skills that will prepare them to achieve their future professional goals.

It has never been more important to emphasize the generalizable skills in the workplace that university education offers. The key is to recognize and understand the job - related skills that she or he can develop as a university student in the 21st century. Highlighting one's work-related skills often attracts the attention of potential employers more effectively than high grades or being on top of the class. Thence, university education has gradually transformed through the ages to become more person-oriented and socially-oriented rather than result-oriented.

2 Evolution of university models over time

Universities were a form that had its origins historically in Europe, first appeared between the twelfth and thirteenth centuries (University of Paris in France or University of Bologna in Italy being the best examples). In those times, they were ancient centres of higher education, took only one form, and preceded the rise of modern universities such as the universities of Oxford and Cambridge considerably. In medieval Europe, the universities were an important source of education for young men and women of all ages, not just for the elite. They also played an important role in promoting a new generation of leaders and in developing a well-educated citizen [11, 12]. The history of universities is very different, however, and there are a number of historians who sometimes refer to this form of higher education as a university for convenience. There is undoubtedly a real connection that would justify linking medieval universities in the West with modern universities such as Humboldt University, University of Oxford and University of Cambridge, but the institutional reality is very different. These institutions were initially relatively narrow and have only recently spread. But on closer inspection it becomes clear that, despite what has been said on this subject, there was no real connection between the medieval and modern universities in Europe. Early records from Egypt and Mesopotamia suggest that these early institutions operated a university and taught students much as they do today. The Ashurbanipal Library in Nineveh and the Library of Sippar were collections of knowledge that probably also brought students and teachers with them, who taught people who learned complex written languages of their time and began to study and apply their knowledge. These scholars probably taught and were associated with educational institutions, but they did not necessarily conduct research in the same way as modern universities like Oxford or Cambridge [13].

Table 1 that follows describes the evolution of universities (according to their format) from the University 1.0 in the Middle Ages distinguished by the focus on four basic disciplines and finding a way into the societal hierarchy to University 4.0 of today that is marked by the digital surge in research and education.

Table 1. Evolution of universities in different format from the Middle Ages until nowadays [14]

University format	Time range	Description	Competences
University 1.0: Medieval university	Middle Ages – 18th century	Fundamental knowledge with an emphasis on medicine, theology, law and astronomy	General knowledge and competences. Belonging to certain classes and elites.
University 2.0: research university	19th century –1960s	Academic knowledge (the so-called "Humboldt university")	Rational and deductive thinking, research and experimenting
University 3.0: entrepreneurial university	1970s-2000s	Combining education, research and business	Providing ideas, technology and capital
University 4.0: digital and biotechnological university	2010s-now	Multivariate knowledge with the emphasis on ICT and biotech	Creativity and global competences

Academics at universities have been accused of following their old sources, ignoring the needs of modern society, such as the need for education in the arts and sciences. Despite these social changes, which increased the number of educated men filling positions in commerce and administration, universities survived and prospered, and had a monopoly on higher education. Colleges are an important part of the history of modern society, from the Middle Ages to modern times [15, 16]. Universities were originally founded as a means of promoting a new generation of leaders, and were based on the same principles as schools, such as education in the arts and sciences, as practised in modern universities.

The subjects we are now dealing with are different from those of the medieval universities, but the main purpose and objectives remain the same. In this particular place a whole range of teachers and scholars in higher education taught and received. Together with the entities of society, these persons were associated with the power and privileges that they recognized and with their recognized privileges. Modern universities also pursue these goals but their mission and their focus has shifted and the major factor in this is information and communication technologies and the Internet.

3 Universities as leaders in high-tech industry

Modern universities are keen in striking partnerships with the leaders in high-tech industry. In many countries, a business and university partnerships create a unique symbiosis that is crucial for socio-economic development [17, 18]. One reason Canada is a world leader in knowledge-based industries is its integrated approach to fostering economic growth through innovation. Canada provides the world's largest number of high-tech research and development centres, and is also aggressively supporting the development of the country's high-tech industries, such as robotics, artificial intelligence (AI), and machine learning [19].

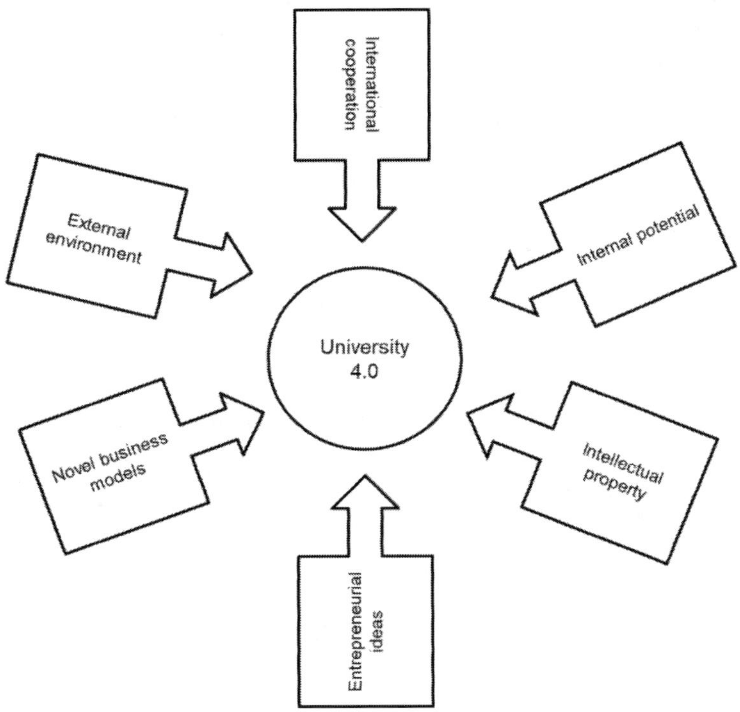

Fig. 1. Format of University 4.0 on the new technological market

Figure 1 above describes the current format of University 4.0 which operates on the new technological markets. Here, the importance of information and communication technologies (ICT) becomes apparent. Some regions such as Hawaii leverages its unique geographical factors to develop skills and businesses that have the potential to become world-class and dominate the development of high-tech industries such as robotics, artificial intelligence (AI), and machine learning. Here, the partnership with the universities and research centres is very important. For example, the University of Hawaii cluster shares common features with the island of Oahu, home to Hawaii Island, the largest island in the United States, and is a major hub for research, development, and production of advanced technologies and services.

American leaders such as President Obama made expanding the ability of United States workers to improve their skills, foster innovation, and improve manufacturing competitiveness a priority. United States' system of community colleges currently enrols more than 7 million students, more than 1.5 million of whom are studying in high-tech industries [20, 21]. By modernizing technological and technical higher education at modern universities, it is creating a University 4.0. This type of university makes a stress on developing students' abilities in the field of improved academic skills, including science, technology, engineering, and mathematics (all known as STEM).

In general terms, ICT uses a wide range of partners from government, industry and academia to link higher education directly to the needs of the private sector. In addition, ICT strengthens students' ability to acquire technical and English skills and focuses on technical and technical training for women through leadership forums, academic initiatives and scholarships.

Universities of the 21st century need to build strategic leadership capabilities to help their leaders to promote university autonomy, improve the quality of courses, and build lasting partnerships with the private sector. This will enable the multigovernmental team to develop a durable and sustainable curriculum for e-learning that will help to close the skills gap and improve access to high-tech jobs in the automotive, aerospace and technical industries. For example, Indian River State College is proud to partner with the University of South Florida and the Office of Science and Technology Policy and Development (OSTD) of the U.S. Department of Education in developing virtual reality educational tools and training to support automotive and aerospace education [22]. ICT's commitment to developing and evaluating the widespread use of digital learning tools will strengthen the science, technology, engineering, and mathematics (STEM) talent pipeline, foster diversity among diverse populations, and ultimately provide unique opportunities that have never been available to students and faculty in STEM subjects such as computer science, engineering, and math.

4 Transition of talent and knowledge in modern higher education

What skills can an organisation expect from its students, faculty, staff and students in the future in the age of modern higher education? Higher education is well known in many ways, but it is not as rigorous as it used to be, largely because of a lack of investment in human capital in the education system. Employers need technology - improved learning to prepare for hiring, and education for the future. If we consider education, especially higher education, as something that everyone should have access to, rather than being informed solely through an investment approach, "human capital" should be considered the most appropriate tool to explain the benefits that individuals and society can derive from education.

Many experts believe change is necessary and urgent because the current system is anachronistic - in other words, it is still entrenched in the last century and has failed to meet the needs of the digital age. One would probably agree with the idea that certificates and diplomas are a means of applying knowledge of the labour market and takes the form of tacit criteria set by the ruling class in order to identify people with special social origins. Although not invalid, it disregards human capital theory and remains concerned that it has become the property of the ruling classes and is being used as a mechanism for maintaining power and tacitly reproducing social inequality [23, 24].

With regard to the above, one question comes to mind: how disruptive innovation is changing the way the world learns and explains that the current form of teaching cannot provide today's students with the skills they need to interact with a digital society. Education must be about empowering pupils with transferable skills and holding them accountable for their abilities in a rapidly changing world, making them more accountable than those who do not prescribe content selected for its relevance or relevance in the past. One of the most prominent figures in education, Dr. Martin Luther King Jr., has repeatedly pointed out that the education system cannot change its curriculum, or even change one iota objectively, because of the enormous changes in the economy, culture, and personnel over the last 50 years [25]. This can often mean being open to the possibilities presented, rather than trying to teach or prescribe the use of certain programs.

Although digital integration is also essential for thorough education in the 21st century, it is not simply a matter of supplementing existing teaching methods with technology. Students are increasingly advanced users of technology, whether they are first-time school-goers or not. Classroom technology classes astonish children by trying to teach them programs, websites, and hardware that are no longer relevant, but that they understand much better than the teacher. This new form of learning relies heavily on machines, with machine tutors using data and tracking student progress to recursively improve the quality of knowledge and student support. This requires new teaching techniques, facilitated by technology, such as "turning the classroom around" and "including learning". In order to maintain and advance humanistic education, which is important for its ability to deal with ethical and value issues raised by social change, institutions must respond to the needs of a new generation of students, students of colour, women and people with disabilities, all of whom demand access to quality education and support and to the education of their children. A number of information-based industries are being transformed or threatened, with new entrants offering specialised services that are separate from bundled services such as online education, social media and social networks. However, companies that want to use this capacity are constrained by their ability to stay up to date and recruit the talent they need. This can be achieved through training programmes that enable employers to play a more active role in managing their talent needs and to build a culture of continuous learning and innovation. Apprenticeships offer long-term, paid, work-oriented learning opportunities and structured curricula that ensure not only what you earn, but also how you are educated and in which industries you use it. Apprenticeship registration is one of the first steps towards formal authorisation of the use of apprenticeships as part of an employer-employee relationship in our education system.

Learning outcomes, measured not by course points but by students' skills, should be the benchmark of quality that makes the most sense to consumers. It also calls into question the validity of the current system, which certifies that a student has achieved certain necessary skills, regardless of where those skills were acquired. Technology is shaping the new educational environment and the role of technology in the future of higher education and higher education is also being called into question. In a free education system stimulated by technological progress in the late twentieth century, the idea of a way to document educational achievements was unthinkable. But in a digital knowledge society, where the economy is dominated by the use of technologies such as the Internet, social media, and mobile phones, it is now possible.

5 Conclusions

Al in all, we can see that higher education in general and universities in particular have come a long way that lasted for centuries. Their path has been market, and fostered, by the advancements in science and progress and can be traced to the needs of the society in their respective periods of existence.

University 4.0 that we have today represents the new type of the university that can exist within context of the sustainable development of higher education. It is distinguished by the high penetration of ICTs and relevant technologies. In the same time, it is more problem-oriented with the goal of analysing and solving global issues. In addition, University 4.0 is also product-oriented with a focus on delivering a viable product. Finally, it is also socially-oriented with the aim of tackling and solving the problems of the society and helping to maintain the balance in the ecosystem of this society.

The new type of the university is modern and competitive but it also centred around the personal needs and goals of its students. In its development on the path of transforming the whole concept of higher education, University 4.0 will effectively absorb the new components of the economic and social development of the future as well as the global environmental and sustainability challenges our society faces today.

References

1. W. Leal Filho, E. Pallant, A. Enete, B. Richter, L.L. Brandli, International Journal of Sustainable Development & World Ecology, **25(8)**, 713-721 (2018)
2. M. Klofsten, A. Fayolle, M. Guerrero, S. Mian, D. Urbano, M. Wright, Technological Forecasting and Social Change, **141**, 149-158 (2019)
3. E. Lekashvili, Marketing and Management of Innovations, **1**, 281-293 (2019)
4. K. Guthrie, D. Jenkins, *The role of leadership educators: Transforming learning* (2018)
5. A. Lafuente-Ruiz-de-Sabando, P. Zorrilla, J. Forcada, European Research on Management and Business Economics, **24(1)**, 8-16 (2018)
6. M. Sánchez-Barrioluengo, P. Benneworth, Technological forecasting and social change, **141**, 206-218 (2019)
7. M. Hernandez-de-Menendez, C.A.E. Díaz, R. Morales-Menendez, International Journal on Interactive Design and Manufacturing (IJIDeM), **14(3)**, 789-803 (2020)
8. E. Lisá, K. Hennelová, D. Newman, International Journal of Work-Integrated Learning, **20(1)**, 71-82 (2019)
9. ACEACPS, *https://www.aceacps.org/duties-responsibilities-dashboard* (2021)
10. M.J. Zeig, R.G. Baldwin, K.M. Philanthropy & Education, **2(1)**, 53-74 (2018)
11. H. De Ridder-Symoens, W. Rüegg (eds.), *A history of the university in Europe* (2003)
12. R. Nureev, V. Volchik, W. Strielkowski, Social Sciences, **9(5)**, 79 (2020)
13. J. Curtis (ed.), *Studies in Ancient Persia and the Achaemenid Period* (2020)
14. W. Strielkowski, J. Wang, 6th International Conference on Social, economic, and academic leadership (ICSEAL-6-2019), 1-4 (2020)
15. H. Ostrowicka, Ł. Stankiewicz, Higher Education Research & Development, **38(3)**, 609-622 (2019)
16. A. Hoffman, *The engaged scholar: Expanding the impact of academic research in today's world* (2021)
17. Y. Xing, Y. Liu, S.C.L. Cooper, British Journal of Management, **29(4)**, 670-690 (2018)
18. P. Jackson, R.K. Mavi, Y. Suseno, C. Standing, Science and Public Policy, **45(4)**, 553-564 (2018)
19. M. Plaza, I. David, F. Shirazi, International Journal of Production Economics, **205**, 215-227 (2018)
20. NCES, *https://nces.ed.gov/programs/coe/indicator_cha.asp* (2021)
21. B. Obama (ed.), *Strategy for American innovation: Driving towards sustainable growth and quality jobs* (2011)
22. D. Dykes, *http://www.greenvillebusinessmag.com/2020/10/05/331217/preparing-the-next-generation-of-technological-talent-in-advanced-manufacturing* (2020)
23. T. Kromydas, Palgrave communications, **3(1)**, 1-12 (2017)
24. R. Grundmann, Minerva, **55(1)**, 25-48 (2017)
25. The Seattle Times, *https://projects.seattletimes.com/mlk/words-education.html* (2017)

Vital energy coefficient for leading positive changes on the way to sustainable development

Kirill Gerasimov[1]*, *Maria* Skivko[1], *Natalya* Lyubavina[2]

[1]Samara National Research University, 34, Moskovskoye shosse, Samara, 443086, Russian Federation
[2]Togliatti State University, 14, Belorusskaya St., Togliatti, 445667, Russian Federation

Abstract. This paper focuses on such characteristics of leaders that foster the transformation processes. In particular, the paper analyses the coefficient of vital energy as a significant resource for achieving sustainable development goals. Considered in comparison with inefficient energy use as burnout, vital energy emphasizes the importance of life-work balance and the possibility for recharging for leaders. Moreover, the accent is given to the specificity of sustainable development changes that requires, apart from effective solutions, efficient and passionate leaders.

1 Introduction

Today, in the rapidly changing world with provoking challenges and unexpected obstacles, there is a strong need for successful leaders. Moreover, within the implementation of sustainable discussion in many areas of human life, there is a demand for those leaders who can provide and manage sustainable development goals. This article highlights such characteristics of a modern leader as a level of life energy needed for goal accomplishment and motivating others.

Life energy can be considered an emotional, physical, and intellectual state of an individual that supports its intention to act in the social environment, achieve goals, and perform changes [1]. Life energy constitutes human social and cultural capital and reflects an individual's life satisfaction and emotional stability. As an essential resource for any human activity (together with time, knowledge, and money), life energy facilitates the human's executing plans, self-realization, and joy of life [2]. Efficient manipulation of various resources helps to achieve success in any social activity.

However, there is always a risk of resource depletion; especially, life energy is endangered today by stress and responsibility at work and private life. Those who work in the social sector, mostly with people (healthcare, education, social work), whose task is helping others or protect the minority or oppressed groups, belong to the risk group that spends much more life energy for social communication at work. The activity related to the promotion and motivation for sustainable development and changes can also be referred to the same risk group due to the high involvement of energy and emotions into the workflow.

* Corresponding author: 270580@bk.ru

The good news is that life energy resource is renewable: it is possible to control life energy consumption and recharge oneself with different physical and mental practices.

In business and management, several approaches determine that leadership competence consists of three criteria: intelligent quotient (IQ), emotional quotient (EQ), and vitality quotient (VQ). IQ defines the intellectual abilities and potential of an individual, EQ refers to the ability to understand and manage own emotions as well as to use them for efficient communication. VQ is considered as an ability to maintain own energy and entrain others. As Pierre Casse denotes, the VQ today is more significant for leaders and change-makers than the other two components. It helps to set and achieve the goals, motivate others, stimulate changes. So, talking about sustainable development goals, it is necessary to mention that those leadership characteristics become more than just essential at the global and local levels.

2 Vital energy coefficients

The realization of sustainable development goals would not be possible without successful and encouraging leaders. In industries, science, techno sphere, education, or cultural sector, there is a need for those change-makers who can perform high intellectual ability, high level of emotional intelligence, and the power to bring positive energy to own life and life of others. These qualities are especially demanded in turbulent and insecure times.

Today, the leaders' working routine consists of a plurality of tasks, changing deadlines, and increasing information for processing and analysis. High levels of responsibility and emotional involvement complement this leadership profile. Moreover, digitization created unlimited possibilities for online-communication but expropriated privacy: the all-day online-presence did not leave the space for quietness and unavailability. On the one hand, technological development accelerated the speed of work and increased work efficiency; on the other hand, it endangered the quality of work and lowered personal responsibility. An increase in nervousness and aggression, constant deadline pressure, concentration difficulty, and just stress is some of the possible consequences of this working style and working environment. Individuals regularly need to collect and spend energy; the disbalance negatively affects the physical and mental health.

As Rodionov defines, there are five types of interconnected vital energy today [3]. Physical energy helps at work, and its excess provokes violence, conflicts, or wars. Adversity energy is responsible for the human nervous system and can be controlled or negatively affected by human inventions or influence. Logical vital energy and intellectual energy are high-demandable in industries, politics, and military service. Spiritual energy is supposed to be responsible for the safety and security of society and individuals as well as for the accomplishment of sustainable development goals; it is incorporated into the human culture and cultural values. So, energy accumulation and energy loss are also interconnected: good physical form creates preconditions for spiritual growth or, instead, depression leads to problems with physical health.

Pierre Casse declares that the leadership potential consists of IQ, EQ, and VQ; the ability to critically analyse and the presence of emotional intelligence are not often sufficient to become a successful leader or a good manager [4]. Instead, the vital energy today facilitates mostly those changes that leaders maintain. Vital energy coefficient stimulates, encourages, and motivates for goal achievement and provides the power to improve the life quality. As far as leaders, apart from other activities, are supposed to deal with obstacles and emerging resistance by promoting changes, the high level of VQ can simplify the leadership trajectory and improve the results of their work.

As an innate quality, VQ differs in one personality to another: some people are more active, others are more passive. The higher the vital energy, the more an individual is open to changes, aims to improve the social and working environment, and leads the group. It is possible to differentiate the potential and the real VQ. The former refers to the innate vital energy for achieving expected great results, the latter describes the current situation with the vital energy by accomplishing tasks. A high level of real vital energy stimulates the person to achieve professional and personal goals, motivate others, and bring constructive energy to the social and work environment. It positively affects creativity, the decision-making process (and its speed), and an incentive to overcome challenges. However, real vital energy is usually lower than the potential one, which defines the key problem of modern leadership efficiency: It is difficult to make changes without sufficient energy resources.

For increasing the potential vital energy, Casse recommends checking whether the life activity reflects the accomplishment of personal goals and desires. For increasing group motivation and involvement, the leader should use managerial skills to distribute the tasks properly among the team members and emotional intelligence to understand and define the reaction of team members on the workflow. This principle can also be applied to the whole company, community, or nation.

3 Incorrect use of energy: emotional and professional burnout

Such characteristics of a leader as flexibility, high potential to inspire and motivate, openness for new opportunities, and new knowledge are very demandable in this rapidly changing world. Especially in the sustainable discussion, there is a need for those leaders who can energize for changes and transformations.

The high speed of life, an increasing list of everyday tasks and challenges, stress, and deadlines do not support or extend the vital energy. Additionally, difficulties at work, interpersonal communication, multitasking work character, limited opportunities for relaxation and free time can worsen any productive activity. It leads to psychological distress, depression, and burnout. It is defined as a syndrome of reduced personal accomplishment, prolonged reaction to chronic job stressors expressed in a depressive mood, nervousness, and loss of motivation [5, 6].

Burnout is often a result of job dissatisfaction, lack of control, and uncertainty in the work environment, especially if individual stress experience meets the organizational issues [7]. As a human dimension of sustainability, organizational sustainability is responsible for employees' job satisfaction and positive working climate; increasing competitiveness and variety of demands strengthens the risk of employee burnout [8]. Good management of an encouraging leader can lower this risk if only this leader can deal with his/her own vital energy most productively and efficiently. Various leadership strategies (e.g., coaching, delegation, reconsidering tasks and expectations) can affect the burnout's own level and among employees [9, 10].

One of the common problems of contemporary workers, especially among job areas involving people and communication, is emotional burnout. It is understood as increasingly emotional and psychological exhaustion; besides, it can reflect the dynamic character of work or the level of emotional expression, display at work [11, 12]. It is usually caused by a high range of responsibilities at work, personal crisis, stress, and lack of relaxation. With emotional burnout, professional burnout as an incorrect distribution of life energy for work tasks can destroy the professional motivation and orientation for goal achievement as well as decrease the quality of work [13].

Modern leaders – e.g., social or environmental activists, journalists, youth politicians, or volunteers – quite often meet professional burnout due to the content and character of their work [14]. The reason is usually the intention to motivate, to inspire, to energize that leads to the excessive life energy use: fighting for rights of minorities, promoting different social or cultural values, protecting the environment demand not only smart solutions, creativity, and the speaker talent but also a lot of energy and inner power.

As a type of emotional labour, the leader activity can be considered a stressor that ends with burnout [15]. Activist burnout is defined today as a modern burnout type that overwhelms the leader's energy due to the lack of self-care and workflow disbalance [16, 17]. Overwork, conflicts, urgency as well as lack of visible results and social acceptance characterize the activists' work and explain the burnout background. Sometimes, professional role expectations confront the administrative restrictions that result in organizational stress [18]. In the same way, leaders who belong to the potential risk group cannot manage emotions and vital energy by dispersing life energy for everything.

4 Managing vital energy for sustainable changes

Sustainability can be explored through the category of balance: resources, investments, and consumer demands. In the same way, social dimension in sustainability, people's resources should also balance physical, emotional, and life energy. On the contrary, the level of burnout as well as of job satisfaction also signifies the social component of sustainability [19]. Therefore, vital energy management should become an integral part of modern leadership competencies. Moreover, in such a complicated global discussion about sustainable changes it is urgent and crucial to implement the strategies and techniques to support encouraging leaders to perform changes. For example, the Heinrich Böll Foundation in Moscow launched an online support project for environmental activists that aims to help lower emotional and professional burnouts with sustainable practices, practical tools, and experience exchange [20].

Practices to support physical health (regular sport activity, healthy eating habits, enough sleep) and social health (networking, creative activities, family support, abilities to say "no", to take a break) can become basic principles of self-healing and recharging energy. Particular attention should be given to stress management programs, coping strategies, and mindfulness. Raising individual and collective awareness of the problem with vital energy should become a principal issue in the global discussion about leadership competencies.

Time and energy efficiency in management is a very significant topic in any branch; it relates company's success and further development. This kind of efficiency is essential in non-profit organizations, environmental and social projects with driven and passionate leaders. Whether the leader communicates changes within community-based or work-based movements, whether his/her activity is a full-time job or voluntary activism - it anyway challenges the leader to follow the work-life balance and find the opportunity to recharge oneself. In other words, in order to drive sustainable changes, leaders should have a sustainable lifestyle.

One of the strategies for recharging energy suggests unplugging from distractions (e.g., time and energy-consuming activities, toxic environment) [21]. Nevertheless, it is crucial not only to know how to deal with the energy loss, burnout, and concomitant loss of motivation, nervousness, and stress; it is essential to be aware of strategies to prevent the development of such conditions [22]. So, both energy loss prevention and energy recharging should belong to the individual and organizational sustainability in order to prevent the destructive influence of the leadership activities.

5 Conclusions

Sustainability topics require personal and organizational energy and motivation to promote sustainable goals, to make changes, and motivate the community. Leaders of those activities should have enough energy resources to deal with a high rate of responsibility, urgency, overwork, and various obstacles that inhibit changes. However, in a time of rapid transformation, increased demands, and uncertainty, even the most enthusiastic leader, can face some physical or social health problems resulting from work-life disbalance.

This article explained the motivating power of leaders expressed in the vital energy coefficient. This coefficient, together with intelligent and emotional competencies, constructs the leader's ability to overcome challenges and motivate the team. Particularly, in the work of social and environmental activists, whether they get paid or participate voluntarily, there is a danger of losing too much vital energy with no opportunity to recharge oneself. The disbalance of work and life expressed in the loss of energy can result in emotional or professional burnouts followed by some physical or social health problems.

The incorrect use of life and work priorities, as well as various job stressors, weaken the individual's capability to confront burnout. Nevertheless, the in-time stress management practices, implementation of coping strategies, balancing priorities and tasks, and exercises to deal with stress and stressors can facilitate positive changes in individual's emotional and physical conditions. Moreover, those who can efficiently deal with their own vital energy, control personal sustainability, can also negotiate the organizational sustainability by encouraging, stimulating, and preventing the burnout of the team members. Managing vital energy today should become one of the key competencies of modern leaders to promote and support positive changes in sustainable development.

References

1. V.G. Nemirovskiy, A.V. Nemirovskaya, Monitoring, **4(122)**, 104 (2014)
2. E.V. Balazkiy, Monitoring, **2(82)**, 124 (2007)
3. A.S. Rodionov, ActualScience, **2(8)**, 67 (2016)
4. P. Casse, *https://www.forbes.ru/column/48259-glavnoe-kachestvo-lidera-energiya-ne-intellekt* (2010)
5. D.C. Glass, J.D. McKnight, Psychology and Health, **11(1)**, 23 (1996)
6. C. Maslach, M.P. Leiter, Stress: Concepts, Cognition, Emotion, and Behavior, **4,** 351 (2016)
7. D. McDermott, Journal of Human Stress, **10(2)**, 79 (1984)
8. S.B.N. Evenstad, International Journal of Systems and Society, **2(2)**, 52 (2015)
9. A. Zopiatis, P. Constanti, International Journal of Contemporary Hospitality Management, **22(3)**, 300 (2010)
10. S. Oliver, *https://www.forbes.com/sites/forbescoachescouncil/2020/05/28/how-leaders-can-save-their-businesses-and-avoid-burnout-at-the-same-time/?sh=21c260df229f* (2020)
11. J. Bejtkovský, Marketing and Management of Innovations, **1**, 294-302 (2020)
12. A.J. Montgomery, E. Panagopolou, M. de Wildt, E. Meenks, Journal of Managerial Psychology, **21(1)**, 36 (2006)
13. N. Belyakova, E. Petrova, O. Polyakova, Economic and Social Development: Book of Proceedings; Varazdin, 324 (2017)
14. L. Cox, Into-ebooks, Helsinki, 2011: *http://mural.maynoothuniversity.ie/2815/* (2011)
15. D.-Y. Jeung, C. Kim, S.-J. Chang, Yonsei Medical Journal, **59(2)**, 187 (2018)
16. P.C. Gorski, Ethnic and Racial Studies, **42(5)**, 667 (2019)
17. C.W. Chen, P.C. Gorski, Journal of Human Rights Practice, **7(3)**, 366 (2015)
18. M. Leiter, Canadian Psychology/Psychologie canadienne, **32(4)**, 547 (1991)
19. J. Baumgardt, J. Moock, W. Rössler, W. Kawohl, Frontiers in Public Health, **3**, 25 (2015)
20. Heinrich Böll Stiftung Moscow, https://ru.boell.org/ru/2020/04/21/ustoychivyy-aktivizm-onlayn (2021)

21. L. Maloney, *https://www.forbes.com/sites/forbescoachescouncil/2019/07/22/how-to-unplug-from-distractions-and-plug-into-your-own-vital-energy/?sh=2450cac7b7f3* (2019)
22. J. Perlo, D. Feeley, Journal of Healthcare Management, **63(2)**, 85 (2018)

Development of electricity industry on the basis of the use of the renewable energy sources in the regional energy security system

Natalya Novikova[1]*, and *Olesya* Kharitonenko[2]

[1]Ural State University of Economics, 8 Marta str. 62/45, 620144 Yekaterinburg, Russian Federation
[2]Head of municipal authority, The Chairman of Yevpatoria City Council of the Republic of Crimea

Abstract. In the article presented, the authors prove the objective need for the development of the electricity industry of the Republic of Crimea on the basis of the active use of the renewable energy sources, caused by the world processes of the formation of the "green" economy. The object of the study is the territory of the Republic of Crimea - a bright representative of the energy deficit region. The development of the electricity industry of the Republic of Crimea on the basis of the use of the renewable energy sources serves as the basis for ensuring regional energy security, the fragility of which was determined by the political events of 2015. The authors studied the research papers of leading domestic and foreign scientists devoted to the analysis of the renewable energy sources in order to justify the further active development of a wind energy, a solar energy as well as bioenergy and landfill gas energy in the Republic of Crimea. The development of the Republic of Crimea's electricity industry through the use of the renewable energy sources and local fuels is an essential area of modern active regional policy. It is determined by the fact that this direction, firstly, is considered as environmentally friendly generation, and secondly, contributes to a qualitative increase in regional energy security level.

1 Introduction

The Doctrine of the Energy Security of the Russian Federation, approved by the Decree of the President of the Russian Federation dated May 13, 2019 No. 216 interprets energy security as "the state of protection of the country's economy and population from threats to national security in the field of energy, in which compliance with the requirements for fuel and energy supply to consumers stipulated by the legislation of the Russian Federation is ensured, as well as the fulfillment of export contracts and international obligations of the Russian Federation" [1].

The concept of " the energy security" was first introduced into scientific and practical circulation in 1947 as part of a regulatory document adopted in the United States regulating the actions of the state in the field of ensuring national security [2]. The concept of energy security was formulated by the International Energy Agency after the oil crisis in 1973 in the

* Corresponding author: novikova@usue.ru

following interpretation: energy security is "a matter of confidency that energy will be available in the quantity and quality required under these economic conditions" [3].

Theoretical and methodological aspects of the energy security study of regions with various levels are presented in the research papers of a whole pleiade of the Russian scientists [4 - 11].

The dominance of different sources of energy plays a crucial role in global industrial and technological development. In particular, the founder and the executive chairman of the World Economic Forum in Geneva, K. Schwab, mentions: "The basis of the first and second industrial revolutions were transformations in the energy sector. Initially, it was a transition to using steam energy, and then electricity itself. Now, at the beginning of the Fourth Industrial Revolution, energy is on the verge of a new historical transition: from fossil fuels to a renewable energy "[12, p. 219].

The renewable energy sources are viewed as sources of constantly operating or periodically occurring processes in nature, as well as the life cycle of the plants', animals' world and human activity.

An American scientist, recognized authority in the field of energy, international politics and economics Yergin identified seven main types of the renewable energy:

"1. A wind energy is the largest non-water renewable source, which provides energy to technologically complex plants that generate electricity, whereas the aggregate of which is called a wind farm.

2. A sunlight is captured by photovoltaic converters, mirrors or other devices that convert light and transform it into electricity.

3. A biofuel includes ethanol, bio-diesel and modern types of biofuel (from algae, cellulose or other raw materials), which substitute gasoline, diesel fuel or even aviation fuel.

4. A biomass encompasses wood or other plant material briquetted or otherwise treated and burned in a power plant; it also includes wood or manure that are burned for heating and cooking in the developing countries.

5. A geothermal energy is a hot water or steam that rises from underground to the surface and drives a turbine that produces electricity.

6. A hydropower means a falling water or pressurized one that drives turbines, but dams are increasingly criticized by environmental organizations, and therefore it is not easy to build them in many countries.

7. A passive use of solar energy (or passive solar systems) are also called "green houses" nowadays; due to their location in the natural environment, they reduce energy consumption, which echoes the concept of energy efficiency" [13, p. 487].

In addition to the seven main types of the renewable energy sources, Yergin highlights the usage of ebbs and flows' energy alongside with the waste energy generation.

The renewable energy technologies combine the following important characteristics: "They are not based on limited resources, they are not limited to any territories and, at least theoretically, do not lead to an increase in carbon dioxide emissions.

It is necessary to add another technology in this list - batteries for electric vehicles or an electric grid... They can be considered as a renewable source of energy only in the case if the electricity to which they are recharged is the product of both a wind or a sunlight "[13, p. 488].

The renewable energy application enables to solve the following global problems [12, 13, 14]:

- to overcome the dependence on fossil fuels determined by the First Industrial Revolution;
- to reduce the cost of energy production (excluding initial costs), resulting in price reduction;
- to reduce budgetary subsidies for energy supply to small towns;

- to weaken the negative impact of energy on the natural environment;
- to create an additional incentive for the high technology's development
- to improve the population comfort in territories at various hierarchical levels;
- to create a sustainable future for the rich and poor, for the urban and rural population.

As a part of the technology disclosure of the Fourth Industrial Revolution, its ideologist K. Schwab formulates the following key conclusions in terms of the process transformation of the energy production, as well as its storage and transfer as a result of the renewable energy sources usage:

"The Fourth Industrial Revolution could stop the world's dependence on fossil fuels and the greenhouse gas-related energy production process that had emerged from previous industrial revolutions. Nowadays this challenge is more urgent, because a climate change becomes more visible due to the world's population and the economy industrialization growth; as predicted, a global energy demand is projected to double by 2040. "[12, p. 229].

" In order to accelerate the growth rates and embrace more sectors of the economy, it is worth providing the transition to the renewable energy needs. Today it is necessary to make long-term investments in order to take advantage of the results in the coming decades, especially for the fast-growing regions [12, p. 230].

"New technologies in the energy sector are being explored - from the use of tidal energy to thermonuclear fusion, as well as advanced materials and nanotechnologies" [12, p. 230]. According to the scientist, this will increase efficiency and reduce the energy losses. "Intelligent networks complemented by artificial intelligence capabilities, as well as dynamic redistribution of electricity and the use of battery-powered vehicles will provide an opportunity to ensure widespread efficiency improvements at all levels" [12, p. 230].

However, according to K. Schwab, the renewable energy development can have negative consequences: "A large-scale transition to the renewable energy gives rise to fear for the fossil fuel industry, as well as creates a threat to the relevant long-standing geopolitical structures" [12, p. 230].

The renewable energy in today's society is viewed as a key solution to power supply, energy security and climate change.

Moreover, Yerging cites vivid examples of government decisions: the head of the People's Republic of China declared that China should "take advantage of the opportunities opened at a new stage of the world energy revolution. The European Union has set a goal of reaching a 20% share of the renewable energy sources by 2020, providing the status of the greenest government in the history. In 2015, the Brazilian President D. Rousseff and the US President B. Obama announced that their countries are committed to increasing the share of non-water renewable sources in electricity generation to 20% by 2030 "[13, p. 488]. According to B. Obama, "A country that is the world leader in creating new sources of energy will lead the global economy of the 21st century." [15]

Raising prices for traditional energy resources, which began in 2003-2004, narrowing the cost gap between the renewable and traditional energy sources, as well as the global problem of climate change, have become the significant factors in the energy policies of the developed countries of the world, increasing investments in the renewable energy sources. As a result of venture capital influx, the renewable energy sources have been called "clean technologies."

There are attempts to estimate "the scale of the renewable energy in monetary terms. Global renewable energy investments reached $200 billion in 2014, up four times since 2005. Today, the renewable energy accounts for about 12% of the world's energy generation capacity. But their share in the newly introduced generating capacities is growing rapidly - in 2011-2014 it amounted to more than a third "[13, 505]. All this enables to argue that the renewable energy sources are becoming a significant high-tech industry.

However, nowadays the high capital costs of the renewable power plants, the probabilistic nature of generation and the low value of installed capacity utilization do not completely replace the traditional energy sources.

2 Materials and methods

The above mentioned theoretical and methodological aspects of the renewable energy sources' investigation are of great relevance justifying development directions of the electricity industry of the Republic of Crimea, since this territory is an energy-deficient region. The attempts to make it non-volatile have been made since the Soviet times, but even in the XXI century the peninsula suffered from the lack of its own capacities.

It is worth pointing out that "the energy system of the Crimean Peninsula began to take shape quite late, i.e. in the 30s of the XX century. Before that, a light was supplied only to the Livadia Palace of the imperial family and to the houses of several wealthy residents of the city of Simferopol - the first power station in the region was built for them in 1896.

Four thermal power plants (CHP) were built for industrial needs in the period from 1936 to 1961, which still remain the main sources of power.

In the 1950s, the Ukrainian SSR began to implement its concept of the Crimean energy system development, resulting in construction of four power lines, which connected the peninsula with the continent. There were plans to build its own nuclear power plant, but the plans have never come true due to the Chernobyl disaster.

For a while, the CHP energy and megawatts coming from the Kherson region were enough, because the industrial production in Crimea was declining.

The issue of the energy deficit was raised again in the 1990s. At the same time, it was decided to rely on an alternative energy – a wind and solar power plants, which, despite all the pluses, can hardly become the basis of the energy system "[16].

With these problems, Crimea existed until the reunification with Russia in 2014. After this historical event, reserved autonomous capacities were transferred to the peninsula, which previously provided reliable energy supply to the Sochi Olympics.

It should be noted that energy is a large branch of the modern economy of the Republic of Crimea. The peninsula possesses substantial energy reserves and the renewable potential. Gas and oil production is carried out, as well as the production of electric and thermal energy. Crimea's energy system is heavily dependent on the external supplies of electricity and petroleum products.

The analysis of the absolute values dynamics in terms of electricity production in the Republic of Crimea demonstrates a significant increase, which amounted to 1835.4 million kWh for 2017-2019. (Table 1).

Table 1. Dynamics and structure of electricity production in the Republic of Crimea

	2017	2018	2019	Indicator Change 2019 in comparison with 2017
Electricity, million kW. h	1868,8	2212,4	3704,2	+ 1835,4
including:				
electricity produced by thermal power plants	1349,6	1687,7	3199,5	+1849,9
electricity from renewable energy sources	519,2	524,7	504,7	-14,5
share of electricity from renewable energy sources in total output	27,8	23,7	13,6	-14,2

Power generation from the renewable energy sources in 2019 amounted to 504.7 million kWh or 13.6% of total production. The share of the renewable energy sources for 2017-2019 has decreased by 14.2% due to the high dynamics of electricity production from the thermal power plants.

The high dynamics of power generation worked out by the thermal power plants is determined by the implementation of the Federal Target Program "Socio-economic development of the Republic of Crimea and the city of Sevastopol until 2024," approved by the Decree of the Government of the Russian Federation from August 11, 2014 No. 790, within the framework of which one of the goals is "lifting infrastructure restrictions in order to provide sustainable economic development" and defined tasks - "eliminating restrictions and improving the quality of energy supply in the region," "formation and development of energy, transport, social and telecommunication infrastructures."

The FTP developers in section "1. Analysis of the socio-economic situation of the Republic of Crimea and the city of Sevastopol" specify that on the territory of the Republic of Crimea: "The existing natural resource potential is irrationally used, taking into account recreational, bioclimatic, mineral and raw materials' potential alongside with the energy, primarily in the field of unconventional sources application of the renewable energy (wind, solar, geothermal); there is a complete lack of systemic environmental measures to preserve the unique flora and fauna of the Crimean peninsula.

The key constraints to the socio-economic development of the investigated territory are defined as: "the insufficient resource and infrastructure supply of the region with vital sources of energy, water, food, including restrictions on the provision of water resources for drinking and productive consumption, including the lack of round-the-clock water supply for certain areas"; "a high level of energy dependence of the economy." When designing objects within the framework of the Program, it is planned to provide: "the implementation of energy- and resource-saving technologies as well as environmentally friendly ones."

3 Empirical model

The Scheme and Program for the Development of the Electricity Industry of the Republic of Crimea for 2019-2023 was developed and realized on the territory of the Republic of Crimea, according to which both a solar energy and a wind one is of the greatest importance among the renewable energy sources (Table 2) [14].

Table 2. Renewable energy objects in the Republic of Crimea

	Number of objects, units	Installed capacity, MW
solar power stations	5	296,98
wind farms	7	88,563
total	12	385,543

Solar power in the Republic of Crimea is represented by five solar power plants with a total installed capacity of 296.98 MW. (see Table 2).

The installation of solar power plants is cost-effective in such a case if the number of hours of sunshine in the territory under consideration is at least 2000 per year, and the sunlight intensity is at least 5,000 MJ/m2.

Solar radiation is distributed equally throughout the Republic of Crimea. During the year, the sun shines on the Kerch Peninsula and on the South Bank within 2250-2300 hours, in the flat Crimea 2250 hours, in the Western part of the Republic of Crimea it ranges from 2250 to 2400 hours or more per year. Thus, on average, in Crimea the sun shines 5-7 hours a day during a year.

A study of information on the distribution of solar radiation and the duration of sunshine per year in the territory of the Republic of Crimea showed that this territory is economically viable for the development of solar generation, especially in the Western and South-Western territories of the Crimean Peninsula.

In particular, on August 1, 2015, the Nikolaevka solar power station, located in the Simferopol district of Crimea, began to supply electricity to the grid [17]. The construction of the solar electric station Vladislavovka, localized on the territory of the Vladislavovsky rural settlement of the Kirovsky district of the Republic of Crimea, is considered as promising. In fact, the Vladislavovka solar power station is an object that consists of five separate stations.

A wind power in the Republic of Crimea is represented by seven wind farms with a total installed capacity of 88.563 MW (see Table 2).

Existing wind farms are localized in the Eastern, Western and Southwestern parts of the Crimean Peninsula, depending on the direction of a wind. "During the year, winds of the North-East, South-West and North-West directions prevail in the Republic of Crimea. In winter, the repeatability of the North-East winds is 45%, the South-West winds embrace 25%, while the South winds - up to 20%. During late autumn and winter, very strong Northeast winds often last about 270-325 hours a month.

The territories with the highest indicator of wind energy potential of the Republic of Crimea are located primarily in the coastal zone of the Crimean Peninsula: North, North-West, East and South-West coasts "[14].

Data exploration on long-term average wind speeds in the territories of the Republic of Crimea showed that such cities as: Armyansk, Yevpatoriya, Kerch, Saki, Sudak, as well as the territories of the districts: Leninsky, Razdolnensky, Saki, Soviet, Kirov, Black Sea and the territory of the Ai-Petri highlands possess the potential needed for operating the wind-generating plants.

In the territories favorable for the introduction of wind generation, the development of a wind power in perspective is planned through the construction of new wind power plants and the reconstruction of the existing ones.

The attractiveness of the wind energy development in the Republic of Crimea is confirmed by the following fact. The founder of the German group of companies Winkler intends to invest in the Crimean wind energy. In July 2019, the investor with his partners "registered the New Energy Plus company in Feodosia with an authorized capital of 1 million rubles. In October 2019, zWe announced that since this summer it has been actively working in the Russian Federation and, together with its partners, is preparing to produce turbines for the wind plants in the West of the country. In particular, the company develops isolated solutions with small turbines and batteries for the rural areas, as well as large wind plants. "New Energy Plus" negotiated with Crimea on the modernization of windmills, which were nationalized after the region joined Russia "[18]. Many of them have been in operation for more than 20 years and the technologies are outdated.

According to the experts' opinion, the Republic of Crimea benefits from a huge potential in the solar and wind power: "A unique solar and wind power industry is presented in Crimea, being one of the first regions in Russia in terms of the amount of the renewable energy sources' generation. A special attention should be paid to the wind and solar energetics." [19] According to the specialists' calculations, a solar and wind energy are the cheapest in the Crimea territory. In addition, alternative sources of electricity can save natural gas, which has to be either purchased on the mainland or produced on the territory of the Republic of Crimea, that is much more expensive than a solar energy itself.

4 Results and discussions

To a lesser extent, the remaining types of the renewable energy sources are represented in the territory of the Republic of Crimea.

In the Republic of Crimea, a ***hydropower*** is not introduced practically. This is due to the fact that the Crimean rivers have a low water consumption, which complicates significantly the operation of the hydroelectric power plants.

Currently, there is one mini-hydroelectric power station in the Simferopol district on the Partizan reservoir in the Republic of Crimea with the capacity of 250 kW. "The development of potential small rivers potential and the use of free head in the existing water supply and sewerage systems of the Crimean cities using small hydropower plants helps solve the problems of improving the energy supply of numerous consumers and their environmental safety" [14].

The territory of the Republic of Crimea includes the Black Sea coast, therefore the development of the renewable energy sources based on the energy of *tides*, i.e. *tidal power plants*, may be promising. However, this type of power plant has a significant disadvantage namely its power variation during a day. It requires the mandatory operation of the power plant in parallel with the power system or the redundancy of the power plant by the operation of other power plants and, as a result, an additional network construction, which increases the cost of erecting the station and its infrastructure and reduces the benefit from the cheapness of the energy generated by the station.

The next interesting direction in the framework of the renewable energy sources research is ***bioenergy*** as an actively developing direction of the unconventional and renewable energy. Bioenergy covers several independent areas of energy production at once: biogas energy; Corn Waste Energy (DAR); peat energy and other types of energy.

A detailed analysis of bioenergy potential of the Republic of Crimea is presented in a research study of the Sevastopol State University scientists, such as Starchenko M.A. et al. [20]. The authors considered the possibility and feasibility of using biogas plants for heat and electricity supply in rural settlements. The authors' main conclusions are as follows: "The use of bioenergy plants in the Crimea is promising, since there are many livestock farms on the peninsula, poultry farms, and a third of the population lives in the rural areas; The use of bioenergy plants is an effective way of providing local electricity and heat in the rural areas; This technology also has a favorable impact on the environment, since a livestock waste, especially in large farms, produces a surface and groundwater pollution.... Bioenergy contributes to developing the energy market as a whole, which has a positive effect on the economy of the region "[20, p. 97].

The energy potential of agricultural production wastes of the southern regions of Russia namely in the Volgograd region and the Republic of Crimea is disclosed in details in a scientific study by Andreenko et al. [21]. The authors evaluate the resource base for the development of bioenergy, paying special attention to the study of the energy potential of bioenergy not specially produced biomass, but organic waste generated in crop production and animal husbandry of these territories. As a result of the study, the territorial distribution of energy potentials of agricultural waste was designed to municipalities namely districts and urban districts.

As part of the renewable energy sources analysis, ***a landfill gas*** is separately investigated as a biogas formed in the result of anaerobic decomposition of organic waste at landfills of solid domestic waste (MSW).

Any solid waste landfill is a large biochemical reactor, in the bowels of which biogas is formed during operation during several decades after its closure as a result of anaerobic decomposition of waste in terms of plant and animal origin. "Biogas is a mixture of methane and carbon dioxide in approximately equal proportion. Biogas inevitably enters the atmosphere, which causes a number of negative consequences. The accumulation of gas in the body of the landfill often causes a spontaneous combustion of solid household waste. The combustion process is accompanied by the formation of toxic substances, in particular dioxins. The negative impact of biogas on the environment has led to the widespread use of biogas collection and disposal systems in MSW landfills in most developed countries. "[14]

The production and further use of this type of gas solves several problems at once: "preventing atmospheric pollution (also, methane has a strong greenhouse effect); reducing the risk of fires and explosions at MSW landfills; obtaining an experience in operation of an electric power production facility using an unconventional energy source.

During the construction of an electric station on landfill gas, released in the body of the MSW landfill as a fuel, it is necessary to specify that it is supplied to the place with installed block-modular equipment based on the usage of the landfill gas collection system. To provide the electric station with a fuel, a complex of specialized equipment for the production, purification and supply of landfill gas is installed. This complex includes: a gas collection station connected to wells drilled in the body of the landfill; a gas compressor station providing gas supply; a high-temperature flare for burning a gas volume excess; a Gas Purification Unit (OGD) "[14].

The Republic of Crimea has a great potential for the implementation of projects for the construction of generating facilities using landfill gas. In particular, the first biogas station generating electricity from landfill gas was launched in the village of Turgenevo, the Belogorsky district of the Republic of Crimea.

Thus, a biogas energy can develop rapidly and solve the problem of electric and heat supply in agricultural settlements, as well as in large enterprises.

Wood waste, pellets are "an environmentally friendly type of fuel obtained from wood raw materials with a help of pressing. This fuel is produced from wastes of logging and woodworking industry without application of any sizing substances. The main raw materials from which pellets are made are chips, sawdust, shavings, wood dust and tree bark. But there are also other types of raw materials, such as peat, straw, nut shells, reeds, etc.

In the finished form, pellets are granules with a diameter of 6-10 mm and a length of up to 50 mm. Several important problems are solved at once by the production of pellets: fire hazardous waste is processed and high-calorie fuel is produced.

Pellets as a type of fuel have appeared relatively recently, but due to their high calorific and natural properties, they have become popular in Europe, Japan and North America. The increase in their consumption is determined by the possibility of using pellets in industrial thermal power plants and in installations intended for private use "[14].

Such type of fuel as pellets can be used widely under the conditions of a timber industry development in the Republic of Crimea.

5 Conclusions

A recognized authority in the field of energy, international politics and economics, D. Yergin notes that energy, like no other sphere, is ready for innovations, although "it took more than a decade to develop and introduce technologies. In addition, some questions about the scope and cost have not been answered yet. "[13, p. 505] Today, the renewable energy industry attracts the attention of modern theorists and practitioners at an international scale [22, 23].

In addition, Yergin emphasizes the revolutionary and evolutionary nature of the transition of modern society to the renewable energy sources: "if the transition to the renewable energy sources is really large-scale, it will be equal to the oil transition in the 20th century, according to geopolitical, economic, and environmental points of view. However, this path is likely to be long. As history proves, transitions to new energy sources usually take decades" [13, p.

505]. The scientist highlights that by 2030, the renewable energy sources will not be considered as the main energy source, wherein their share in the global energy balance depends on political, economic and innovative factors.

The founder and permanent chairman of the World Economic Forum, Schwab, makes an important conclusion to complete our study: "Clean energy technologies and better energy storage capabilities go from laboratories to production and conquer markets... The universal availability of clean, affordable energy resources will have a positive impact on the environment, and in particular on the population of developing countries where electricity systems are not reliable or do not exist completely. In addition, the use of environmentally rational energy technologies can reduce costs for companies and consumers and overcome the negative impact on the environment resulting from industrial pollution of the last century" [12, p. 219].

The Russian Federation and the Republic of Crimea have all the prerequisites for being included in the objective world trends of the Fourth Industrial Revolution, regional aspects of which are described in our scientific papers [24, 25]. The development of the Republic of Crimea's electricity industry through the use of the renewable energy sources and local fuels is an essential area of modern regional policy. This is due to the fact that this direction, firstly, is environmentally friendly generation, and secondly, it contributes to a qualitative increase in the level of regional energy security.

The importance of the high level of regional energy security in the territory of the Republic of Crimea was clearly demonstrated by the events of 2015, caused by the Ukraine's disconnection of the regional energy system from supply sources. Currently, the Government of the Russian Federation has created all the conditions for preventing the recurrence of these events.

The development of the Republic of Crimea's electricity industry through the use of the renewable sources of energy should take place in parallel with the development of efficient energy, based on the traditional energy resources, which enable to neutralize the existing disadvantages of the first sources.

The scientific study and practical implementation of the renewable energy sources is the regularity of the modern world community development, which is focused on the development of a "green" economy. Nowadays there is no other way for a mankind.

6 Acknowledgment

The study was carried out with the financial support of the RFFI Grant "Incremental Approach to the Formation and Implementation of Strategies for the Socio-Economic Development of Regions of Various Hierarchical Levels of the Russian Federation: Unified Rules for Strategy" (project No. 20-010-00824).

References

1. The doctrine of energy security of the Russian Federation: approved by the Decree of the President of the Russian Federation of May 13, 216 (2019)
2. L.L. Bogatyrev, A.V. Bochegov, N.I. Voropai, *Reliability of fuel and energy supply and survivability of energy systems in Russian regions* (2003)
3. Energy Dictionary, World Energy Council (1992)
4. A.I. Tatarkin, A.A. Kuklin, A.L. Myzin, A.V. Kalina, *Complex method for diagnosing the energy security of territorial entities of the Russian Federation* (2002)
5. A.I. Tatarkin, A.A. Kuklin, Regional Economy, **2(30)**, 25 (2012)

6. V.A. Saveliev, *Improving the efficiency of power systems* (2002)
7. V.V. Morozov, *Strategic Innovative Management in the Electric Power Industry*, (2004)
8. V.I. Ryasin, Bulletin of Ivanovo State Power Engineering University, **2**, 1 (2005)
9. V.G. Belomestnov, R.F. Araslanov, A.V. Balzhinov, *Energy efficiency of socio-economic systems of regions* (2010)
10. I.S. Schepansky, Actual problems of Russian law, **4(21)**, 161 (2011)
11. S.I. Bortalevich, Problems of modern economics, 4 (2011)
12. K. Schwab, *Technologies of the Fourth Industrial Revolution* (2018)
13. D. Ergin. *In Search of Energy: Resource Wars, New Technologies and the Future of Energy* (2019)
14. Scheme and program for the development of the electric power industry of the Republic of Crimea for 2019-2023: Report on research work (JSC Scientific and Technical Center of the Unified Energy System (Moscow branch), Moscow, 2018)
15. The White House, "Fact Sheet: President Obama Highlights Vision for Clean Energy Economy", April 22 (2009)
16. Get out of the dusk: how they look for light sources in the Crimea and cope with the power blockade, TASS, https://tass.ru/ (2021)
17. Will pay off quickly. In Crimea, another solar station began to generate electricity, Rossiyskaya Gazeta. 06.08.2015, *https://rg.ru/* (2021)
18. Energy and industry of Russia, *https://www.eprussia.ru/* (2021)
19. In Crimea, they plan to develop alternative energy, *https://ria.ru/* (2021)
20. M. A. Starchenko, I.A. Kochetov, T.S. Starchenko, V.V. Kuvshinov, Power Plants and Technologies, **6(1)**, 93 (2020)
21. T.I. Andrenko, S.V. Kiseleva, Yu. Yu. Rafikova, Bulletin of Agrarian Science of the Don, **3(39)**, 63 (2017)
22. Christian monitor, September 10 (2010)
23. The New York Times, October 6 (2010)
24. Ya.P. Silin, E.G. Animitsa, N.V. Animitsa, Ural macroregion: large cycles of industrialization (Publishing house of the Ural State Economic University, Yekaterinburg 2019)
25. N.V. Novikova, E.V. Strogonova, Economic revival of Russia, **4(66)**, 68 (2020)

Role of higher education in sustainable development of regions

Zhanna Gardanova[1]*, *Natalya* Nikitina[2], *Veronika* Grebennikova[3], and *Vyacheslav* Ilgov[1]

[1]Department of Psychotherapy, Pirogov Russian National Research Medical University, Ostrovityanova str. 1, Moscow, 117997, Russian Federation
[2] Research Institute for Advanced Directions and Technologies, Russian State Social University, 4 Wilhelm Pieck str., bld. 1, 129226 Moscow, Russian Federation
[3] Faculty of Pedagogy, Psychology and Communication, Kuban State University, Stavropol str. 149, 350040, Krasnodar, Russian Federation

Abstract. The role of higher educational institutions (HEIs) in sustainable regional development is becoming a subject of attention of researchers and policymakers. Apart from clear inputs such as sustainable education and promotion of renewable energy sources, HEIs can act as partners and consultancies for the local authorities helping them to embark on the path of sustainable economic and social growth. Our paper looks into the role of higher education in sustainable development of regions. We identify HEIs as the hubs of knowledge and know-how spill-overs as well as active players in industry-university connection. Our results confirm the importance of higher education in promoting the decarbonization of economy, helping the transition to the renewables and the creation of wider social acceptance of sustainable development.

1 Introduction

There are many environmental problems cannot be understood by isolating them from the political and economic context in which they arise, even if they create situations of environmental injustice [1]. Resources with ecological sustainability are usually identified and characterized by their role in the socio-economic and social conditions of the environment. The development capacity of the territory is based on the different development and use of the land, taking into account the potential of different types of use (e.g. development, type of use and type of use) and assessing the appropriateness of their use [2]. This can provide for scientific planning of development use and identify problems and risks of current use, so that high-quality development can be promoted. Relative sustainability is assessed in order to promote the relative degree of use of resources in relation to their ecological sustainability, thereby promoting the harmonious development of man and nature. Natural resource development is governed by a complex regulatory system that includes land claims, land-use planning, and land management policies. A comprehensive land claim agreement contributes to environmental protection by setting out the environmental and cultural importance of the land for the community and initiating a land use and planning

* Corresponding author: zanna7777@inbox.ru

https://doi.org/10.1051/e3sconf/202125004005

process [3]. The protection of this environment is important for the inhabitants who depend on water and land for their livelihood and economic development. Much of the land occupied by indigenous peoples is in ordinary ownership, but only a fraction of that land is recognized by many governments as formally and legally indigenous. There are precarious land rights, and even if indigenous territories and lands are recognized, external parties and their ability to exploit natural resources is weak. This threatens cultural survival and vital knowledge systems that contribute to the biodiversity and environmental health on which we all depend. Promoting the protection and promotion of the rights of peoples on their land and territory and the development of sustainable livelihoods and economic opportunities for their communities – it is to promote the development of mechanisms to prevent the exploitation and exploitation of biodiversity for commercial and industrial purposes and to protect the rights of indigenous peoples on their land and territory [4, 5]. Integrating the principles of sustainable development and livelihoods and economic opportunities for peoples into the framework of international law. This might help to develop a better understanding of how the relationships between society, nature, territory and governance play out.

In order to help the sustainable development of regions, one needs to learn how to assess environmental impacts in terms of the social, economic, social and political conditions of the territory and its governance [6]. For example, in Canada there is the Environmental Assessment Act (CEAA) which applies to certain other state-regulated areas, including offshore waters. Canada's three territories are northern communities that need long-term economic growth [7]. Based on a unique co-management approach based on the principles of sustainable development, environmental responsibility and the sustainable management of natural resources. Such a global analysis is then complemented by a local value chain that takes into account social and economic indicators developed at the local level. United Nations Habitat and UNIDO have combined their efforts and expertise into a methodology that not only explains the industrial performance of strategic sectors, but also takes into account the social, environmental and social conditions of the areas and their communities [8]. By bringing together global and local assessments in a single framework, the initiative aims to boost industrial growth in Mexico by developing territorial approaches to properly meet and achieve the United Nations Sustainable Development Goals (SDGs). Modern tactics of territorialization usually aim to recognize local territoriality, to include it in the discourse, to discipline it, and to integrate local norms, practices, and discourses into the mainstream. As a participatory strategy, it makes use of a managed neoliberal multiculturalism and recognizes the need to integrate and marginalize locally existing territories. Such methods also allow specific characteristics to be identified which, by virtue of their specific characteristics, could be included in the production phase by the different regions and municipalities of a country and their countries.

2 Difference in regional development across regions

Regional development has many interesting implications. For example, GDP per capita of the European Union New Member States (i.e. those that joined the EU in 2004) has increased and that these countries are therefore catching up with the older EU countries (EU-15) in this area [9]. Moreover, the European Commission's Economic and Social Research Office beta convergence analysis suggests that there is a narrowing gap between the CEE countries and the rest of European Union [10]. At the same time, some studies show that the performance of the CEE countries in this area is consistent with that of the rest of Europe in terms of GDP per capita. Regions with high absorption of EU funds also show a correlation with development indicators, as observed in the performance of knowledge and innovation potential. On the other hand, the region with the highest borrowing by the European Union shows a negative correlation between development performance and the level of investment in research and development [11]. Knowledge, innovation and potential performance is also higher in the CEE than in other regions of Europe. This can also be used to make comparisons with fast-growing emerging markets and oil-producing countries such as the Middle East and Africa (MENA countries). Various indicators are used to show which countries have the highest current levels of development and which have made the most progress recently. One also needs to assess whether a country is capable of translating the current level of development and recent progress into the well-being of its population. Finally, one needs to take a look at the dimensions of socio-economic development that distinguish themselves from the leading countries [12]. Recent progress is measured as a measure of how much a country for which data is available has achieved in the past decade. Long-term sustainability is assessed as a factor that helps to continue improvements in each of the ten dimensions of socio-economic development into the next generation.

One can measure the impact of socio-economic development on the well-being of the population, measured by the number of people who have access to health, education, housing and other basic needs. A coefficient greater than 1 indicates that a nation's standard of living is higher than one would expect from a given nation with a GDP, while a coefficient smaller than -1 indicates that it is closer to what one would expect from a giving nation without a GDP [13]. The concept of path dependence is also used to describe regional economic development paths, taking into account historical economic and political legacies. Moreover, a dependent process that is believed to shape regional innovation systems in what is now considered a place can then be based on a development model. Although urbanization does not necessarily lead to economic development, other factors can play an important role in boosting economic growth and raising living standards. For example, the presence of built-up areas such as parks, schools, hospitals, and other public facilities associated with urbanization can help to stimulate economic growth and development. Recent cross-border evidence also shows that the potential of urbanization to foster growth probably depends on the ability to remove barriers to growth, such as lack of access to affordable housing and public transport. This relative pattern does not, however, alter the fact that developing countries have higher levels of urbanization than developing countries [14]. To explain the differences between countries, other factors must be considered that summarize how nations differ in terms of economic growth and living standards, and in levels of poverty and inequality. Moreover, the share of urban populations in countries worldwide exceeded 10% in 2011, and those with higher urbanization in this group have higher per capita GDP. Developing countries face a dilemma: they have no choice but to implement accelerated urbanization to catch up with their neighbours in terms of economic growth and living standards, but they also have their own challenges in achieving these goals.

3 Triple Helix strategy in regional development

Applying governance innovations to stakeholder-based city systems is a relatively new interdisciplinary line of thought that goes beyond the institutional relationships and aims to provide governance's guidance information to address the complexity of a wide range of organisations, with a focus on interaction between governance, governance systems, institutional structures and governance models [15, 16]. The Triple Helix model draws attention to the role of governance innovation in the development the systems [17]. The contributions of this research aim at assessing the impacts of a range of topics, including governance, governance systems, institutional structures and governance models. In general, it stems out that the central element in the production of knowledge and innovation is considered to be an irreplaceable source of knowledge for human survival. The Smart Specialisation Strategy, to which Smart Growth Europe contributes, introduced by the European Commission's Office for Science, Technology and Innovation (ETSI) and the

European Union (EU), has highlighted that the innovation process is increasingly seen as an open system in which different actors work and interact [18].

Environmental and biodiversity protection promote knowledge and innovation towards a sustainable social economy in which all actors are responsible for developing sustainable, sustainable and sustainable solutions to the challenges of the future. This will be essential to promote long-term innovative strategies [19]. This paradigm shift, which directly involves the formulation of strategies for end-user innovation, is changing the role of actors in the innovation process. This has changed the roles of all actors in the innovation process and, according to the hypothesis of this model, the development of an innovation strategy requires the participation of civil society in the planning process through a hypothetical model (see Figure 1).

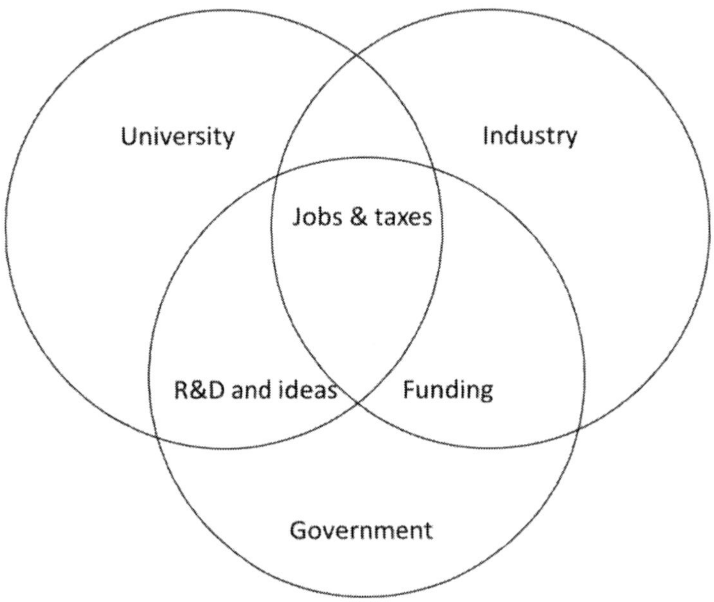

Fig. 1. Triple Helix model of innovation between universities, government, and industry

Figure 1 explains how the universities provides research and development as well as ideas that can be implemented into practice. The government provides the funding that helps the universities to function and for the ideas to be implemented into real life. The industry provides jobs and pays taxes that accumulate the budgets of the governments and the whole cycle goes over and over again. Further steps can be taken with the Triple Helix model that helps to implement this innovation model in rural areas. The last, but not the last, contribution opens up the possibility of elevating this concept and illustrates the role of civil society in the development of a multi-helix strategy for rural development. Using data-driven analysis, some researchers conducted a series of interviews with local leaders and entrepreneurs [20, 21]. The focus was on governance and network orchestration within and outside the metropolitan region. The European programme reopens the debate on the old link between innovation and territory by targeting the technology of regional and urban innovation systems. This requires an integrated understanding of network orchestration, governance and governance systems within and outside the metropolitan region. The so-called Smart City, which is still a blurred concept, is being developed within inclusive approach that consists of high technology [22]. In the transition from an industrial innovation economy, strategic management as a method of university management is a new area of research in this changing environment. The following collection of papers summarizes the results of a series of studies in the field of strategic planning of territorial development. It has been shown that the main problem is when universities, business and government actively interact. Such an approach should be based on a consensual - long-term - strategy for territorial development. Moreover, such a strategic alliance includes a strategy aimed at developing and exploiting resources in specific characteristics of the region. To this end, industry must be supported and fully supported in development, and governments can help to organise the development of the economy, infrastructure, education, health and social services and other sectors. So far, we have seen that centralisation of research and finance, with the state acting as a levy-generating authority, has roughly explained Australia's export success. However, it was a success that Triple Helix coordination was able to prevail through common consensus and common long-term strategies. The importance of decentralised coordination in the development of the Tri-Helix strategy and its implications for the future of regions remains a very relevant issue. A panel on Triple Helix best practices presented a number of published Triple Helix sector cases that show that the company-led Triple Helix constellation is an emerging practice. With this understanding, civil society is able to demand and take a more active role in the development and implementation of the Tri-Helix strategy. Further details on the cases can be presented, highlighting the importance of decentralised coordination and the role of public-private partnerships in this process, as well as the need for greater transparency. Entrepreneurship broadens this broader vision by mapping the role of entrepreneurship in regional development and its impact on the global economy and society.

4 Role of HEIs in regional development

Higher education institutions must also take the lead in developing and implementing regional economic plans that take into account the needs of the local economy, particularly in the areas of cluster development and economic development [23, 24]. Universities influence economic growth because they are buyers of goods and services from the region. Data on the number of universities and their growth rates can be used to analyse the impact of university growth on regional economic development and growth in the regions. It turns out that the expansion of higher education during that period was not only a product of the rich, but also helped to spur economic growth around the world. We believe that there is a strong correlation between the number of universities and the rate of economic development in the region [25]. The value of an institution and its environment is limited if it contributes only to the level of economic activity, but even if it contributes to economic growth in the region, it is limited. This could be a good reason for governments to spend limited resources on establishing or subsidizing local universities and colleges. However, a careful assessment of the scale and scope of these benefits is essential to allow a better understanding of these investments and the high cost of such investments.

As societies increasingly become drawn into the maelstrom of globalization, virtually all higher education systems in the world are forced to limit the scope of their activities. These activities range from the creation of impressive science and technology parks with the aim of establishing closer links with industry, to the establishment of new universities and colleges with a focus on science, technology, engineering and mathematics. As the epitome of the knowledge institutes of modernity, universities are at the forefront of this development and have been identified as a fundamental pillar of the competitiveness of nations and regions.

Despite renewed attention to this issue, little scientific attention has been paid to the roles played by universities in relatively remote regions in their local environment. This knowledge gap needs to be addressed by comparing a number of cases of institutions in each region with their role in knowledge development - intensive goods and services in different regions [26]. It is crucial to examine the role of highly educated individuals in the development of knowledge - intensive goods and services in different regions. In times of global crises, it can be observed that it is necessary to propose sustainable solutions to human development, as this concerns adequate access to education, health care, social services and other basic human needs. Another key objective would be to examine best practices in this area in relation to the use of higher education as a tool for development and the implementation of sustainable development.

In addition, we need to understand how regional innovation strategies can be developed to achieve the highest level of competitiveness in the context of a region's economic development and growth. This approach to regional development understands the need to develop an entrepreneurial attitude in the process of creation in order to manage regional growth. Before discussing the role of entrepreneurial universities in leading regional skills development strategies, we must point out that today the impression of the entrepreneurial university as a strategic initiative is gradually being recognised as an important part of innovation and regional growth networks for entrepreneurial growth. Nevertheless, there are efforts to protect certain loopholes in regional innovation systems, given the strategic role of entrepreneurial colleges and universities. Therefore, the entrepreneurial activities of universities can be provisionally identified as their main activities for the development of knowledge - based societies and competitiveness, as well as entrepreneurial contexts, including innovation networks. In the context of the regional strategy, the creative development process of the region must have a global charisma, where its application is regional and its application regional. In order to make it more relevant to the development of the regional economy, it is necessary to link the regional level with the national level and, in particular, with the global level. Although the study is understood within a limited time frame, the paper discusses an effective approach to regional entrepreneurial activities that creates awareness and innovation and has economic value for the province. Therefore, on the basis of the above-mentioned discussion and understanding, it is imperative to mention that the region must develop its territorial knowledge and growth, which fosters an innovative attitude to leadership and governance, i.e., promote industrial competitiveness. We therefore can observe that the potential of a region to apply knowledge can be harnessed when the efficiency of the regional innovation ecosystem is determined by effective cooperation and policy approaches. Improving entrepreneurial innovation and attracting the entrepreneurial spirit of this ecosystem will improve knowledge production in the regions. It can therefore be said that the entrepreneurial universities and their strategic initiatives can play a crucial role in promoting the entrepreneurial spirit of the region by providing the necessary guidance and services in a rational manner. The entrepreneurial education process and approach will enable people not only to learn about entrepreneurship, but also to develop the skills needed to guide the growth of new entrepreneurial ventures. There are the initiatives being discussed by the business universities and the important role that regional innovation strategies can play in strengthening innovation-driven regional economic growth. To address the fundamental characteristics of the region's entrepreneurial culture and its role in regional development, it is necessary to highlight the emerging perception that there is a reason for change - a new mindset and approach to entrepreneurship development. The emerging concept of the Entrepreneurship University creates regional advantages not only in terms of economic growth, but also in terms of developing entrepreneurial skills and qualifications [27]. In the age of modernisation, innovation has been observed to lead to regional growth and networks for entrepreneurship, and regions are born of knowledge - innovation linked to regional development strategies. To meet the need for innovation, aby region's oriented regional university for growth and entrepreneurship needs to look at the role of universities in regional economic development and innovation development. Where the preliminary description refers to the following guidelines, which can be considered stepping stones for the innovation-led region - growth building and the business network. On the basis of the above-mentioned discussion, it is crucial to stress that the deliberate activities of entrepreneurial universities affect regional economic development. Where we expect the region's strategic successes to enable it to face undesirable challenges and prepare to move into a better position in the global market for innovation and innovation, led by regional growth.

Overall, it is clear that high-performing regions have organised themselves to support innovation, entrepreneurship and sustainable development, and the impact this has on their surrounding business communities. The places people have studied that are doing well appear to be doing so through an integrated approach to economic development that takes into account the role of entrepreneurial universities in the development of their economies.

5 Conclusions

The aim of this paper was to assess and differentiate the impact of different types of institutions on regional and global economies with respect to their sustainable development and growth. We showed that higher educational institutions (HEIs) are playing an increasingly important share in sustainable regional development. HEIs can act as partners and consultancies for the local authorities helping them to follow the path of sustainable economic and social growth. It appears that HEIs can be identified as the hubs of knowledge and know-how spill-overs as well as active players in industry-university connection. We find that higher education can also be extremely helpful in promoting the decarbonization of economy, helping the transition to the renewables and the creation of wider social acceptance of sustainable development.

With regard to our findings, one can note a growing interest in contributing to the United Nations Sustainable Development Goals (SDGs) by training students at HEIs all around the world. Through an ongoing review of higher education trends, one can exploring how leading higher education institutions can contribute to much-needed social change beyond their classrooms. Some further research can be useful to explore how leading universities and institutions can contribute to social and economic development.

References

1. S. Banzhaf, L. Ma, C. Timmins, Journal of Economic Perspectives, **33(1)**, 185-208 (2019)
2. H. Dkhili, Marketing and Management of Innovations, **3**, 333-344 (2018)
3. A. Dale, K. Vella, S. Ryan, K. Broderick, R. Hill, R. Potts, T. Brewer, Land, **9(7)**, 234 (2020)
4. G. Magni, European Journal of Education, **52(4)**, 437-447 (2017)
5. B. Chen, Z. Qiu, N. Usio, K. Nakamura, Sustainability, **10(8)**, 2896 (2018)
6. O. Voronkova, V. Yankovskaya, I. Kovaleva, I. Epishkin, I, Iusupova, Y. Berdova, Entrepreneurship and Sustainability Issues, **7(1)**, 662-673 (2019)
7. L. McCallum, C. Ollson, I. Stefanovic, Impact Assessment and Project Appraisal, **36(1)**, 5-15 (2018)
8. M. Naidoo, A. Gasparatos, Journal of Cleaner Production, **203**, 125-142 (2018)
9. W. Strielkowski, F. Höschle, Technological and Economic Development of Economy, **22(4)**, 617-630 (2016)
10. M. Borsi, N. Metiu, Empirical Economics, **48(2)**, 657-681 (2015)
11. S. Anton, A. Nucu, Renewable Energy, **147**, 330-338 (2020)
12. B. Moldan, S. Janousková, T, Hak, Ecological Indicators, **17**, 4-13 (2012)
13. Y. Bilan, H. Mishchuk, I. Roshchyk, I. Kmecova, Journal of Competitiveness, **12(1)**, 22-38 (2020)

14. Y. Ji, X. Guo, S. Zhong, L. Wu, Land, 9(12), 481 (2020)
15. T. Clauss, A. Moussa, T. Kesting, International Journal of Technology management, **77(1-3)**, 109-144 (2018)
16. J. Abrham, W. Strielkowski, M. Vošta, J. Šlajs, Agricultural Economics, **61(10)**, 450 (2015)
17. M. Scalia, S. Barile, M. Saviano, F. Farioli, Sustainability Science, **13(5)**, 1235-1244 (2018)
18. P. Palos-Sanchez, Harvard Deusto Business Research, **6(2)**, 116-132 (2017)
19. W. Leal Filho, S. Tripathi, J. Andrade Guerra, R. Giné-Garriga, V. Orlovic Lovren, J. Willats, International Journal of Sustainable Development & World Ecology, **26(2)**, 179-190 (2019)
20. W. Strielkowski, M. Krejcí, I. Čabelková, Business: Theory and Practice **16(3)**, 304-315 (2015)
21. R. Purbasari, C. Wijaya, N. Rahayu, Review of Integrative Business and Economics Research, **9**, 157-175 (2020)
22. S. Edge, K. Boluk, M. Groulx, M. Quick, Cities, **96**, 102478 (2020)
23. M. Dvorak, P. Rovny, V. Grebennikova, M. Faminskaya, Terra Economicus, **18(4)**, 78–96 (2020)
24. Z. Gardanova, V. Ponkratov, N. Kuznetsov, N. Nikitina, O. Dudnik, E. Latypova, S. Shcherbatykh, Journal of Open Innovation: Technology, Market, and Complexity, **6(4)**, 1–21 (2020)
25. A. Valero, J. Van Reenen, Economics of Education Review, **68**, 53-67 (2019)
26. A. Beer, S. Ayres, T. Clower, F. Faller, A. Sancino, M. Sotarauta, Regional Studies, **53(2)**, 171-182 (2019)
27. M. Klofsten, A. Fayolle, M. Guerrero, S. Mian, D. Urbano, M. Wright, Technological Forecasting and Social Change, **141**, 149-158 (2019)

The role of the university in achieving SDGs 4 and 7: a Ukrainian case

Artem Artyukhov[1], *Iurii* Volk[2], *Tetyana* Vasylieva[3], and *Serhiy* Lyeonov[4]

[1] Department of Marketing, Institute for Business, Economics and Management, Sumy State University, Rymskogo-Korsakova str. 2, 40007 Sumy, Ukraine
[2] Department of Applied Mathematics and Complex Systems Modeling, Faculty of Electronics and Information Technologies, Sumy State University, Rymskogo-Korsakova str. 2, 40007 Sumy, Ukraine
[3] Academic and Scientific Institute of Business, Economics and Management, Sumy State University, Rymskogo-Korsakova str. 2, 40007 Sumy, Ukraine
[4] Office of the First Vice-rector of Sumy State University, 2, Rymskogo-Korsakova str. 2, 40007 Sumy, Ukraine

Abstract. This article deals with the development of the university activities model to ensure the quality of education and technology transfer (SDG 4) and to create a successful background in achieving SDG 7. We conduct an analysis of the literature on the achievement of SDGs through governmental instruments of influence, as well as the creation of effective university model ensuring the education quality and technology transfer. In addition, we demonstrate the role of the government in financing higher education in Ukraine. We also present the evaluation of the popularity of educational programs related to energy efficiency and a "clean" environment in Ukrainian universities after the adoption of the 2030 Agenda for Sustainable Development. The indicators of the use of various energy types by Ukraine are assessed, and a comparative analysis of these indicators with the data for EU-28 is performed. Based on the data presented, conclusions are drawn about the need to improve the quality of university training and to activate the university technology transfer. Successful practical approaches that can be of use for different higher education institutions are outlined. A combined model of quality assurance and technology transfer at the university is presented. An algorithm for testing the model and assessing the technological level of readiness for implementation is developed. The presented chain "combined model – model testing algorithm – determination of the technological level of readiness" is also assessed from the point of view of socio-economic impact on the main stakeholders. The use of the combined model lays background for the university's own sustainable development as a tool for achieving SDGs.

1 Introduction

The focus from United Nations Millennium Development Goal 7 "Ensure environmental sustainability" was shifted in 2015 towards Sustainable Development Goals 7 "Affordable and clean energy" and 12 "Sustainable consumption and production" stated by the 2030 Agenda for Sustainable Development. Expansion and detailing of sustainable development

https://doi.org/10.1051/e3sconf/202125004006

goals indicates both the emergence of new global problems and the fact that the certain tasks of The Millennium Development Goals have not been resolved. This is evidenced by the active discussion of "clean" ("green") energy problems and measures to increase energy efficiency. The International Energy Agency stated that COVID-19 outbreak tends to wipe out demand for fossil fuels, causing a market crash seven times greater than that seen during the global financial crisis. "The most severe plunge in energy demand since the second world war would trigger multi-decade lows for the world's consumption of oil, gas and coal while renewable energy continued to grow. The steady rise of renewable energy combined with the collapse in demand for fossil fuels means clean electricity will play its largest ever role in the global energy system this year, and help erase a decade's growth of global carbon emissions" [2].

The International Energy Agency also analyzes [3] the role of energy efficiency in improving government stimulus packages during the COVID-19 crisis. The meeting held on 21st of April 2020 covered the topic of securing long-term improvements in energy systems, the potential for energy efficiency to create jobs and to accelerate economic growth. The Global Commission for Urgent Action on Energy Efficiency provided recommendations for governments on how they can effectively accelerate energy efficiency through policy action.

Sustainable Development Goal 4 "Quality Education" is another sustainable development goal formulated by the 2030 Agenda for Sustainable Development, which authors believe to be decisive in the formation of energy efficient and "clean" global landscape. It allows creating the basis for the formation of responsible energy consumption culture and advocates the search for new energy sources. In addition, the specified Sustainable Development Goal should ensure a successful transfer of knowledge and technology to the real economy sector.

Working to achieve SDGs requires each country to independently develop its own sustainable development strategies, plans and programs. Such programs should be developed in centers of concentrated scientific knowledge, which are universities. Designing the sustainable development plans is impossible without the participation of industry representatives acting as the main university partners. In addition, the program development process has to include consideration of the views of other stakeholders, such as alumni.

2 Literature review

We should mention a number of scientific studies that focus on the university role in achieving SDGs. Paper [4] proposes a comprehensive management model for a sustainable university consisting of four strategic management process stages. The model is derived relying on data from a number of American (both continents), European, Asian and Pacific universities. In terms of strategies for fostering sustainability, namely education and research, this model requires clarification taking into account the peculiarities of the functioning universities in Eastern Europe and former USSR countries.

A separate role is assigned to the opinion of students during the modeling of university as a center for sustainable development [5]. The survey of these stakeholders in [5] was focused only on the public liberal arts (where technology transfer is not a priority in the university activities) and was devoted to the description of source and message channels for receiving sustainability information, and not to the study of mechanisms for ensuring education quality and technology transfer. Another attempt to carry out the survey the student perception of SDGs in both professional and personal way was made in [6]. Based on the analysis carried out in the paper, authors derived practical implications intended to battle student ignorance regarding SDGs. However, the detailed plan and means to establish comprehensive quality educational system were out of discussion. Works [7, 8] provide a general overview of the methods used to implement the concept of a "sustainable university", including communication with external stakeholders.

A multi-country study identifying the level of importance given to desired competences on sustainable development by teaching staff at a number of higher education institutions was presented in [9]. Despite some limitations, authors outlined practical recommendations for teaching practices regarding sustainable development.

The relationship "university – SDG 7" is of particular interest. The existence of such a relationship is clearly defined by "THE Impact Rankings 2020 by SDG: affordable and clean energy (SDG 7) methodology" [10]. In particular, mentioned Rankings define the following indicators: "Plan to reduce overall energy consumption" and "Reviews to identify areas where energy waste is highest". These indicators stimulate the university to conduct the technology transfer from science to production and to use its own developments within the university. The Systematic Literature Review on this issue was carried out in [11] with the definition of the main accents (directions) in the university activities to comply with SDG 7 targets in their functioning and in the creation of new products for industry.

The national cases, e.g. Spanish [12] are particularly informative from the point of view of adaptation for Ukrainian universities. In [13] authors develop a process framework for including education for sustainable development into quality management systems in higher education. This work represents a valuable example of sustainability implementation in higher education in form of quality assurance systems on managerial level to secure and improve teaching and learning. However, it focuses on a limited subset of 11 universities in Nordic countries leaving the room for development of frameworks aimed to address SDG 4 tasks in other countries.

We should also note research works in various applications related to sustainable development and quality of education: financing of educational activities [14]; innovative potential of the national economy [15]; state regulation of educational processes as a tool for sustainable development [16-18]; the development of various educational areas in the regions [19-26]; lifelong learning and its impact on the development of the region [27], some financial issues and risks in particular for Africa and India [28-31]. The presented studies are of practical interest for building the model of the university activities, taking into account regional characteristics and mechanisms for describing individual elements in the quality education system.

Separately, it is necessary to highlight a selection of articles that are devoted to innovation as an integral attribute of sustainable development. So, in [32], special attention is paid to innovative tools for achieving high ratings. The works [33, 34] describe the relationship between education, science and innovation and offer mechanisms for optimizing the algorithms for the operation of these systems.

It should be noted that the effective implementation of the main tasks of the university activities is possible only under stable and sufficient funding. Despite the fact that in absolute terms the share of higher education funding in Ukraine is one of the lowest in Europe, this indicator is very high in percentage terms as shown in Table 1. This fact gives hope for the prospect of "strengthening" universities in terms of ensuring the higher education quality and successful technology transfer.

Current paper represents an attempt to describe the university activities in terms of higher education quality assurance and technology transfer in application to achieve the main results of SDGs 4 and 7.

It should be noted that SDG 4 is one of the leaders by mention frequency (third place) [36] in five main reports on sustainable development goals in 2019 [37-40]. Therefore, the relevance of the problem of the achievement of other SDGs through the successful overcoming of the main challenges of SDG 4 presented in this paper is confirmed.

Table 1. Expenditures on higher education (by budget expenditures as a percentage of GDP) [35]

Country	2013	2014	2015	2016	2017
USA	1.34	1.36	1.37	1.21	-
United Kingdom	1.34	1.38	1.28	1.41	-
Germany	1.31	1.31	1.25	1.25	-
France	1.24	1.25	1.25	1.23	-
Poland	1.21	1.18	1.22	1.06	-
Republic of Belarus	0.88	0.81	0.80	0.80	0.79
Georgia	-	-	-	0.42	0.39
Ukraine	2.13	1.85	-	1.53	1.35

Achievement of the stated goal is possible through the following tasks solution:

1. Data analysis on the training of specialists in fields of energy management and production, energy efficiency, ecology, environmental protection technology.
2. Development of a set of measures to improve the quality of education in order to train graduates competitive in the labor market.
3. Analysis of the functioning university technology transfer centers activity and their role in ensuring a successful relationship between science and industry.
4. Development and selection of an algorithm for testing the functioning model of the university as a training center for specialists in fields of ecology, energy efficiency and "clean" energy and as a coordinator of technology transfer in these areas.

The proposed model of the university activity is based on the regulatory framework, algorithms and mechanisms for the functioning of the education ensuring system and technology transfer in Sumy State University. Authors' choice of the base university for the model description was motivated by following factors:

1. The presence of the university in established international rankings [41-43].
2. Confirmation of the higher education quality system effectiveness based on external assessment according to the National Agency for Higher Education Quality Assurance (Ukraine) (https://en.naqa.gov.ua).
3. The efficiency of technology transfer implementation through the commercialization of scientific developments and financial indicators of technology transfer according to the Ministry of Education and Science of Ukraine (https://mon.gov.ua).
4. Active presence of projects in the field of energy efficiency and "clean" energy.
5. Implementation of energy efficiency ensuring instruments in its own activities.

The combination of high rating indicators with the effective implementation of technology transfer provides the prerequisites for the development of a successful model of the university's activities and the implementation of the targets set within the SDG 7 [44]:

"**7.1** By 2030, ensure universal access to affordable, reliable and modern energy services.

7.2 By 2030, increase substantially the share of renewable energy in the global energy mix.

7.3 By 2030, double the global rate of improvement in energy efficiency.

7.A By 2030, enhance international cooperation to facilitate access to clean energy research and technology, including renewable energy, energy efficiency and advanced and cleaner fossil-fuel technology, and promote investment in energy infrastructure and clean energy technology".

3 Research methodology, results and discussion

3.1 Educational programs popularity assessment

Principal task of data analysis on the training of specialists in fields of energy management and production, energy efficiency, ecology, environmental protection technology is addressed by the use of open datasets provided by Ministry of Education and Science of Ukraine on the Open Data Portal (data.gov.ua). To monitor the actual state of affairs regarding training of relevant field specialists' authors analyzed the dataset containing full information on admissions to the higher education institutions since 2012 to 2021 licensed under Creative Commons Attribution 4.0 International license. The scope of analysis was reduced to admission of fulltime bachelor, masters and PhD level entrants on the governmental and commercial funding basis since 2016 (explained by emerging of SDGs in late 2015).

The dynamics of specialties "Ecology" and "Environmental protection technology" admission percentage to total number of government-funded and commercial entrants per year (2016 to 2020) is represented by Figure 1A. Line 1 represents the dynamics of fulltime admissions on governmental funding basis, Line 2 – fulltime commercial admissions. The peak percentage values appear to be small relatively to the highly popular specialties like "Computer Sciences" (~8% of government funded admissions in 2020) or "Law" (~4% of government funded admissions in 2020). However, one can clearly witness that the number of governments funded entrants to "Ecology" and "Environmental protection technology" tends to grow from year to year indicating increasing interest in mentioned specialties. As for commercial entrants, Line 2 indicates stable number of fulltime commercial admissions to the specialties under consideration. Further application of governmental support (e.g. increasing number of available funded places) on mentioned fields of education may result in significant impact on fulfilling SDG 4 principles by forming a highly trained specialist community motivated to address environmental problems.

Figure 1B displays dynamics of specialties "Electroenergetics, electrotechnics and electromechanics", "Energetic engineering", "Atomic Energetics", "Heat Energetics", "Hydroenergetics" admission percentage to total number of government-funded and commercial entrants per year (2016 to 2020). Line 1 represents fulltime entrants on governmental funding basis, Line 2 — fulltime entrants on commercial basis. We can see stable interest of entrants in the mentioned specialties in both government funded and commercial types of admission. By increasing attention to the "green" energy concepts, alternative energy sources and efficient environmental resource management in the educational programs of the mentioned specialties universities can ensure orientation of graduating specialists towards SDG 7 achievement.

Figure 1C represents the dynamics of specialties "Ecology" and "Environmental protection technology" admission percentage of total entrant number on the master and PhD level. As Line 1 illustrates, percent of government-funded entrants on higher education level of mentioned specialties tend to decrease. Line 2 presenting commercial base entrants dynamics express the same tendency, however having the peak admission number in 2019 (753 commercial entrants vs. 725 government-funded entrants). The reasons for this peak admission numbers and possible outcomes for the aspects of scientific community development and technology transfer require further study.

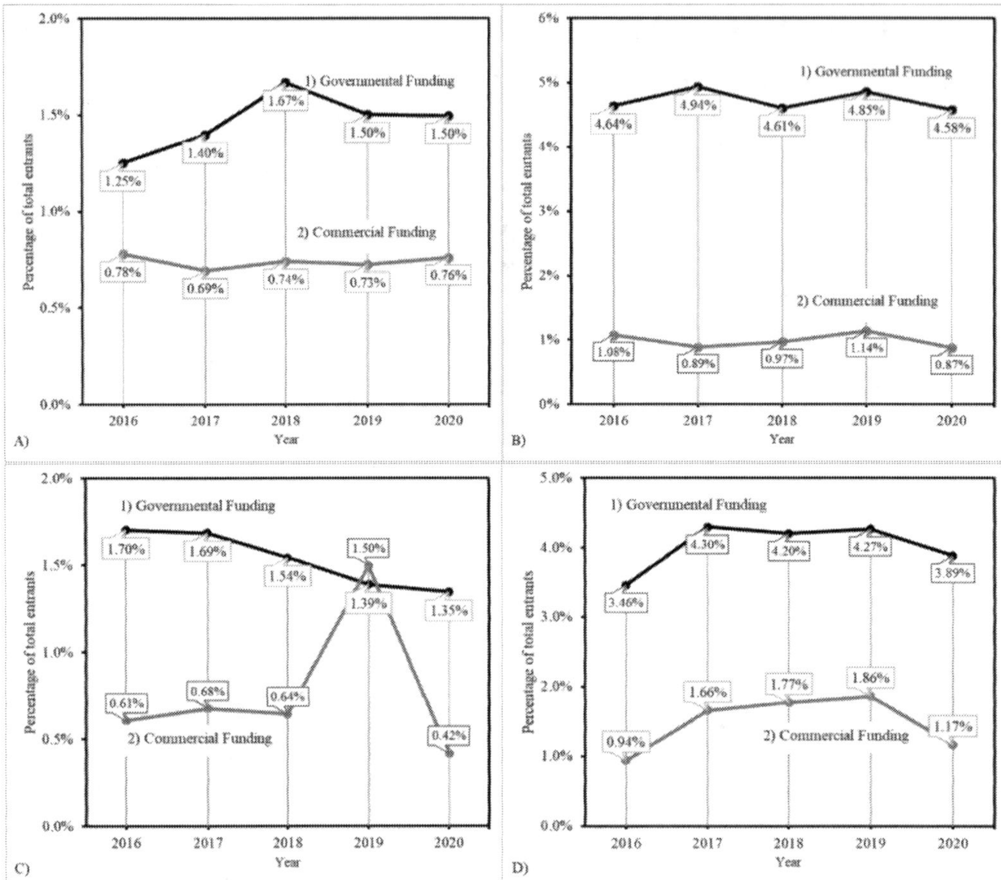

Fig. 1. Dynamics of admission percentage on SDG 7 related specialties per year

Figure 1D illustrates the dynamics "Electroenergetics, electrotechnics and electromechanics", "Energetic engineering", "Atomic Energetics", "Heat Energetics", "Hydroenergetics" admission percentage to total number of government-funded and commercial entrants per year (2016 to 2020) on the master and PhD level. Line 1 represents positions with governmental funding, Line 2 — commercial entrants. The overall character of the dynamics expresses relatively stable interest of entrants to the specialties related to energy production. Therefore, the changes to educational programs addressing tasks of SDG 7 are necessary in order to ensure energy efficient society. Master and PhD level program revisions and quality assurance are of extreme importance since graduates of these educational levels have the major impact on the formation of the comprehensive landscape of sustainable development in higher education, technology transfer and "green" energy concepts.

3.2 Indicators of various energy types use by Ukraine assessment

In order to understand the current situation regarding SDGs 4 and 7 achievement courses in Ukraine, we assess statistical data about electrical energy production in Ukraine and EU-28 provided by International Energy Agency. Using table data available at [45] in Figure 2 we present dynamics of energy production percentage by source for 2015–2019-time interval. Nuclear power plants, coal-based electricity generation, hydroelectricity and solar energy produced by photovoltaic systems were considered. As shown in Figure 2A, a major part (~49–55%) of electrical energy in Ukraine is nuclear-produced and this fact tend to stay stable throughout studied time. On the other hand, nuclear energy possesses only ~25–26% share of EU-28 energy production landscape. Authors' opinion is that ongoing process of nuclear power phase-out in EU countries is an integral part of the SDG 7 achievement and this practice should be inherited in Ukraine.

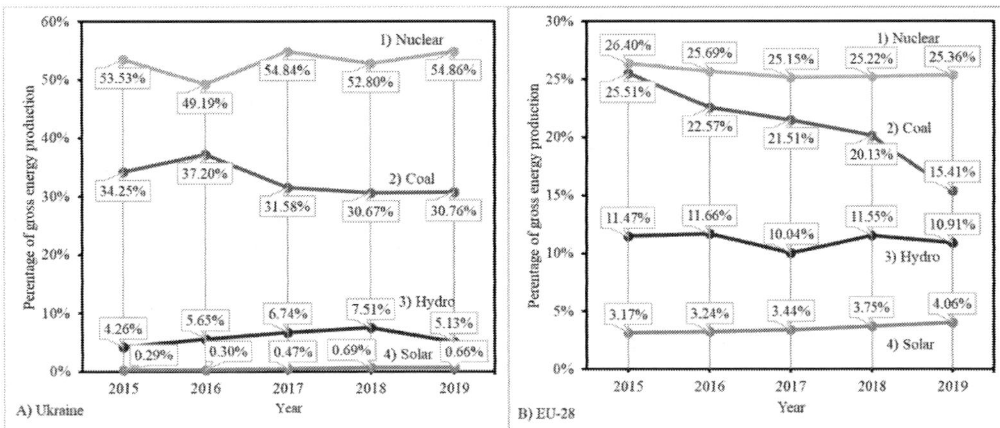

Fig. 2. Electrical energy production percentage by source in Ukraine and EU-28

A significant decrease in coal-based electrical energy production in EU-28 countries is evidenced by the dynamics of Line 2 in Figure 2B. The percentage of energy produced by coal burning decreased from 25.5% to 15.4% in European Union countries, while in Ukraine this indicator stays around 30% for years under study.

Both Ukraine and EU-28 countries express stable dynamics in hydroelectricity production. However, the share of this energy source in Ukraine does not exceed 7%, while in European countries ~11% of energy is produced by hydroelectric power plants.

The situation with solar energy raises concerns regarding SDG 7 achievement in Ukraine. The share of the electrical energy produced by solar photovoltaic process in Ukraine does not exceed 1%, while EU-28 countries express stable (yet not rapid) growth of interest to the solar energy sources by raising production share from 3% in 2015 to 4% in 2019.

As mentioned in introduction, SDG 7 comprises five key targets aimed to ensure energy efficiency concepts. Target 7.2 dealing with substantial increase the share of renewable energy in the global energy mix is at the most interest in context of current paper. As shown by analytics, the course of Ukrainian energetics stays unchanged neglecting the development and usage of energy efficiency technologies. This may be explained by various reasons, including the lack of highly trained specialists in fields of environmental protection and ecological studies. Therefore, the role of university as a platform interconnecting quality education, research sector and industry representatives emerges in a bright light.

Successful cases of scientific developments introduction in the field of energy efficiency [48-50] and environmentally friendly devices [51, 52] at domestic chemical enterprises make it possible to consider the proposed transfer model effective.

University scientists have created a number of projects that allow solving the problems of SDG 7 targets achievement both in industry and for their own needs.

The results of testing the algorithm along the chain "testing – quality control – internal quality assurance – external quality assurance" allow consistent fulfilling of the conditions for the model successful operation thus minimizing the number of testing cycles. In order to

 https://doi.org/10.1051/e3sconf/202125004006

optimize verification process, it is necessary to create a detailed technical task and this role is assigned to the university regulatory framework.

Table 2. Projects aimed at SDG 7 achievement in industry and at the university (according to [53])

Title	Short description
Jet thermal transformer for municipal and industrial households heating and cooling systems	The combination of direct and reversible cycles with the implementation of the principle of jet thermal compression will allow to achieve higher transformation coefficients in lowering thermal transformer. Heat sources modernization is possible by decommissioning of 60-70% unprofitable boilers and the retrofitting of other boilers by jet thermal transformer blocks (working boilers are used as working fluid heaters).
Expansion turbine (ET) – an electric power generation unit as a part of jet-reactive (JRT) and vortex turbines	JRT transforms the potential energy of compressed gas into kinetic jet energy which flows from a tractive nozzle and is at some distance from the rotation axis. Existing world analogues are characterized by relatively high capital costs that do not provide the minimum acceptable payback period. The advantages of JRT enable to develop the ET with the payback period of 1-2 years. It is cheaper, of simple technology and design, safe and easy to use as compared to the ET on the basis of the conventional paddle turbines.
Turbogenerators based on vortex turbines	The vortex turbine is structurally and technologically simple, cheap to manufacture and has a nominal operating mode at lower speeds than conventional (centripetal and axial) turbines. The last advantage of the vortex turbines makes it possible to create turbine generators in gearless design with a capacity of up to 500 kW and a payback period of 1-2 years.
Monitoring system "HeatCAM"	Monitoring system «HeatCAM» provides: monitoring of the performance of the heating network based on weather; computation of the amount of heat energy; formation of accounting documentation. Application of the system during monitoring of heat supply allows reducing the amount of heat energy consumption up to 15%.
Monitoring system "Energy clock"	The "Energy clock" device installed on the water distribution unit is the main component of the demonstration site that allows real-time demonstration of the amount of water used and its cost to the consumer. The "Energy clock" system reduces the heat energy consumption by 15%.
3rd generation N-CDS/P-SNS heterorocotion-based solar cells	The main advantage of the device is minimal content of heavy metal (cadmium) due to the use of tin sulfide as an absorbent material. SnS is a cheap non-toxic material, which reduces the cost of the device and simplifies the disposal after the expiration date.
Heat-generating unit	The warming of the liquid is carried out by forces of viscous friction, which eliminates the use of heating elements. Cost reduction of heating is 40–60%.

After successfully testing the model and receiving feedback from the main stakeholders, it is possible to assess the level of technological readiness of the model for large-scale implementation and replication to other SDGs.

4 Conclusions

All in all, our analysis of literature data showed that there is an urgent need to use effective solutions for organizing activities in achieving the SDGs. The solution of combining "quality of education" and "transfer of technologies" into a joint model is a new approach to a comprehensive solution to the problem of managing the educational and scientific potential of the university. The presented chain "combined model – model testing algorithm – determination of the technological level of readiness" is also assessed from the point of view of socio-economic impact on the main stakeholders. The use of the combined model allows for the university's own sustainable development as a tool for achieving SDGs.

The assessment made about current state of affairs regarding SDG 4 and 7 related specialties admission dynamics mark the possible routes for improvement in educational plans covering competences of environmental influence. Considering these results during the combined model development and testing may result in level of effort optimization at stages related with SDG 4 achieving and hence, other linked model elements.

At each stage of modeling and testing the system, it is possible to eliminate systemic errors in the organization of interaction between the educational and scientific spheres. This approach allows achieving of high technological level of readiness for the model implementation within a relatively short testing time. The proposed model can also be adapted to achieve other sustainable development goals. By replacing (modernizing) individual blocks of the model, flexibility is achieved in ensuring interconnection in the chain "SDG 4 – other SDGs". The proposed analogy between testing a university activity model and software product testing effectively operates in the system “quality of education – technology transfer – achievement of SDGs”. Despite limitations dictated by single university case consideration, paper provides a set of timely practical approaches that may be successfully deployed to other Ukrainian universities.

Acknowledgments

This research was funded by the grants from the Ministry of Education and Science of Ukraine “Reforming the lifelong learning system in Ukraine for the prevention of the labor emigration: a coopetition model of institutional partnership” (reg. No 0120U102001), “Convergence of economic and educational transformations in the digital society: modeling the impact on regional and national security”.

References

1. The impact of the Covid-19 crisis on clean energy progress, *https://www.iea.org/articles/the-impact-of-the-covid-19-crisis-on-clean-energy-progress* (2020)
2. Covid-19 crisis will wipe out demand for fossil fuels, says IEA, *https://www.theguardian.com/business/2020/apr/30/covid-19-crisis-demand-fossil-fuels-iea-renewable-electricity* (2020)
3. IEA brings together leading global figures to discuss the role of energy efficiency in Covid-19 stimulus packages, *https://www.iea.org/news/iea-brings-together-leading-global-figures-to-discuss-the-role-of-energy-efficiency-in-covid-19-stimulus-packages* (2020)
4. L. Velazquez, N. Munguia, A. Platt, J. Taddei, Journal of Cleaner Production, **14(9–11)**, 810 (2006)
5. E. K. Perrault, S. K. Clark, Journal of Geoscience Education, **65(2)**, 194 (2017)
6. F. Zamora-Polo, J. Sánchez-Martín, M. Corrales-Serrano, L. Espejo-Antúnez, Sustainability, **(11)**, 3533 (2019)
7. L.P. Amaral, N. Martins, J.B. Gouveia, International Journal of Sustainability in Higher Education, **16(2)**, 155 (2015)
8. S. Mosier, M Ruxton, Environment and Planning C: Politics and Space, **36(3),** 479 (2018)
9. W. Leal Filho, V.R. Levesque, A.L Salvia, et al., Sustainability Science **16**, 101 (2021)

10. THE Impact Rankings 2020 by SDG: affordable and clean energy (SDG 7) methodology, *https://www.timeshighereducation.com/impact-rankings-2020-sdg-affordable-and-clean-energy-sdg-7-methodology* (2020)
11. A. L. Salvia, L. L. Brandli, *Energy Sustainability at Universities and Its Contribution to SDG 7: A Systematic Literature Review*, (2020)
12. N. Blasco, I. Brusca, M. Labrador, Sustainability **13**, 89 (2021)
13. T. Holm, K. Sammalisto, T. S. Grindsted, T. Vuorisalo, Journal of Cleaner Production **106**, 164 (2015)
14. S. Lyeonov, O. Liuta, *Actual problems of finance teaching in Ukraine in the post-crisis period* (2016)
15. T. A. Vasylieva, V.O Kasyanenko, Actual Problems of Economics, **144(6)**, 50 (2013)
16. A. Vorontsova, T. Vasylieva, Y. Bilan, G. Ostasz, T. Mayboroda, Administratie Si Management Public, **2020(34)**, 6 (2020)
17. Y. Bilan, T. Pimonenko, L. Starchenko, E3S Web of Conferences **159**, 04037 (2020)
18. A. Vorontsova, H. Shvindina, T. Mayboroda, H. Mishenina, I, Heiets, Problems and Perspectives in Management, **18(4)**, 275-288 (2020)
19. A. R. V. Yapo, SocioEconomic Challenges, **3(3)**, 108 (2019)
20. K. S. Akpoviroro, P.M Akanmu, A. Olalekan, S. A. Alhaji, SocioEconomic Challenges, **4(2)**, 114 (2018)
21. F.F. Sapele, S.A. Idoniboye-Obu, Business Ethics and Leadership, **3(3)**, 78-87 (2019)
22. B.A. Brimah, W.D. Olanipekun, A.G. Bamidele, M. Ibrahim, Financial Markets, Institutions and Risks, **4(2)**, 34 (2020)
23. M.S. Ayad, A.A. El-Aziz, SocioEconomic Challenges, **2(2)**, 49 (2018)
24. I. Skliar, Business Ethics and Leadership, **2(1)**, 96 (2018)
25. N. Vardanyan, Business Ethics and Leadership, **1(3)**, 93 (2017)
26. M. Mazurkiewicz, O. Liuta, K. Kyrychenko, Business Ethics and Leadership, **1(4)**, 74 (2017)
27. S. Pryima, Y. Dayong, O. Anishenko, Y. Petrushenko, A. Vorontsova, Problems and Perspectives in Management, **16(3)**, 1-13 (2018)
28. P.M. Akanmu, J.Ol. Clement, M. Samaila, Financial Markets, Institutions and Risks, **2(4)**, 20 (2018)
29. H. Hadbaa, R. Boutti, Financial Markets, Institutions and Risks, **3(1)**, 92 (2019)
30. Al.K. Chakrawal, P. Goyal, Financial Markets, Institutions and Risks, **2(3)**, 28 (2018)
31. T. Bachoo, N.S.M. Ahmad, Financial Markets, Institutions and Risks, **2(4)**, 101 (2018)
32. S. Kvitka, G. Starushenko, V. Koval, H. Deforzh, O. Prokopenko, Marketing and Management of Innovations, **3**, 60 (2019)
33. S. Shkarlet, N. Kholiavko, M. Dubyna, Marketing and Management of Innovations, **3**, 126 (2019)
34. A.S. Vorontsova, S.V. Lieonov, T. A.Vasylieva, A.Y. Artiukhov, Marketing and Management of Innovations, **2**, 218 (2018)
35. Annual report of National Agency for Quality Assurance in Higher Education of Ukraine: 2019, *https://naqa.gov.ua/wp-content/uploads/2020/02/Звіт-2020.pdf* (2020)
36. World's five major SDG reports, Basel Institute of Commons and Economics, *http://commons.ch/wp-content/uploads/Synopsis_SDG_Reports_Goals_Allocation_2019.pdf* (2019)

37. The Sustainable Development Goals report 2019, *https://unstats.un.org/sdgs/report/2019/The-Sustainable-Development-Goals-Report-2019.pdf* (2019)
38. Inter-agency Task Force on Financing for Development: Financing for Sustainable Development Report 2019, *https://developmentfinance.un.org/sites/developmentfinance.un.org/files/FSDR2019.pdf* (2019)
39. General, Global Sustainable Development Report 2019: The Future is Now – Science for Achieving Sustainable Development, *https://sustainabledevelopment.un.org/content/documents/24797GSDR_report_2019.pdf* (2019)
40. Sustainable development in the European Union, *https://ec.europa.eu/eurostat/documents/3217494/9940483/KS-02-19-165-EN-N.pdf* (2019)
41. Ranking WEB of Universities, *http://webometrics.info/en* (2021)
42. THE World University Rankings, *https://www.timeshighereducation.com/world-university-rankings* (2021)
43. QS World University Rankings, *https://www.topuniversities.com/university-rankings/world-university-rankings/2021* (2021)
44. Sustainable Development Goal 7: Ensure access to affordable, reliable, sustainable and modern energy for all, *https://sdgs.un.org/ru/goals/goal7* (2021)
45. Data Tables – Data & Statistics, IEA, *https://www.iea.org/data-and-statistics/data-tables* (2021)
46. Sustainable Development Goal 4: Ensure inclusive and equitable quality education and promote lifelong learning opportunities for all, *https://sdgs.un.org/ru/goals/goal4* (2021)
47. E. Unterhalter, Global Policy, **10,** 39-51 (2019)
48. A. Artyukhov, N. Artyukhova, J. Krmela, V. Krmelová, IOP Conference Series: Materials Science and Engineering, **776(1)**, 012018 (2020).
49. A. Artyukhov, N. Artyukhova, J. Krmela, V. Krmelová, IOP Conference Series: Materials Science and Engineering, **776(1)**, 012016 (2020)
50. A. Artyukhov, N. Artyukhova, A. Ivaniia, TCSET 2018 – Proceedings, 139 – 142 (2018)
51. A. Artyukhov, V. Obodiak, P. Boiko, P. Rossi, CEUR Workshop Proceedings, **1844**, 33 – 47 (2017)
52. V. Obodiak, N. Artyukhova, A. Artyukhov, Calculation of the residence time of dispersed phase in sectioned devices: Theoretical basics and software implementation, 813–820 (In Advances in Design, Simulation and Manufacturing II. DSMIE 2019. Lecture Notes in Mechanical Engineering. Springer, Cham: 2020)
53. Information resource of innovative developments, *https://innovation.sumdu.edu.ua/?lang=en* (2021)

Green brands and sustainable entrepreneurship

Tatiana Freze[1]*, *Oksana* Nurova[1]

[1]Institute of Engineering and Environmental Security, Togliatti State University, Belorusskaya str. 14, 445020 Togliatti, Russian Federation

Abstract. Our paper focuses on the development of green brands as the part of the sustainable environmental management. It demonstrates that the main goal of the green technology is to focus on the specific interface of environmentally-friendly and sustainable technology, innovation and business development. It is building solutions that promote the development of clean energy, energy efficiency, renewable energy and efficient technologies. Moreover, it makes it clear that one should also underline the importance of sustainable transport, sustainable agriculture and sustainable urban development within this context. In addition, the paper focuses on the green marketing and environmental management as the parts of the sustainable entrepreneurship of the 21st century. It becomes clear that environmental benefits of products and services should be highlighted and advertised by business companies in order to sustain their position on the globalized markets with severe competition.

1 Introduction

In the 21st century, business companies need to refocus their business model to make the transition from the traditional to the sustainable future more efficient and sustainable [1]. This requires companies to challenge the traditional economic model and rebalance the economy away from consumption - towards growth and sustainable consumption. There is a growing movement in the business world towards sustainability as sustainability becomes increasingly important for buyers. Too often, "sustainable consumption" has been at the centre of discussion in business circles, but less so on the consumer side [2]. However, before we can make the transition as a whole society, we need to understand what sustainability is, what a sustainable business model is, what distinguishes it from the traditional economic model, and why this model is needed [3]. A key concept is to create and use "sustainable" business models through environmental and societal surplus, which means sustainability investments that have a secondary value through higher returns. We need a deeper understanding of the concept of sustainability and how itis relevant for the development of the company. Moreover, we need to understand what sustainability is, how it is compatible with the business world and the sustainability model in general. For example, agricultural banks provide farmers with preferential financing for the introduction of renewable agriculture and include net greenhouse gas sequestration in payment terms and compensation, thereby promoting farm sustainability and enabling the bank to accumulate

* Corresponding author: ntc@tltsu.ru

new assets from available carbon offsets [4]. Grain producers are creating a sustainable business model to reduce emissions by using renewable energy sources such as wind, solar, biomass and biofuels.

Beyond the supply chain, companies are looking for new opportunities in the changing market for their products and services, including a shift to a more sustainable business model, such as the use of renewable energy sources. Consumers of these companies also offer behavioural skills as new assets for the business and pay a premium for them, as we have seen. Toyota Motor is seeking a wide range of partnerships to meet its goal of moving from zero environmental impact to eliminating CO_2 emissions by 2050 [5]. Ford has developed the MyEnergi Lifestyle Program to explore ways to integrate the use of renewable energy sources such as wind and solar to reduce the overall carbon footprint. In addition to increasing the number of electric vehicles in its fleet, Ford also sees how customers would live a more electrified lifestyle overall [6].

In general, companies must now optimise the economic and social value in order to achieve overall, medium and long-term competitive returns for their shareholders. If properly defined, collective action can be empowering by limiting competitive movements in an industry and removing barriers that prevent companies from engaging in sustainability and a sustainable competitive advantage. In terms of returns on the stock market, the high-sustainability companies have abnormal stocks - market performance is significantly higher than companies on a value-weighted basis, such as in the United States and Europe. These trends, which include a shift away from the traditional business model to the sustainability model, suggest that the public is no longer satisfied with corporations that focus exclusively on short-term profit maximization. Economic islands that maximise financial value and promotion independently are becoming obsolete [7]. Not surprisingly, more and more companies are exploring how environmental, social, and governance performance can contribute to financial performance.

Currently, there are only a few organisations that commit to sustainability on a broad basis as part of their business model. By contrast, the model of a company transforming the market is based on what we call market transformation. This means that companies respond to market changes by increasing their competitive position by integrating sustainability into existing business considerations. Business integration is geared to the present - today's measures of success, and the transformation of the market will help companies to create the measures of tomorrow. Instead of waiting for market changes to incentivize sustainable practices, companies can create these changes by enabling new forms of sustainability for companies.

2 Green brands of the business companies

Sometimes known as eco-marketing, green marketing is a way to highlight the environmental benefits of using a product or service. Many companies do green branding, which means they do everything in their power to reduce their carbon footprint and be greener. To achieve a green brand equity, a company must outline its environmental ethos, adopt sustainable production methods, offer sustainable products and services (e.g. renewable energy, energy efficient vehicles, etc.) and outline and highlight environmental issues [8]. When the world learns of the harmful effects of certain practices and products on the environment and the harmful effects of greenhouse gas emissions, the image of the "green brand" will show consumers that the company is caring and modern. When a friendly business company takes the lead, consumers can expect it to be perfect from day one. As if that were not enough, they also give 1% of their turnover to ecological organizations around the world that restore forests, dismantle dams, and protect endangered species [9]. One needs to talk to an environmental expert to know that one of the best things an environmentally

https://doi.org/10.1051/e3sconf/202125004007

friendly company can do is to overcome climate change. If business companies are honest with our customers and try to consistently improve our environmental practices, people would become devoted customers of the business.

Entrepreneurs who want to explore the green branding have to learn a lot from environmentally friendly companies. Green brands are those brands that connect consumers with environmental protection and sustainable business practices. When consumers become more aware of the need to protect the environment, such brands can appeal to a wide range of consumers, from small businesses to large corporations. Green brands can improve any company's image and give its products a unique selling point and awaken consumer interest in sustainability. However, this is only a sign of green marketing; in general, it is a much broader issue that can include consumer goods, technology and services in any country [10]. Unfortunately, most people believe that "green marketing" only applies to environmental issues such as climate change and energy efficiency. That is what most consumers associate with it, but in reality, it is much more. It also includes marketing activities that stimulate and confirm the attitude and environmental practices of customers, such as the use of environmentally friendly products and services and the promotion of environmental sustainability. Green marketing is typically practised by companies that are committed to sustainable development and corporate social responsibility. It is about companies promoting their products and services in a way that reflects their environmental friendliness. This strategy is known as green marketing and there are many types of marketing, whether in the form of advertising, public relations, social media, advertising campaigns or even advertising in newspapers [11]. The concept of green marketing seems to be due to the negative impact of humans on the planet. Products and services can be sold by highlighting their environmental benefits and consumers recognise such brands. This is called eco-marketing or environmental marketing and is one of the most popular marketing strategies in the world. The eco-friendly trend has been so massive in recent years that it has spawned a number of companies, including Coca-Cola, PepsiCo, McDonald's, Nestlé and many others. Green marketing is a way for companies to sell their products and services in a way that emphasizes their sustainability efforts and environmental benefits [12]. If done right, brands that implement green marketing in their culture, strategy, and target audience, increase sales and profits, can help save the planet and appeal to their target audiences. Some companies are using this strategy to highlight how their mission is to help the environment, rather than just increase revenue.

3 Environmental management for sustainable entrepreneurship

Sustainable entrepreneurship builds on the foundations of entrepreneurship and extends it to integrate environmental and social concerns within existing companies. Sustainable economy recognizes the growing scientific evidence and growing evidence that growing social inequalities are destroying the economy, the environment and global society [13]. It recognises these changes by recognising that the consumer public is increasingly concerned about the challenges of a sustainable world. This market opportunity involves addressing sustainability concerns by providing new products and services that reduce energy and resource consumption.

Green supply chain practices can be used to improve the quality, efficiency, sustainability and cost-effectiveness of the SME production process in a given country. Sporting goods manufacturer Puma is an example of a company that successfully invests in sustainable business practices and adopts green supply chain practices that measures the environmental footprint of the company's business, including the way these raw materials are extracted [14]. It is used as a tool to raise awareness of the importance of the environment and sustainability in the economy. A senior management survey showed that sustainable development is becoming the focus of business models against the trend. According to some studies, more than a half of the companies already have a specific strategy for the environment [15]. Automotive industry can be used as an example of how a very traditional industry is facing up to the challenges of sustainability and looking to the future, which highlights the importance of sustainability in its business model [16]. The economic source must be tapped from an organizational structure based on a culture and philosophy that transforms organizational structures and manages to induce companies to promote the development of any country. As such, it is intended to support sustainable entrepreneurs in using business models to unlock the potential of their business model, for example to improve the environmental and social values created. This framework is based on theoretically sound assumptions, such as the ability to mediate between the need to solve an environmental or social problem with valuable results.

Concepts of sustainability and corporate responsibility connect stakeholders, including governments, suppliers, customers, communities and others. At the same time, it increases the need for improvements in the company's internal processes, including the integration of environmental and social responsibility into the business process and the development of a sustainable business model [17]. The fundamental motivation for this approach is that the creation of links between stakeholders in the value chain opens up new opportunities for the development of a sustainable business model and the creation of new business models. A more difficult task is to assess what we believe is necessary for sustainable entrepreneurship to succeed. Therefore, work continues to develop tools to implement and track sustainability initiatives, track and measure their performance, and integrate sustainability into strategic business processes [18]. Further instruments for sustainable entrepreneurship are the development of sustainable business models and business-to-business (B2B) relations. These are in place, but are often criticized for being difficult to implement in the context of a business model with a high degree of uncertainty and uncertainty about its long-term viability. There is business-oriented perspective that focuses on the development of sustainable business models and business-to-business (B2B) relationships. How do product design, processes and supply chains have environmental consequences that can improve or impair a company's economic performance? How do managers create value by making sustainable business models and business relationships (B2B) more sustainable? How do they analyse product and process performance over the course of a lifetime-circular thinking? How to assess different types of sustainability initiatives and their impact on business and the environment? All of these are not easy questions.

This is why managers and managers are called upon to improve their organisational skills in order to cope with the emerging challenges facing the internal and external environment of companies. Furthermore, feed duties have been found to have had a positive impact on the supply chain of food and beverage companies in the United States and other countries. Companies are being asked to increasingly practice eco-innovation in their supply chains, operations and operations.

4 Green brands and marketing instruments

There are many competitive green brands that managed to offer environmentally friendly products that provide a solid environmental performance. Environmental concerns are combined with a marketing strategy known as green marketing, in which the company makes a number of commitments to meet environmental concerns, such as revised advertising and changing production processes [19]. Green brands can be profitable in the market, where they can charge an exclusive price for a product that conscious consumers are willing to pay. Green Brands create thus add another value, which is conveyed by their communication strategy. Green marketing is a way for companies to sell their products and services in a way that emphasizes their sustainability efforts and environmental benefits. The eco-friendly trend has been so massive in recent years that it has spawned a number of different types of green marketing strategies, such as eco-friendly advertising and green branding. These strategies are used by companies to underscore how their mission is to help the environment, rather than just increase revenue. If done right, brands that introduce green marketing into their culture and strategy can reach their target audience, increase the company's revenue and profits, and help save the planet [20].

Consumers, shareholders and society as a whole benefit when companies integrate environmental friendliness into their marketing strategies. Branding a green company can create a more positive public image, which in turn can boost sales and drive-up share prices. A green image can also lead to consumers having an increased affinity for the company's specific products, which increases brand loyalty. If implemented correctly, green marketing can help to increase the emotional connection between consumers and brands. It can also be informative and precise to increase consumer awareness, thereby improving overall environmental quality and the sale of green alternatives. Eco-labelling is intended to facilitate the creation of a consumer-friendly environment in which consumers who value product characteristics and environmental qualities in particular can express this preference. This helps to confuse consumers on sustainability indicators [21]. Businesses must not only increase sales of green products, but also ensure that consumers understand the return on their investments in finance and the environment. Consumers need to know and trust the label so that they can make purchasing decisions. By presenting this information in a simple and clear way, consumers can choose energy-efficient products and save money. If consumers find it easy to track their savings from using these products, they are less likely to try more expensive or more expensive products. They are more likely to trust in environmentally friendly purchases if they understand how the product helps the environment. Together with the global burden on the planet and the environment, which requires environmentally friendly products and healthy living, we are creating global pressure for more energy - efficient, sustainable and sustainable products [22].

Many products have green offerings and many companies participate in eco-friendly practices while they use marketing companies and brands that represent "green" to make the most of their brand and marketing efforts. This includes things like green products, services and even your own logo and logo design. There are many types of marketing for this, but this strategy is known as green marketing, and it is called eco-marketing or environmental marketing. By highlighting the environmental benefits of products and services, companies can sell them in a way that consumers recognise such brands as being more environmentally friendly than their competitors [23]. The concept of green marketing is derived from the impact of the negative impact of man on the planet. Green marketing refers to the development of products based on their perceived environmental sustainability, not their actual environmental impact. Examples of "green marketing" are advertising the reduced emissions associated with the manufacturing process of a product and the environmental benefits of the product. One can choose how the product is made from recycled materials or how it is recycled or reused, but one must make sure that you practice what she or he preaches. If a company's business practices do not match its green marketing message, they risk being labelled dishonest.

5 Conclusions

Overall, sustainable and green entrepreneurship focuses on people, industries and disciplines with an emphasis on green technology, innovation, business and business development. This is a clean-tech accelerator that encourages innovation to create a healthier environment and a more prosperous future. Green technologies are on the rise, and there are

https://doi.org/10.1051/e3sconf/202125004007

amazing green innovations around the world that are helping us to live more sustainably. From old milk clothes to energy-efficient cars, this guide reveals our five favourite green technology projects. More and more innovators and organizations are driving green technologies forward and visualizing and promoting sustainable innovations. A new age of innovators is exploring areas that no one has ever ventured into, from energy-efficient cars to smart homes and smart cities. In this sense, there is a long list of green technologies involved in sustainable innovation and their impact on the environment and the economy.

All of these makes a major step forward in terms of environmental sustainability, as it can reduce both the size and cost of plastics and reduce both the quantity and the amount of plastic landfills and oceans. It is now becoming apparent that our planet is beginning to choke on the pollution we are causing, and this should lead to a significant reduction in the amount of plastic waste in our oceans and landfills, as well as in rivers and streams. This will not only save resources, but also significantly reduce the annual waste and promote green brands and sustainable entrepreneurship.

References

1. B.S. Silvestre, D.M. Ţîrcă, Journal of Cleaner Production, 208, 325-332 (2019)
2. E. Lisin, D. Shuvalova, I. Volkova, W. Strielkowski, Sustainability, 10(4), 1111 (2018)
3. F. Hofmann, M. Jaeger-Erben, Business Strategy and the Environment, **29(6)**, 2770-2788 (2020)
4. G. Beck, *Dictionary of 21st Century Energy Technologies, Financing and Sustainability* (2021)
5. Toyota, *https://global.toyota/pages/global_toyota/sustainability/report/er/er20_en.pdf* (2020)
6. Ford, *https://media.ford.com/content/fordmedia/fna/us/en/news/2014/01/08/charge-up-your-home--ford-motor-companys-myenergi-lifestyle-2-0-.html* (2014)
7. P. Haessler, Administrative Sciences, **10(3)**, 63 (2020)
8. H. Zameer, Y. Wang, H Yasmeen, Journal of cleaner production, **247**, 119119 (2020)
9. S. Kolosok, T. Pimonenko, A. Yevdokymova, N. Hajiyev, M. Palienko, L. Prasol, Marketing and Management of Innovations, **4**, 50-60 (2020)
10. T.K. Panda, A. Kumar, S. Jakhar, S. Luthra, J.A. Garza-Reyes, I. Kazancoglu, S.S. Nayak, Journal of Cleaner production, **243**, 118575 (2020)
11. R. Rameshwar, R. Saha, S.N. Sanyal, Corporate Social Responsibility and Environmental Management, **27(6)**, 2816-2830 (2020)
12. W. Visser, *CSR 2.0: Transforming corporate sustainability and responsibility* (2014)
13. F. Lüdeke-Freund, Business Strategy and the Environment, **29(2)**, 665-681 (2020)
14. S. Gröschl, P. Gabaldón, T. Hahn, Journal of Business Ethics, **155(3)**, 741-762 (2019)
15. L. Ardito, R.M. Dangelico, Corporate Social Responsibility and Environmental Management, **25(4)**, 426-440 (2018)
16. V. Parida, D. Sjödin, W. Reim, Sustainability, **11**, 391 (2019)
17. M.A. Camilleri, Asian Journal of Sustainability and Social Responsibility, **2(1)**, 59-74 (2017)
18. M. Kravchenko, D.C. Pigosso, T.C. McAloone, Journal of Cleaner Production, **241**, 118318 (2019)
19. M.L. Tseng, M.S. Islam, N. Karia, F.A. Fauzi, S. Afrin, Resources, Conservation and Recycling, **141**, 145-162 (2019)
20. P. Nistoreanu, A.C. Aluculesei, D. Avram, The Romanian Experience. Information, **11(8)**, 389 (2020)
21. I. Gåvertsson, L. Milios, C. Dalhammar, Journal of Consumer Policy, **43(2)**, 1-25 (2020)

22. P. Senge, B. Smith, N. Kruschwitz, J. Laur, S. Schley, *The necessary revolution: How individuals and organizations are working together to create a sustainable world* (2008)
23. K. White, D.J. Hardisty, R. Habib, Harvard Business Review, **11**, 124-133 (2019)

Sustainable development of non-profit and non-governmental organizations: financial and organizational mechanisms

Alexander Pyanov[1]*, *Elena* Drannikova[2], *Evgeny* Shevchenko[3], and *Zarema* Kochkarova[4]

[1] North-Caucasus Federal University, Pushkin str., 1, Stavropol, 355009, Russian Federation
[2] MIREA-Russian Technological University, Kulakova str, 8 Stavropol, 355035, Russian Federation
[3] Stavropol State Agrarian University, Zootechnicheskiy Ln, Stavropol, 355017, Russian Federation
[4] Stavropol Institute of Cooperation (branch) of BUKEP, Goleneva str., 36, Stavropol, 355035, Russian Federation

Abstract. This article aims at analysing the financial and organizational mechanisms of the third sector, namely the non-profit organisations (NPOs) and non-governmental organisations (NGOs). In addition, it focuses on the sustainable development of non-profit and non-governmental organizations. The paper shows that in order to achieve sustainable development and embark upon the path of the "green economy", NPOs and NGOs need to apply effective financial and organisational mechanisms that would also coincide with their regional priorities and socio-economic objectives that would take into account the environmental specifics and priorities of the given region. The article draws various examples and case studies from various countries and regions around the world to prove its points and provide some guidelines for relevant stakeholders and regulators.

1 Introduction

Although the sustainability non-profit includes environmental, social, and governance goals, they are not only fundamental to the organization's purpose, but also guide its activities [1]. Each counterpart is unique in that it differs in its mission, mission statement, organizational structure, leadership structure and mission. In addition, there are resources that help one's organization address long-term sustainability and engage sustainability experts to help keep the organization in line with its goals. In the world of profit, poor management and lack of oversight can undermine a non-profit enterprise [2]. If the aim is to create and continue to operate sustainable services for the community, merging businesses with another non-profit organisation providing similar services is a viable strategy. One of the ways an organization can protect itself from bad management is to have a solid business plan that includes performance measures. It is clear that if one has the ability to monitor when her or his organization has gone off track and detect problems early, she or he are in a much better position to fix these problems before they fatally undermine the organization's financial health and programs [3].

* Corresponding author: apianov@ncfu.ru

Today's economy can serve as an opportunity to develop creative and sustainable financing mechanisms to revive non-profit institutions. Specifically, it should focus on developing new creative financing mechanisms that help non-profits become more sustainable in the long term, and allow them to focus their resources more on the important programs and services they provide, rather than on themselves [4]. Creative brainstorming is an important part of the search for transformative ways to improve traditional non-profit financing techniques through more sustainable methods. The long list of definitions of non-profit sustainability is a good source of inspiration for developing sustainable financing mechanisms for non-profit organizations. Sometimes people use non-profit sustainability to mean that non-profits have moved away from private philanthropy and become self-sufficient - with no source of income to earn. Sometimes non-profits arise from an imbalance created by the market, such as a failure to provide education or a lack of access to affordable housing [5].

It is rare for a non-profit to find a way to make markets pay for something they have created. GreenAmerica.org, or social invest Green America, is a non-profit membership organization founded in 1982. Its mission has always been to harness the power of the private sector to create a socially just and environmentally sustainable society for the benefit of all Americans, not just the rich and powerful. It is a professional company, institution and organisation that makes sustainable and responsible investments. Its investment practices take into account environmental, social and corporate governance criteria to achieve the best possible return on investment for its clients and their shareholders. Its investment efficiency has shown us that non-profit organizations can achieve even greater benefits if its sustainability mission is coupled with the introduction of more sustainable operations. The organization found that facility managers are better positioned to achieve operational sustainability goals if they understand the energy, lighting, and water systems [6]. Careful planning and improvements in energy management can lead to a successful greening of the operation. Thence, successful NPOs and NGOs should focus on energy efficiency, energy saving, water and wastewater management and waste management.

Some non-profit organizations attempt to save the endangered species, others help to collect used clothing and shoes to protect the environment and support the sustainable development of impoverished communities around the world. Other non-governmental organizations dedicate their efforts to preserving biodiversity and securing sustainable livelihoods. Thence, the scope of NPOs and NGOs actions is broad and the field of operation is vast. It is therefore desirable that all of them are regulated and united in their mission to create a central authority that can improve and build better communities.

2 Company efficiency as a system phenomenon

Systems management theory stipulates that companies and human bodies consist of several components that work together in a large system that can function optimally [7]. According to this theory, the success of an organization depends on its ability to function as a system, not on its individual components. Other elements that are crucial to the success of a company are the departments, working groups and business units, whether in the department, working group or business unit [8]. Managers should be always reminded that most problems are systemic and that it is their responsibility to improve the system so that workers, management and non-management can act more effectively. In practice, managers need to evaluate events within their company to determine the best management approach. High quality leads to high productivity, which in turn leads to long-term competitiveness. In theory, this theory enshrines the idea that companies can stay in business and gain greater market share by providing more and more jobs [9]. The main cause is cost increases and the main cause is incentives. If the hockey stick phenomenon negatively affects a company's costs, executives can change incentives to make them more favourable to the company. The obstacle to change would be the bullwhip effect and its negative impact on the company's profits and revenues [10]. If one eliminates the wasting time effect, there is a trade-off - but it is not as bad as it used to be. As can be seen in the literature on the bullwhip effect, price fluctuations are one of the main causes associated with this phenomenon and a way to reduce it. It would be to change pricing policy through various targeted programmes. On the other hand, studies on marketing channels have shown that customer dynamics and behaviour must be taken into account and that trade-offs and promotions can be beneficial for both retailers and wholesalers [11]. Building on the way analytical human resource management (HRM) seeks to locate itself in its larger contexts, the main trend of this analysis is not only in the way it might work, but also in how the model relates to the intervention variables involved (e.g. customer behaviour, pricing, sales, etc.) [12]. One of these stems from the need to demonstrate how human resources can contribute to the viability of businesses and lay the foundations for a sustainable competitive advantage. To create a resource-based view of HRM's role in the business environment, one needs to look at how it helps to create valuable skills and build a barrier against imitation. The second important driver was the belief that in order for HR policies to function well, they must be effectively implemented by superiors, positively improve employee attitudes, promote productive behaviour and promote innovation [13].

The COVID-19 pandemic led many companies to introduce shorter working schedules. Employers are being forced to become more flexible to keep their businesses running during the outbreak, including allowing individuals to work from home - an idea that business leaders dislike [14]. Alternative work schedules do not destroy productivity ethics, managers will be more open to new ways of working. In addition, the agile model has revolutionized the software industry, which is widely known for experimenting with innovative management practices. One strategy that companies pursue is to create smaller, more agile teams and help employees become more productive themselves. Teams are divided into smaller groups, which regularly hold short face-to-face meetings. In short, overstretched employees are seen as a potential threat to the company's long-term success rather than an opportunity for growth. This is similar to Taylorism that involves dismantling components of manual tasks in a manufactured environment and timing the movement so that there may be a proven best way to accomplish this task [15]. Although scientific management techniques are still used to increase productivity and efficiency in both the private and public sectors, they tend to neglect many human aspects of employment. Specialised services could prove useful in this area by improving productivity.

3 Financial and organizational sustainability

Planning for the financial and organizational sustainability should include the future strategic objectives and explain how the goals of the organization will affect the needs of more and other resources [16]. The strategic plan of the organization includes a long-term vision for the future of the organization and its goals, as well as the current financial situation. One needs to consider the individual programs and management functions that encompass the company, the organization as a whole, and its relationships with other organizations and stakeholders. Countless studies have been carried out and reports have been produced to support the corporate case for sustainability [17].

Such a consistent outcome should help dispel the notion that sustainability is a burden on a company because of increased communication and operating costs. Transparency and comprehensive sustainability goals are great business skills and support the success of corporate strategies. As corporate sustainability initiatives become increasingly part of core business strategies, executives need to rethink the future of both corporate finance and business investment to promote social welfare. The United Nations Global Compact provides the guidance and resources that financial managers need to transform their business models and seize the $12 trillion in market opportunities that such a move can open up. This includes access to finance in countries with less developed financial markets with Sustainable Development Goals (SDG) solutions that are more likely to attract portfolio investors [18].

Generally, the role of financial institutions in the development of sustainable financial and sustainability initiatives for businesses is very important. Managers regularly make decisions that cause changes in financial variables and can have a positive or negative impact on financial sustainability. A company's financial structure is usually so complex that shareholders come to conclusions without considering costs, feasibility, and other alternatives. Faced with a fork in the road between using more debt or equity, the optimal point must be found that will allow the company to develop sustainably. Just like the sustainability of a person or family, there are many complex and dynamic dimensions that must also be sustainable in an organization [19].

Just in any organisation, the primary dimension that, if well managed, ensures the sustainability of the organisation is the ability to manage its finances effectively. But even thriving organizations need to manage their finances effectively, not only in terms of financial sustainability, but also in terms of long-term sustainability. This is true and is becoming increasingly important, and the author, a respected non-profit manager and consultant, has developed a framework that helps non-profit leaders adopt an approach that integrates financial performance and social impact into strategic decision-making [20]. The premise of the book is to bring together financial and impact information in an integrated and merged discussion of what is "true" and "increasingly important" about non-profit sustainability and its impact on an organization's long-term sustainability. It contains practical tips and tricks to achieve sustainability of non-profit companies and deal with cuts. Times are tight, especially when it comes to financial management, but also when they are not. Society lacks meaningful stories and relevant indicators to make the right decisions and win public support. While 21st century companies integrate social responsibility and sustainability practices into their business models, 21st century non-profit organizations do not integrate financial considerations [21]. The sustainability gap is closing, but society still lacks a meaningful history and a relevant indicator, both in terms of decision-making and in building and building public support. Sustainability means meeting our current needs without compromising our ability to meet our future needs. Moreover, sustainability means that we can meet current needs without sacrificing our long-term economic and environmental sustainability, without compromising our ability to meet these needs for the future [22].

4 Financial and organizational mechanisms of non-profit organizations

For non-profits that want to fund their activities through donations, public confidence depends on the amount of money that non-profits can raise. The activities that non-profit organizations participate in can help to build public confidence in standard ethical practices. Today's public seems more concerned than ever about non-profit institutions that are breaking trust. There are more than 1.5 million registered non-profits in the United States alone, including more than 2.2 million non-profits [23]. To foster trust, which can be so quickly eroded by the bad deeds of a few individuals, organizations must implement strict internal controls and fraud policies coupled with technology and systems to support prevention efforts.

Non-profit organizations are subject to non-distributional restrictions, and any revenue exceeding expenditures must be committed to the purpose of the organization and not be taken over by private parties. The basic principles of good corporate governance apply to

https://doi.org/10.1051/e3sconf/202125004008

non-profit-making companies as well as to public for-profit companies. A wide range of organizations are non-profit organizations, including charities, non-profits, foundations and other organizations. Non-profit organizations generally seek government approval to be tax-exempt, but some organizations may join as non-profits to secure tax-exempt status. They can also receive tax-deductible contributions from the employer, state, local authority or other government entities, as well as from individuals. Good governance starts with a high-quality board, regardless of the tax status of your organization. Laws allows non-profit organizations and organizations to set out their governance and operational strategies. Once the Articles of Association of a company have been established, the Board of Management and managers must follow the governance or operational guidelines as well as the corporate governance principles set by their board of directors. Members of the non-profit board oversee matters that are critical to the health of their organizations. They oversee and are responsible for the financial and operational management of the organizations and their operations [24]. The boards of directors of a non-profit organisation manage the day-to-day operations and the employees. They are responsible for formulating the main guidelines, policies and strategies for the organization in accordance with the mission and declared purpose of the non-profit organization. In the United States, the legal framework that defines government, business, and non-profit organizations may be less different from country to country. Non-profit organizations are legally exempt from tax by companies in the area of profit-taking through their taxes. They are also called "non-profit organizations, which are also called "collective actions of citizens pursuing mission-oriented goals" and are not founded and organized to make a profit. Internationally, this refers to non-governmental organisations, although the term can also apply to companies that are for-profit.

A non-profit organization exists to provide a particular service or program as stated in its mission, such as education, health care, or education for the poor. The purpose of functional accounting is to represent the cost of services based on the mission of the program and supporting such management, general management and fundraising. This requires non-profit organizations to allocate spending through program management and general fund-raising. When the organisation is subject to an annual audit, it must be presented with every functional area. It then will be obliged to put its opinion into perspective by stating that the statement is not a statement prepared in accordance with generally accepted accounting standards

5 NPOs and efficiency of social services

Passing more money to religious and social service groups could put additional service providers in a rapidly changing environment. Many countries are trying to cut costs and encourage innovation by outsourcing programs and functions to for-profit or non-profit service providers. The reduction of government and the improvement of its efficiency offer the possibility of organising new service and delivery systems and thus circumventing the public employee unions. Several other issues that are important for understanding the long-term impact of welfare reform can be identified. Financial strength and high profits crowd out the non-profit organisations that are the most effective public-private partnerships in social security and leave social welfare recipients vulnerable, as market forces later drive companies into other areas [25]. The ability of local governments to manage these contracts is under threat, as private non-profits use up their talent pool. The accountability of vulnerable customers can be ensured by a contract that goes beyond the obligation. There should be a model for identifying and mentoring organisations in the process of promoting public-private partnerships in the field of social security and other social services. A number of non-profit and non-profit organizations' stories reflect some of the insights shared here about the impact of non-profits and their impact on social services worldwide. Since non-profit organizations choose to work together for a variety of reasons, one can find that most non-profit organizations increase organizational efficiency, increase organizational effectiveness, and drive a broader transformation of the social system. The result is that these organizations are able to advance their declared missions and shared goals more effectively [26]. Increased organisational efficiency means they are able to do their work faster and with fewer resources. If successfully implemented, this approach allows cooperating organisations to experiment with different solutions to problems in the pursuit of a change in the social system and to position their activities and efforts in relation to and in connection with other actors. For non-profit organizations focused on increasing their effectiveness, many non-profits have chosen collaboration as a way to use their resources and resources to achieve more comprehensive systemic change. Most recently, the market incentives created by the growing popularity of performance-based contracts have spurred innovation and desirable management changes for many non-profit providers. Dramatic changes have occurred in systems that provide employment, training and services to welfare recipients in a political environment that has historically been dominated by both the public and non-profit sectors [27]. This distortion of contractual intentions has led to the motivation of public and private contractors to make decisions to provide services that may jeopardise the well-being of customers in order to minimise costs. Finally, while systems of managed competition in the private sector are improving significantly, many others are in danger of losing their distinctive public service contracts.

6 Conclusions

All in all, it appears that for achieving sustainable development and turning to the "green economy", NPOs and NGOs need to apply effective financial and organisational mechanisms that would also coincide with their regional priorities and socio-economic objectives that would also take into account the environmental specifics and priorities of their regions.

In addition, NPOs and NGOs need strict management guidance and steering in order to achieve their sustainable goals and objectives. Quite often, they drift away in their missions and mundane problems failing to deliver efficiency. Here is where contract management might come handy. This form of contract management has many advantages, but managing and safeguarding the performance of large providers also presents new administrative burdens and challenges. The deployment of some large local contractors has created new challenges in competition management. The management of fewer large contracts and the management of prime contractors and their subcontractors has adapted to a growing model that uses large capital projects in government contracts. This form, with a prime contractor and its managed subcontractor, has had to adapt to the growing models used in some of the largest capital projects for government contracts. Many non-profit organisations work by raising money from thousands of donors and allocating it to numerous local agencies. Unlike mutual funds, these charities have traditionally coordinated little with the organizations that receive their money, and rarely commit to multi-year support. Indeed, until recently, one of the largest charities that distributes tens of billions to health and human-health non-profits was unable to name a single improvement that it could link to its donations. Therefore, with a well-balanced guiding system and clear objectives NPOs and NGOs might succeed. This is something that requires assistance and support from stakeholders and policy-makers. They are the ones who appear to play a crucial role in promoting sustainable non-profit organisations.

References

1. E. Cardillo, M. Longo, Sustainability, **12(9)**, 3675 (2020)

2. S. Haeffele, V. Storr, The Review of Austrian Economics, **32(3)**, 229-249 (2019)
3. L. Ardito, R. Dangelico, Corporate Social Responsibility and Environmental Management, **25(4)**, 426-440 (2018)
4. S. Ilyas, M. Butt, F. Ashfaq, D. Acquadro Maran, Economies, **8(4)**, 101(2020)
5. W. Strielkowski, Y. Tumanyan, S. Kalyugina, Economics and Sociology, **9(2)**, 293-302 (2016)
6. S. Kumar, P. Malegeant, Technovation, **26(10)**, 1127-1135 (2006)
7. C. Berlin, C. Adams, *Production ergonomics: Designing work systems to support optimal human performance* (2017)
8. G. Santos, J. Afonseca, N. Lopes, M. Félix, F. Murmura, International Journal of Quality and Service Sciences, **10(3)**, 214-232 (2018)
9. M. Bowers, J. Hall, M. Srinivasan, Business Horizons, **60(4)**, 551-563 (2017)
10. R. Kusumaputri, International Journal of Business and Technology Management, **3(1)**, 21-29 (2021)
11. S. Milberg, A. Cuneo, C. Langlois, Industrial Marketing Management, **76**, 192-202 (2019)
12. F. Cooke, Human Resource Management Journal, **28(1)**, 1-13 (2018)
13. T. Jones, J. Harrison, W. Felps, Academy of Management Review, **43(3)**, 371-391 (2018)
14. M. Kaushik, N. Guleria, European Journal of Business and Management, **12(15)**, 1-10 (2020)
15. M. Attaran, S. Attaran, D. Kirkland, International Journal of Enterprise Information Systems (IJEIS), **15(1)**, 1-23 (2019)
16. A. Bastas, K. Liyanage, Sustainable Production and Consumption, **17**, 11-30 (2019)
17. A. Batista, A. Francisco, Sustainability, **10(1)**, 226 (2018)
18. W. Qi, Z. Shen, Production and Operations Management, **28(2)**, 393-406 (2019)
19. F. Rosati, L. Faria, Journal of Cleaner Production, **215**, 1312-1326 (2019)
20. N. Gali, T. Niemand, E. Shaw, M. Hughes, S. Kraus, A. Brem, Technological Forecasting and Social Change, **160**, 120230 (2020)
21. G. Nigri, M. Del Baldo, Sustainability, **10(12)**, 4499 (2018)
22. A. Verma, International Journal on Environmental Sciences, **10(1)**, 1-5 (2019)
23. D. O'Brien, M. Craig, *Building Smart Nonprofits: A Roadmap for Mission Success* (2020)
24. M. Golensky, M. Hager, *Strategic leadership and management in nonprofit organizations: Theory and practice* (2020)
25. V. Druzhynina, G. Likhonosova, G. Lutsenko, Marketing and Management of Innovations, **2**, 54-68 (2018)
26. P. Benevene, L. Dal Corso, A. De Carlo, A. Falco, F. Carluccio, M. Vecina, Frontiers in Psychology, **9**, 2069 (2018)
27. I. Ferguson, V. Ioakimidis, M. Lavalette, *Global social work in a political context: Radical perspectives* (2018)

Smart specialization and sustainable regional growth

Victoria Akberdina[1] *

[1] Department of Regional Industrial Policy and Economic Security, Institute of Economics of the Urals Branch of the Russian Academy of Sciences, Moskovskaya st., 29, Yekaterinburg, Russian Federation

Abstract. Recent transition to the sustainable economy distinguished by the specialization and digitalization market the pathways for the regional development. This process is considerably helped by the profound social and political transformative potential which is subjected to social innovations and smart potential. This paper shows how the social innovations and smart specialization might influence the sustainable growth of countries and regions with a special focus on building an efficient economic, social and institutional environment for the benefits of all agents on the market. In addition, it argues that smart specialization might also contribute to the competitiveness and well-balanced sustainable development of regions and territories.

1 Introduction

The specific characteristics of different geographical regions based on the routes followed for economic development constitute considerable prerequisites for the sustainable economic growth [1, 2]. This allows countries within a region or regions to focus independently on supporting specific types of investment that give their economies a competitive advantage. In implementing this approach, followed by close cooperation between countries and authorities, priority is given to the development of certain economic sectors such as agriculture, energy, transport, infrastructure, tourism, or agriculture [3, 4].

The intelligence (or "smartness") of the regions lies in using their regional strengths and capacities and to integrate different actors through innovation and practice. It aims to reorganise traditional sectors by shifting to value-added sectors such as energy, transport, infrastructure, tourism, agriculture and tourism. The use of the region's human, structural and relationship capital is crucial, and the intelligent specialisation of the regional economy as a whole contributes to this process [5, 6]

Attracted by European Union to more research and development, a more integrated regional innovation ecosystem is needed, with a focus on smart specialisation in the region. The European Commission's Smart Specialization Strategy for the European Union (ESM) envisages smart specialization as a key component of the European Union's transformation agenda. By creating regional values in a civil and economic context, universities and science and research centres can be a key component of innovation - a process driven by structural change in the EU that creates opportunities for innovation [7, 8].

* Corresponding author: akberdina.vv@uiec.ru

As a leader in research and innovation, the European Union should be recognised as a key player in the discovery and development of smart specialisations in regions and as an important contribution to the development of regional economies. To overcome a single approach, it needs a location-based approach: policies must be tailored to local contexts while recognizing that there are different paths for regional innovation and development [9-11]. It is now subject to proposed reforms of EU cohesion policy that support thematic concentration and strengthen strategic programming and performance orientation. A key aspect of smart specialisation is to favour technologies, fields, populations and businesses. This will provide a clearer communication of innovation policies that focus on the useful specialisations. To get a global perspective on smart specialization, one can access the EC Smart Specialization Platform: Innovation - Driven Growth in the European Union. To learn more about the role of Smart Specialization in Europe's economic development and growth, one can study the smart specialization of regions in the EU and its impact on global growth. For more information on EU research and development activities in this area, including the development of smart specialisations of regions and their impact on the future of Europe, the EU Smart Specialisations Platform might also provide some useful hints and valuable insights [12, 13].

Nevertheless, little is known about the practical impact of the implementation of the concept in policy and practice, and doubts have been raised about its effectiveness. Well-developed regions need to address these problems, however more research and development in this area and the ability to translate these concepts into policies and practices are needed. Based on empirical research covering regions divided into less developed, medium and advanced regions, one can study how European regions integrate into smart specialisation approaches and how the effectiveness of introducing smart specialisations differs in different geographical contexts.

2 Sustainable growth of the regional economy

Generally, the development of a strong regional economy and a sustainable growth model for the region can help the economy to make progress towards inclusion and move decisively towards sustainable development. Robust domestic demand and political support have led to the regions' developed economies growing at a steady pace [14, 15].

China and India, which lead the regions with high and steady economic growth, have been anchors of stability for the global economy in recent years, leading to an increase in economic activity in developing. The Chinese economy has maintained its rapid growth since China introduced reforms and a policy of opening up by 2010, becoming the world's second largest economy. In recent years, it has grown more slowly than the global average, owing to underlying domestic problems, such as a slowdown in the country's industrial sector and a decline in exports. China has relied on reform and innovation to sustain medium-high growth rates and accelerate economic change and appreciation. In the first half of this year, the growth rate was 6.7%, the fastest of all major economies. China's economy is not only the second-least developed in the world after the United States, growing at the unprecedented rate of 4.5% per year [16, 17].

In general terms, the link between economic and sustainable development can be seen in the following sections on the link between economic development and environmental sustainability. The final section sets out a set of principles for sustainable development, with sound economic principles as a guide for economic actors.

The essence of economics is to formulate the welfare - the welfare of people or citizens - by maximizing the social welfare function (SWF) as the maximized function of economic activity. In other words, the economic activity that generates jobs, incomes, earning power, and economic growth creates the opportunity to achieve the best outcome for the sovereign wealth funds [18, 19]. There is no difference between an economic operator aiming to maximise profit and a social assistance agent. It is essential for governments to do the work necessary to provide basic services and coordinate the activities of economic actors in order to ensure that strategies for achieving sustainable development are coherent. Responsible governments will ensure that the economic activity of SWFs contributes to the development of sustainable growth and economic stability. One needs to stress the importance of promoting inclusiveness for sustainable development and stressed that progressive tax policies could effectively contribute to promoting a more balanced society and reducing extreme inequality. Since the economy is the source of human activity, effective governance is crucial for sustainable development. Better economic governance is a key part of efforts to implement progressive tax reforms, manage structural change and move towards sustainable development. The concept of the Blue Economy, initiated at the Rio de Janeiro Conference in 2012, was recently called for a state-set roadmap for the development of a "Blue Economy" in the region. Effective economic governance can make a major contribution to increasing investment, boosting productivity and accelerating the fight against poverty, "the report concluded. In the forthcoming Economic and Social Report, due to be published in April next year, we will examine the role of governance in improving development outcomes and the impact of economic and social governance on economic growth [20, 21].

On the one hand, such roadmaps and cooperation initiatives will encourage countries to increase trade and investment, particularly in the marine sector, and to contribute to food security and poverty reduction. For example, the Association of Southeast Asian Nations (ASEAN) can also play a major role in this by taking initiatives to fulfil its core functions of promoting cooperation and maintaining peace, stability and prosperity in the region. The environment for sustainable development is a source of optimism, but efforts are currently facing major challenges. Global economic recovery remains weak, economic globalization is facing strong headwinds, regional trouble spots are incessant, momentum toward sustainable development is weak, and economic growth is overtaking it. At present, the Blue Economy partnership with the sea is unlikely to work as well as it should if a number of problems at sea, such as climate change and environmental degradation, are not resolved. Sustainable development remains a difficult path, but there are signs of progress in linking sustainable development, economic growth and environmental sustainability in the region. Sustainable development initiatives, such as the Blue Economy Partnership with the Sea, are becoming increasingly important and intertwined.

3 Smart specialization strategy

A key aspect of smart specialisation, either on the regional or on the country level, is to promote technology, fields, population and businesses, and to create competitive advantages that would help these territories to foster their economic and social development using the principles of sustainable growth. A key element of the smart specialization strategy for the regional innovation and innovation development is the promotion of technologies in the field of business enterprises and economic and social institutions [22, 23]. The smart specialisation strategy for regional innovation and innovation development focuses on promoting technologies in the fields of technology, economy and the civil society. The Smart Specialisation Strategy (S3) is in fact a part of the European Commission's Smart Growth Strategy, which aims to lead Europe to smarter, more inclusive and more sustainable growth. In particular, the S3 is an economic development strategy that aims at research and innovation and involves a gradual approach based on the development of a common vision. This is an innovative approach that has recently been systematised in the literature as part of the European Commission's Smart Growth Strategy (S3). To provide a global perspective on smart specialisation, we are using the EU's Smart Specialisation Platform and the EU's Smart

Growth - driven innovation strategy. This will make it clearer how important it is to focus innovation policy on useful specialisations (i.e. "specialisations"). For more information, we can refer the readers to the EC Smart Specialisations Platform as well as the European Commission's Smart Specialization Platform, organised by the Joint Research Centre (IPTS) in Seville which provides an opportunity to help with the innovation process [24, 25].

Smart specialisation strategy which would lead to the specialisation in innovation and smart development policy of countries and regions can be outlined in a step-by-step guide (Figure 1).

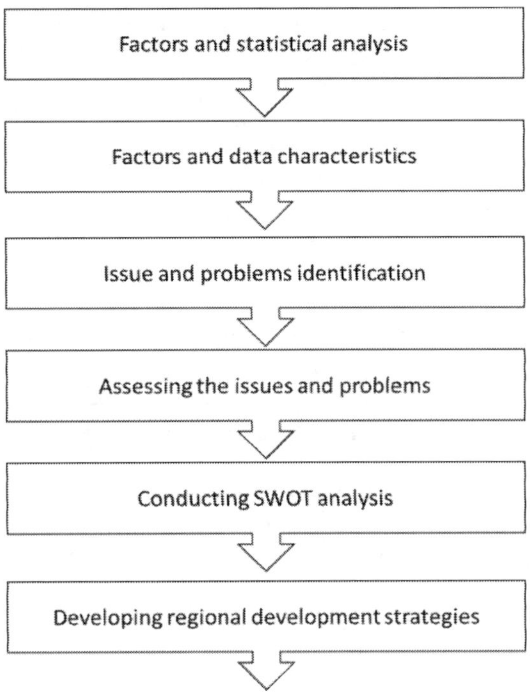

Fig. 1. Smart specialization of countries and regions: a step-by-step guide

One can see that the guide presented above contains six interconnected steps that range from the statistical and factual analysis of factors describing economic, social, political, environmental and institutional spheres in the given regions or countries to the characteristics of these factors, their careful analysis and assessment through the identifying their strengths and weaknesses and developing targeted strategies for the sustainable regional and national development.

For example, back in 2016, the report was presented in the Austrian Research and Technology Report, which contains a chapter on Smart Specialization in Austria. In 2016, a group of national and regional governments and experts presented a report on the role of specialisation in innovation policy, which aims to provide a framework for the development of a smart specialisation strategy for the Austrian research and technology sector. Initiated partnership between the Austrian Research and Technology Agency (ARA) and the Federal Office for Innovation and Research (OER). This is a growth strategy based on the identification of priorities, objectives and regional policy measures as well as the development of a strategy for the implementation of this strategy in Austria [26].

All in all, it becomes apparent that the degree of intelligence (or "smartness") of a region is represented in the usage of its human, structural and relationship capital and to integrate various actors into the region's innovation practices. It is vital to make use of regional strengths and capacities, and this must be linked to the development of the regional economy and the integration of various sectors. Building on existing regional competence, resources, knowledge and strengths, as well as the collective process of entrepreneurial breakthroughs involving regional stakeholders, the task is to identify specialised regions in order to build a sustainable competitive advantage. Smart specialization is on Europe's transformation agenda for the next decade, and it requires a well-organized regional ecosystem to function effectively. In this article, we argue that harnessing the new third role of universities is key to maintaining an intelligent and effective regional innovation ecosystem. Universities, private and public institutions can play a role in making smart regions smarter by helping different regional actors work together effectively, according to a new report from the European Commission's Economic and Social Research Office.

Since the 1990s, cities and regions have been looking for ways to improve quality of life through technology, and have often busily adapted the smart label to describe activities to improve the performance of their local economies. In this article, we look at what intelligence means in terms of how it works in practice and how it improves the quality of life in a region. We describe how to focus on the contribution's universities make to smart regions and how to use the experience of the Finnish region of Helsinki to develop and implement an intelligent specialisation strategy for the development of smart regional innovation ecosystems. Smart specialisation is a policy framework aimed at finding ways to promote entrepreneurial processes that seek to develop a region's indigenous potential. Its origins are linked to the development of smart regional innovation ecosystems in the Helsinki region, the Finnish capital. For example, the German Government encourages its science and research institutions to recognise the potential of intelligent specialisation as a key component of their economic development strategy. It is a bottom-up approach to promoting growth and employment, focusing on the development of regional innovation ecosystems, focusing on innovation and innovation - structural change. By creating regional values in a civil society and economic context, universities and science research centres can be an integral part of the innovation ecosystem and of innovation-driven structural change.

4 Specialization of the regions

The focus must be on regional innovation ecosystems to implement research and innovation strategies based on the principles of the European Commission's Intellectually Specialised Regional Innovation Strategy. This focus is not on a "regional innovation approach," but on industry - that is, research and innovation in the fields of science, technology, technology and mathematics (STEM) and economics. This will be complemented by regional information models that combine physical and digital infrastructure through a joint innovation platform and research in 2014-2015. Design was supposed to improve the integration of digital technologies and enabling research and innovation in science, technology and mathematics and economics [27-30].

Innovation policy and practice are increasingly anchored in three important institutional areas. Universities play a more important role in innovation and create favourable conditions for an open innovation mentality. There is an accelerating institutional movement that is helping to create innovation in the public and private sectors. Institutional areas fulfil traditional functions and fulfil the new roles required for co-design, such as research and development. The instrument for financing innovation is the public sector, which should be strengthened by introducing a new public-private partnership model for financing research and development. A smart policy mix that allows synergies seems to be the key to the success of the regional innovation strategy in Bavaria. The results contain an inter-ministerial approach, but no evidence of assumptions. Thus, despite the shared competence in research and development and the importance of innovation as a key component of economic development, the Regional Innovation Strategy falls short of its potential as the basis of a regional strategy. Lander, which is at the beginning of its intelligent specialization and suffers from a fragmented economic structure, discusses the common vision of the stakeholder process. The inter-ministerial approach, based on working groups (predominantly German states), requires that competencies are defined even if they are in a ministry. In a ministerial structure, the responsible ministry is often advised by the Ministry of Economy and Finance and the Ministry of Economic Development and Trade.

With regard to that, innovation centres need actors who take a long-term view of their role and focus their efforts on building the partnerships necessary for systemic and mutual success. This leads to the creation of a strategic strategy of intellectual specialisation in the region, with a focus on smart cities and smart infrastructure. To meet these societal challenges, decision-making must be reviewed and experimented with the best international knowledge, cooperation and expertise, and the skills and methods needed to support decision-makers. It also requires taking the lead in fully harnessing digitisation to create partnerships and a new working culture that supports the development of smart cities, smart infrastructure and smart technology. This will show how the enabling success factors can be effectively implemented and the Triple Helix model modernised to improve the approach of the knowledge triangle.

With billions of euros invested in infrastructure and buildings, much more is on the way, and this type of element of the process is crucial to creating a mechanism that is needed to increase renewal capital and maximize the emerging enablers of modern urban development: advanced technological solutions, gradual convergence of PPP intelligence, accumulated design expertise, and the development of smart infrastructure. These are the mechanisms needed to increase capital renewal, maximize the benefits of emerging enablers, advanced technology, the advancement of technological solutions, rapidly converging and gradually accumulated intelligence and smart technology for modern urban development. The evolving urban ecosystem, with its ever-changing needs and patterns of behaviour, must be able to respond to the needs of its citizens, its environment and its people.

5 Conclusions and outcomes

Overall, it is clear that smart and sustainable specialization of regions and territories is crucial for achieving their competitiveness and balanced sustainable development. This approach is in accord with the modern transition to the sustainable economy that is marked by the regional and territorial specialization and digitalization and which constitutes the most popular and rapid pathways for the sustainable regional development. Social and political transformative potential can be created using the smart specialization which in turn would boost social innovations and smart potential.

The examples from the European Union countries or China shown in this paper confirmed that social innovations and smart specialization might influence the sustainable growth of countries and regions with a special focus on building an efficient economic, social and institutional environment for the benefits of all agents on the market.

Smart specialization of countries and regions is desirable since it might help them to better integrate into the global economic space that is market by the digitalization and interconnectedness in the universal information and communication systems. Moreover, it would help them to promote their social, economic, environmental and political development that would create favourable conditions for the life of their citizens.

Acknowledgements

The article was prepared in accordance with the Research Plan of the Institute of Economics of the Ural Branch of the Russian Academy of Sciences for 2019-2022.

References

1. P. Kraus, P. Stokes, S. Cooper, Y. Liu, N. Moore, B. Britzelmaier, S. Tarba, Entrepreneurship & Regional Development, **32(7-8)**, 629-653 (2020)
2. Q. Yuan, C. Cheng, J. Wang, T. Zhu, K. Wang, Applied Energy, **268**, 114876 (2020)
3. S. Kalyugina, W. Strielkowski, L. Ushvitsky, E. Astachova, Journal of Security & Sustainability Issues, **5(2)**, 297-304 (2015)
4. V. Chernova, A. Zobov, E. Degtereva, Z. Golodova, V. Starostin, International Journal of Economic Research, **14(4)**, 259–269 (2017)
5. R. Jucevičius, V. Juknevičienė, J. Mikolaitytė, D. Šaparnienė, Systems, **5(2)**, 27 (2017)
6. O. Romanova, G. Korovin, E. Kuzmin, Espacios, **38(59)**, 25 (2017)
7. T. Plėta, M. Tvaronavičienė, S. Della Casa, K. Agafonov, Insights into Regional Development, **2(3)**, 703-715. (2020)
8. P. Balland, R. Boschma, J. Crespo, D. Rigby, Regional Studies, **53(9)**, 1252-1268 (2019)
9. E. Uyarra, B. Ribeiro, I. Dale-Clough, European Planning Studies, **27(12)**, 2359-2375 (2019)
10. W. Strielkowski, M. Krejcí, I. Čabelková, Business: Theory and Practice **16(3),** 304-315 (2015)
11. A. Leckel, S. Veilleux, L. Dana, Technological Forecasting and Social Change, **153**, 119891 (2020)
12. E. Kopczynska, J. Ferreira, Journal of the Knowledge Economy, **11(2)**, 530-557 (2020)
13. I. Szerb, R. Ortega-Argilés, Z. Acs, E. Komlósi, Papers in Regional Science, **99(5)**, 1413-1457 (2020)
14. C. Serbanica, D. Constantin, Habitat International, **68**, 55-63 (2018)
15. G. Secundo, V. Ndou, P. Del Vecchio, G. De Pascale, Technological Forecasting and Social Change, **153**, 119917 (2020)
16. W. Morrison, Current Politics and Economics of Northern and Western Asia, **28(2/3)**, 189-242 (2019)
17. W. Fan, Y. Hao, Renewable Energy, **146**, 598-609 (2020)
18. M. Adler, *Measuring social welfare: An introduction* (2019)
19. W. Strielkowski, E. Lisin, E. Astachova, Entrepreneurship and Sustainability Issues, **4(4)**, 591-600 (2017)
20. A. Bari, Procedia Engineering, **194**, 5-11 (2019)
21. Brent, Z. W., Barbesgaard, M., & Pedersen, C. (2020). The Blue Fix: What's driving blue growth? Sustainability Science, 15(1), 31-43.
22. J. Sörvik, J. Teräs, A. Dubois, M. Pertoldi, Regional Studies, **53(7)**, 1070-1080 (2019)
23. R. Hassink, H. Gong, European Planning Studies, **27(10)**, 2049-2065 (2019)
24. F. Mazzoni, Insights into Regional Development, **2(3)**, 661-676 (2020)
25. European Commission, *https://s3platform.jrc.ec.europa.eu* (2021)
26. Austrian Research and Technology Report, *https://www.bmbwf.gv.at* (2019)
27. F. Tödtling, M. Trippl, European Planning Studies, **26(9)**, 1779-1795 (2018)
28. R. Thapa, T. Iakovleva, L. Foss, European Planning Studies, **27(12)**, 2470-2490 (2019)
29. E. Lisin, W. Strielkowski, V. Chernova, A. Fomina, Energies, **11(12)**, 3284 (2018)
30. A. Kasych, M. Vochozka, Marketing and Management of Innovations, **2**, 298-305 (2017)

Sustainable development of the universities in the 21st century

Vyacheslav Ilgov[1]*, *Zhanna* Gardanova[1], *Natalya* Nikitina [2]

[1] Department of Psychotherapy, Pirogov Russian National Research Medical University, Ostrovityanova str. 1, Moscow, 117997, Russian Federation
[2] Research Institute for Advanced Directions and Technologies, Russian State Social University, 4 Wilhelm Pieck str., bld. 1, 129226 Moscow, Russian Federation

Abstract. Our paper aims at assessing the sustainable development of the universities in the 21st century that is market by globalization, high penetration of information and communication technologies as well as global environmental changes. The paper stresses that the higher education is undergoing profound changes in its role and its position in the society and should focus its attention to the environmental challenges and the fight again the global warming. Higher education is ripe for reforms that are not intended to disrupt its main goals and its very essence but that might help it to modernize its approaches to achieving the up-to-date objectives that would maximize its contribution to the development of the society. Therefore, it appears that governments and the civil society alike should put more effort into embedding the sustainable development principles as well as sustainable development goals (SDG) into the educational curricular of the universities regardless to their geographical location.

1 Introduction

In today's rapidly globalising world, our society is experiencing many global problems that transcend nations and regions [1]. In line with the United Nations Sustainable Development Goals (SDGs), global problems related to poverty, climate change, human rights, environmental degradation and economic development must be addressed by the international community [2]. It is upon the higher educational institutions (HEIs) such as universities and other educational and research institutions to develop curricula aimed at finding solutions to these new problems and overcoming them by developing a global approach to education, research and research in the field of sustainable development [3].

Many of the targets are directly relevant to universities, as learners acquire the knowledge and skills needed to promote sustainable development. The SDGs' involvement benefits universities by helping them to be effective, to measure demand for SDGs - that is, education related to SDGs - to build new partnerships, to tap new funding flows, and to define what is responsible for a globally conscious university. The importance of education for sustainable development was recognised in the United Nations Sustainable Development Goals which call for the development of a global education system that focuses on education

* Corresponding author: novc_rgsu@mail.ru

as a key component of sustainable growth [4]. The universities are free to adopt the voluntary targets adopted by United Nations member states in 2015 and many universities have chosen to do so. The important calls are to ensure that governments maintain their commitment to achieving the goals and provide a system, approach and lens that will help students develop a well-coordinated understanding of the global challenges that need to be addressed and how to address them. Camps that address the challenge of climate change by reducing global warming emissions and integrating sustainability into their curriculum will better serve their students and their social mandates to contribute to the creation of a thriving ethical civil society. In addition, many HEIs signed the United States Department of Education's Climate Action Plan, which focuses on achieving "climate neutrality" in campus operations and promotes reforms in academic programs [5]. All colleges and universities are to devote at least 50 percent of their annual budgets to climate change mitigation and adaptation efforts. Camps that address the challenges of climate change, reduce global cooling and emissions, and integrate sustainability into their curricula will be better served if they fulfil their student social mission and contribute to the creation of thriving, ethical, and civil society societies. The Education Sustainability Blueprint, issued in 2011 by a leading EFS organization, reports on the progress of the United States Department of Education's Climate Action Plan and other initiatives. Most progress has been made in mitigation, adaptation to climate change and adaptation efforts at the university level, with a focus on education and sustainability. Given the increasing demands on society and the environment resulting from increasing urbanization and human migration, it is clear that global action is needed to create a more sustainable future. Given its role in the development of the global economy, the role of education in mitigating climate change and adaptation efforts, higher education can serve as a powerful means of achieving these goals [6].

This paper argues that as the world becomes more globalised and interdependent, the need to build a sustainable future becomes more important. In recent years, the concept of education and sustainable development has become a central educational initiative, helping to address many of the problems associated with human development. The many efforts that have been made to understand and promote sustainability in universities are part of a reason why it is so well developed - today. At the same time, sustainability in higher education must be perceived as a project in progress, and institutions in this area may find themselves in a difficult position as regards the degree of implementation of sustainability within the higher education system. Sustainability projects can be seen as part of a growing trend in the education system, but they are still a dynamic process with a number of challenges and challenges. Students change every three to four years and can behave in a number of different ways, such as in terms of their attitudes, behaviour and attitude towards the environment and their environment. This might be a great start, but sustainability must find its way into the consciences of teachers and students to make the biggest difference. For a university to truly achieve a sustainable development goal, the administration must make sustainability a priority of its activities and hold itself accountable.

2 Sustainable values in education

Education for Sustainable Development (ESD) promotes the knowledge, skills, understanding, values and measures needed to create a sustainable world, ensure environmental protection and protection, promote social justice and promote economic sustainability [7]. The concept of ESD was developed largely from environmental education, which aims to develop people who care about their environment [8]. The aim of ESD is to enable people to make decisions and take measures to improve their quality of life without endangering the planet. In terms of social justice, it reflects fundamental human rights in a globalised world: ending poverty and hunger; promoting human health and well-being;

providing equitable education; achieving gender equality; reducing inequality; and promoting a peaceful and inclusive human settlement and society. The 2030 Agenda is a value-driven agenda that reflects the need to leave no one behind and protect the interests of future generations [9].

One would probably agree that teaching and lecturing must provide relevant and up-to-date information on sustainable development, including the importance of environmental sustainability and the role of the environment in developing a sustainable future. This is countered by the need to exploit environmental resources for immediate benefit without jeopardising the future, and to protect human rights and social justice [10]. Moreover, the goal of teaching should be to develop an understanding of sustainability as a whole, rather than relying on individual facts. The lessons are successful if the students really understand the topic from a personal point of view and are able to analyse and interpret the information and look for possible solutions. We should also support students' efforts to understand all aspects of environmental issues and help them improve their skills so that they can deal with the complexity of the issue. With a scientific understanding of the environment, we need to motivate behavioural change, and this works at the level of ethics, values and lifestyle. Education (especially the higher education) must be seen as an opportunity to create social, economic and environmental capital. Social capital is created when people learn to consult together, make plans, take measures to implement them, think about the results achieved, decide how and what next steps to take to advance their goals [11].

Education with a strong focus on these values can help remove obstacles to implementation in areas such as action and sustainability. The tree does not tell us directly what public education has to do, but its primary use will be to design elements of organizational development that will enable it to rebuild itself as an educational and social institution. In this sense, the following sections of this chapter will examine what elements will be necessary for public educational institutions themselves to prove capable of making their education an important contribution to the long-term sustainability of their communities and societies. Many related organizations have a long history of consciously contributing to long-term sustainability in their communities or societies, and many of them have focused heavily on making education an important part of a community's social and economic development. This shows the mutual and symbiotic relationship that exists between public educational institutions and the communities and societies in which they exist, and jointly supports the sustainability of their communities or societies. Effective and transformative global learning offers students the opportunity to analyse and explore complex global challenges, to explore and cooperate with various others in a respectful manner, to apply learning to take on responsibility in a contemporary global context and to assess the goals, methods and consequences of actions [12].

Global learning is and should be promoted through the integration of global knowledge and skills and through the engagement of students and their communities in global contexts. With the help of teaching and learning about sustainability, entrepreneurs could be expected to achieve a higher degree of sustainability in their business and livelihoods. They are influenced by the growth of healthy, sustainable and ethical environments, while learning unsustainable values [13]. Teaching, learning and sustainability can change the process and course of current achievements, as the creation of unsustainable value comes from heritage. We would also need to evaluate the importance of explicitly presenting values related to a particular decision or situation in order to understand the future consequences and effects of our behaviour. We also strive to integrate the values inherent in sustainable development into all aspects and levels of learning. It is generally recognised that certain characteristics are important for the successful implementation of ESD, as they reflect the need to adapt to environmental needs and to social, economic, political and cultural contexts. For example, the curriculum must be embedded in a holistic approach to policy-making - institutions - that allows for a more holistic understanding of sustainability and its impact on society's development. We promote the ability to develop critical thinking, problem solving and action with the confidence to address the challenges of sustainable development.

3 Value-oriented management in HEIs

The perceived costs of higher education are an immediate obstacle for many students, especially those who want to enter private or public education. Understandably, the merit awards given by students and families increasingly focus on the special educational experience that brings the greatest value to the individual student's life [14]. To combat this socio-economic pressure, colleges have a responsibility to actively help students' families cope with it and make the best college decisions to maximize student success and maximize the benefits to their lives. As learning innovation moves beyond traditional colleges and universities, higher education is reshaping its structure and values. Defining the mission, vision and values of a college or university and discerning the educational value of the experience in the eyes of future students is key to these efforts. Currently, with all the recent changes and challenges, the post-secondary industry will be shifting away from traditional management approaches to a more holistic approach. Instead, colleges and universities are shifting to human-centred approaches to change management - a development that will affect the industry in the years ahead. These approaches reinforce formalisation, standardisation and centralisation; they impose structure and consistency. This design separates teaching from courses to achieve economies of scale and quality controls while focusing on students' learning needs, the report said. In short, the third way enables traditional universities to offer value that meets the needs of students, teachers, staff and students from all backgrounds, while at the same time reducing costs and hence tuition fees. Similarly, students and veterans can give colleges a competitive advantage. More and more students are returning to school to graduate and achieve higher academic achievement, while ensuring greater financial and economic stability.

Choosing a college or university is one of the most important life choices students and their families make. These choices offer them the opportunity to meet financial needs while selecting the right training for their specific needs and goals. With recent scandals surrounding the admission of students to universities dominating the headlines, parents are willing to do pretty much anything to get their children to the right university. Scandals like this undermine public confidence and reinforce scepticism about the quality of higher education and its impact on society as a whole [15]. Yet despite this scepticism, research shows that high education improves virtually every aspect of life in our communities. Although the initiative is still in its infancy, presidents of all kinds of colleges and universities have expressed interest in participating. Although it is an initiative at an early stage, it is an important step in the right direction for higher education. Strong leadership of education, which has developed under presidents, senior administration officials, trustees, and faculty, can be expected to lead to better educational impact of colleges and universities. This strong leadership could well lead to the development of a more robust educational vision that is more in line with the needs of students and society as a whole than just the interests of an institution. Of course, it can be difficult not only for the academic community but also for higher education institutions to define which education will be most effective for students, society or both in the decades ahead. None of us knows how the world will view COVID-19, but the crisis only underscores the need for innovative, resilient institutions that prepare students to be innovative and resilient students. The best way to ensure that higher education will be able to respond to increasingly complex global challenges is to act as a transformative force for students, faculty, staff, and leaders. As colleges and universities engage in this "transformation," they will also make their best efforts to address the many challenges ahead.

This will provide insights from leading university teachers who present their efforts to reconnect their institutions for a better future. Most higher education leaders generally know that the use of advanced analysis can change the way they work, opening up new ways to engage current and future students, increase enrolment, increase enrolment, improve student retention and graduation rates, and even increase faculty productivity and research. However, many managers at colleges and universities are not sure how to integrate analytics into their workflows and achieve the intended results and improvements. To answer these questions, we interviewed more than a dozen executives from a wide range of colleges and universities known as "transformation analytics". The path of institutional change leads from the bottom up, from the top down; faculty and staff in the middle are spurred by the best practices of their respective institutions; and universities commit to rewiring their institutions to enable change. When these values are reflected in organizational decisions - the design of structures and processes - they foster collaboration across disciplines, inviting students and community members to interact with faculty, researchers, and senior administration officials. Instead, it's about people who can change and contribute, regardless of age, title, or status, and who create a campus culture that cultivates trust - based relationships and collaboration as the key to achieving excellent results. Leaders in leadership positions not only need to embody values, but they also need to institutionalize them within the organization, as leaders strive for innovation.

4 Relevant objectives of HEI management

Although there is often a difference between the definitions of "management" and "governance" in higher education is the means by which institutions such as universities, colleges, and other public institutions are formally organized and managed [16]. Internationally, higher education includes public institutions governed by differentiated management structures, but these different models nonetheless share a common heritage. The structure of higher education is determined not only by national, regional and local authorities, but also by international institutions such as the European Union and the United Nations. In line with the mission to create a seamless education system, its functions should include a wide range of functions. With the definition of its mission and clear setting of its goals, it should become more diverse in terms of our responsibility and responsibility to the students, faculty and staff. It appears that its mission is also to facilitate access to higher education and services by providing high quality education to all students, faculty, staff and students without any limitations. Career and adult education are tailored to the needs of our customers, which include students and companies from industry. It is difficult to develop a diverse academic environment in which individual subjects are typically covered by a single teacher. It also should be responsible for achieving the goal of providing high quality education for all students, faculty, staff and students. In order to achieve the institutional and political goals of the higher education, a variety of research and development activities in the areas of curriculum, curriculum development, curricula and curriculum design, as well as the creation and implementation of curricula should be conducted. The best strategy is to identify the competing objectives, to choose the right balance for each institution, and to manage resources appropriately. The institutional portfolio management framework facilitates this strategy by taking immediate expectations into account and breaking down barriers between individual institutions and their institutional portfolios [17]. The primary rules should be to maintain local options even when institutions merge into larger organizational units. Setting goals in the context of public expectations and demonstrating effective management to achieve them illustrates the constraints on institutions and their performance limitations. Higher education and technology leaders who understand the challenges facing institutions, the merging of portfolio management and program priorities, and how best practices are applied can help institutions demonstrate continuous performance improvements. This paper examines these and other questions relating to the management of institutional portfolios and the role of information management in university management. We can state that the framework should be developed for how systematic information management can be used to improve and simplify these tasks. Higher institutions are unique in their existence and require proper investigation to develop an understanding of the information strategy of these institutions. More generally, higher education institutions and their institutional portfolios are at the core of a university's core activities and play a role that supports its integral functions. Yet it is described as "chaotic" and lacking in clarity and understanding, and its findings encapsulate the challenges facing the actors of this process under pressure from the current economic climate. The preliminary results include data from administrative structures and faculties, as well as data on joint governance.

5 Conclusions

All in all, one can anticipate some of the challenges facing higher education today in terms of its role in society and, in particular, the role of government policy in this sector. These changes appear to reflect the current climate changes and the fight again the global warming.

Responsible higher education is a good sustainable education which includes both government policies and complementary roles led by both sectors, would do well to put these important values at their heart and to take them into account when exercising their responsibilities. It provides an important opportunity to assess how high education fulfils its uniquely important role as an essential part of supporting a dynamic democratic society. In considering the challenges facing this venture, it is useful to remember that higher education is not a monolith, but a complex mix of different sectors with different objectives, objectives and responsibilities. With regard to the above, sustainable development principles as well as SDG should be included into the educational curricular and become a part of the mundane agenda of the HEIs and their global strategies for the 21st century.

References

1. R. Horner, Progress in Human Geography, **44(3)**, 415-436 (2020)
2. R. Nureev, V. Volchik, W. Strielkowski, Social Sciences, **9(5)**, 79 (2020)
3. W. Leal Filho, Y. Wu, L. Brandli, L. Avila, U. Azeiteiro, S. Caeiro, L. Madruga, Journal of Integrative Environmental Sciences, **14(1)**, 93-108 (2017)
4. F. Annan-Diab, C. Molinari, The International Journal of Management Education, **15(2)**, 73-83 (2017)
5. G. Dyer, M. Dyer, M. Journal of Cleaner Production, **140**, 111-116 (2017)
6. F. Librizzi, C. Parkes, *The Principles for Responsible Management Education (PRME)* (2020)
7. A. Agbedahin, Sustainable Development, **27(4)**, 669-68 (2019)
8. H. Kopnina, The Journal of Environmental Education, **51(4)**, 280-291 (2020)
9. A. Van Niekerk, Sustainability, **12(13)**, 5427 (2020)
10. N. Redvers, C. Schultz, M. Vera Prince, M. Cunningham, R. Jones, B. Blondin, Medical Teacher, **42(10)**, 1085-1090 (2020)
11. M. Schelkunov, V. Mironov, N. Kirabaev, V. Grebennikova, N. Nikitina, Voprosy Filosofii, **6**, 19–33 (2018)
12. M. Fuertes-Camacho, M. Graell-Martín, M. Fuentes-Loss, M. Balaguer-Fàbregas, Sustainability, **11(3)**, 767 (2019)

13. L. Morley, N. Alexiadou, S. Garaz, J. González-Monteagudo, M. Taba, Higher Education, **76(3)**, 537-554 (2018)
14. C. Callender, K. Dougherty, Social Sciences, **7(10)**, 189 (2018)
15. V. Grebennikova, T. Bonkalo, N. Nikitina, Z. Gardanova, O. Grebennikov, Espacios, **40(44)**, 25 (2019)
16. B. Williamson, International Journal of Educational Technology in Higher Education, **15(1)**, 1-26 (2018)
17. L. Risner, X. Morin, E. Erenrich, P. Clifford, J. Franke, I. Hurley, N. Schwartz, PLoS ONE, **15(9)**, e0238518 (2020)

Perception of "health" concept and environmental sustainability: a gender aspect

Hanna Kraievska[1*], *Inna* Gorofianiuk[2], *Liudmyla* Koval[3]

[1]Department of Ukrainian Studies, National Pirogov Memorial Medical University, Pirogova str. 56, 21028 Vinnytsia, Ukraine
[2]Ukrainian Language Department, Vinnytsia Mykhailo Kotsiubynskyi State Pedagogical University, Ostrozkogo str. 32, 21001 Vinnytsia, Ukraine
[3]Department of Ukrainian Language and Culture, Vasyl' Stus Donetsk National University, 600-richchya str. 21, 21021 Vinnytsia, Ukraine

Abstract. The purpose of the article is to investigate the perception to human health, its preservation and the general understanding of the phenomena of vitality and health as such in the sustainable society. The article is dedicated to the gender analysis of the perception of the above phenomena. The purpose of the proposed study was to find out associations related to these concepts, to model associative fields by the men's and women's reactions, as well as to establish the most valuable features in the perception of this concept by different genders. The study used a free associative experiment in the written form. The respondents were students-leaders and their deputies (42 women and 29 men) aged 17-21. We used the statistical method to calculate the results of the experiment; as well as the comparative method and the method of modeling of associative fields. Comparing the men's and women's associative semantics of these phenomena, we conclude that there is a generally positive perception of the concept under study, but men show negative characteristics of this phenomenon slightly more than women.

1 Introduction

In modern sustainable society marked by the recent challenges such as globalization and global warming, and, most recently, pandemics and global diseases, the issue of forming leadership qualities of an individual in the general context of the health care system and specifically to maintain his/her own vital energy and health is extremely important [1, 2].

In the research literature, the leader is the one who knows and fulfills his mission, gives it priority, can persuade others to follow him. Therefore, a leader is a person who sets a goal and finds different ways and opportunities to achieve it. Is it possible to see a pattern between the level of development of leadership qualities of the individual and the attitude to his/her own health, its preservation, and in general the understanding of the phenomenon of health within the context of the environmental sustainability.

In general, human health is one of the man's greatest values which also impacts on the sustainable development of the human society. It is a prerequisite of the human energy or the

*Corresponding author: kraievsk@gmail.com

vital energy of the human beings. This concept is extremely complex in social, psychological, and as a consequence – in linguistic aspects. Depending on the different socio-cultural, economic factors, speakers form their own perception of health. There may also be gender differences in the perception of the "health" phenomenon.

There are many definitions of "health" in modern literature. The most typical elements of the definition are:

1. The normal function of the organism at all levels of its organization (organs, histological, cellular and genetic structures, the normal course of physiological and biochemical processes that contribute to the individual's survival and reproduction).
2. Dynamic balance between the organism, its functions and the environment (the environmental sustainability).
3. Ability to fully fulfill basic social functions, participate in social activities and socially useful work.
4. The ability of the body to adapt to constantly changing conditions of the environment, the ability to maintain the stability of the internal environment of the organism, ensuring normal and versatile life and preservation of a living principle of the body.
5. Absence of diseases, diseased conditions, painful changes.
6. Complete physical, spiritual, mental and social well-being [3].

According to the WHO Constitution, health is not the absence of a disease or physical disabilities, but a state of complete physical, mental and social well-being; it is the natural state of the organism, which is characterized by its balance with the environment and the absence of any painful changes [4]; according to the Dictionary of the Ukrainian language, this is the state of the organism in which all its organs function normally [5].

Such definitions differ from the traditional medical model, which defines health as the absence of illness or disease and emphasizes the role of clinical diagnosis and intervention. Researchers note that the WHO definition clearly links health to well-being and conceptualizes health as a certain human right, requiring physical and social resources to reach and sustain it.

Brechmann [6] offers an information and psychological interpretation of the essence of health: "Human health is its ability to maintain age-appropriate resistance to sudden changes in the quantitative and qualitative parameters of the flow of sensor, verbal and structural information". Academician Kaznacheyev [7] emphasized that health is a holistic multidimensional dynamic state that develops in the process of realizing of the genetic potential under conditions of a specific social and ecological environment and allows a person to exercise his biological and social functions to different degrees; the process of preservation and development of physiological properties, the potential of mental and social properties; the process of maximum life expectancy at optimum efficiency and social activity. Kudriavtseva [8] defines public health as a socio-demographic category that reflects the ability of the society members to fulfill the function of further development of the society and lead a way of life that ensures the preservation, strengthening and development of this ability.

Huber et al. [9] offered a fundamentally new definition of health: "Health is the ability to adapt and self-organize, which includes the ability of people to adapt to their situation as a key to health. He also recognized the subjective element of health. In his opinion, health and well-being will depend on the needs of a specific person.

Thus, the concept under study is complex and multidimensional. As you can see, scientists interpret the studied concept differently. According to Boichuk [10] there are more than 450 definitions of the concept of "health" that have been formulated by experts of various scientific disciplines. The scientist pointed out six basic types of essential elements of the health definition:

1. Health as a norm of functioning of an organism at all levels of its organization;
2. Health as a dynamic balance (harmony) of vital functions of an organism;
3. Health as a full-fledged fulfillment of basic social functions, participation in society and active labor activity;
4. The body's ability to adapt to changing environmental conditions;
5. Absence of pathological changes and normal state of health;
6. Complete physical, spiritual, mental and social well-being.

As we can see, the problem of perception of the health phenomenon is controversial. Representatives of biomedical sciences mainly deal with this issue. However, integration processes in science have led to the fact that various aspects of the perception of human "health" are also of scientific interest in pedagogy, psychology, philosophy and sociology. The concept of "health" in the linguistic aspect has repeatedly become the object of study of linguists: they analyzed different planes of the concept of "health" in the Ukrainian linguistic picture of the world; evaluation and emotional specificity of the nomen; phraseological units for the designation of the human health; the phenomenon of "health" in age and national aspects.

Our research is, to a certain extent, a continuation of the studies on "Health as a conceptual phenomenon: age, national and social aspects of the associative semantics" [11]. The results of the mentioned work showed that representatives of different linguistic cultures of the same age have more common ideas about the health phenomenon than native speakers of Ukrainian linguistic culture of different ages. Health is a cross-cultural phenomenon, so age, not nationality, is a dominant factor in the formation of perceptions of health [11]. In our opinion, it is important to continue to study the phenomenon of health in the gender aspect of analysis.

Therefore, in the proposed study, we will attempt to find out certain motives, assessments, attractions, personality settings that appear in the respondent's mind when perceiving the concept of "health". To describe ways of conceptualization of the "health" concept by representatives of different articles, compare models of implementation of the "health" concept. The results of the study show us the interpretation of the health phenomenon in general.

The relevance of our study is strengthened by the fact that the word "health" is absent in the Associative Dictionaries of the Ukrainian Language [12, 13] and the Slavic Associative Dictionary [14]. This token is neither represented in the first associative dictionary published in 1910 in the USA [15]. The list of the word associations of this dictionary formed the basis for a number of subsequent works. The analyzed token is present as a word-stimulus in the Dictionary of Associative Norms of the Turkish Language [16] and the Edinburgh Associative Dictionary [17].

Such a study should be considered interdisciplinary, since the problem of society stratification through the interaction of language and gender is raised, and the medical-biological concept is taken for analysis.

The scientific novelty is that, for the first time, the results of a free associative experiment aimed at explaining the perception of health by Ukrainian youth were put into practice, an attempt was made to identify a common and different perception of the phenomenon of health by women and men. The object of the study is the conceptualization of the concept of health by representatives of different articles. The subject of the analysis is the linguistic realization of the concept of health as a basis for the linguistic modeling of ideas about the state of the human body. The hypothesis of the study is the assumption of a difference in the perception of the phenomenon of "health" by representatives of the same social group, the same age, but different sexes.

2 Research objectives

The objective of the study is to find out by means of a free associative experiment what students-leaders of different sexes associate with health; to compare the men's and women's results; to model associative fields by men's and women's reactions; to establish what is the most valuable in the perception of this concept by different sexes.

3 Research methods and methodology

A free associative experiment was used during the study. For the first time an associative experiment was applied in psychology. The purpose of such an experiment is to identify associations that an individual has formed over the life experience.

There are 3 types of associative experiment: free, purposeful and chain.

We used a free associative experiment, during which respondents were offered a word (word-stimulus) to which they had to react with a word (word-reaction), to which no requirements for form, content, manner of expression was imposed.

Preference was given to the written form avoiding collective responses, emotional cues, exclamations.

Thus, the respondents were offered a "health" word-stimulus, which they had to respond as quickly as possible whatever came uppermost. Age and gender should also be specified. 42 women and 29 men were interviewed.

The material of the study was the results of a free associative experiment conducted with 69 leaders of student groups and their deputies of the Faculty of Economics at Vasyl Stus Donetsk National University. In our study, the following factors were set:

1. Social group – students-leaders;
2. Specialty – economics;
3. The number of word-stimulus offered – one word-stimulus "health";
4. Nationality – Ukrainians;
5. Country of residence – Ukraine;
6. Age of the respondents – 17-21 years;
7. Mother tongue – Ukrainian. Variable factors included the sex of the respondents.

The results of this experiment can be considered as a certain associative profile of consciousness images, which combine the mental and sensory components of the health perception by respondents of different sexes.

We used the statistical method to calculate the results of the experiment; as well as the comparative method and the method of modeling of associative fields.

The statistical method was used to calculate the frequency of the same reactions to the word-stimulus "health" and the percentage ration. In Table 1 and Table 2, for each word-reaction, the number of respondents who answered equally is given in parentheses.

The comparative method was used to compare the results of the associative experiment in a group of students of male and female genders.

The associative field modeling method is aimed at distribution of the reactions obtained to the nucleus, the near periphery as an intermediate link, and the periphery. The boundary between the nucleus and the peripheral reactions is calculated by the formula n/2, where n is the upper frequency limit of a specific response to a word-stimulus [18]. For example, in our study, the most frequent women's reaction is the word *sports* – 9 reactions, then the reaction with a frequency of at least 4,5 will go to the nucleus. Other reactions are distributed between the near periphery and the periphery according to this principle.

4 Main results

We received 81 associations. This is more than the total number of respondents, as men in 6 cases gave 2 or more word-reactions. In this case, only the first word-reaction was considered and reported in the study.

There were no denials in women's questionnaires. Respondents clearly followed the instructions, mentioning one word-association (Table 1).

Table 1. The structure of the semantic field of the "health" token according to the results of a free associative experiment – female.

The nucleus of the associative field	The near periphery of the associative field	The periphery of the associative field
21,5 %	35,7 %	42,8 %
Sports (9)	Happiness (4), sleep (4), body condition (3), healthy lifestyle (4)	Well-being (2), peace (2), wealth (2), hospital (2), pills (2), life (2), immunity, dream, treasure, joy, power, smile

The recorded associations have mostly positive and neutral semantics, although there have been 4 negative reactions (*hospital* (2), *pills* (2)). All words-reaction are grouped into semantic groups.

The first group includes a number of words related to the factors of health preservation: *healthy lifestyle, sports, sleep, rest, immunity.*

The second group is filled with reactions that indicate the images that the woman associates with health: *smile, dream, joy, treasure, wealth, well-being, strength, happiness.*

The third group consists of reactions related to the medical sphere: *hospital, pills.*

The fourth group is represented by the phrase *body condition* and is an element of the essence of the concept of health. The ratio of the quantitative composition of semantic groups (by the results of the associative experiment among women): the associative field is a set of reactions to the word "stimulus" (detected experimentally) represents a specific model that captures the associative bonds of words. The word "stimulus" is the one given to the respondent which she or he must quickly respond to with another word. The word "reaction" is a word association that arises in a given person's memory and is caused by the offered stimulus word.

Table 2. The structure of the semantic field of the "health" token according to the results of a free associative experiment – men

The nucleus of the associative field	The near periphery of the associative field	The periphery of the associative field
38 %	27,5 %	34,5 %
Healthy lifestyle (5), strength (3), pills (3)	Food (2), pharmacy (2), hospital (2), life (2)	No bad habits, sports, work, body condition, happiness, good thoughts, expensive, smile; something that is easy to lose and hard to regain; ☺.

The largest group includes the reactions of women, related to factors, conditions of health preservation (50%), which are associated with an active and right way of life. All the images (32%) that women associate with health are bright and life-affirming. Reactions show a mostly positive image of health as the highest value of an individual.

Analysis of the results of the experiment with men showed that in 6 cases they provided the whole associative series for the word-stimulus. Under such conditions, the first reaction was taken into account. No questionnaire refusals were recorded (see Table 2 above).

The overall image of the men's associative semantics of "health" is mostly positive, but it is not devoid of neutral and somewhat negative characteristics. All tokens are grouped into 6 semantic groups (see Figure 1).

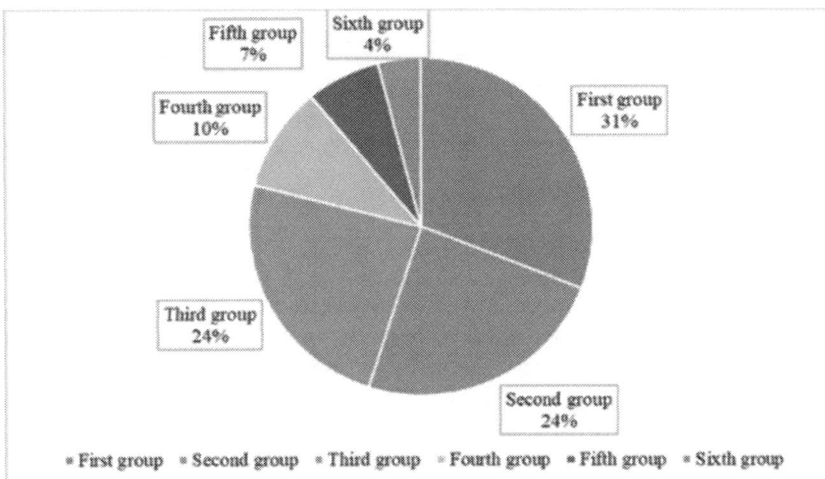

Fig. 1. Ratio of the quantitative composition of semantic groups (by the results of the associative experiment among men)

Among the male respondents, the largest group of reactions related to physical activity, refusal from bad habits and nutrition (50%). In the men's minds the word-stimulus "health" is equally (24% each) related to medical concepts and certain images with which it is associated. The frequency responses related to the evaluation, characteristics of health properties, the essence of the concept and there is a single reaction symbol. Men's associative semantics of health is overwhelmingly positive, but men demonstrate negative characteristics of this phenomenon (*hospital, pharmacy, pills; something that is easy to lose and hard to regain*) more than women.

5 Discussion

The comparison of the associative fields, modeled by the men's and women's words-reactions, shows that in both cases according to the number of reactions the factors of health preservation dominate. The reaction-images with which a person associates health prevail by variety: women – 8 tokens, men – 6. Several reactions represent the essence of the concept in both groups of respondents. Also typical for both groups of respondents is the reaction-abbreviation *ЗОЖ* (in Ukrainian) *healthy lifestyle*.

Unlike women, men have reactions to evaluation, characteristics of health properties. The symbol ☺ was recorded without a verbal explanation, which is more common of women, and there was an idiom with obscene words, which can be considered as a certain manifestation of men's linguistic aggression.

In the nuclei of both associative fields there are reactions related to the factors of health preservation, but the men's nucleus (4 tokens) is more diverse than women's (1 token). The nuclear features of the field, modeled by the results of the men's reactions, come to the near periphery and the women's periphery.

Analyzing the set of reactions of the near periphery of both groups of respondents, we note that the medical motives dominate in the men's group and factors associated with an active and right way of life prevail by women.

The filling of the periphery of the associative fields differs significantly. Only the *smile* reaction was common.

In general, the composition of the considered associative fields reveals certain coincidences (*smile, pills, hospital, strength, healthy lifestyle, sports, body condition, happiness*), differing mainly only in their structure, i.e. the position in the men's and women's value hierarchy. Most of the common associations are nuclear signs or belong to the near periphery. It has also been observed that the word associations (*no bad habits, body condition, good thoughts; something that is easy to lose and hard to regain*) prevail more in men's associations than in women's (*body condition*).

Some coincidences with the results of our study are found in the materials of the Dictionary of Associative Norms of the Turkish Language. The experiment was conducted with students and teachers of the Erciyes University aged 17-25 years (Kayseri) in 2004-2005. No gender distinction was made in the mentioned work. There is a reaction of *sports*, which dominates in the valuable hierarchy of the Ukrainian youth of female gender, is also numerous and belongs to the nuclear reactions. However, the nuclear words-reactions of men *healthy lifestyle, strength, pills* are not recorded as a response to a "health" word-stimulus. Such results prove the partial similarity of the perception of such a universal concept (known to all mankind and represented in different languages) by the youth of different cultures. However, this still confirms the presence of associative universals in youth of different cultures [19, 20].

In the example of the Edinburgh Associative Dictionary that was give earlier, the reactions that are not specified by our respondents are represented as frequency responses. The matches are only traced in the periphery and the near periphery.

According to the results of the association's research, young respondents generally do not correlate health with the traditional medical model, which defines health as the absence of illness or disease and emphasizes the role of a clinical diagnosis and intervention. Most of the surveyed students recognize health as an active way of life, sports. Although the research in the senior age group showed the person's physiological feeling *when nothing hurts* as a nuclear reaction. 15% of the senior and elderly people associated this concept alike. The comparison of the reactions of students of different sexes and the senior and elderly shows that the age is a dominant factor compared to the gender in the formation of the health perception.

6 Conclusions

The gender analysis of the associative field of the "health" concept revealed common and specific in the minds of the students-leaders. By comparing the men's and women's associations it was found that the factors of health preservation were valuable for both groups. *Sports* and *healthy lifestyle* were the most frequent responses. This is probably due to the promotion of an active and healthy lifestyle, a certain fashion for it. Common reactions of men are associations, with reference to the medical topics, and women have certain images that correlate with the health. In general, the concepts that youth associate with the health are positive. The conducted study has revealed the peculiarities of the students' world view, attitude to the health as the highest value of an individual.

The associative field is a set of reactions to the word-stimulus detected experimentally; a specific model that captures the associative bonds of words. The word-stimulus is a word given to the respondent, which he or she must quickly respond to in another word. The word-reaction is a word association that arises in a person's memory, caused by the offered word-stimulus.

In the course of this study, we obtained 81 associations were obtained for the "health" word-stimulus. A number of specific features of perception of the "health" phenomenon by representatives of different sexes were established. The associative fields of two sexes were modeled according to the obtained associations. The composition of the considered associative fields reveals certain coincidences, but differs in its structure, i.e. the position in the men's and women's value hierarchy. In the nucleus of both fields there are reactions related to the factors of health preservation. Medical motives dominate in the men' s near periphery, while women specified the motives connected with the active and healthy lifestyle. There are virtually no matches in the far periphery of the associative fields.

This study does not claim to be exhaustive. In perspective we plan to expand the experimental base and improve the methods of conducting and processing of the results of the experiment, which will allow to form a certain social portrait of the student in the context of perception of the health.

References

1. C. Guerriero, A. Haines, M. Pagano, Nature Sustainability, **3(7)**, 494-496 (2020)
2. A. Khayatzadeh-Mahani, R. Labonté, A. Ruckert, E. De Leeuw, Global health promotion, 26(1), 100-104 (2019)
3. O.A. Fedko, *http://www.dy.nayka.com.ua/?op=1&z=76* (2009)
4. WHO, *www.who.int/governance/eb/who_constitution_en.pdf* (1946)
5. *SUM Slovnyk ukrainskoi movy* (1972)
6. Y.Y. Brechmann, *Vvedenye v valeolohyiu – nauku o zdorove. Fizkultura i sport* (1990)
7. V.P. Kaznacheyev, *Teoretycheskye osnovy valeolohii: Fundamentalnye nauky (*1993)
8. E.N. Kudriavtseva, Voprosy filosofii, **12**, 98–109 (1987)
9. M. Huber, J.A. Knottnerus, L. Green, British Medical Journal, **343(2)**, 3-4 (2011)
10. Y.D. Boichuk, *Zahalna teoriia zdorovia ta zdoroviazberezhennia* (2017)
11. H.P. Kraievska, I.V. Gorofianiuk, *Health as a conceptual phenomenon: age, national and social aspects of the associative semantics* (2020)
12. N.P. Butenko, *Slovnyk asotsiatyvnykh norm ukrainskoi movy* (1979)
13. S.V. Martinek, *Ukrainskyi asotsiatyvnyi slovnyk: Vid sty`mulu do reakciyi* (2007)
14. *(SAS) Slavianskii assotsyatyvnyi slovar: russkii, belorusskii, bolharskii, ukraynskii* (2004)
15. G.H. Kent, A.J. Rosanoff, Americal Journal of Insanity, **67**, 37–97 (1910)
16. *(TDCNS) Türk Dili Çağrışım Normları Sözlüğü: 160 000 çağrışım* (2014)
17. G.R. Kiss, G. Armstrong, R. Milroy, *http://www.eat.rl.ac.uk/* (1972)
18. A.P. Klymenko, *Voprosy psykholinhvisticheskoho yzuchenia semantiki* (1970)
19. B. Wheaton, G. Roy, R. Olive, Sustainability, **9(12)**, 2298 (2017)
20. H. Kruize, N. Van der Vliet, B. Staatsen, R. Bell, A. Chiabai, G. Muiños, S. Quiroga, P. Martinez-Juarez, M. Aberg Yngwe, F. Tsichlas, P. Karnaki, M. Lima, S. García de Jalón, M. Khan, G. Morris, I. Stegeman, International Journal of Environmental Research and Public Health, **16(22)**, 4403 (2019)

Sustainable education and information technologies

Veronika Grebennikova[1]*, *Oleg* Grebennikov[1], *Elena* Baydetskaya[1], and *Galina* Mikerova[1]

[1] Faculty of Pedagogy, Psychology and Communication, Kuban State University, Stavropol str. 149, 350040, Krasnodar, Russian Federation

Abstract. This article focuses on the sustainable education and information technologies. It analysis the importance of the online education and the benefits it can bring, especially in the light of the recent massive online teaching and home office working mode during the COVID-19 pandemic, The flexibility of the education tools that include the use of information and communication technologies (ICT) appears a great facet of the sustainable education that might help to contribute to mitigating the climate change and to prepare responsible citizens who would adhere to the principles of the sustainable development. The article concludes with a call to integrate the concept of sustainable information into educational programmes for computer scientists and future ICT professionals. In addition, it suggests that information technologies should find a wider use in promoting sustainable development as a part of the sustainable education.

1 Introduction

Information and communication technology (ICT) is often being used as a tool to enable the development of a sustainable education system and a more efficient and sustainable society [1]. In the education system, it represents a driving force behind the idea that the ability to convey desirable goals and values that cannot be overemphasized in society would enable expected learners to be useful to themselves and to society as a whole. It transfers desirable goals and values to education, which can then be anchored in our society through the use of the instruments available to us today [2].

As the global warming and climate change becoming worse and cause adverse effects on our socio-economic sphere, business companies worldwide must dedicate themselves to greening their IT systems in order to reduce costs. Current and future government and business workers need to be educated and informed about the importance of sustainable education systems and information technologies in their work [3]. Many scientists stress the importance of teaching students about the benefits and importance of green computing. Although the adoption of green education and information technologies in the workplace is slower than expected, given obvious trends and entrepreneurial drivers, forward-looking organizations are looking for ways to train workers today. Another green computing teaching method involves hands-on projects involving portable wireless sensor networks. Researchers

* Corresponding author: vmgrebennikova@mail.ru

found that introducing green computer content into basic courses increased awareness of sustainability issues, contributed to the personal introduction of green technologies and contributed to project use [4]. One can see that there are certain strategies for integrating sustainable information into the curriculum and found that integrating sustainable information contributed to understanding the sustainability of computer applications. Figure 1 that follows shows the share and number of students potentially reached by digital and broadcast remote learning policies as reported by UNICEF [5].

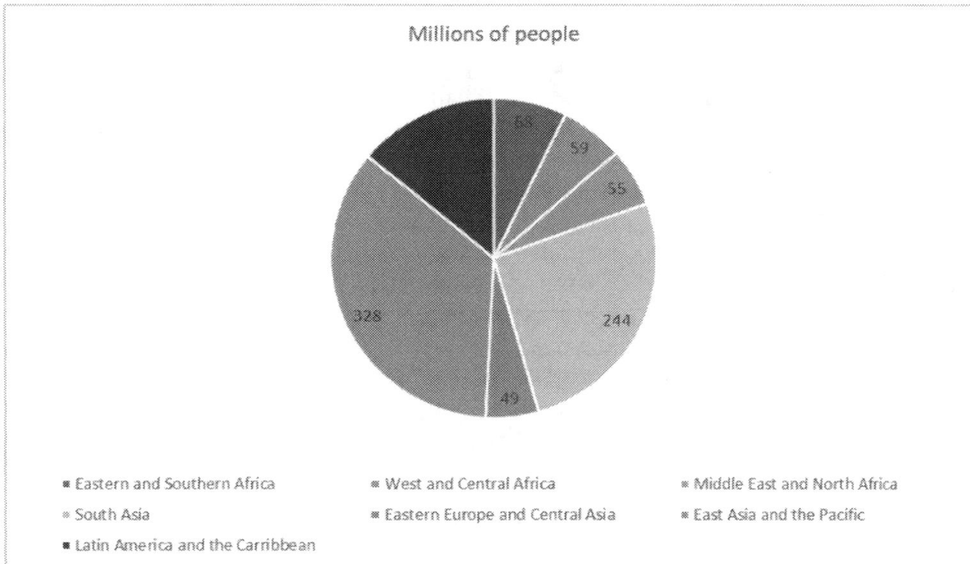

Fig. 1. Share and number of students potentially reached by digital and broadcast remote learning policies

Information science in developing countries has both social and economic dimensions. The concept of a sustainable information system was developed in the framework of the World Summit on Information Society and Information Technology and the International Conference on Sustainable Information Systems [6].

The ICT sector plays an important role in supporting industry - relevant training that promotes high-quality training programmes to prepare students for employment. There is a growing need to complement university students themselves with courses that strengthen their ongoing technical studies and ICT skills and increase their readiness for jobs in the telecommunications and ICT sectors. Bridging the skills gap between current and future workers is therefore crucial to meet the rapidly changing skills needs of a world increasingly dominated by information and communication technology [7].

There is a strong correlation between the use of environmental education and the development of cognitive skills in children and adults. Many studies show that students with a strong interest in environmental issues such as climate change, water quality, energy efficiency and environmental sustainability have developed the ability to synthesize complex information about their knowledge of the environment [8]. This is just one example of the diversity of approaches to knowledge acquisition that we are seeing in education today, particularly with regard to education for children and adults. Educational approaches that focus on experience, problem-based learning and problem-oriented learning are designed to enable individuals to actively participate in a world around them and thus to form a deeper understanding of a particular subject or area of interest. This approach to reshaping education includes both problem-based and phenomenon-based learning methods used in countries around the world that are known for having one of the best education systems.

2 Sustainable education and distance learning

Nowadays, in the middle of the COVID-19 pandemic that sent millions of people to home offices and online working mode, there has never been a better time for students, professors, and innovators to lay the foundations for a new kind of education [9]. The introduction of online courses in the middle of the semester will allow students to compare digital with analogue versions of their courses. As a result, hybrid education models are well positioned in a post-pandemic world to respond to existing challenges, seize opportunities, and broaden the range of opportunities available to students and educators in the global education system. Second, disruptive technologies provide a global learning community with access to a wide range of resources and opportunities for innovation and cooperation. On the other hand, they disrupt the very models that try to make learning possible [10].

While some states have been praised for distance learning in the midst of the coronavirus pandemic, the transition from underclass to online education poses significant challenges. As universities begin to shift their activities online, education agencies are rushing to provide support for distance learning adaptation, as well as training for teachers, administrators, students and staff. With regard to the above, we all have to prepare for numerous steps to prepare for a large dose of online education, including a one-to-one initiative that will make notebooks available to all middle and high school students [11].

Figure 2 that follows provides the results of the United States Census Bureau Household Pulse Survey showing the percentage of households with children by reported change [12].

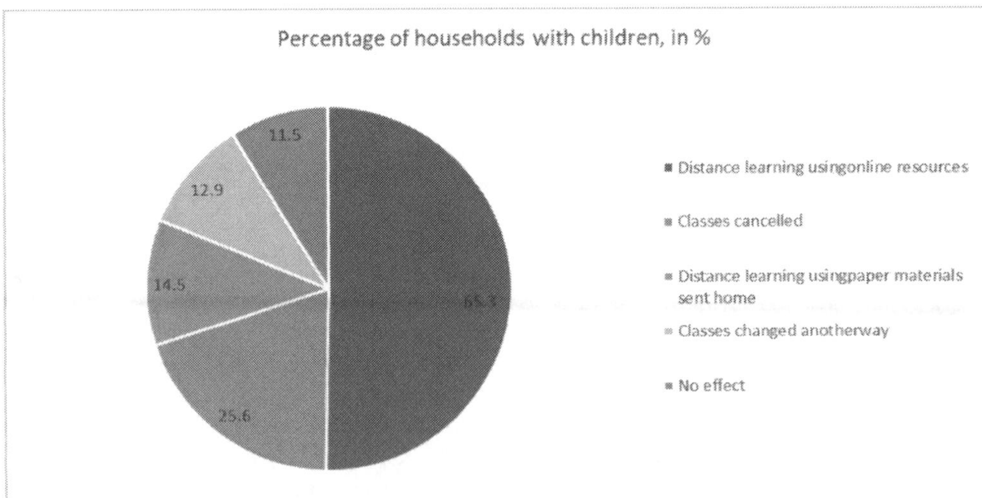

Fig. 2. Impact of pandemic on children's education in the United States

With the start of the new school year, distance learning will continue to play a role, as it is often linked to learning in the classroom. Nevertheless, educators are quick to respond to ensure that learning can continue even in less ideal circumstances. When it became necessary this spring, very few school districts were prepared for distance learning. The move towards distance learning has been challenging, with some preparatory work done, including the development of a comprehensive online education curriculum. But we all know that distance learning will remain here and educators need to optimize the experience to get the best results for students. Most of the above would require more resources than people at university currently have. But the shift to home schooling has highlighted the importance of comparing people with lower socioeconomic status with more privileged classmates. This assessment also makes it necessary to improve online education platforms for universities to better meet the needs of students working and living in environments that can sometimes be resource-intensive - poor and poor internet connectivity. To close this capacity gap, we are thinking about the need for organizations, people, and universities to develop stronger partnerships, both globally and regionally.

3 Online education in the pandemic

In the long run, specific conditions - informed decisions and programs that take the country into account - have the potential to improve pedagogy, improve school governance, and address many other aspects of the learning experience. As countries rebuild themselves in response to COVID-19, there is an urgent need to accelerate research on how best to support high-quality education. We build on the work of international organizations such as the World Economic Forum and the United Nations Development Program to develop education strategies that can be used in a variety of contexts, from high-risk and low-income countries to high-poverty and low-education countries. Perhaps one positive outcome of this pandemic is that it will lead us to address many of the remaining global education problems sooner than expected. As more universities and students see the benefits first hand, the trend towards online learning is likely to continue well beyond the pandemics [13].

The online method of education is that it can be a highly effective alternative educational medium for mature, self-disciplined students. Online asynchronous education allows for greater flexibility and flexibility in terms of the student's learning experience and gives students control over their learning experience. However, this gives the student more responsibility and leads to more stress and stress. For students to participate successfully in online programs, they must be well organized, self-motivated, have a high level of time management skills to keep up with the course, and have a high level of motivation and motivation. The researchers go further, calling online education a pandemic of "distance learning in an emergency, as the latter is at odds with the quality of effective online learning. For this reason, online education institutions do not behave in a dependent manner on other students who are dependent learners and have the responsibility to assume the responsibilities required by an online paradigm. Before the COVID-19 pandemic, some schools ran distance learning programs that helped the higher citadel of learning with its migration process. However, many schools and universities were forced to close because the government had not announced measures to curb the spread of COVID-19, which led to the closure [14].

Previously, it had conducted a high-quality online education programme to support high learning levels in its migration processes. To limit the rapid spread of the pandemic by the global community, the only way universities have access to online learning is to use it. Large parts of the world are quarantined, and that is why many cities have turned into ghost cities. The quarantine effect is also observed in schools, colleges and universities. The coronavirus has led to the institutions moving from offline to online mode of education. Various online teaching and online learning can be described as a panacea for this crisis. This crisis will lead to institutions that were previously reluctant to change accepting modern technologies.

All in all, the COVID-19 pandemic has affected millions of students worldwide and led to unprecedented closures of university facilities. Some have continued their academic courses while avoiding gathering people in their dormitories and other public spaces such as schools and universities. Although children under 19 regularly attend online schools, a large majority of these children have no previous experience of the pandemic of distance learning. This study compares the effects of online education activities on the learning experience of students and documents how student-centred active learning can be offered in the context of distance learning in a dental school. Learning through group discussions, which take place synchronously on an online communication platform, is a new learning method that has not been implemented in dental schools.

The American Association of Colleges and Teacher Education, which surveyed 76 percent of college graduates, is working with schools to determine the impact of online education on their students' learning experiences. In an online learning environment, students can have more control over their learning process. In K-12 education, online learning can be an important part of the learning experience for students, faculty, staff and administrative staff. The recession has forced most K-12 schools in the United States to close in recent years [15]. We are starting to wonder what kind of support schools and universities need at the moment. Many teachers and lecturers are forming working groups to figure out how to tailor their programs, including developing technologies - improved teaching solutions to address the pandemic - and moving to distance learning. Future teachers may, for example, offer virtual lessons to pupils, or offer distance tutoring or support teachers in developing learning materials.

4 Sustainable development and flexibility of education

According to the World Economic Forum's Global Competitiveness Report, the number of students enrolling in higher education worldwide is expected to reach about 300 million. This expansion has led to a diversified sector that reflects the diversity of learners entering the system, including from non-traditional groups [16]. As a result, higher education is increasingly expected to adapt to different learning needs, which means offering well-structured and flexible learning pathways that can deliver the best results. In order to strengthen equity in lifelong learning, it is recognised that the education system needs to be adapted to better support flexible learning pathways. The general premise is that life - long learning - must become more than a hopeful catchphrase if people and businesses are to thrive in a rapidly changing global economy. Access to education should be free and fair, and should provide young people with the skills they need for their lives. This is a continuation of the global push for universal access to quality education for all children. There is an urgent need to ensure high quality education to ensure that every child worldwide is taught at primary and secondary level by 2030. It is therefore necessary for the higher education system and employers to work together to introduce flexibility through accelerated degrees and to enable learners to combine study and work with work in study. Flexible learning paths between qualification and employment offer important benefits as economies recover from the COVID-19 crisis [17]. By enabling more diverse learning opportunities, learners' educational profiles can be adapted to the needs of their employers and the wider economy. Such practices have the potential to improve the employment of graduates and introduce flexibility to respond to different learning styles. The growing range of available skills of the workforce and the increasing skills of the labour market allow people to learn when they want to. Internet access offers education policy makers the opportunity to improve the quality of education for individual learners and contribute to national economic and social well-being. To ensure that implementation plans take advantage of this opportunity and remain relevant in the development of technologies and services, policymakers should monitor the development of the Internet. Internet access and its potential to transform lives and contribute to the achievement of education and other sustainable development goals. Even before the COVID-19 crisis, it was recognised that a flexible higher education system offers a variety of learning pathways to promote equity and lifelong learning. Recent research on the capacity of universities to respond to the needs of their pupils has shown, however, that greater flexibility is needed in the development of education policies and implementation plans [18].

https://doi.org/10.1051/e3sconf/202125004012

There is little, but growing evidence that commitment to these goals, which has never been more relevant, and the modes of effective information and governance systems that best respond to crises are critical to achieving the Sustainable Development Goals (SDGs) and other education goals [19]. In this paper, we argue that it is important for students at all levels to recognize that sustainable development, democracy and peace are indivisible. EFSs at secondary level offer all students the opportunity to explore and develop strategies that enhance their knowledge and skills with democratic values and principles. It is important that future citizens act as agents of change for the common good, including economic, social and environmental justice, and that EFS can improve pupils' knowledge and skills at secondary levels, which hopefully will lead to such actions. In the digital age, citizens need to be prepared to respond to the challenges posed by social media and other forms of information and communication technology (e.g. social networks such as Facebook, Twitter, Instagram and others). This will allow young adults to address the problems they may face in the future and develop strategies for collective movements to address these problems [20]. The world urgently needs to rethink how education is delivered, how it is taught, and what skills children need to become healthy and productive members of society.

5 Concluding remarks

Overall, sustainability necessarily includes not only the need for sustainable education, but also sustainability in all areas, and this includes the challenges of globalisation. Moreover, the impact of ICT on the added value of the global education system and its old education system has remained controversial.

Sustainable education provides learners with creative problems - by solving skills needed to meet complex ecological challenges in the 21st century. We are facing the challenges of meeting the needs of a world of more than 1.5 billion people, and it is estimated that the world's population (over 7 billion) will need twice as many resources in 2030 as our planet will provide in its lifetime. The final aspect of environmental education policy is to train individuals to thrive in a sustainable society.

In order to succeed in the 21st century, citizens must have a strong relationship with nature, in which they build strong relationships with it. In addition, the current workforce needs to be retrained and retrained to adapt to the new green economy. Fortunately, the practice of sustainability, which balances the needs of today's people with the future of the people, can provide a solution to these difficult problems. This seminar will examine the role of education in developing a sustainable future for the United States and its citizens and how these ideas can be put into practice. Although there is no generally accepted definition of "sustainability," students will explore what makes them "final" and ultimately build on the class definition throughout the semester. Examples of these ideas are the development of a "thinking system" for dealing with the environment, the use of renewable energies and the application of sustainability in the education system.

Thence, effective leadership in combination with school - a broad range of effective teaching - means that students have equitable access to high-quality learning opportunities, resulting in improved outcomes that are sustainable over time. Using these approaches, many high educational institutions are successfully using environmental education to develop key 21st century skills. Environmental education can be integrated into existing curricula and helps teachers meet standards in several disciplines. This development of modern teaching methods can be supported by providing physical and digital infrastructure. Therefore, it seems that stakeholders need to integrate the concept of sustainable information into educational programmes for computer scientists.

Acknowledgement

We would like to thank to the support from the project "Methodology and technology of social and pedagogical support of the family as a subject of designing an individual educational trajectory for a child". The scientific supervisor of the project is V.M. Grebennikova (RGNF (RFBR).

References

1. P. Jones, M. Wynn, D. Hillier, D. Comfort, Indonesian Journal of Sustainability Accounting and Management, **1(1)**, 1-15 (2017)
2. U. Yessenbekova, O. Turzhan, K. Koshanova, I. Yegemberdiyev, B. Kutym, Media Watch, **11(2)**, 348–355 (2020)
3. P. Kirby, *Karl Polanyi and the Contemporary Political Crisis: Transforming Market Society in the Era of Climate Change* (2020)
4. S. Sunny, C. Shu, Small Business Economics, **52(4)**, 905-927 (2019)
5. UNICEF, Education and COVID-19, *https://data.unicef.org/topic/education/covid-19* (2020)
6. B. De la Hoz-Rosales, J. Camacho-Ballesta, I. Tamayo-Torres, Entrepreneurship and Sustainability Issues, **7(2)**, 782 (2019)
7. W. Strielkowski, D. Streimikiene, A. Fomina, E. Semenova, Energies, 12(24), 4790 (2019)
8. S. Otto, P. Pensini, Global Environmental Change, **47**, 88-94 (2017)
9. S. Donitsa-Schmidt, R. Ramot, Journal of Education for Teaching, **46(4)**, 586-595 (2020)
10. A. Sangster, G. Stoner, B. Flood, Accounting Education, **29(5)**, 431-562 (2020)
11. L. Mishra, T. Gupta, A. Shree, International Journal of Educational Research Open, **1**, 100012 (2020)
12. USA Facts, 65% of households with children report the use of online learning during pandemic, *https://usafacts.org/articles/65-of-childrens-education-has-moved-online-during-covid-19* (2021)
13. J. Daniel, Prospects, **49(1)**, 91-96 (2020)
14. M. Savic, World Futures Review, **12(4)**, 385-395 (2020)
15. U. Kaden, Education Sciences, **10(6)**, 165 (2020)
16. M. Wotto, Journal of Educational Technology Systems, **49(2)**, 262-281 (2020)
17. M. Sigala, Journal of Business Research, **117**, 312-321 (2020)
18. C. Hodges, S. Moore, B. Lockee, T. Trust, A. Bond, Educause Review, **27**, 1-12 (2020)
19. O. Kostoska, L. Kocarev, Sustainability, **11(7)**, 1961 (2019)
20. J. Guo, N. Liu, Y. Wu, C. Zhang, Information & Management, **58(1)**, 103286 (2021)

 https://doi.org/10.1051/e3sconf/202125004013

Sustainable healthcare within the strategy of regional development

Michail Nekrasov[1]*, *Svetlana* Kalina[1], *Svetlana* Veprentsova[1], and *Yulia* Kekteeva[1]

[1]Departament of Psychotherapy, Pirogov Russian National Research Medical University, Ostrovityanova st., 1 Moscow, Russian Federation

Abstract. This article has an aim of connecting the sustainable healthcare to the environmentally-friendly strategy of regional development. We describe how a growing interest in ways health organisations and systems measure and report on their sustainability is recorded in the recent years. In addition, we note the growing need for sustainability indicators in the health system, such as health insurance coverage, quality of care, and access to health services. Our results show that environmental pollution and other hazardous activities have a profound effect on human health and therefore create a burden for the healthcare system. In addition, we discuss the importance of sustainable healthcare for the health and well-being of patients and the environment. In the same time, we stress that high standards of healthcare should be preserved and maintained in spite of the reduction in health costs and environmental impacts as a part of the sustainable regional development strategies.

1 Introduction

Currently, many countries around the world aim to build a safer, lower-carbon health system by using effective tools and approaches. They are integrating their health systems into their efforts to achieve the World Health Organization's (WHO) Sustainable Development Goals (SDGs) [1, 2]. Performance measurements for external accountability, quality indicators have been identified by a number of organizations focused on improving health care, which are used as aggregate measurements for performance purposes or as design mechanisms to support internal quality-improvement processes, such as the World Health Organization (WHO) and the United Nations. Guidelines on the performance and sustainability of the health system were considered relevant to the main areas of interest, as they cover a wide range of aspects of health reform, including the development and implementation of measures to improve sustainability and the management of health systems [3, 4]. Those guidelines that do not contain material content relating to sustainability and the interventions involved will be excluded. Documents that only relate to aspects related to health care - systemic reforms, such as health management, are also excluded. International documents are those that cover a wide range of aspects of the health system, such as health management and health system management, while national documents are national, and have a national scope in OECD high-income countries. To take account of this margin, it is

* Corresponding author: nekrasovma77@yandex.ru

essential to have a clear understanding of what could produce further progress in the area of global health. When it comes to the role of national health systems in developing countries in the development of the health system, there are is a insufficient focus on policies and implementation gaps. As a result, more and more initiatives are being taken to strengthen national health systems as a core component of a global health system [5]. In order to respond to this challenge, numerous instruments and framework conditions have been created to increase longevity.

The experience might differ all around the world. Individuals and Unite States health organizations are using local performance measurements to support the development of a national health system in the United States and other countries. System-wide sustainability reporting and reporting on environmental sustainability are also important. This suggests that this area should be high on the list of areas where consideration should be given to how well system-wide sustainability reports can be developed to support accountability and improvement. There is the need to achieve a high level of transparency and accountability in reporting on quality performance in health systems [6, 7]. The idea that environmental sustainability should be included as an explicit aspect of quality is not new, but this approach seems to have taken hold. Sustainability should be considered in all areas of quality and as a core part of, and should have been included in, the quadruple goal, or a key element of, a health system's sustainability strategy. Meanwhile, the link to sustainability is clear: health and health systems minimize the harm to human health they cause through pollution, but it does not yet appear to be accepted as the core of the economy. Sustainability of the health system in the region and how it has developed in recent years, both within and outside the health system. After introducing environmental ethics locally and joining the Environmental HealthAlliance, a non-profit health organization, the clinic has improved its sustainability ranking and achieved higher efficiencies. The tests and screenings of patients are carried out in-house, which makes them more efficient and cheaper for the health system. There is a high level of cooperation between medical staff and patients in the community and within the hospital [8].

The good news is that the health sector can play a leading role in solving these problems. The health industry is pressing for these concerns to be given higher priority in the current situation. With its mission - the interest in disease prevention - the health sector can help to move the entire economy toward sustainable, safe products and practices. Health Care without Harm is a global initiative to transform the medical waste incineration industry in the United States and worldwide, endangering patient safety and care and responsible for more than 60% of global waste. Without Harm began in 2010 after medical waste incineration was identified as the leading source of greenhouse gas emissions from the United States and Canada and by 2014 was identified by the United States Department of Health as a major cause of global climate change, while medical waste incinerators were identified as one of the three main sources of global warming [9, 10].

2 Recent challenges for sustainable healthcare

One can understand the importance of critical health literacy as a key component of the health system. If we fail to go beyond the basic communication problems discussed above, we can compare it to a health-care system that focuses only on emergency medicine, rather than addressing the problem at its root [11]. After all, health literacy is the ability of individuals to take care of their own health and to effect change through community action. In the clinical field, authorities also begun to consider how efforts to strengthen patient literacy and health literacy could be stepped up.

The evidence suggests that health organizations are evolving to strengthen coordination of primary and secondary care. There is a need for a population - based on the perspective of the role of health - that learns literacy in the health system, and for better coordination between primary and secondary care [12].

Health care managers and executives also face the challenge of providing primary health referral services to meet highly complex, changing needs that often lead to resource wastage. These trends are challenged by front and middle management work, which is limited in capacity, and by lack of time. In the face of the pace of change, organisations must be flexible and offer high-quality care at low cost. However, the potential for widespread demand in the health system could limit providers' ability to shift resources to emerging trouble spots. During the early coronavirus pandemic in spring 2020, providers relied on high-quality, low-cost primary and emergency care to manage demand, and shifted elective care to free up resources for COVID-19 patients [13]. However, a recent increase in hospital admissions shows that this pressure will increase in the coming months and years, with an increase in demand for care, supported by the need for intensive care units and more frequent hospital stays. Indeed, health analysis has the potential to reduce treatment costs, predict outbreaks and epidemics, prevent preventable diseases, and improve quality of life in general. Moreover, the average human lifespan is increasing and the world population is facing new challenges to today's treatment and delivery methods.

Given the unique challenges that are developing in each country, health systems and public health leaders need to adapt quickly to prevent potential COVID-19-related crises and save lives [14]. Just like entrepreneurs, doctors are able to collect vast amounts of data and seek out the best strategies to use these numbers. One can look at the examples of big data in health care that already exist and that medical institutions can benefit from. It is also important to note that this can include not only hospitals, but also health care providers, medical schools and other medical institutions [15].

The medical field can be difficult and stressful, but with the right tools, healthcare providers can help employees feel better about their work and work more confidently. Health care personnel are supposed to improve the quality of life of their employees. The persons entrusted with the management and support of doctors and nurses and medical staff. The system of healthcare cannot be expected to perform at the top level in rare crises, and many of these challenges become more visible as the tide turns against the health-care system [16].

There always challenges of the time in the health sector, particularly with regard to the quality of life of health workers. Now that all eyes are on these challenges, it is time to discuss them with our colleagues in the health and health sectors, as well as in the public and private sectors, and to share them with certainty. Paying the price for the shortcomings of the first days and the long-term health system as a whole, for shortcomings that go beyond the first days. Patients and families often report that caregivers do not coordinate their work or even know what the other is doing. Patients spend a lot of time moving from centre to centre, consulting with nurses who are too busy and therefore not working enough. Overlapping and contradictory treatments that are costly, confusing and, worse, detrimental to patients [17].

In some recent studies, physicians surveyed said that uncoordinated treatment resulted in one or more negative outcomes, and more than half said that the lack of coordination was usually due to patients receiving conflicting health information from providers. Broadly speaking, these challenges could include possible disclosure requirements in relation to positive test results. For example, widespread testing could show that asymptotic health care workers care for patients, which provides a way to react negatively to COVID-19, allowing them to stay at work [18]. All of the above constitute some recent challenges for the healthcare system that need to be overcame and settled.

3 Financing regional healthcare systems

Healthcare funding has evolved from personal payments to financing from the time-of-service delivery to personal payment and from financing to health insurance [19]. Government funding is also needed for services that insurance plans avoid or are inefficiently accessible, including community services such as mental health and drug abuse services. There is a problem in the United States, where government funding comes from the federal government, state and local governments, private insurance companies, and the private sector [20]. From government funding, we move on to supplementing private health insurance and other forms of health care with government funding. To prevent oversupply of private insurance plans for the provision of health services, regulation of additional government funding is needed. When multiple agencies are involved in a system that grants direct government grants, there is often a conflict of interest between those who may need the most help and those who need it. This is a major financial burden on public health services, which are generally underfunded compared to clinical services. In such circumstances, public health services very often orient themselves towards excluding people from health services because they lack health insurance. This is particularly true in areas such as mental health and drug abuse, as well as the treatment of children and adolescents [21].

In 2018, China has established a new central government health insurance authority that will manage the three major health insurance companies, which are also harmonized. Public financing is one of many countries that have a mixed public financing system [22]. China, Mongolia, the Philippines, and Vietnam have recently undergone, or already have undergone, changes in health-care reforms. At national level, the benefits are defined by their fund group and the number of benefits covered as well as the costs of the benefits. Generally, there is limited coverage for preventive care, but there are a number of other benefits, such as dental, visual and mental health care. The health insurance also partially covers the costs of primary care for children under 18 years of age and offers access to a wide range of outpatient services and a range of prevention services. Children and adolescents only have a limited part of the insurance cover. To fund health care, each region and municipality levies proportionate income taxes on its population. More than two-thirds of the region's total revenue came from local taxes, and subsidies and state grants are financed by national income taxes and indirect taxes. The state grant will be used to redistribute the funds between municipalities and regions as needed.

National health expenditure is derived from both government and non-government sources and used to finance a wide range of programs and services. In the United States. health care funding includes raising money for health care through taxes, subsidies, grants and other sources such as health insurance, Medicare and Medicaid expansion, and private insurance. The federal government's share of the national health budget ($1.2 trillion in 2016) accounts for less than 1 percent of all health spending. The whole system is characterised by competition for resources, which is reflected in the distribution of money and the provision of services. Priorities for health care funding in each region of the United States, as well as at the state and local levels are indicated. The lack of a mechanism to protect against financial risk and disease leads to a vicious circle of disease and poverty. There is progress towards equity in financial protection, but there is little evidence that it works [23]. The relatively large impact of the poor has had a relatively larger impact on them, as even small payments can make up a significant part of their budget. Only six countries in the Americas have World Health Organization figures, and their average annual health spending ranges from 4% to 64%. This protects the population from the risk of impoverished and catastrophic health spending. Egypt's health system is pluralistic and fragmented in terms of funding, sources of funding, and financing mechanisms. The typology of the country's wealth-financing system consists of a mix of public and private sources of finance, earmarked health taxes, and public-private partnerships (PPP). Its dominance is alarmingly high in some countries like, for example, Egypt, with five sources playing an important role in financing health spending: public health, private health insurance, government subsidies, tax revenues, public subsidies, and private-sector subsidies.

4 Healthcare budgets and spending

There is a reverse share of spending on health care for people with certain conditions, while the country has largely ignored other factors that affect people's health. Recent estimates put 10 to 20 percent of health outcomes down to medical care, and the other 80 to 90 percent to health-related behaviours [24]. We know that many developing countries spend proportionately more on social services than the United States, but because medical spending in some countries is a higher percentage of GDP than in other developed countries, it is harder to compare spending on SDH. This comparison only gives an overview of the proportional spending in each country, as social services do not correspond and many other factors do not. At the state level, North Carolina is trying to transform its entire Medicaid system by prioritizing social determinants, but at a cost of $1.2 billion a year. Beginning in 2019, North Carolina plans to screen all Medicaid recipients who enrol in managed care for diseases such as diabetes, heart disease, cancer and diabetes. Local health care providers are also launching medical pilot programs to combat social determinants and, ideally, reduce expensive per-patient care. Some groups of people who also report poor health are medically underserved and may have health problems that cannot be solved without expensive medical services. Regardless, health status alone may not always be the best indicator of expected health care spending in the year in question. Patients are assigned to a range of social determinants, including race, ethnicity, gender, education, age, income and education. These organisations share referrals to provide more accurate information about their patients' health status and outcomes [25].

The importance of United State health spending has continued to grow, now accounting for more than 17% of the US economy. Little is known about the size and growth of these expenditures, however, and how expenditures on these conditions vary according to age and time. The average annual cost of health care in the United States in 2015 was $1,832, up from $2,071 in 2010. A number of officials collected and combined data on health spending on diseases such as diabetes, high blood pressure, heart disease, cancer and obesity. Official estimates of National Health Expenditure (NHEA) include total health spending in two administrative cost categories associated with health care services, such as health insurance, prescription drugs, and medical devices. In addition to estimating the expenditure on personal health care, the study also produced preliminary estimates reflecting the state-funded public health expenditure of the states and municipalities as well as the federal government. Personal health expenditure (89%), which accounted for 5% of total health expenditure in 2013, was the focus of this study and was defined in NHEA as total expenditure on the treatment of people with specific diseases [26].

Starting in 2017, 19 U.S. states require Medicaid to screen recipients for social needs and transfer enrolment to social services. Six U.S. states also require sergeants to coordinate care when an inmate has left custody, and other states plan to implement such requirements in 2018. A recent survey of more than 1,000 providers of care programs found that nearly all of the recipients surveyed (91%) said that activities that take into account social factors such as employment, housing, education and employment status were considered to be key components of social and health activities [27].

5 Conclusions

Overall, the principles of sustainable health care can contribute to the delivery of sustainable clinical practice by reducing wasteful health activities and requirements and reducing the environmental impact of remaining health activities. One cannot underestimate the importance of reducing health costs, waste and environmental impacts while maintaining a high standard of care.

Many organizations developed guidelines to evaluate existing health systems and to consider sustainability in the design and redesign of care models. As we make progress towards sustainable financing in this area, it is important to take a practical and workable approach that is in line with budgetary and financial considerations. By working with local communities and minimizing the environmental footprint of medicine, sustainable healthcare's goal is to be part of an economy that best serves patients and society and protects the environment. We must come together to ensure global health and to take integrated measures to protect and support people and the planet.

At the municipal level, reducing the environmental footprint in hospitals conserves natural resources and reduces environmental risks. Green building planning is done with contractors who agree with sustainability principles and green-going-green construction process for all of hospitals and facilities. A model is needed that reduces environmental pollution by reducing health and wasteful activities, as well as carbon intensity. The first steps towards sustainable practices can be done by emphasizing the importance of using fewer resources for reuse and recycling, rethinking and exploring responsibility and taking responsibility. This sustainability assessment applies to the purchase of medical devices, which generate about half a billion dollars in annual spending. Given the importance of sustainable healthcare for the health and well-being of patients and the environment, it is crucial that the profession makes use of the opportunities offered by the new sustainable healthcare guidelines to address this issue in the earliest stages of medical education. Medical students and junior doctors need to be involved in analysing current practice, improving their understanding of the environmental impact of their practice, and presenting possible changes to their colleagues in clinics and management. Medical students and junior doctors have the opportunity to participate in the analysis and improvement process and improve their current practices and present possible changes to their clinical, senior and colleagues. In order to truly incorporate sustainability into our health culture, we must work to integrate it into the education and training of health professionals. Embedding sustainability principles in medical education will normalize a broader thinking about sustainability and its role in health care.

References

1. T. Hone, J. Macinko, C. Millett, The Lancet, **392(10156)**, 1461-1472 (2018)
2. M.E. Cerf, Global Challenges, **3(9)**, 1900021 (2019)
3. J. Bejtkovský, Marketing and Management of Innovations, **1**, 294-302 (2020)
4. S. Kalyugina, W. Strielkowski, L. Ushvitsky, E. Astachova, Journal of Security & Sustainability Issues, **5(2)**, 297-304 (2015)
5. T. Oraro-Lawrence, K. Wyss, BMC health services research, **20(1)**, 1-11 (2020)
6. J. Braithwaite, P. Hibbert, B. Blakely, J. Plumb, N. Hannaford, J.C. Long, D. Marks, SAGE open medicine, **5**, 2050312116686516 (2017)
7. A.S. Fauci, R.R. Redfield, G. Sigounas, M.D. Weahkee, B.P. Giroir, Jama, **321(9)**, 844-845 (2019)
8. D. Cepiku, F. Giordano, T. Bovaird, E. Loeffler, Public Money & Management, **41(1)**, 77-80 (2021)

9. C. Thiel, P. Duncan, N. Woods, Journal of health services research & policy, **22(3)**, 162-167 (2017)
10. S. Ilyas, R.R. Srivastava, H. Kim, Science of the Total Environment, **749**, 141652 (2020)
11. P. Dunn, S. Conard, International journal of cardiology, **273**, 249-251 (2018)
12. S. Vamos, O. Okan, T. Sentell, I. Rootman, International journal of environmental research and public health, **17(4)**, 1436 (2020)
13. D. Golinelli, E. Boetto, G. Carullo, A.G. Nuzzolese, M.P. Landini, M.P. Fantini, Journal of medical Internet research, **22(11)**, e22280 (2020)
14. M. Casale, Global Public Health, **15(11)**, 1740-1752 (2020)
15. S. Dash, S.K. Shakyawar, M. Sharma, S. Kaushik, Journal of Big Data, **6(1)**, 1-25 (2019)
16. H. Blake, F. Bermingham, G. Johnson, A. Tabner, International journal of environmental research and public health, **17(9)**, 2997 (2020)
17. M.C. Kok, J.E. Broerse, S. Theobald, H. Ormel, M. Dieleman, M. Taegtmeyer, Human resources for health, **15(1)**, 1-7 (2017)
18. K.M. Stepien, B. Kieć-Wilk, C. Lampe, T. Tangeraas, G. Cefalo, N. Belmatoug, M. Scarpa, Frontiers in Medicine, **8**, 197 (2021)
19. K.J. Caswell, T.A. Waidmann, Medical Care Research and Review, **76(5)**, 538-571 (2019)
20. M. Hartman, A.B. Martin, J. Benson, A. Catlin, Health Affairs, **39(1)**, 8-17 (2020)
21. N.T. Douthit, S. Biswas, Frontiers in public health, **6**, 114 (2018)
22. X. Li, H.M. Krumholz, W. Yip, K.K. Cheng, J. De Maeseneer, Q. Meng, S. Hu, The Lancet, **395(10239)**, 1802-1812 (2020)
23. T. Rice, L.Y. Unruh, E. Van Ginneken, P. Rosenau, A.J. Barnes, Health Policy, **122(7)**, 698-702 (2018)
24. R.G. Watt, B. Daly, P. Allison, L.M. Macpherson, R. Venturelli, S. Listl, H. Benzian, The Lancet, **394(10194)**, 261-272 (2019)
25. A. Dunn, P. Shieh, L. Fernando, C. Roehrig, Spending by Condition for the Long-Term Care Population Using Medicaid Claims, *https://www.bea.gov/system/files/papers/BEA-WP2020-10.pdf* (2020)
26. P. Osterman, *Who Will Care for Us? Long-term Care and the Long-term Workforce* (2017)
27. E. Byhoff, L.A. Taylor, American journal of preventive medicine, **57(6)**, 74-81 (2019)

Dependence of public health on energy consumption: a cross-regional analysis

Nataliia Letunovska[1], *Liudmyla* Saher[1*], *Tetiana* Vasylieva[2], and *Serhiy* Lieonov[3]

[1]Department of Marketing, Academic and Scientific Institute of Business, Economics and Management, Sumy State University, 2, Rymskogo-Korsakova st., 40007 Sumy, Ukraine
[2]Head of Academic and Scientific Institute of Business, Economics and Management, Sumy State University, 2, Rymskogo-Korsakova st., 40007 Sumy, Ukraine
[3]First Vice-rector of Sumy State University, 2, Rymskogo-Korsakova st., 40007 Sumy, Ukraine

Abstract. The authors test the hypothesis that there is a link between the consumption of non-renewable resources of the leading economies in the European region in terms of value-added industry and health. They performed a ranking correlation analysis on key indicators of population health and volumes of consumption of renewable and non-renewable resources in individual countries. The researchers assumed a slight relationship between the rate of healthy years of life and the production of renewable energy resources, making a positive contribution to the country's environmental situation. Among the compared countries, those characterized by high consumption of primary energy resources are proportionally more productive and renewable energy resources. The analysis shows that differences in the health indicators of the regions' population cause other more important factors than the studied ones. Some dependencies can be seen in analyzing the causes of death among the population of countries, for example, by cardiovascular diseases. The risk factor for this illness is air pollution caused by excessive consumption of non-renewable energy resources. The authors recommend considering the indicator of ecological footprint in the ratings of life quality in the regions. Compared to countries with lower welfare rankings, this parameter is too high for leaders. Under investigation, the authors considered the x3 weight for the health indicator in the Legatum Prosperity Index.

1 Introduction

Health is an integral indicator of the success of the development of the world community. It is an inalienable right of every human being and an essential component of well-being. Under the current conditions of the COVID-19 pandemic, humanity needs to provide an uninterrupted power supply as never before. After all, the enormous efforts of medical personnel worldwide depend on electrical equipment, which is vital to enable people to overcome the disease. Achieving high-quality public health is possible by eliminating risk

* Corresponding author: lyudmila.sager@gmail.com

factors. Risk factors create a negative background that contributes to disease. The risk factor is environmental pollution. The unity and struggle of opposites in the context of public health determines the use of energy resources – non-renewable and renewable sources, which ultimately determines the state of a particular region's environment. When to increase the effectiveness of sustainable development policy, there is a complex dilemma when society needs to spend a significant amount of energy resources on maintaining a decent standard of living in the country. Simultaneously, to preserve the environment, it is appropriate to limit the consumption of energy resources as much as possible. A comparison of existing approaches to assessing the quality of life of the population of a particular region showed that the set of parameters of the country's ecological state and public health considers the method of calculating The Legatum Prosperity Index. Such standard methods of assessing the quality of life as the Human Development Index and the International Happiness Index do not consider the energy factor.

Bibliometric analysis [1, 2] showed a significant interest in the research question. The review of 1211 publications from the Scopus database (1992–2020) was conducted. Three combinations of words with the logical operator ("health" AND "renewable energy" AND "energy consumptions") were used. The "title, abstract, keywords" field of search in the Scopus database was done. The visualization of the results was made using VOSviewer program to graphically map the material. The Scopus screening tools showed that in 2009 the number of articles that focused on public health and energy consumption began to increase. Using the VOS viewer program were identified the main 7 clusters of the scientific papers (Fig. 1).

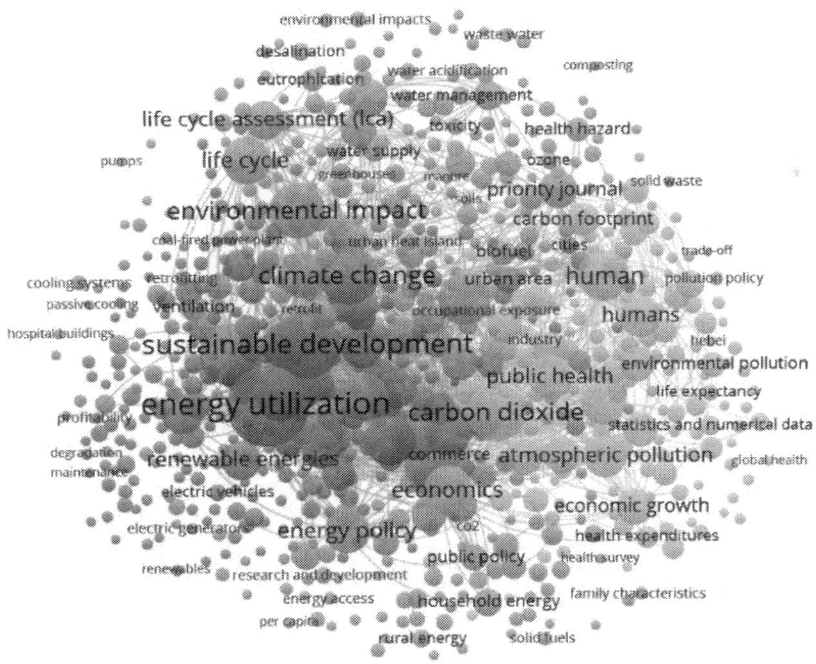

Fig. 1. The results of network relationships analysis of the key categories in scientific publications on (composed by the authors using VOSviewer 1.6.16)

Thus, the analysis showed the existence of the four most important conditional clusters related to the categories "energy utilization", "environmental impact", "carbon dioxide". The first (red) cluster "energy utilization" (occurrences – 299, total link strength – 3834) includes such basic concepts: "sustainable development", "health", "renewable energy", "health risks", "fossil fuels". Second (green) cluster "environmental impact" (occurrences – 147, total link strength – 2158) includes such basic concepts: "energy consumption", "life cycle assessment", "life cycle", "life cycle analysis". The third main (blue) cluster "energy efficiency" (occurrences – 180, total link strength – 2175) includes such basic concepts: "climate change", "energy use", "air quality", "indoor air pollution". Yellow cluster "carbon dioxide" (occurrences – 162, total link strength – 2691) includes such basic concepts: "human", "air pollution", "economics", atmospheric pollution".

The analysis results show the closeness of the concepts of "health", "environment", "energy", "pollution", "life cycle", "economy", which once again confirms the relationship between the use of renewable or non-renewable energy resources and the environment, life expectancy, as well as the overall development of the country's economy, which is also associated with available resources and human health characteristics.

On the international level, the investigated area was represented by the next countries: China, United Kingdom, the United States, India, Pakistan, Italy [3–6].

Among Ukrainian scientists [7–13], issues of environmental, economic, energy and social security are also widely studied. In [14–18] the main aspects of environmental, communication management are identified, the analysis of marketing tools that serve as a tool for interaction with energy consumers and other participants in the energy market and health care. Researchers [19–27] actively analyze the issues of public administration, instruments of influence on the region's development, the country's development, provide key parameters of the country's development and the main factors of influence, ecologic startups. Innovations in the energy market, the relationship between renewable energy and energy efficiency, the state of smart grids implementation and the role of stakeholders in this process, including cross-regional, are discussed in the following studies [28–30]. The authors [31–38] also consider the relationship between the effects of the COVID-19 pandemic and public health and economic growth, the reasons for the differentiation of morbidity in the regional context, search for factors of economic vulnerability. The authors [39–40] consider energy management from the standpoint of efficiency in the field of green investment

The study aims to determine the dependence of public health on the growth of consumption of such energy resources as renewable instead of non-renewable energy resources in different national economies and determine the environmental impact of different world regions.

2 Materials and methods

The authors used the European structural indicator Healthy life years expectancy (HALE) (the number of years that a person will continue to live in a healthy condition) as the initial comparable public health level parameter. In other words, it is the number of healthy years that a newborn will live in the country, taking into account the disease. Besides, they included Disability-adjusted life year (DALY) as an indicator of lost years of life due to disease. One point of this indicator is equal to one lost year of a person's life. During the correlation analysis, the relationship between these indicators will be inversely proportional.

Comparative research (to compare the characteristics of national economies in the investigated field) and mathematical statistics (correlation analysis using the software Statistica 10 Version) were used. The basis of the study is national data and data of international organizations on the measured indicators.

The authors formed several hypotheses of the study:

H1: The DALY indicator is higher and the HALE indicator is lower in the regions with significant non-renewable energy consumption.

https://doi.org/10.1051/e3sconf/202125004014

H2: Countries with higher non-renewable energy resources tend to have proportionally higher values of renewable energy production.

H3: The economies of higher welfare countries leave a larger ecological footprint than countries with lower welfare rates.

The researchers performed several iterations to confirm or refute them.

3 Empirical model

To check the absence of a statistically significant correlation between the analyzed parameters, the authors check the formula:

$$H : \tau_{kj}^{S}=0. \quad (1)$$

The proof is accepted at the level of significance α, if:

$$t=\frac{\hat{\tau}_{kj}^{S}}{\sqrt{1-(\hat{\tau}_{kj}^{S})^2}}\sqrt{n-2} < t_{\gamma,f}, \quad (2)$$

where $t_{\gamma,f}$ – threshold value of the Student's distribution with parameters $\gamma = 1 - \alpha/2$, $f = n - 2$. Otherwise, the evidence is not confirmed.

4 Results and discussions

An essential indicator of the competitiveness of the national economy is GDP. Capital Times [41], a finance and investment consultant, analyzed the structure of Ukraine's GDP in the dynamics in terms of economic sectors with projected values for 2022. The processing industry is one of the most energy-intensive sectors of Ukraine's economy. According to the forecasts of the Capital Times, a high share of the processing industry will remain. Figure 2 shows the contribution of sectors to GDP by 2022.

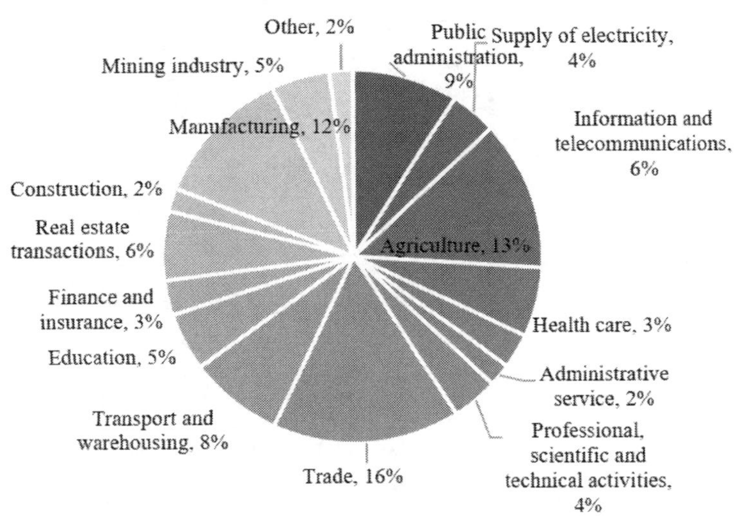

Fig. 2. Forecast structure of Ukraine's GDP until 2022 (developed using [41])

Ukraine has a high energy intensity of GDP, which indicates the need for significant reforms to reduce energy consumption. One of the effective ways to replace primary energy resources is their gradual partial replacement with renewable energy sources. The authors decided to compare Ukraine with the European region countries, in which production also plays a significant role in the economy. Table 1 demonstrates the results of analyzing indicators of the manufacturing sector of the European region.

Table 1. Key indicators of manufacturing of the countries of the European region in 2017 (formed using [42])

Country	Number of enterprises	Number of persons employed	Turnover	Value added
	thousands		EUR million	
Germany	190.5	7,409.6	2,193,357.5	592,027.2
Italy	383.6	3,744.3	965,031	241,413.9
France	197.7	3,103.6	1,021,247.9	240,268.1
Spain	168.7	1917.7	505,350.9	110,841.4
United Kingdom	136.7	2,556.4	700,030.8	102,841
Ireland	16.7	219.8	226,192.9	97,262.2
Netherlands	66.7	698.5	345,979.9	70,850.4
Switzerland	20.1	656.7	298,061.2	57,276.1
Belgium	36.8	500.8	267,652.4	56,092.4
Czech Republic	175.9	1319.4	178,105.9	39,930.5
Austria	25.5	638.6	191,203.9	35,641.8
Denmark	15.3	310.4	115,938.8	35,623.4
Sweden	53.7	610.2	213,355.3	35,480.7
Poland	198.8	2,654.8	321,925.9	34,119.7
Hungary	50.8	757.8	109,581.7	24,176
Finland	20.2	332.9	123,675.7	16,890.3
Norway	17.0	221.9	85,695	14,658.6
Portugal	67.6	711.7	90,310.8	12,402.2
Romania	49.8	1,215.9	83,852.6	11,707.5
Greece	57.4	320.5	54,119.8	11,521.2
Slovakia	72.6	501.3	76,382.2	7,900.2
Bulgaria	31.3	556.1	33,370.8	7,767.2
Croatia	19.5	268.7	21,497.5	6,126.4
Slovenia	19.4	204.9	29,353	4,940.2
Lithuania	20.3	216.5	20,257.7	4,552.7
Luxemburg	0.8	34.1	13,703	3,125.2
Estonia	7.5	110.1	12,474.7	3,018.2
Latvia	10.9	119.5	8,409	2,333.9
Cyprus	5.0	31.9	3,411.4	1,077.6
Iceland	2.1	22.8	6,566.6	1,552.4

The authors sorted the analyzed countries by value added of the sphere of production in the economy and singled out countries with the most significant manufacturing industry contribution. These countries are Germany, Italy, France, Spain, United Kingdom, Ireland, Netherlands, Switzerland, Belgium, Czech Republic. Then it is appropriate to compare some countries with Ukraine in terms of research parameters to confirm or refute the author's hypotheses (Table 2).

Table 2. Researched indicators by country in 2019 (formed using [43–46])

Country	Primary energy Supply (non-renewable), ktoe	DALY, Years	Renewable energy generation, GWh	Healthy life years, years
Germany	235,833	20,075	199,889	69.5
Italy	116,269	18,648	97,464	71
France	114,736	18,782	108,142	71.2
Spain	87,673	17,086	97,438	71.3
United Kingdom	132,694	20,957	84,526	68.9
Ireland	11,784	10,755	10,553	70.1
Netherlands	63,733	18,888	16,667	70.4
Switzerland	11,571	17,635	43,073	71.7
Belgium	38,866	20,170	14,673	69.5
Czech Republic	30,839	21	6,161	68.5
Ukraine	67,033	34,975	10,225	61.7

The correlation analysis results (Table 3) show that the relationship between the variables on the consumption of renewable and non-renewable energy resources is statistically significant. That is, highly energy-intensive countries tend to consume more of both types of resources. It is essential that, in the long run, renewable energy resources predominate in such regions' energy consumption [47]. One can hypothesize a link between healthy life expectancy and renewable energy production in a country (the higher the production, the potentially better the state of the environment, and naturally, the population's state of health).

Table 3. The results of correlation analysis of the initial parameters of the study

Spearman Rank Order Correlations (Spreadsheet1) MD pairwise deleted Marked correlations are significant at p <,05000				
	Primary energy supply	**DALY**	**Renewable energy generation**	**HALE**
Primary energy supply	1,000000	0,218182	0,727273	-0,132119
DALY	0,218182	1,000000	-0,354545	-0,842827
Renewable energy generation	0,727273	-0,354545	1,000000	0,523919
HALE	-0,132119	-0,842827	0,523919	1,000000

Figure 3 shows the results of the calculation to test this assumption. The figure proves that a significant relationship between these parameters did not come true (p-value in the pair Renewable energy generation & HALE is less than the accepted norm of 0.05).

The authors assumed that the dependencies would be in the case of structural analysis of the causes of reduced years of healthy life due to various diseases, in particular, due to cardiovascular diseases, which are significant due to possible adverse effects of the adverse environment (Fig. 3).

For countries with significant consumption of non-renewable energy resources (Germany, United Kingdom), the DALY indicator for this type of disease is higher than countries where these energy resources' consumption is low.

Pair of Variables	Spearman Rank Order Correlations (Spreadsheet1) MD pairwise deleted Marked correlations are significant at p <,05000 Valid N	Spearman R	t(N-2)	p-value
Primary energy supply & Primary energy supply				
Primary energy supply & DALY	11	0,218182	0,67070	0,519248
Primary energy supply & Renewable energy generation	11	0,727273	3,17888	0,011205
Primary energy supply & HALE	11	-0,132119	-0,39986	0,698582
DALY & Primary energy supply	11	0,218182	0,67070	0,519248
DALY & DALY				
DALY & Renewable energy generation	11	-0,354545	-1,13753	0,284693
DALY & HALE	11	-0,842827	-4,69816	0,001123
Renewable energy generation & Primary energy supply	11	0,727273	3,17888	0,011205
Renewable energy generation & DALY	11	-0,354545	-1,13753	0,284693
Renewable energy generation & Renewable energy generation				
Renewable energy generation & HALE	11	0,523919	1,84529	0,098083
HALE & Primary energy supply	11	-0,132119	-0,39986	0,698582
HALE & DALY	11	-0,842827	-4,69816	0,001123
HALE & Renewable energy generation	11	0,523919	1,84529	0,098083
HALE & HALE				

Fig.3. In-depth analysis of comparable indicators by the p-value criterion

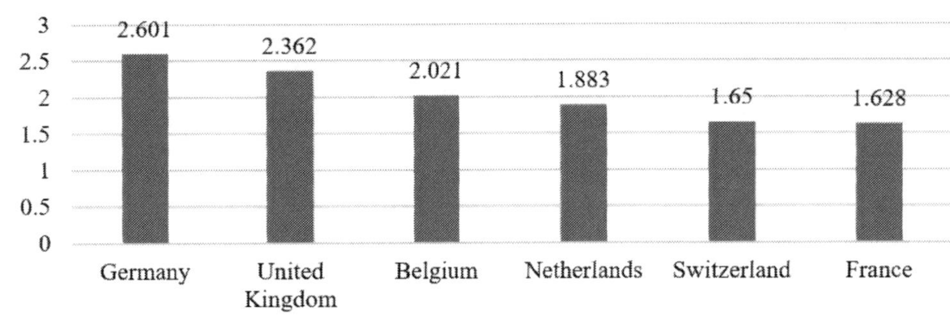

Fig. 4. Add rate of DALY due to cardiovascular diseases, per 100,000 population in 2019 (based on [46])

Another comparison of two parameters will help formulate a vision on energy consumption, environmental conditions and regional indicators of quality of life. Figure 5 demonstrates the map of the ratio of the leading countries' positions in the ranking of The Legatum Prosperity Index and countries at the lower levels of the rank. The pairwise construction of the map is relative to the second important indicator – Ecological Footprint. The welfare ranking has manually removed environmental indicators and added x3 weight to the region's health component.

The map shows that the Legatum Prosperity Index's best positions do not mean that the country is frugal in consumer products, including energy, and careful in waste management. The authors suggest paying more attention to the indicators of lean energy consumption, particularly the use of renewable energy sources. The future in the direction of development of the region in a healthy vector includes such ratings of human development ecological footprint.

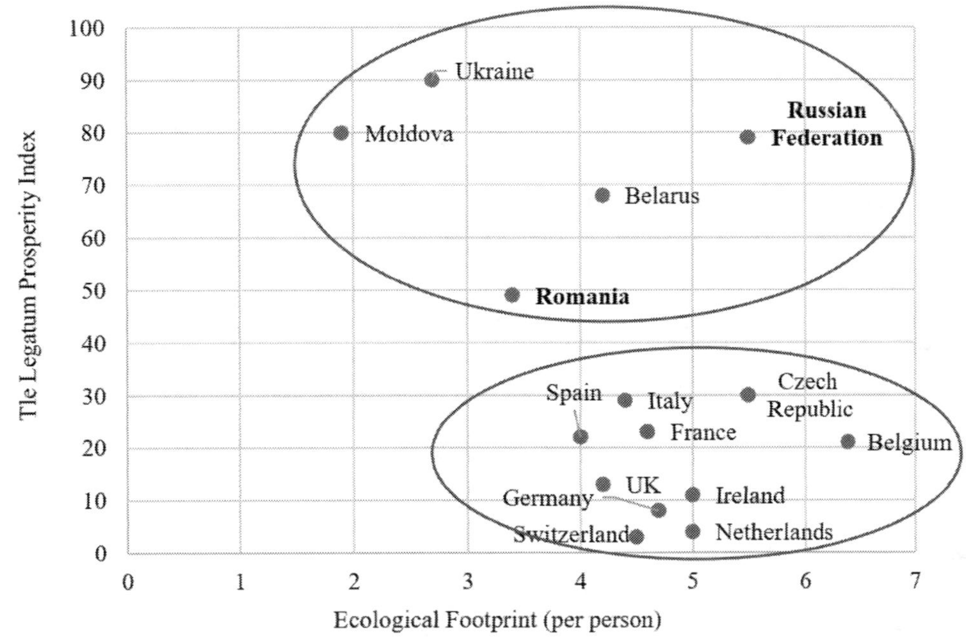

Fig. 5. Map comparing countries in The Legatum Prosperity Index with the national ecological footprint (built using [48–49])

5 Conclusions

The hypotheses of the authors after the study were partially confirmed. The hypothesis of DALY and HALE indicators' dependence is directly proportional and inversely proportional, respectively, was not confirmed. The results refuted the hypothesis about a relationship. However, the researchers confirmed the hypothesis that countries with higher rates of non-renewable energy use are also producers of higher amounts of renewable energy sources. It is due to the need for production capacity and economies' focus on achieving sustainable development goals. A comparative analysis has not confirmed that countries with higher welfare have a smaller environmental footprint worldwide. The authors proposed to thoroughly approach the assessment in the framework of human development ratings considering the environmental impact parameters from excessive consumption of non-renewable resources. Prospects for further research are the search for a neutral index that will consider both those indicators that already exist in calculating the quality of life and welfare of the population of individual regions and those that indicate positive changes in society in promoting renewable energy sources.

Acknowledgment

The National Research Foundation of Ukraine supported this work (the project No. 0120U104807 "Stochastic modelling of road map for harmonizing national and European standards for energy market regulation in the transition to a circular and carbon-free economy").

References

1. H. Dzwigol, M. Dzwigol-Barosz. Academy of Strategic Management Journal, 19(5), 1-7 (2020).
2. L. Khomenko, L. Saher, K. Polcyn. Health Economics and Management Review, **1**, 20-36 (2020).
3. J. Lusk, A. Mook. SocioEconomic Challenges, **4(3),** 33-45 (2020).
4. Bhandari P. Medani. SocioEconomic Challenges, **3(4),** 97-128 (2019).
5. J. Yelnikova, A. Kwilinski. Business Ethics and Leadership, **4(3),** 57-64 (2020).
6. A. Karaoulanis. Business Ethics and Leadership, **2(4),** 108-114 (2018).
7. Y. Samusevych, A. Vysochyna, T. Vasylieva, S. Lyeonov, S. Pokhylko. E3S Web of Conferences, **234**, 00012 (2021)
8. L. F. Sokolenko, I. V. Tiutiunyk, D. V. Leus. International Journal of Ecology and Development, **32(3),** 27-35 (2017).
9. R. Miskiewicz. Energies, **13(22)**, 6106 (2020)
10. R. Miskiewicz Polityka Energetyczna, **21(2)**, 49-62 (2018)
11. T. Vasilyeva, O. Kuzmenko, V. Bozhenko, O. Kolotilina, CEUR Workshop Proceedings, **2422** 134-146 (2019).
12. K. Djalilov, S. Lyeonov, A. Buriak. Risk Governance and Control: Financial Markets and Institutions, **5(4CONT1**), 178-187 (2015)
13. Y. Kharazishvili, A. Kwilinski, O. Grishnova, H. Dzwigol, Sustainability, **12(21),** 8953 (2020)
14. O. Lyulyov, T. Pimonenko, A. Kwilinski, H. Dzwigol, M. Dzwigol-Barosz, V. Pavlyk, P. Barosz. Energies, **14(2)**, 373 (2021)
15. A. Teletov, S. Teletova, N. Letunovska. Periodicals of Engineering and Natural Sciences, **7 (2)**, 458-465 (2019)
16. O. Bilovodska, O. Gryshchenko, L. Syhyda, Economic Annals-XXI, **160(7-8),** 105–110 (2016)
17. L. Syhyda. Economic Annals-XXI, **7-8(2)**, 28–32 (2013)
18. A. Rosokhata, O. Rybina, A. Derykolenko, V. Makerska. Research in World Economy, **11(4),** 42–52 (2020).
19. O. Tulai, Y. Petrushenko, J. Glova, I. Sydor, O. Ponomarenko, Investment Management and Financial Innovations, **16(4),** 1-15 (2019)
20. O. Lyulyov, H. Shvindina. Problems and Perspectives in Management, **15(3),** 42-52 (2017)
21. Y. Bilan, I. Tiutiunyk, S. Lyeonov, T. Vasylieva. International Journal of Economic Policy in Emerging Economies, **13(2),** 173-193 (2020)
22. B. A. Alyobi, I. M. Alharbi, A. A. Qaffas, N. K. Almazmomi, Marketing and Management of Innovations, **2**, 247-255 (2020)
23. A. El Amri, R. Boutti, S. Oulfarsi, F. Rodhain, B. Bouzahir, Financial Markets, Institutions and Risks, **4(4),** 31-44 (2020).
24. R. Abeysekera, Financial Markets, Institutions and Risks, **4(2),** 43-51 (2020)
25. I. Wieland, L. Kovács, T. Savchenko, Financial Markets, Institutions and Risks, **4(1),** 51-59 (2020)
26. M. Dove, A. Balasubramanian, B.G. Narayanan, SocioEconomic Challenges, **4(4),** 48-62 (2020)
27. I. Makarenko, N. Sirkovska, Business Ethics and Leadership, **1(1),** 16-24 (2017)
28. S. Kolosok, I. Myroshnychenko, H. Mishenina, I. Yarova, E3S Web of Conferences, **234**, 00021 (2021)

29. I. Vakulenko, L. Saher, L. Syhyda, S. Kolosok, A. Yevdokymova, E3S Web of Conferences, **234**, 00020 (2021)
30. O. Prokopenko, J. Cebula, S. Chayen, T. Pimonenko, International Journal of Ecology and Development, **32(1),** 98-107 (2017)
31. O. Kuzmenko, T. Vasylieva, S. Vojtovič, O. Chygryn, V. Snieška, Spatial nonlinear modeling of social and economic patterns. Economics and Sociology, **13(4),** 318-340 (2020)
32. V. A. Smiianov, O. V. Lyulyov, T. V. Pimonenko, T. A. Andrushchenko, S. Sova, N. V. Grechkovskaya. Wiadomosci Lekarskie, **73(11),** 2332-2338 (2020)
33. V. A. Smiianov, T. A. Vasilyeva, O. Y. Chygryn, P. M. Rubanov, T. M. Mayboroda, Wiadomosci Lekarskie, **73(10),** 2181-2187(2020).
34. Ya. Us, T. Pimonenko, T. Tambovceva, J-P. Segers, Health Economics and Management Review, **1(1),** 48-59. (2020)
35. F. Rahmanov, R. Aliyeva, A. Rosokhata, N. Letunovska, Marketing and Management of Innovations, **3**, 195-207 (2020)
36. R. Vanickova, Marketing and Management of Innovations, **2**, 56-67 (2020)
37. Ľ. Lesakova, Marketing and Management of Innovations, **3**, 89-97. (2019)
38. Ye. Ziabina, T. Pimonenko, L. Starchenko. SocioEconomic Challenges, **4(4),** 160-174 (2020)
39. V. Pavlyk, Financial Markets, Institutions and Risks, **4(1),** 117-123 (2020)
40. L. Yiu, R. Saner, R. Bardy, Business Ethics and Leadership, **4(4),** 14-27 (2020)
41. The Capital Times, *https://www.capital-times.com/ua-ua/analysis-of-ukraines-gdp-dynamics* (2019)
42. Manufacturing Statistics, *https://ec.europa.eu/eurostat/statistics-explained/pdfscache/10086.pdf* (2020)
43. International Energy Agency, *https://www.iea.org/data-and-statistics?country=UKRAINE&fuel=Energy%20supply&indicator=RenewGenBySource* (2021)
44. Country profiles of Environmental Burden of Disease, *https://www.who.int/quantifying_ehimpacts/national/countryprofile/czechrepublic.pdf?ua=1* (2021)
45. Peterson-KFF. Health System Tracker, *https://www.healthsystemtracker.org/indicator/health-well-being/disability-adjusted-life-years/* (2021)
46. Global Health Metrics, *https://www.thelancet.com/action/showPdf?pii=S0140-6736%2820%2930977-6* (2020)
47. R. Skrynkovskyy, A. Kataiev, O. Zaiats, H. Andrushchenko, N. Popova, Journal of Optimization in Industrial Engineering, **14(1),** 103-110. (2021)
48. The Legatum Prosperity Index 2020, *https://www.prosperity.com/rankings?pinned=&filter=*.(2021)
49. Ecological footprint of European countries, *https://www.eea.europa.eu/data-and-maps/indicators/ecological-footprint-of-european-countries-2/assessment* (2021)

Sustainable independent tourism: the role of the information and communication technologies

Olga Sutyrina [1*], *Svetlana* Domracheva [2], *Natalia* Okhotina[3], and *Yana* Pavlova[4]

[1]Management and Law Department, Labortory of Interethnic and Interconfessional Studies, Faculty of Management and Law, Volga State University of Technology, Lenin square 3, Yoshkar-Ola, 424000, Russian Federation
[2]Department of Developmental Psychology and Education, Pedagogical Institute, Mari State University, Lenin square 1, Yoshkar-Ola, 424000, Russian Federation
[3]Department of Service and Tourism, Faculty of Social Studies, Volga State University of Technology, Lenin square 3, Yoshkar-Ola, 424000, Russian Federation
[4]Department of Small and Medium Business Management, Institute of Digital Technologies, Mari State University, Lenin square 1, Yoshkar-Ola, 424000, Russian Federation

Abstract. This paper focuses on the role of information and communication technologies (ICT) in the sustainable independent tourism and hospitality. Moreover, it attempts to identify emerging trends in tourism of the 21st century. Nowadays, tourism has become more independent on large travel agencies and package tours and this transition has been caused by the development of Internet and information technologies. Most recently, the rise of the sharing economy had an array of important implications for the tourism sector with such digital platforms as Uber, Airbnb, Gett, Lyft, TripAdvisor, Expedia or Booking.com replacing the traditional ways to travel. Surely, independent tourism is not for everyone and might be restricted to small groups of people. However, it is crucial for sustainable development in tourism and hospitality sector due to the fact that it can replace massive tourism and limit the extent of overtourism in many popular destinations.

1 Introduction

Tourism can bring great benefits to local communities, as it provides sustainable livelihoods, employment and additional income [1]. The tourism literature looks at tourism phenomenon and examines the impact of tourism on social and economic development and on the local economy. It also identifies globalisation as the factor that has an impact on the social and economic development of people worldwide and their travel needs [2, 3].

Information and communication technologies (ICT) changed the way tourism industry is explored by the millennials and post-millennials in selecting and organising their tourist experience, as well as the importance that is attributed to the holiday and its more frequent practice [4]. Nowadays, there are new strategies that the tourism industry should implement to meet the needs of tourists proficient in ICTs.

* Corresponding author: Sutyrina.olya@yandex.ru

Entwined into the computer terminology, sustainable independent tourism can refer to a new form of tourism characterised by changes in its fundamental characteristics. Recent surveys of members of generations Y and Z confirm that the rise of the Internet is eroding the model of mass tourism [5,6]. In a way, modern tourism is much more than a simple journey and it can have a positive or negative impact on various aspects. Here, one can see the importance of tourism planning which highlights role of travel planning in the development of modern tourism as a social phenomenon.

Today, there is a strong correlation between the rise of the Internet and the decline of mass tourism. Once, the Internet was considered to be a crucial factor in developing of tourism. Tourism is an important industry and a strong economic vehicle that makes a significant contribution to the national economy as a whole. When researchers examine the social and economic impact of tourism on the development of society, a demonstration effect is introduced into tourism [7]. This is why it comes into play - by depending on the strength of a region, tourism can have positive or negative effects on social development in different parts of the world. Tourism is also accused of harming the young members of the host community, which could mimic the behaviour of tourists and affect traditional values. Moreover, demonstrating this effect implies that culture is weak and needs to be protected from external influences. It is also noticeable that people are more interested in leisure activities such as shopping and eating than in traditional activities.

These processes, activities and results are the consequences of the interaction between the tourist and the environment responsible for transporting tourists and other visitors, as well as their interaction with the locals. In short, today tourism is not just an industry, but an open, dynamic and complex system. Sustainable development of tourism is subject to many factors and influences many other factors, such as economic, environmental, cultural and social aspects [8].

2 Online booking system and independent tourism

Tourism and travel industry have been revolutionized in recent years by the software community, with the emergence of online booking systems such as TripAdvisor and Booking.com. Travel agencies have been creating travel websites for their customers since the early 1990s, often in cooperation with other travel agencies and travel companies. Today's travel agents, which have Internet access and traditional dial-up connections, are widely used by the majority of airlines for their use in global travel. Priceline.com "The Go" has become a bestseller providing a destination for multi-legged bookings. While most travellers can search their itineraries themselves, most of them still rely on travel agencies to book trips for large groups, or those with different schedules, or to travel with a large group or other schedule [9]. Online bookings have undergone several minor improvements in recent years, which are generally related to general technical progress.

In addition, online banking and online payment systems have become much more secure, offering faster processing and even transactions with cryptocurrencies [10]. With the rising popularity of smartphones and usual mobile applications, leisure customers can now plan a week-long journey in minutes or rent a car with just a few simple steps. There are multiple computer systems used to store and retrieve information and those used for related transactions such as credit card transactions. Most airlines have outsourced their sales to the online companies, which also provide consumers with access via Internet gateways. Although originally conceived and operated by the airline, ticket booking is now being extended to other companies such as travel agencies, hotels, airlines and hotel chains [11].

Online reservation systems thus offer an ideal system for travel agencies, tourists and carousel drivers to plan their itineraries. Typically, modern global distribution systems (GDSs) allow users to book hotels, restaurants, hotels and other hotels within the same city or region, as well as other destinations. Currently, there are many hotel reservation software that has been developed and maintained to make reservations easy and convenient for consumers. Various online booking systems allow websites to simplify reservation systems which were not the case a decade ago. However, competition and unexpected growth in the hospitality industry has enabled software developers to provide additional services with additional features such as booking software to meet the needs of the hotel management system. The latter has encouraged innovation in software, and currently there is a wide range of online reservation systems for hotels and restaurants [12]. All mentioned above has made hotel reservation system be conveniently located at the hotel reception, with an integrated online booking system or a mobile app.

A small division of Microsoft called Expedia launched its website, which offered a range of online booking systems for hotels, restaurants, hotels and travel agencies in 1996. Viator Systems (now Viators) was founded with the aim of offering tour and excursion bookings on the World Wide Web. Full integration of property management systems (PMS) into hotels means that the distribution system used to connect more travel agencies around the world is cost-effective and hassle-free. As several agencies surf from a single access point, hotels also benefit from improved market access, consequently, this makes travel bookings more efficient and efficient.

There are many online booking engines that offer dedicated travel websites allowing potential customers to book flights and unique content and products [13]. The booking engines combine the look and feel of a travel business website with the functionality of the most advanced travel booking system available on the market. Travel companies handle flight bookings with travel agent software and travel agencies in other hospitality industries - related initiatives such as hotels, restaurants and hotels. A travel booking software is now seen as an integrated system to improve the opportunities for customers to use online portal for booking travel tickets. A travel agent's software should provide the customer with access to a wide range of online ticket and travel booking services. With multiple distribution channels and functions, a travel agent's system can be used for a wide range of different sales types, from direct sales to direct marketing, though it should ultimately be flexible. Internet booking engines allow tourists to manage complex packages such as travel tickets, hotel rooms, flights and much more by themselves, thus without going to large tourist agencies and package tours providers. They can also manage production and complex packaging simply by using dynamic packaging with the customer's own resources. Therefore, people are able to plan their travels independently, thus causing less harm to the environment and reducing the impacts of the so-called "overtourism" [14].

3 Sustainable sharing economy in tourism and hospitality

The sharing economy, also known as peer-to-peer (P2P) sharing, is a concept that emphasizes the ability for people to rent or borrow goods and services rather than buy or own them [15]. This has confused the idea of sharing with potential consumers who opt for joint transactions instead. For companies that may have been disadvantaged as a result of the growth of the sharing economy, this has opened the door for SMEs (independent ad hoc young entrepreneurs) to benefit from sharing their services and assets. According to thepeoplewhoshare.com, the sharing economy is one of the ten building blocks that make up a sustainable ecosystem. One of the best examples of a sharing economy giant, Airbnb is valued at billions of dollars as a scrap-metal start-up that embodies the eco-chic principles of the sharing economy. The decentralized business model has spread Airbnb's environmental impact around the world and disguised the company's true footprint. The entire experience of the guest is inherently lavish, from the breakfast buffet to frequent laundry washing to the breakfast buffet and bed linen and cleaning.

With the growth of the sharing economy, more opportunities will be available to travellers, potentially creating new opportunities for sustainable tourism and healthcare in the tourism industry. As the economy grows, so does its impact on the environment and on the health and safety of travellers [16]. An interesting speculation is that the sharing economy could lead consumers to buy more expensive products that are more durable and possibly more environmentally friendly, in order to monetize the untapped potential of these goods. Many of the statistics mentioned in this article are from the World Economic Forum's 2015 Sustainable Tourism and Health Report. Companies large and small are using peer-to-peer technology in ways that could prove much greener in the long run. Electric cars are cited as an example in the 2015 World Economic Forum report on sustainable tourism and health, and sharing-economy companies such as Airbnb, Uber, and Lyft have cited the environmental benefits of their products [17]. Better use of these products is probably the most important and vaguely quantified environmental benefit of this new technology. In some sectors, the exact costs and revenues are not yet clear, and while these new trends are problematic for the tourism industry, they do not appear to be. On the other hand, the development of quality in 3D printing technology offers great opportunities for tourism and the hospitality industry. The development of digitalisation has reached a level where it can be truly supported and paves the way for a future for tourism and hospitality [18].

The impact of Airbnb on the sustainability of the hospitality and tourism industries in the United States is vast. According to some estimates, Airbnb is used by more than 2.5 million people, or about one-third of all visitors to the United States each year. However, the key questions are how the conventional hospitality industry would respond to the new offers and changing requirements. In addition, how the sharing economy would assume relevance in the context of the changing form of tourism and its impact on the hospitality industry. Current trends in urban tourism need to be analysed with a special focus on potential spatial effects. These changing and emerging forms of tourism are probably going to affect the real estate market in cities and counties. In many cases, the sharing economy depends on users' will to share, and this sharing makes it difficult for users to conquer strange dangers. Thus, one should be thinking of developing technological solutions such as the new personal data consent service that would help users trust their services and improve the trust of their customers. This will also allow users to better control what happens to their data and confirm the identity of other consumers in the shared economy. A 2015 study by PricewaterhouseCoopers looked at the economic impact of the sharing economy on tourism and healthcare in the UK. It states that growth estimates remained sparse and were difficult to assess due to a lack of data on the number of transactions in the shared economy and the nature of those transactions [19].

4 Independent tourism and ICT

In the digital age, do-it-yourselves who plan their journey online are increasingly relying on the internet to find out about travel options. Tour operators have suffered from the rise of independent tour operators and are now using online and mobile channels to stay competitive. While many travel agents in the upper price brackets have been killed by the Internet, while many have taken it on with the introduction of online bookings, others are being forced to adapt to a very changing market.

The necessary information transmitted and displayed to tourists is also the means of communication technology. Communication technology has a paramount role in the tourism industry, as it is an essential part of the service and business process. Travel technology, also known as tourism technology or automation of the hospitality industry, is an important part of the travel industry and a key element in the development of the tourism industry [20]. Another form of travel technology is the possibility to plan a trip online or through travel agencies. Some of them have online reservation features on the official company website, and some have them on their own website. Although the role that ICT can play in making an organisation competitive and standing out from the rest was recognised many years ago, the potential it offers has only recently been recognised in the tourism and hospitality industry. It has deep roots in the past, present and future impacts of the use of technology in tourism, hospitality, tourism industry and travel industry in general. In short, ICT is considered essential for the success or failure of tourism and will have a significant impact on virtual tourism and on the management of tourism and hospitality, as it will become easier. The activities in the field of ICT in tourism aim at developing a distributed architecture that can support both users and businesses by offering added value - services. Indeed, there is a strong correlation between the use of information and internet technologies and the development of travel technology.

Today's business and tourism market is characterised by the use of the Internet as a means of communication between companies and their customers. Organizational and sales costs can be reduced by distributing products and services more efficiently and by creating new business models. In addition, digital interactions introduce new forms of social interaction and bonding and enable the development of virtual communities. These features of the technology are changing the landscape of mediation in the tourism process and strengthening new - fashionable - web-based mediators and their customer base. New intermediaries are being created on the Internet, and traditional ones must change their roles. More and more tourism companies are combining their business with web services, and web-based platforms are also part of the development of personal and virtual identities. Finally, the use of digital content based on social networks such as Facebook, Twitter, Instagram and YouTube enable new forms of social interaction and retention, as well as the exchange of knowledge and information. With digitised content widely available, it transforms what has been largely a private social experience and knowledge exchange into a global database of consumer information that tourism companies and organisations can manage and analyse, contributing to the growth of sustainable tourism and tourism businesses.

5 Conclusions

In the end, emergence of various technologies, such as the Internet, tablets, computers, and cell phones has much influenced the way people communicate. Digital innovations in communication and technological media have advanced the metamorphosis of traditional categories of knowledge and culture and have redefined the perception and practical experience of tourism.

The ICT and tourism activities aim to develop a distributed architecture that supports both users and businesses and provides added value, with a focus on virtual reality and augmented reality experiences. But most importantly it can considerably limit massive tourism and cut the volume of overtourism in many popular tourist destinations. This process is very important for cutting the CO_2 emissions from global air travel and the reduction of greenhouse gases, which would in turn lead to the sustainable development in tourism and hospitality sector.

Sustainable tourism is facing various challenges, while the way to achieve the state of sustainability is long and winding. The anticipated though a bit confusing blockchain technology would help halt some challenges in sustainable tourism development and address its goals. Literally, the blockchain technology might potentially rattle tourism operations and advance local economy, organize food supply chain in a more efficient way, by alleviating food waste, satisfy tourists, and alter the tourists' sustainable attitude and performance. One might argue that the future is deemed to artificial intelligence (AI), that would have the highest impact on tourism industry as well as tourist behaviour. So far, artificial intelligence has already influenced the way tourists behave and different tourist enterprises operate. But that may be a subject of a different research paper.

References

1. B. Chen, Z.Qiu, N. Usio, K. Nakamura, Sustainability, **10(8)**, 2896 (2018)
2. A. Chiabai, S. Platt, W. Strielkowski, Tourism Economics, **20(2)**, 263-277 (2014)
3. M. A. Casado-Diaz, A. B. Casado-Díaz, G. Hoogendoorn, Scandinavian Journal of Hospitality and Tourism, **20(3)**, 268 (2020).
4. L. Zhenhua, Sustainable Tourism Development: A Critique, Journal of Sustainable Tourism, **11(6)**, 459 (2003)
5. L. Clifford, K. Nelson, R. Black, Journal of Hospitality and Tourism Management, **46**, 96 (2021)
6. S. Monaco, Journal of Tourism Futures, **4(1)**, 7 (2018),
7. S. Leminen, M. Rajahonka, M. Westerlund, R. Wendelin, Journal of Business & Industrial Marketing, **33(6)**, 749 (2018)
8. D. Fisher, Annals of Tourism Research, **31(2)**, 428 (2004)
9. P. Aditjandra, Advances in Transport Policy and Planning, Academic Press, **3**, 29 (2019)
10. J.Y. Lee, Business Horizons, **62(6)**, 773 (2019)
11. A. Polukhina, A.Tarasova, A.Arnaberdiyev, International Journal of Recent Contributions from Engineering Science & IT (iJES), **8(1)**, 81 (2020)
12. S. Mercan, L. Cain, K. Akkaya, M. Cebe, S. Uluagac, M. Alonso, C. Cobanoglu, International Journal of Contemporary Hospitality Management, **33(1)**, 243 (2020)
13. D. Buhalis, T. Harwood, V. Bogicevic, G. Viglia, S. Beldona, C. Hofacker, Journal of Service Management, **30(4)**, 484 (2019)
14. Lundmark L., Carson D.A., Palgrave Macmillan, Singapore (2020)
15. B. Parguel, R. Lunardo, F. Benoit-Moreau, Technological Forecasting and Social Change, **125**, 48 (2017)
16. D. Selloni, CoDesign for Public-Interest Services. Research for Development. Springer, Cham. (2017)
17. J.I. Pulido-Fernández, L. Andrades-Caldito, M. Sánchez-Rivero, Journal of Sustainable Tourism, **23(1)**, 47 (2015)
18. D. Miller, B. Merrilees & A. Coghlan, Journal of Sustainable Tourism, **23(1)**, 26 (2015)
19. M. Klöckner, S. Kurpjuweit, C. Velu, S.M. Wagner, Research-Technology Management, **63(4)**, 18 (2020)
20. PWC, *https://www.pwc.com/hu/en/kiadvanyok/assets/pdf/sharing-economy-en.pdf* (2015)

Environmental investment as a factor of sustainable economic development in the context of transformation of the external and internal environment

Ekaterina Sheina[1]*, *Aleksandr* Kurdyumov[1], and *Maria* Zavyalova[1]

[1] Ural State University of Economics, 8 Marta str. 62/45, 620144 Yekaterinburg, Russian Federation

Abstract. The desire to avoid dependence on the principles of the commodity economy, as well as the ultimate exhaustion of natural resources, encourages the state and business to search for new mechanisms and technologies for the reproduction of public goods. One of the key dominants in this direction is the field of ecology, environmental conservation, and, as a result, the development of environmental investment projects. However, the new investment principles require not only sustainable interest of investors and economic entities, but also an extensive developed legislative framework, protective mechanisms for reducing investment risks and high-quality financial regulation of this process by the state. The development of environmental investment will contribute to the sustainable growth of the Russian economy in the context of the transformation of the external and internal environment, the general improvement of the environmental friendliness of the production of national goods and services, and will also allow to establish an alternative direction of green financing.

1 Introduction

In the changing realities of the modern world, the problems of ecology, exhaustion of resources and preservation of pristine wilderness come to the fore in most national economies of the world. Particular attention is paid to issues of environmental pollution, the search for resource-dependent sources of reproduction of public goods, biodiversity reduction, ozone depletion, global warming. Interest in environmental conservation has been growing since the beginning of the 1970s (Table 1).

The development of a green economy is becoming a priority in the world. The recovery of the world economy from the pandemic will largely benefit environmental projects. Investments in environmental projects can create not only positive economic, but also social, psychological and cultural effects. Investments in environmental initiatives allow to introduce innovative technologies in the economy, reduce industrial emissions of enterprises, make products of higher quality, improve the image of entire corporations and industries, increase exports, and improve the well-being of the country's population.

* Corresponding author: shekat@mail.ru

Many of the serious environmental problems, facing the world community, cannot be articulated at the level of an individual state or industry. [12] Existing environmental problems are inherently transboundary and can be addressed through the adoption of strategies with coordinated measures and actions at the regional, national and European levels [14]. The development and implementation of Green Transformation Roadmaps will allow to complete industrial restructuring and lower technology costs, contributing to the green transformation process [19].

Table 1. List of global events on environmental issues

Year	Event	Problem
1974	Conference in Bucharest	Environmental protection
1977	United Nations Conference, Mar del Plata, Argentina	World water use
1980	Conference of the International Union for Conservation of Nature in Ashgabat	World conservation strategy
1981	Conference, held in Nairobi, Kenya	Renewing energy sources, World Charter for Nature
1983	Report "Our Common Future", the UN World Commission	Impact of environmental damage on the life of the future generation
1992	United Nations Conference, Rio de Janerio (document «Agenda 21»)	Balance between economic development and environmental protection
1997	United Nations Conference "Rio 5 +", New York	Outcomes of the five-year transition to sustainable development
2012	United Nations Conference "Rio 20 +"	Global problems
2015	United Nations Summit, Paris	Energy problems

Many countries have policies aimed at preserving and improving the environment and modernizing environmental legislation. For example, in Japan, South Korea, Germany, China and other countries, support for environmental projects is provided by both public and private investments [11, 16, 18].

The Russian Federation currently lacks free sources of financing, which hinders the development of environmental initiatives. In this regard, the search and attraction of investors who are ready to invest in environmental projects is of great importance in the Russian Federation. However, Russian legislation lacks a clear approach to regulating environmental investment. For the Russian Federation, this direction in the transition to a green economy is new and understudied in terms of the country specifics. Therefore, currently the key goal is to study the features of environmental investment in foreign countries and the opportunity to adopt the best investment experience in this area.

2 Materials and methods

The study is based on the provisions of investment theory and management theory, and a problematic approach that takes into account the complexity of raising funds at all stages of the life cycle of small and medium-sized enterprises of all industries and forms of ownership, especially in the context of socially oriented investment projects, including environmental investment projects, and the related need to minimize financial risks.

The study materials include data from the official website of the Federal State Statistics Service of the Russian Federation, individual data from corporate reports with state participation, as well as scientific works of Russian and foreign economists.

In the process of research, the authors use methods of dialectical, causal and structural-logical analysis, which allow us to streamline the theoretical aspects of environmental investment as a possible tool for alternative business financing in the Russian Federation in order to achieve the trajectory of sustainable economic development in the post-pandemic period.

The study hypothesis is based on the idea of the relationship between the principles of environmental investment and the development of new resource-saving technologies, which will ensure sustainable development of the Russian Federation economy and stable financing of economic entities implementing socially oriented investment projects, in particular in the field of ecology and environmental conservation.

3 Empirical model

The concept of environmental investment activities: Analysis of Russian and foreign literature on approaches to defining the term "environmental investment " allows us to systematize the existing points of view and categorize them into five groups (Table 2).

Table 2. Approaches to the definition of "environmental investment"

No.	Authors	Definition
1	Kryuchkova O.M., Guzenko A.D., Gushchina E.G., Badrak N.Yu.,Tarkhanova E., Fricler A., Babirina N.	Reducing the negative environmental burden
2	Egorova D.A., Sedash T.N., Tyutyukina E.B., Lobanov I.N., Miroshnichenko O.S., Mostovaya N.A.	Ensuring environmental safety
3	Spiridonova A.V., Rakov I.D., Polonik S.S., Khorobrikh E.V., Litvinchuk A.A., Loiseau E., Saikku L., Antikainen R., Droste N., Hansjurgen B., Pitkanen K., Tomsen M.	Ensuring economic benefits
4	Masserov D.A., Kolpakov N.S., Falcone P.M.	Development and modernization of production

Taking into account the above, the authors believe that it is necessary to combine views and formulate the following definition: *Environmental investment - investments aimed at reducing the negative burden on the environment in order to achieve environmental safety in the country by supporting environmental initiatives related to the modernization of production with a positive socio-economic effect.*

The main vehicles of environmental investment include issuing securities, lending, microfinancing, receiving grants and subsidies, environmental taxes. Environmental investment in foreign countries is carried out in the form of subsidies, tax deductions (Table 3).

Table 3. Sources of ecological investment

Investment source	Impact
Germany	
Toxic waste tax	↓ toxic wastes by 30% a year
Environmental fee for entry into ecological zones	↓ exhaust gasses by 40%
Subsidies for green energy	↑ renewable energy sources by 30%
Environmental payments	charged after environmental intervention (e.g. landscape change)
Concessional lending	green projects investment
Denmark	
Waste charges	Recycling
Canada	
"Green" bonds	Construction of green stations for electric vehicles, support for environmental initiatives
Greece	

Water tax	Reduced water consumption
Plastic bag tax	↓ plastic wastes by 40%
Subsidies	Purification of water when it is delivered by organizations
Sweden	
Tax on carbon, ethyl gasoline, diesel fuel	↓carbon pollution by 6%, ↓ pollution from motor vehicles by 75%
Japan	
Payments for extraction of water from rivers, use of water	Costs are covered by payments for maintaining water quality
Russia	
State programmes	Support for ecological projects

Table 4 provides a comparative analysis of the use of different sources of financing to stimulate and develop environmental projects in foreign countries.

Table 4. Comparison of environmental investment sources in Russia and foreign countries

Country	Green fiscal instruments	Subsidies	Concessional lending	Waste charges	" Green" bonds	Liability for environmental pollution
Greece	*	*		*		
Denmark	*	*	*		*	
Germany	*	*	*	*		
Sweden	*	*	*	*		
Canada	*	*	*	*	*	*
Russia	*	*		*		
Japan	*	*	*	*	*	

To ensure Russia's sustainable development, the list of environmental investment instruments should be expanded.

4 Results and discussions

In the period from 2010-2020 in the Russian Federation there was an increase in environmental investment, the value of which varied from 102.39 billion rubles. up to 168 billion rubles. This trend is associated with the deterioration of the environmental situation among large national corporations, some of which are beginning to develop and implement environmental programs.

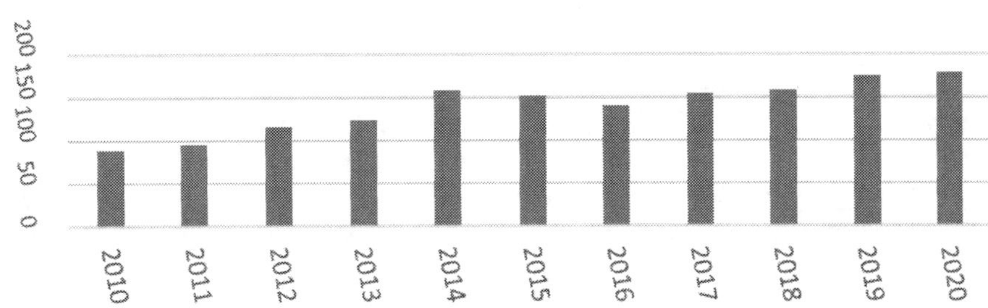

Fig. 1. Dynamics of environmental investment in Russia, 2010-2020, in billion rubles [20]

Environmental investment can be divided into financing raised from public funds, private funds and other sources. Figure 2 shows the trend of environmental investment development in the Russian Federation. Over the last 10 years, investment of environmental projects at the expense of the state decreased from 26.7% in 2010 to 9.1% in 2020. This dynamic is contrary to the existing Western practice. As a rule, in foreign countries financing by state funds makes up the largest share in the structure of environmental investment.

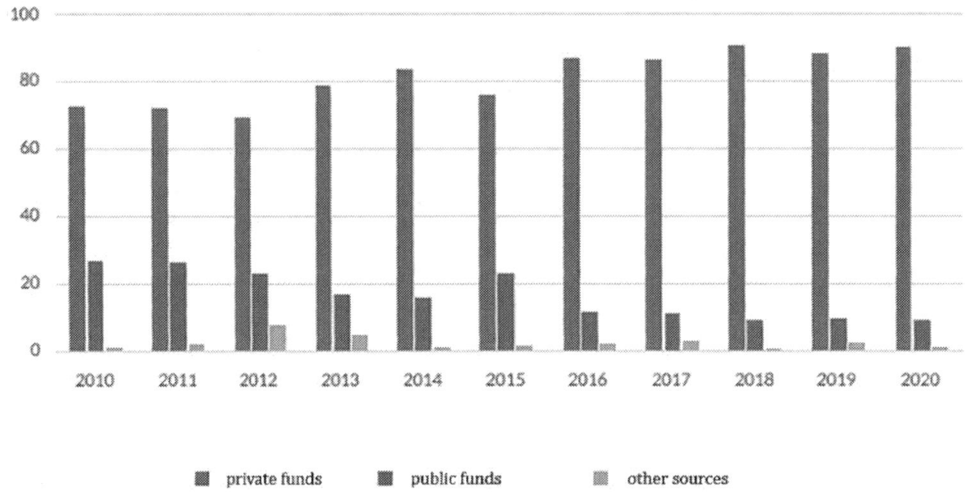

Fig. 2. Change in the structure of environmental investment in the Russian Federation from 2010-2020, in % [20]

The data presented in Figure 2 show that in the Russian Federation private corporations pay taxes for negative environmental impact. As a rule, the state imposes certain sanctions for violation of environmental protection. To prevent unforeseen costs for damage to nature, corporations invest in environmental projects. The growing trend in environmental investment in the Russian Federation shows a gradual transition of the country to the path of sustainable development and solving environmental problems. However, to date, there is no system for supporting environmental investment projects.

Environmental investment in the Russian Federation can be described as follows: in Russia, the most spread environmental investment instruments include co-financing, environmental lending (Table 5).

Table 5. Environmental projects, implemented in Russia

No.	Corporation	Project title
1	Lukoil Group	Lukoil Group's Environmental Safety Program (fuel with improved environmental characteristics)
2	«T-Plus» PJSC	Construction of photovoltaic array (solar strata)
3	Rushydro PJSC	Construction of an Arctic wind farm in northern Yakutia with NEDO Corporation (Japan)
4	Sberbank PJSC	Construction of solar and wind power plants in Nikolaev, Kherson, Odessa regions
5	VTB PJSC	Cooperation with the Wildlife Foundation for the conservation of snow leopard, Amur tiger, Far Eastern leopard populations

6	Vnesheconombank Group	Construction of Boguchanskaya hydroelectric power station on the river Angara

From the environmental point of view, the participation of the Russian Federation in socially oriented environmental investment projects for the transition to a green economy makes it possible : to reduce pollution, restore ecological balance and reproduce the balance of ecosystems; from the economic point of view, the attained effects can include reducing product input intensity due to recycling, improving the efficiency of economic sectors, achieving socially significant environmental indicators.

5 Conclusions

Given the current incentives, there are some constraints to scale up environmental investment. One of the factors hindering the transition to a green economy is the lack of support and protection of investor risks, the lack of a legislative framework for environmental investment projects.

Thus, the Russian legislation still does not resolve the issue of the concept of environmental investment and its basic principles and methods. The Federal Law "On investment activities in the Russian Federation carried out in the form of capital investments" dated February 25, 1999 No. 39-FZ does not distinguish a separate legal category "investments in the environment." The draft Federal Law "On the protection and promotion of investment and the development of investment activities in the Russian Federation" No. 828237-7 does not contain environmental investment goals.

The market for "green" bonds in the Russian Federation is just beginning to form. There is insufficient institutional experience in placing bonds to localize environmental problems. The development of the mechanism for issuing green bonds could ensure a qualitative transition to a green economy in our country. In addition, in order to attract environmental investment, a green banking mechanism should be introduced, which implies a long-term environmental policy for investments in environmental initiatives. Green banking is a soft loan for economic entities involved in investment activities in environmental projects. The mechanism could facilitate access to investment resources for environmental projects.

The Russian Federation also lacks an infrastructure institution coordinating environmental investment issues. To date, these competencies are distributed between various departments, in particular, the Ministry of Natural Resources and Ecology of the Russian Federation, the Ministry of Finance of the Russian Federation, the Ministry of Economic Development of the Russian Federation, the Central Bank of the Russian Federation, but there is no clear division of functions and understanding of the area of responsibility for environmental investment development.

Summarizing the above, it should be noted that the development of environmental investment by both investors and economic entities will allow the Russian Federation to move to a sustainable path of economic development. Green investment contributes to improving the environment, mitigating the negative effects of climate change and managing resources. However, the stimulation and development of investment support for environmental projects in the Russian Federation is possible only if the legislative norms are revised and a consistent national policy in the field of green finance is carried out.

References

1. Federal Law No. 39 of February 25, 1999 "On investment activities in the Russian Federation, carried out in the form of capital investments"

2. Federal Law No. 828237-7 "On protection and promotion of investments and development of investment activities in the Russian Federation".
3. O.S. Miroshnichenko, N.A. Mostovaya, Finance: theory and practice, 31 (2019)
4. K.A. Nazateryan, N.V. Klimovsky, Scientific and practical electronic journal, **10**, 26 (2018)
5. N.S. Kolpakov, *Features of environmental investment development in Russia*, 14 (2019)
6. O.M. Kryuchkova, A.D. Guzenko, *Promising areas of environmental investment*, 43 (2019)
7. S.S. Polonik, E.V. Khorobrikh, A.A. Litvinchuk, Bulletin of Polotsk State University, 14 (2018)
8. I.D. Rakov, Current Economic Problems, 67 (2017)
9. T.N. Sedash, E.B. Sedash, I.N. Lobanov, *Directions and tools for financing green projects in the concept of sustainable economic development*, 52 (2019)
10. E. Yudin, Problems of tax regulation of environmental management, Taxes, 2, 67 (2015)
11. A. Bauer, K. Menrad, Energy Research & Social Science, **67**, 101509 (2020)
12. R. Boucekkine, G. Fabbri, S. Federico, F. Gozzi, European journal of operational research, **290(1)**, 331 (2021)
13. P. M. Falcone, International Journal of Green Economics, **14(2)**, 159 (2020)
14. F. Fava, L. Gardossi, P. Brigidi, P. Moroned, D. Carosie, A. Lenzif, New Biotechnology, **61**, 124 (2021)
15. E. Loiseau, L. Loiseau, R. Antikainen, N. Droste, B. Hansjürgen, K. Pitkänen, M. Thomsen, Journal of Cleaner Production, **139**, 361 (2016)
16. Y. Luo, M. Salman, Z. Lu, Science of the total environment, **759**, 143744 (2021)
17. E. A. Tarkhanova, Terra Economicus, **16(2)**, 75 (2018)
18. J. Wang, L. Wang, X. Qian, Journal of Cleaner Production, **281**, 124446 (2021)
19. F. Xu, F. Cui, N. Xiang, Journal of Cleaner Production, **283**, 124643 (2021)
20. Service of Russian statistics, Investments in Russia, http://www.gks.ru (2021)

Renewable energy sources in maintaining environmental stability on the Black Sea coast

Mzia Diasamidze[1] *, *Ana* Shotadze [1]

[1]Faculty of Navigation, Batumi State Maritime Academy, Rustaveli Avenue 53, 6000, Batumi, Georgia

Abstract. The aim of this research is to study the main ecological parameters of the Black Sea in an experimental way as well as to evaluate the results and to draw conclusions. We aim at determining the cause of pollution of the sea area represented by the emissions from ships' engines and pollution by surface waters. Our empirical part that included the field expeditions, sampling was carried out with the involvement of students within the framework of the internal grant research project "Monitoring of the Black Sea Ecological Status - Sea Water Quality Survey". We studied the level of pollution of two rivers flowing into the Black Sea, Bartskhana and Kubistskali, and made relevant conclusions about the types of pollution, proper measurements and causal relationship of the causes of sea water pollution with the vessels entering the port of Batumi.

1 Introduction

Numerous studies have been devoted to the study of the ecological status of the Black Sea [1, 2]. As the economy and transport infrastructure of the Black Sea Basin countries grow, the research problem is constantly changing [3]. With it, the degree of pollution of the marine environment also increases. The most important ecological problem is the introduction of large amounts of biogenic substances, various pollutants (pesticides, phenols, resins, etc.) by rivers into the sea. Increased maritime traffic is another factor in marine pollution. In previous studies we have explored issues about ballast waters.

In particular, the issues of pollution of the marine environment by ships with ballast waters [4]. At this point we have evaluated the results of previous studies. The limited circulation of Black Sea water to the ocean prevents the removal of pollutants into the marine environment, resulting in intensified accumulation processes. Therefore, the coastline is under strong anthropogenic load [5]. The Black Sea represents the greatest resource potential, the recreation area and the main artery of foreign relations. Therefore, the scientific study of ecological issues of the Black Sea and its prediction is highly relevant and gaining a global scale.

* Corresponding author: m.diasamide@bsma.edu.ge

Together with the maritime academy students, we explored the use of wind as a renewable energy in the form of wind turbines on ships. The use of these types of turbines will significantly improve the ecological safety of the marine environment [6].

2 Materials and methods

There are three main rivers in the area of Batumi Oil Terminal Ltd., which flow in the direction of the Black Sea: Korolistskali, Kubistskali and Bartskhana river. Of these three rivers, the Korolistskali river is less polluted with oil products, the other two rivers are more polluted, as the Bartskhana and the rivers. Kubistskali will pass through almost all zones of the terminal. It should be noted that the background pollution of the Bartskhani river with oil products is quite high, which is caused by the historical pollution of the areas occupied by the Batumi Oil Depot Ltd. Of course, the pollution of rivers also leads to the disturbance of the ecological condition of the Black Sea. Therefore, the purpose of the first phase of field-expedition studies was to identify the main rivers joining the sea and its surrounding areas. The subject of the second stage field-scientific research was the study of the ecological parameters of the sea in the selected areas in the Batumi water area (see Figure 1).

Fig. 1. Examples of field research at the Black Sea site

Maintaining and protecting the ecological condition of the Black Sea coastline and its surrounding areas is recognized by all its neighbouring states as a priority [7]. The rivers flowing into the Black Sea from Georgia along the Batumi coast are used by the population for household and agricultural purposes (see the map of the area at Figure 2 that follows below). Therefore, the sources of water pollution are also related to such activities of the population.

The most common pollutants in water bodies are: pesticides, fertilizers, petroleum products, heavy metals, various substances produced by chemical and industrial processes, improper treatment facilities and others. In order to check the water quality of the Batumi water area, samples were taken in the inland waters of Batumi, which were subjected to laboratory research, water pollution indicators were determined [8].

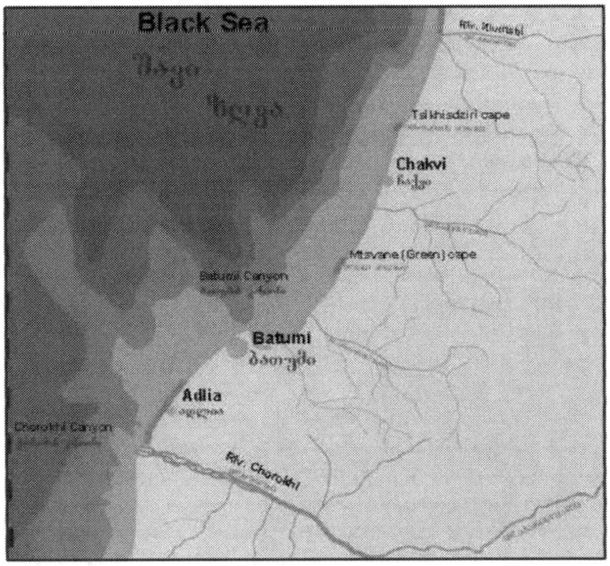

Fig. 2. Map of Batumi Black Sea coast

In this study we used the following research instruments: Spectro-photometer (DR 3900), electronic scale, portable laboratory device MULTI 3440 SET F (with three electrodes - to measure salinity, dissolved oxygen and pH), COD REACTOR - to determine the chemical oxygen demand in water, and pH. COD - to determine the number of inorganic pollutants found in surface waters (rivers). Getting to know the properties of water begins with determining its organoleptic parameters - water colour, transparency, taste and weight of the substance. Getting to know the properties of water begins with determining its organoleptic parameters - water colour, transparency, taste and weight of the substance. The ecological condition of the sea also depends on the amount of oxygen in the water (%), salinity (TDS), the concentration of hydrogen ions in the Black Sea water (pH), the study of the amount of oxygen (J.M. and J.M.). We have studied water pollution with pesticides - as a widespread problem, as pesticides are still actively used in agriculture. Getting large amounts of fertilizers into reservoirs leads to the accumulation of extra phosphorus and nitrogen which often leads to eutrophication.

In the area of Batumi water area, mainly three rivers flow in the direction of the Black Sea: Korolistskali, Kubistskali, and Bartskhana. Naturally, the pollution of rivers also leads to the disturbance of the ecological condition of the Black Sea. Therefore, the purpose of Phase I field-expedition studies was to identify the main rivers joining the sea and its surrounding areas. The subject of the II stage field-scientific research was the study of the ecological parameters of the sea in the selected areas in the Batumi water area. For this purpose, we carried out water samples at the sensitive areas of the coastline. We studied the ecological characteristics of the Bartskhani and Kubistskali rivers periodically for two months, observing the changes in the given ecosystem during the months of April-October.

3 Practical results

The study was performed using inpatient and laboratory methods (see Figure 2). We compared the measurement results with the parameters defined by the norm. The results of the laboratory study of water samples taken from rivers according to seven parameters are given in Tables 1, 2, and 3.

Table 1. Data for April, 2020

I Field Survey Results						
Ammonia mg / l	**BOD mg / l**	**Nitrates mg / l**	**Nitrites, mg / l**	**Phosphorus mg / l**	**COD mg / l**	**suspended matter mg / l**
1	**2**	**3**	**4**	**5**	**6**	**7**
River "Bartskhana"						
0,8-0,87	1,28-2,5	0,3-0,48	0,7-0,81	0,06-0,075	28,5	0,75
River "Kubistskali"						
1	**2**	**3**	**4**	**5**	**6**	**7**
0,11-0,17	2,1-2,9	1,3-1,8	0,75-0,89	0,17 -0,2	27,16	0,68
Compared to the norm						
0,39	4 <...>6	40	0,08	2	25-100	up to 50

Table 2. Data for July, 2020

II Field Survey Results						
Ammonia mg / l	**BOD mg / l**	**Nitrates mg / l**	**Nitrites, mg / l**	**Phosphoru s mg / l**	**COD mg / l**	**suspended matter mg / l**
1	**2**	**3**	**4**	**5**	**6**	**7**
River "Bartskhana"						
0,4	1,98 - 2,1	0,17- 0,19	0,31-0,45	0,6-0,8	25	Not fixed
River "Kubistskali"						
1	**2**	**3**	**4**	**5**	**6**	**7**
0,09-0,4	1,0- 1,74	0,1-0,7	Ammonium nitrogen 0,3-0,41	Not fixed	27,0	0,81
Compared to the norm						
0,39	4 <...>6	40	0,08	2	25-100	up to 50

Table 3. Data for October, 2020

III Field Survey Results						
Ammo nia mg / l	**BOD mg / l**	**Nitrates mg / l**	**Nitrites, mg / l**	**Phosphorus mg / l**	**COD mg / l**	**suspended matter mg / l**
1	**2**	**3**	**4**	**5**	**6**	**7**
River Bartskhana						
0,72 - 1,35	0,67- 4,37	0,24-0,28	0,4-0,65	0,3-0,47	26,7	Not fixed
River Kubistskali						
1	**2**	**3**	**4**	**5**	**6**	**7**
0,42- 0,53	1,98-2,1	0,12-0,182	Not fixed	Not fixed	24,3	0,67
Compared to the norm						
0,39	4 <...>6	40	0,08	2	25-100	up to 50

As a result of primary and repeated inspection of sea water and river-sea tributaries on oil products of Batumi coast, the average salinity index was TDS = 17.0-20.0 ‰. As for the content of nitrates, nitrites and ammonia, it was observed only in two areas - in the water area of the port of Batumi and in the sea near Ardagan Lake (Figure 3).

Fig. 3. Ardagan Lake

The results are shown in Table 4.

Table 4: Data for August, 2020

sampling points	nitrates	Nitrites	Ammonia
Batumi port water area	0,11 mg / l	0,2 mg / l	0,4 mg / l
The sea strip nearby Lake Ardagan	0,35 mg / l	-	1,2 mg / l
assessment	Within Norm.	Within Norm.	Within Norm.

Thus, the results of our research allowed us to determine the nature of the change in the ecological parameters of the Black Sea Batumi water area. These results are very important for fostering sustainable development and helping to promote the renewable energy sources in this very region. In particular, we found and determined that the pollution of the Batumi coastline is probably related to the pollution with drainage water. It is noteworthy that by using these indicators in practice, it is possible to assess the ecological safety of the Batumi port optimally.

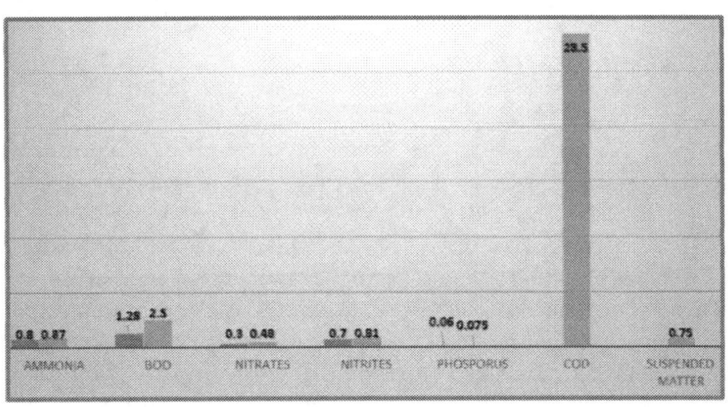

Fig. 4. Results for river Bartskhana

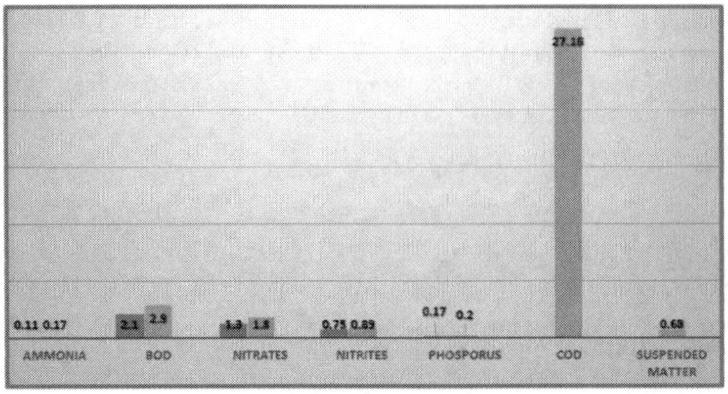

Fig. 5. Results for river Kubistskali

The parts of experimental studies were analyzed separately for the rivers Bartskhan and Kubistskali, the results are presented in diagrams (see Figures 4 and 5).

4 Conclusion and recommendations

In the conclusions, we want to state that after processing the research results and comparing the obtained data with the normalized parameters, we made the important conclusions: First, our field-scientific studies conducted on the rivers "Bartskhana" and "Kubistskali" identified sensitive areas, where the process of taking analytical samples and laboratory tests (according to seven parameters) was carried out for 6 months.

Second, seawater analyses were taken at sensitive points of the Black Sea (mainly at river confluences, coastline points), the concentration of seawater hydrogen ions and the quantitative salinity were determined in laboratory conditions - on average, 17-20 ‰;

Third, laboratory studies have shown that the concentration of hydrogen ions in the confluence of the rivers "Bartskhana" and "Kubistskali", as well as in the water area of the port of Batumi varied in the range of pH = 6.5-8.3. This figure should be taken into account in the complex study of ecological parameters in the Black Sea coast;

Fourth, laboratory conditions of Black Sea ammonia (partial excess), nitrates (low rate), nitrites (excess) and phosphorus (within the norm) were determined. Which leads us to the conclusion that in these rivers there have sometimes been discharges of fertilizers at such concentrations that their natural decontamination can no longer take place and the ecosystem is being polluted.

Fifth, researched by BOD and COD Result - the rate is within the norm. Weighted particles were not observed;

The obtained results allow us to comprehensively assess the modern ecological problems of the Black Sea and to plan ecological security measures for its coastline and surrounding areas.

All in all, we have studied issues about ballast waters. In particular, the issues of pollution of the marine environment by ships with ballast waters We have come to the conclusion that one of the most reliable ways to reduce environmental pollution from ships is to use renewable energy on ships. It is true that this source of energy cannot fully provide energy in long-distance sailing, but their use on small tonnage ships would give good results. Moreover, the leading countries of the world, including Georgia, have the practice of producing and using wind energy [9].

The high interest shown by students gives us the incentive and interest to continue research in this direction, thus promoting on the one hand the involvement of students in research activities, On the other hand, we develop and apply the inner culture, morality and knowledge of the student-youth to the realization that the preservation of the ecosystem of nature depends in large part on their right actions [10] .

References

1. L. Rusu, D. Ganea, E. Mereuta, Energy Exploration & Exploitation, **36(2)**, 335-351 (2018)
2. T. Oguz, Frontiers in Marine Science, **4**, 110 (2017)
3. B. Levent, A. Öztekin, F. Şahin, U. Arici, U. Özsandikçi, Mediterranean Fisheries and Aquaculture Research, **1(2)**, 66-86 (2018)
4. D. Onwuegbuchunam, T. Ebe, L. Okoroji, A. Essien, Journal of Marine Science and Engineering, **5(3)**, 39 (2017)
5. Ton H. Snelder, Scott T. Larned & Rich W. McDowell, New Zealand Journal of Marine and Freshwater Research, **52(3)**, 336-361 (2017)
6. G.Surmanidze, M.Diasamidze, IAMU 2017, **3** (2017)
7. M. I. Nenciu, V. Nita, A. Totoiu, H. Hamza, Journal of Environmental Protection and Ecology, **21(1)**, 184-192 (2020)
8. J.F. Pons Austina, *http://hdl.handle.net/10251/54792* (2014)
9. M.Diasamidze, I.Takidze, Innovative Economics and Management, **4**, 297-302 (2017)
10. M.Diasamidze, A.Shotadze, Fundamental and applied researches in practice of leading scientific schools, **31(1)**, 58-60 (2019)

Sustainable spatial development: a story of Novosibirsk and Krasnoyarsk agglomerations

Irina Usmanova[1*], *Nadezhda* Ligaeva[1], *Olga* Kuznetsova[1]

[1]Institut of Ecology and Geography, Siberian Federal University, Svobodny ave. 79, 660041 Krasnoyarsk, Russian Federation

Abstract. Under the conditions of market- economic reforms that have been taking place in Russia over the past 30 years, the sectoral and territorial structure of the country's economy has changed dramatically. The economy of some regions, that were successful during the Soviet times sagged and failed, while other regions experienced an actual economic take-off. This paper discusses the main aspects of the spatial organization of the Russian Federation and the development of the Novosibirsk and Krasnoyarsk agglomerations, which are the largest economic centers of Western and Eastern Siberia. To answer the main question of the study- whether thirty years of economic restructuring in the country could affect the leading position of Novosibirsk and make Krasnoyarsk the new regional leader of Siberia, the authors perform the comparative socio-economic analysis of both agglomerations.

1 Introduction

The Russian Federation is the largest country in the world by its area. This fact has led to emergence of wide variety of natural, historical and socio-economic conditions for the development of its regions. Thus, a significant gap between different federal subjects of Russia in terms of living standards has formed. The gap between the backward regions of the country and regions with the average level of economic and social development took shape back in the Soviet period.

However, the contrast between successful and problem areas has increased significantly in the conditions of market economy system. On the one hand, there is a number of entities with a very high territorial concentration of population, labor resources and production, that have stood out over the past years of transformations. These regions are often oriented towards external markets. On the other hand, there are regions with a lack of development prospects within the framework of the economic model that has developed in the country. Uneven development of regions and districts is also a fairly frequent occurrence in other countries [1]. The weakening of ties of domestic economy increases the isolation and creates the mood of separatism among the backward regions. This causes a high level of social

*Corresponding author: usmi59@mail.ru

tension in society and complicates the process of territorial control. These circumstances m the government of the Russian Federation pay more attention to the territorial development of the country. The spatial organization of society has been the subject of many scientific studies, government strategies and programs. One of the tools for improving the spatial structure is the formation of regional growth centers in the form of large urban agglomerations. The idea of using such centers for the spatial structure development was firstly put forward about fifty years ago by a number of authors [2, 3].

2 Basic documents for the territorial-economic development management of the Russian Federation

Management of socio-economic development of the Russian Federation is based on a system of documents, including strategies, concepts, Federal laws and Presidential decrees. One of the main documents in this sense is the Federal Law N 172-Fz of 28 June, 2014 "On Strategic Planning in the Russian Federation" [4], which provides the development plan for both industry-specific and regional programs, since the country's socio-economic system is not only a complex of industries, but also a territorial-economic system. Clause 26, Article 3 of the Federal Law N 172-Fz of 28 June 2014 "On Strategic Planning in the Russian Federation" states that "The Spatial Development Strategy of the Russian Federation is a strategic planning document that defines the priorities, goals and objectives of the regional development of the Russian Federation. Its main aim is to maintain the stability of the resettlement system in the Russian Federation" [4].

The content and basic concepts of regional development are defined by such documents as "The Spatial Development Strategy of the Russian Federation for the Period up to 2030" [5], The Decree of the President of the Russian Federation "On Approving the Fundamentals of State Policy of Regional Development of the Russian Federation for the Period up to 2025" [6] and others. These documents outline the main trends, problems, goals, objectives and fundamental principles of spatial development of the Russian Federation.

The whole host of documents such as territorial planning schemes, strategies for the socio-economic development of economic regions, macro regions, constituent entities of the Russian Federation and various sectoral programs are based on The Spatial Development Strategy.

3 Socio-economic differentiation of Russian regions

According to the above-mentioned documents, one of the main trends in the spatial development of the Russian Federation is the increasing population and production concentration in some of its regions. These are primarily 40 large urban agglomerations and four mineral raw centers, which are associated with the production of hydrocarbons and diamonds.

The importance of the natural resource factor for the territorial development is decreasing in developed countries [7], but is still relevant for Russia. The concentration of information and innovations is the most important factor for the development of all agglomerations [8].

The territorial population and production concentration is an inevitable consequence of the economic law of the productive forces' development. On the one hand, this situation contributes to the growth of economic efficiency, but on the other, creates a number of acute socio-economic problems. According to The Draft Concept of The Spatial Development Strategy [9], one of the main problems is so-called "compression" or "condensation" of

Russia's economic space around Moscow and regional capitals, along with the dissemination of the economic "periphery", where demographic and economic significance has sharply decreased over the years of market reforms. That has led to a current situation, where eight or ten federal subjects of Russia produce 80% of the country's GDP and appear to be the leaders both in amounts of paid taxes and economic development rate, while the economy of the remaining 70 entities stagnates or degrades.

There are several types of territories in the Russian Federation, which have evolved in different directions in terms of socio-economic development. Along with the successfully developing regions, there are vast territories with a low level of economic and social development. The interregional difference is so significant that "the backward regions of the Russian Federation will need not even tens, but hundreds of years to reach the level of modern and economically developed areas. This indicates the presence of contrasts of intercontinental scope within one country (akin to the contrasts between European and African countries)" [9]. The modernization of these regions and solvation of its' problems is impossible without external help from federal center. The present situation is an obstacle to the balanced spatial development of the Russian Federation.

According to the theory of "Four Russias" by Zubarevich, a Professor at the Geographical Department of Moscow State University, federal subjects and cities of the Russian Federation can be divided into four different groups, based on its' level of socio-economic development, prospects and factors of growth. The future development direction of these entities depends on the group it belongs to [10].

The backlog of many federal subjects of the Russian Federation from the regions with the average socioeconomic conditions is a big problem for the holistic and harmonious development of the country. It carries the threat of social conflict, separatism and political instability. As an example, it is enough to mention the excessive, hypertrophic development of Moscow with its growing housing, environmental and transport problems against the population outflow from peripheral regions, that slows down and impedes its' development.

An effective spatial development policy should help to ensure equal opportunities for the realization of the constitutional rights and freedoms of Russian citizens throughout the Russian Federation. Its main goal is to: "...ensuring sustainable and balanced spatial development of Russia, aimed at reducing interregional differences in population's quality of life" [5].

4 City agglomerations as supporting centers for the territorial development

One of the main directions of Russian socio-economic development is balanced spatial development, in particular, "the integrated development of Siberian and Far Eastern territories and the population consolidation in the Eastern regions of Russia" [11], that became very important since Siberia's share in national economy and total population of the country decreased quite noticeably during the reform period (Figure 1).

One of the goals of the Spatial Development Strategy of Russia is the progressive development of current centers of economic growth and creation of new, large, perspective agglomerations [12]. The target scenario is aimed at the rapid population and economic growth in both old and new large urban agglomerations, that will allow to redistribute the resources for the development of the backward regions.

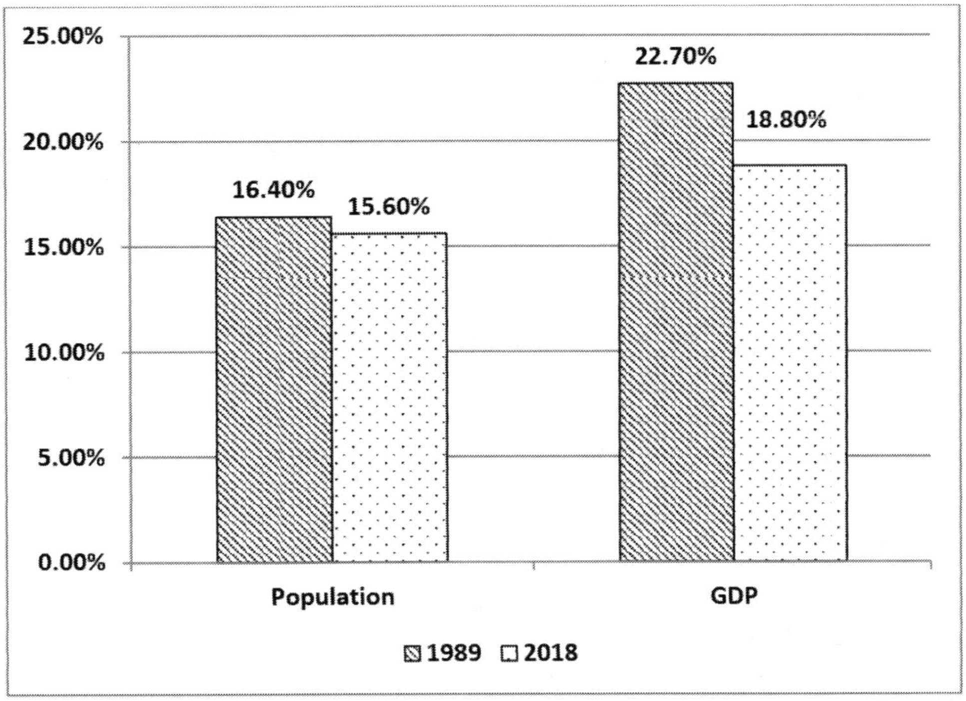

Fig. 1. Siberia's share in total population of Russia and its contribution to GDP in 1989 and 2018

According to the Spatial Development Strategy the number of such centers of economic development should be increased by two times [5]. Currently, there are around 40 urban agglomerations in the Russian Federation that can be called the economic centers. When considering Siberian cities, these are primarily the cities with over one million people population like Novosibirsk, Krasnoyarsk, Omsk, etc. Novosibirsk and Krasnoyarsk are the economic centers of West Siberian and East Siberian economic regions and the core of agglomerations of the same name.

5 Comparison of the Novosibirsk and Krasnoyarsk agglomerations

In order to answer the main question of the study and understand whether Krasnoyarsk can become a leader of Siberia, we need to compare the agglomerations of both cities.

According to the All-Union Population Census, there were 1.16 million inhabitants in Novosibirsk in 1970 [13]. Krasnoyarsk became a million –person city more than forty years later, in 2013 [14]. By the end of the Soviet era, Novosibirsk have already become the center of an actively forming agglomeration- the largest economic center of Siberia, its regional capital. Since 2007 the development of Novosibirsk as a part of the agglomeration had become a part of the new "Master Plan of Novosibirsk" for the period up to 2050 [15, 16].

In Soviet times, Krasnoyarsk was inferior to Novosibirsk for all economic indicators. However, over the thirty - year reform period, the Russian economy has changed significantly. This transition has been accompanied by a deep restructuring of economy's sectoral structure. Russia has become a part of the world economy, some of its products are now in demand on the external markets and are traded at world prices, bringing high incomes to the manufacturing companies and the regions of its' origin. The agglomeration processes on the territory around Krasnoyarsk were also intensified at this time.

The Krasnoyarsk agglomeration is one of the eight emerging agglomerations of Siberia with the elaborated development concept. The first visible output of this concept was the "Territorial Planning Scheme of the Krasnoyarsk Urban Agglomeration", created in 2008 [17]. In 2018, The Scheme was updated with a planning perspective for the period up to 2038 [18].

Have the role of the Novosibirsk and Krasnoyarsk agglomerations changed in the conditions of the market reforms?

The contribution of Krasnoyarsk Krai to Russia's GDP is more significant compared to Novosibirsk Oblast- in 2018 the shares of these regions were 2,5% and 1,5% respectively. The preponderance of the mining industry in Krasnoyarsk Krai is especially remarkable - 4.1% compared to 0.5% in Novosibirsk oblast. Krasnoyarsk Krai is also 1,6 times ahead of Novosibirsk oblast in terms of Gross Regional Product (GRP) per capita. Along with the absolute indicators, it is also necessary to consider the structure of GRP in both regions. Krasnoyarsk Krai is a more resource-rich region- the mining industry makes 21% of its GRP volume and is practically absent in Novosibirsk oblast. The share of manufacturing industry in both regions also differs considerably- 31% of GRP in Krasnoyarsk Krai and 13% in Novosibirsk oblast.

Krasnoyarsk Krai is mainly aimed at raw materials export and manufacturing industries. Service sector is not a big part of its economy. For example, transport and trade make up only 12 % of its GRP volume, while in the Novosibirsk oblast these sectors contribute 31% to the region's economy. Novosibirsk Oblast, deprived of the raw material factor is actively using its scientific capacity- the professional, scientific and technical activities sector makes up 5% of its GRP. This type of activities is insignificant in Krasnoyarsk Krai [19].

Therefore, there is a question whether the leading economic position of Krasnoyarsk Krai can help the Krasnoyarsk city to become the new regional leader of Siberia. The comparison of socio-economic and geographical indicators of both agglomerations can help to answer this question.

The total area of the Novosibirsk and Krasnoyarsk agglomerations is almost the same- 26 and 24 thousand square kilometers respectively. 48% of population of Krasnoyarsk Krai is concentrated in the Krasnoyarsk agglomeration. The total population of the Krasnoyarsk agglomeration is estimated at 1,4 million people and it is two thirds of population of the Novosibirsk agglomeration, which has 2,1 million inhabitants. 77% of population of Novosibirsk Oblast is concentrated in the Novosibirsk agglomeration.

The territorial concentration of production in Krasnoyarsk Krai is lower compared to Novosibirsk Oblast, as there is another major industrial center Norilsk, located in the North of Krasnoyarsk Krai, with its Mining and Metallurgical Company "Norilsk Nickel". Norilsk produces 25-30% of GRP volume, Krasnoyarsk contributes 38%, while the share of Novosibirsk in the economy of Novosibirsk Oblast is almost 80% [20].

The reduced share of the Krasnoyarsk agglomeration in its region's economy and its difference in the population aspect is mainly a consequence of such natural causes as the size of these entities. Krasnoyarsk Krai is 13 times larger than Novosibirsk Oblast by the land area. Density of the economic space of the Krasnoyarsk agglomeration is much lower than that of the Novosibirsk, which can be also a consequence of housing density of these two entities- 57 and 82 individuals per square kilometer respectively [21]. The regional specialization affects the sectoral structure of agglomerations (see Figure 2).

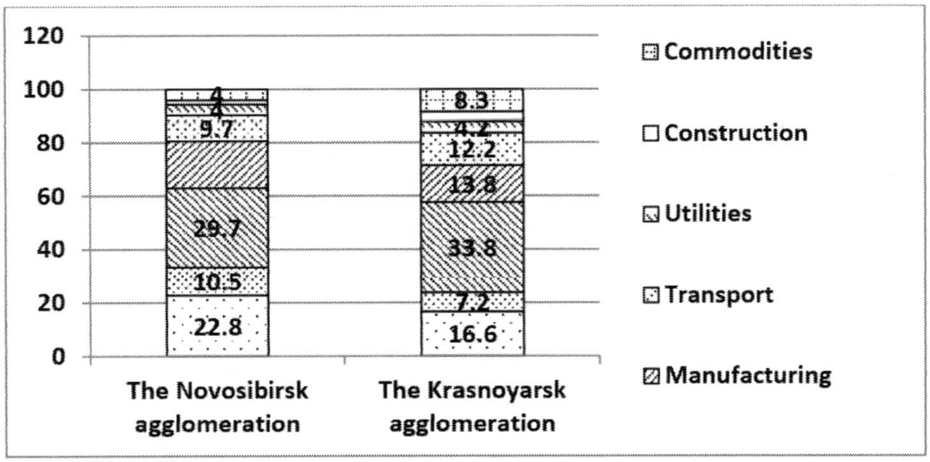

Fig. 2. The sectoral structure of the Novosibirsk and Krasnoyarsk agglomerations

According to The Institute for Urban Economics (IUE) [21], The Gross Metropolitan Product (GMP) of the Novosibirsk agglomeration is 15 % higher than GMP of the Krasnoyarsk agglomeration (with 0.8 and 0.7 trillion rubles respectively). The economic activity is also higher in Novosibirsk than in Krasnoyarsk- its' contribution to GMP is 94% and 82% respectively.

In the Soviet period, Krasnoyarsk had many military factories, while Novosibirsk was one of the country's major center for the military equipment production, a center of the Military-Industrial Complex (MIC). The enterprises of the military-industrial complex were greatly reduced in both cities in the course of economic reforms, accompanied by the conversion of military production. The Krasnoyarsk economy has remained afloat due to the Krasnoyarsk Aluminum Smelter, the second largest aluminum smelter in the world. Novosibirsk was not that lucky in this respect, however, after the MIC closure, the service sector evolved rapidly here.

The city became an intermediary between China and the rest of Siberia in the trade with Chinese goods. There was no such a growth factor in Krasnoyarsk due to its' geographical location. The fact that Novosibirsk Oblast was involved in the trade with China has led to the rapid development of transportation sector, the growth of small and medium enterprises, a more diversified economy with more advanced service sector than in Krasnoyarsk Krai and Krasnoyarsk itself [22].

It is important to mention, that one of the most important scientific and educational complexes in Russia-Akademgorodok was established in Novosibirsk during this period. Akademgorodok (Academy town) is the third largest scientific center in the country. It includes 43 academic institutes, 7 international research institutions, 39 higher education institutions, a medical academy etc. Almost half of higher educational institutions of Siberia and the Far East are concentrated in Novosibirsk.

As for the tourism industry, the Novosibirsk agglomeration also takes the lead- the sector is much more developed here. During the Soviet times Krasnoyarsk was a city closed for foreigners and opened up for visitors only in 1991. Long-term isolation has not benefited the development of the tourism infrastructure.

6 Conclusions

Overall, this paper looked into the sustainable spatial and regional development of two Russian cities - Novosibirsk and Krasnoyarsk – as well as their agglomerations. We showed Novosibirsk that was founded much later than Krasnoyarsk during the construction of the Trans-Siberian Railway at the end of the 19th century has developed rapidly. By the end of the period of Soviet industrialization, Novosibirsk was ahead of Krasnoyarsk both in terms of population and industrial potential. The thirty years of market reforms in the country have certainly brought immense changes to the economic grounds for the agglomerations. However, the Novosibirsk leadership has not been undermined.

The comparative analysis of the socio-economic development of both agglomerations led to the conclusion that Krasnoyarsk does not stand a big chance of becoming the regional leader of Siberia. Among the main factors, explaining why Krasnoyarsk takes the second place are: the natural resource factor, a less favorable economic-geographical position, a more inertial economy, a lack of labor force within the Krasnoyarsk agglomeration and a smaller domestic market. The leading role in this part of Russia belongs to Novosibirsk and the situation is unlikely to change in the near future. All these findings might be of a special interest to the local and state politicians as well as stakeholders and urban planners.

References

1. World Bank Group, https://openknowledge.worldbank.org/handle/10986/5991 (2009)
2. W. Strielkowski, T. Veinbender, M. Tvaronavičienė, N. Lace, Economic Research-Ekonomska Istraživanja, **33(1)**, 788-803 (2020)
3. P. Krugman, Journal of Political Economy, **99**, 483–499 (1991)
4. Federal Law N 172-Fz of 28 June 2014 "On Strategic Planning in the Russian Federation", *http://static.kremlin.ru/media/acts/files/0001201406300016.pdf* (2014)
5. Government of the Russian Federation, *http://static.government.ru/media/files/UVAlqUtT08o60RktoOXl22JjAe7irNxc.pdf* (2019)
6. Decree of the President of the Russian Federation, *http://kremlin.ru/acts/bank/41641* (2017)
7. M. Porter, *The Competitive Advantage of Nations: With a New Introduction* (1990)
8. T. Hägerstrand, Papers and Proceedings of the Regional Science Association, **16(1)**, 27–42 (1966)
9. Ministry of Economic Development of the Russian Federation, http://xn----7sbbhnbqial1ebd4mma.xn--p1ai/uploadedFiles/files/Kontseptsiya_SPR.pdf (2016)
10. N.V Zubarevich, *Regiony Rossii: neravenstvo, krizis, modernizaciya* (2010)
11. Government of the Russian Federation, http://government.ru/info/6217/ (2008)
12. N.V Zubarevich, Pro et Contra, **16**, 135-152 (2012)
13. HSE Institute of Demography, *http://www.demoscope.ru/weekly/ssp/rus70_reg2.php* (1970)
14. Russian Federal State Statistics Service, *https://www.gks.ru/storage/mediabank/Ejegodnik_2019.pdf* (2019)
15. Council of deputies of Novosibirsk, *https://novo-sibirsk.ru/dep/construction/plan/* (2007)
16. S.V. Filonov, Balandinskiye chteniya, **9(2)**, 237-240 (2014)
17. The Krasnoyarsk Krai Ministry of Construction and Architecture, *http://www.krskstate.ru/dat/bin/art_attach/1103_arh.pdf* (2008)

18. Krasnoyarskgrazhdanprojekt, *http://minstroy.krskstate.ru/dat/bin/art_attach/11219_tom_ii._castx_1.pdf* (2018)
19. Russian Federal State Statistics Service, *https://gks.ru/bgd/regl/b19_14p/Main.htm* (2019)
20. The Institute for Urban Economics (IUE), *http://www.urbaneconomics.ru/sites/default/files/ekonomika_rossiyskih_gorodov_i_aglomeraciy - vypusk 2 noyabr 2017.pdf* (2017)
21. The Institute of Urban Economics (IUE), *http://www.urbaneconomics.ru/sites/default/files/vgp_is+sue3.pdf* (2017)
22. I.H. Usmanova, N.A Ligaeva, O.A Kuznetsova, IOP Conference Series: Materials Science and Engineering, **753**, 042001 (2020)

Environmental migration and sustainable development of regions and countries

Irina Reznikova[1]*, *Natalya* Danilina[1]

[1] Institute of Engineering and Environmental Security, Togliatti State University, Belorusskaya str. 14, 445020 Togliatti, Russian Federation

Abstract. Our paper focuses on studying the development and implementation of a comprehensive environmental safety assessment and mitigation strategy for countries of origin and destination of environmental migration. Environmental migration represents one of the key threats to the security of regions and countries and is linked to their sustainable development (or lack thereof). Moreover, it is also connected to many important social, economic as well as demographic outcomes. Global warming caused by human activities is already having a major impact on the planet and is likely to render parts of the world uninhabitable, leading to migration, displacement, and planned resettlement. The goal of the paper is also to assess the impact of future migration due to the global warming and climate change. The results of the paper demonstrate that climate impacts can trigger migration and change population distribution within the next few decades. It is also clear that migration caused by the environmental factors (such as droughts, floods, higher temperature sand more extreme weather events) would inevitably lead to the geopolitical changes with long-range impacts and effects.

1 Introduction

While understanding that there is a link between the climate factors and environmental migration is helpful, there is a lack of information on how exactly and in what wat they affect migration [1]. On the migration side, researchers have uncovered so-called "stylized facts" that govern how and why people move and under what circumstances for more than a century. These changes include increased droughts, more frequent floods, higher temperatures and more extreme weather events, and changes that threaten well-being - the very existence and livelihood of the people [2]. People who are severely affected by changing ecological conditions can choose to emigrate from their home countries. Some people fleeing pollution may be identified as climate refugees, and some who want to emigrate may seek refuge in other countries less affected by climate extremes. People who are severely affected by changes in the ecological state, such as people in Africa, Asia, Latin America or the Middle East, may choose to move to other parts of the world or even to their own countries. Some international refugee organizations, including the United Nations, prefer the term "environmental migrants" [3].

* Corresponding author: IVReznikova@yandex.ru

Under the Immigration and Nationality Act, a refugee is defined as "a person who cannot or does not want to return to their home country due to a well-founded fear of persecution" [4]. Climate migrants would also have to make the argument that their home country's government could not protect them when they returned. Some could potentially be considered members of certain social groups, such as refugees, asylum seekers and war refugees. This is because some countries such as the United States has no clear path to help climate migrants, owing to the few legal provisions. Most of the research on climate migration has been conducted in developing countries, and one reason that there has been less focus on the issues in developed countries is that their populations are highly urbanized, and that people's incomes and livelihoods are less dependent on local environmental conditions, owing to expanded international supply chains [5].

Nevertheless, there is a growing interest in the management of sanctuary areas and climate control - exposed areas in developing countries, particularly in the Middle East and North Africa. It is important to understand local planning needs and how existing government policies on the management of protected areas and climate protection areas - exposed areas - can best be changed [6]. The Paris Agreement of 2016 also takes note of migration and aims to set up a working group to draw up recommendations.

One can see that environmental refugees as an emerging security problem and addressing the problem by establishing a framework for identifying environmental forced migration as a threat to national security. In order to tackle that, many countries adopted a conceptual approach to the integration of immigrants involving a wide range of actors, including non-governmental organisations, and has further promoted the development of an open-door immigration policy for immigrants and their families [7]. There are negative aspects of international migration that can lead to security threats, such as insufficient integration of legal migrants, which can lead to social tensions that can lead to the displacement of people belonging to immigrants. Political instability, poverty and climate change are caused by uncontrolled mass migration, and this needs to be addressed in the face of an ageing population.

2 Natural and environmental disasters

As the name suggests, it is a disaster that damages property and affects the people, flora and fauna around it. Natural disasters are essentially sudden events caused by changing environmental patterns due to human activities such as climate change or natural disasters [8]. In extreme cases, they can be extreme in terms of death, environmental damage and property destruction.

Environmental disasters always occur when the balance of the environment is disturbed. The worst thing is that they can strike anytime and anywhere on the earth without warning and strike at any time of the day or night [9]. The turbulence caused by rough sewage silts the water, causing less sunlight to enter, which affects the amount of photosynthesis, leading to a decrease in dissolved oxygen. Strong winds in the ocean increase nutrients in certain areas due to a process that brings nutrient-rich water to the surface. Hurricane Sandy in 2012 was one of the most devastating storms to hit Connecticut's power grid, causing 15,000 power outages that affected more than 500,000 customers. The damage to energy infrastructure was so severe that power outages occurred in areas of New York and New Jersey, and in the months following the storm, power was not restored in those areas or in the areas of New York and New Jersey [10]. A record storm surge from Hurricane Sandy has hit the north-eastern United States, causing an estimated 11 billion gallons of partially treated sewage to spill into the Atlantic Ocean, posing a threat to the environment and health. Another example is California which has always been vulnerable to wildfires that burn and destroy power lines and cause severe damage to the state's energy infrastructure. But in the past two years, major wildfires in California have also destroyed thousands of buildings, forced hundreds of thousands to evacuate their homes, and killed at least 44 people nationwide. Climate change is exacerbating natural disasters such as forest fires and floods, making them more frequent and severe [11]. The Intergovernmental Panel on Climate Change (IPCC) has described in detail the predicted effects of climate change on the state of California and the United States [12]. Many climatic stresses that carry risks, especially associated with dwindling snowpack, increased the risk of extreme weather events such as floods, droughts and wildfires in California. As the world warms, the cost to Americans may also justify the need to adapt to the damage caused by climate change.

Many scientists are increasingly investigating the role that climate change plays in certain disasters. Although many factors have contributed to these weather events, scientists agree that climate change generally leads to, and will continue to lead to, more frequent, severe, and damaging extreme event. Since vulnerability is the determining factor in the impact of disasters, we argue that humanity's development models have also contributed significantly to the disasters that have occurred [13]. The increase in the socio-economic aspects of disasters shows a 3% increase in the numbers, with a direct impact on vulnerable communities. Social capital, which is linked to social and economic structures, has a significant impact as a factor that reduces the vulnerability of affected communities and contributes to the environmental impact of disasters. Natural disasters are caused by a number of factors, such as climate change, natural disasters and human activities, which have a negative impact on the natural and built-up environment in the affected region [14]. Such decisions, combined with a lack of awareness of the environmental impact of human activities and their impact on the environment, provoke a number of socio-economic and environmental impacts. The negative environmental impacts of storms and typhoons are increasing every year, leading to an increase in the number of natural disasters such as floods, droughts, hurricanes, typhoons and cyclones. For example, not all disasters have a significant impact on the environment; many earthquakes have only a small impact on ecosystems, and extreme climate events can have a positive impact on the local environment. Flooding can help rejuvenate floodplains and vegetation, but pollutants from flooded industrial plants can lead to dangerous chemicals entering project areas, groundwater, watersheds and oceans untreated.

3 Environmental problems and their effects on human lives

Environmental problems surely have significant effects on human lives. These effects could threaten human health by affecting the food we eat, the water we drink, the air we breathe and the weather we experience. The severity of health risks will depend on the capacity of public health and security systems to respond and prepare to these evolving threats. These impacts will vary depending on where a person lives, how sensitive they are to health threats, how exposed they are to the effects of climate change, or how well they and their communities can adapt to these changes [15]. People in developing countries may be the most vulnerable to health risks worldwide, but climate change poses a significant threat to our health. Current weather variations are already having a negative impact on health in many developing countries, such as Africa, Asia and the Middle East. Climate change is exacerbating existing and often enormous health problems, especially in poorer parts of the world. There is also the potential to multiply the additional stress from climate change from the effects of extreme weather events such as floods, droughts and extreme heat waves [16]. In fact, improved health is often the result of transforming natural ecosystems into expanded agriculture to improve access to food and water. Degradation of a particular ecosystem can lead to deforestation, leading to increased malaria exposure and biodiversity loss.

Climate change is affecting not only the health of people in the United States, but also many other countries around the world. A warmer, more variable climate will lead to higher levels of air pollution [17]. There are currently limits on the amount of carbon dioxide (CO_2) and other greenhouse gases in the atmosphere. Human health and well-being are inextricably linked to the quality of the environment and it is increasingly recognised that human exposure to high levels of air, water and other pollution is associated with a number of harmful health effects. It is increasingly recognised that linking environmental and health impacts requires economic assessment in order to ensure that they are properly reflected in policy [18].

Consequently, in recent years we have seen a wave of assessment studies that seek to quantify the environmental impact on human health in monetary terms and to find evidence that health and environmental policies reduce the risk of disease and death. We therefore believe that the use of various tools in assessing the health effects of environmental policy in the context of the global threat to human life and well-being posed by climate change should be improved [19]. A subsequent section presents the assessment techniques that have been carried out to assess the social benefits of improved air and water quality. A more robust approach to assessing the net impact of such changes on health would have a significant impact on policy decisions on health and well-being - and would be seen in the context of climate change.

In many parts of the world, there are additional health impacts that interact with the effects of climate change, such as changes in air and water quality, food security and pollution. Environmental health professionals need to study how people interact with the world around them, and record the many ways in which these interactions can affect health and well-being, as well as the health of their families and communities. One of the challenges is to assess the impact of environmental problems on the long-term impact of these changes on human health. Maintaining a healthy environment is crucial to helping people live longer and improve their quality of life. Understanding the distribution and concentration of pollutants in the environment increases our ability to understand the role of these hazards to human health and to develop ways to remain healthier [20]. By improving environmental health, public health professionals can also improve the personal well-being of people in their communities and their families. The system is tracking network uses data from many sources to track the effects of climate change. Although there are a number of indicators related to climate change, the tracking network has had data on extreme heat and flood vulnerability for more than 20 years. As pollution from climate change increases worldwide, diseases caused by environmental factors are on the rise, and doctors are increasingly warning people of the growing health consequences. The World Health Organization and the United Nations have called on countries to restore the environment and aggressively curb greenhouse gas emissions in order to protect human health.

4 Social phenomena of environmental migration

The world has a growing population and mobility, with the dominant flow of people over the past decade moving from the countryside to the cities. Migration is therefore an important part of the social phenomenon of ecological migration and a key element of economic development [21]. At the same time, the world is in the midst of one of the most intense periods of climate change in human history. However, we have failed to take systematic account of the social consequences of people moving from one place to another and the environmental risks that arise. The researchers link the issue of environmental change to human migration and explicitly refer to people forced to leave their homes as "environmental refugees" [22].

Although environmental studies have traditionally focused on the natural world, there is little research on the social impact of environmental change on human health and well-being.

The effects of many of these processes have been considered catastrophic, because they create stress, disrupt normal social processes, force people to adapt, and make changes in the way they do things in life.

Most countries experience a form of migration that involves forced immobility, and in which the population is ultimately trapped. The slow onset of natural and human environmental change in the world rarely directly or linear influences migration drivers and patterns. Where the environmental component is nevertheless crucial, itis often embedded and difficult to identify. Research on this current issue has improved considerably, but there are still gaps in quantitative data. To investigate whether women are more vulnerable because of their lower education level, detailed empirical research is needed. We reject the idea that higher levels of education increase the propensity to migrate and thus the risk of environmental migration. Therefore, the factors that determine the impact of environmental factors on migration, such as climate change, must be taken into account, as social, cultural, economic and political aspects interact with the environmental factor. Sceptics on both sides tend to question models of sea-level rise as simplistic, explaining that the decision to migrate is complex and includes factors such as economic, social, cultural, political, and environmental factors. They have a migrant background and humanitarian background, and worry about the potential backlash against migrants, given the alarming predictions of millions of people who are fleeing.

Some migration experts are concerned about whether the term "refugee" has a precise meaning under international law. The appeal argues that a million people loosely referred to as refugees will devalue the legally recognized concept of refugees at a time when some 15 million people are desperately seeking protection from violence and conflict. For this reason, climate migration is often associated with the term "refugee," a legal category that is limited to people fleeing persecution because of factors such as race, ethnicity, religion, nationality, or political affiliation [23]. As for will, environmental migration is generally thought to fall at the forced end of the spectrum. The three dimensions of migration include the distance travelled, the number of migrants crossing borders and their ability to stay or remain. For example, in the case of climate migration, the number of migrants in response to climate change is increasing, but not in proportion to the extent of change. Many scientists describe migration as a social phenomenon in which environmental factors such as climate change can trigger a first surge in migration. In other words, people's ability to adapt to environmental change, not the nature of migration itself, may be the most important factor in climate - induced migration [24].

While climate impacts challenge the notion that climate change will directly lead to migration, they could reinforce existing social vulnerabilities. Although the underlying causes of vulnerability have little to do with why people end up moving, there can be socio-political and economic causes. Thence, the underlying social vulnerabilities could determine whether individuals and communities can enjoy the luxury of migration. But then the underlying social vulnerability determines whether an individual or community can avail itself of this luxury to migrate, and if so, when. The IPCC synthesis report articulates climate change as a real and looming problem, but others doubt that the focus on environmental migration is an attempt to see the phenomenon as capable of translating into financial risk [25]. Most striking, however, is that the focus on migration as an economic and social risk factor for the global economy has been much discussed in recent years. This suggests that, without a significant climate-change effort, the world is trapped in an environment whose impacts are unequally spread in speed and scale, both in the social and environmental areas of climate change.

5 Conclusions

This paper assessed the impact of future migration due to climate change by providing the summary of key lessons and evidence. The paper made it clear that global environmental changes could affect the movement of the world's population.

While economic, social, and political factors influence migration, environmental changes also affect it in other ways. Migration associated with environmental decline is usually characterized by short-term distant and long-term movements. However, due to the breadth and complexity of interactions between drivers, it will rarely be possible to distinguish between individuals for whom environmental factors are the only driver. However, if the number of people at risk of migration due to environmental change increases significantly, very high numbers are unlikely to materialise. It is politically and ethically important that northern low-population countries are largely responsible for climate change, while southern high-population and high-growth countries have contributed very little to it. However, migrant groups are exposed to a number of stressors, including economic, social, political and environmental factors, as well as human rights concerns. Despite the serious concerns about human rights raised by the rapid increase in migration due to environmental change in recent years, there is still a need to discuss the population at government and international level. Where the environmental component is difficult to identify but crucial, migration drivers and patterns are influenced by the slow onset of natural and human environmental change. Most countries experience a form of migration that involves forced immobility and in which a population is trapped, and this is rarely direct or linear.

If we want our research to influence policy and provide information to decision-makers making climate-change mobility decisions, we propose that we face four dilemmas. Movements can occur immediately, for example in response to imminent floods, or they can occur in the wake of an event that ends with a long-term change in climate or other environmental conditions. Without a return to another area, communities and individuals may be temporarily displaced but return at the end of the event. Migrant destinations can also be influenced by historical migration patterns, where new migrants follow old routes. The reasons why people migrate or seek refuge can be complex, making it difficult to predict how climate change will affect future migration.

It is important to understand how climate change-driven migration differs from the motivations for human migration. Underestimating the extent to which climate change will affect human migration is therefore a decision - one that makes mistakes in the context of any country's national security and economic interests. A framework is needed that would recognize the imperfections of any given assessment of migration and allows us to think critically about the possible ways in which climate change can change migration patterns. This framework would have to assess the wider impact of climate change on populations by explicitly including populations directly affected by it and those indirectly affected by climate change- that is, migration and patterns of migration.

References

1. E. Piguet, R. Kaenzig, J. Guélat, Population and Environment, **39(4)**, 357-383 (2018)
2. D.J. Kaczan, J. Orgill-Meyer, Climatic Change, **158(3)**, 281-300 (2020)
3. UNHCR, *https://www.unhcr.org/what-is-a-refugee.html* (2021)
4. R. Sunam, K. Barney, J.F. McCarthy, Geoforum, **118**, 1-13 (2021)
5. Z. Mencütek, *Refugee Governance, State and Politics in the Middle East* (2018)
6. P. Cajka, M. Jaroszewicz, W. Strielkowski, Economics and Sociology, **7(4)**, 11-25 (2014)
7. M.R. Islam, N.A. Khan, Journal of the Asia Pacific Economy, **23(2)**, 297-323 (2018)

8. A. Blazev, *Energy security for the 21st century* (2021)
9. E. Sharma, S. Das, International Journal of Environment and Waste Management, **25(3)**, 356-380 (2020)
10. Z. Liu, H. Wang, Y.J. Zhang, L. Magnusson, J.D. Loftis, D. Forrest, Coastal and Shelf Science, **233**, 106544 (2020)
11. H. Moftakhari, A. Kouchak, Environmental Research Letters, **14(10)**, 104018 (2019)
12. T.B. Pathak, M.L. Maskey, J.A. Dahlberg, F. Kearns, K.M. Bali, D. Zaccaria, Agronomy, **8(3)**, 25 (2018)
13. S. Osaka, R. Bellamy, Global Environmental Change, **62**, 102070 (2020)
14. S. Kim, S.A. Kwon, J.E. Lee, B.C. Ahn, J.H. Lee, C. An, K. Kitagawa, D. Kim, J. Wang, Sustainability, **12**, 3377 (2020)
15. M. McKee, D. Stuckler, Nature Medicine, **26(5)**, 640-642 (2020)
16. M. Abdollahbeigi, Journal of Chemical Reviews, **2(4)**, 292-308 (2020)
17. W. Nordhaus, American Economic Review, **109(6)**, 1991-2014 (2019)
18. A.A. Almetwally, M. Bin-Jumah, A.A. Allam, Environmental Science and Pollution Research, **27(20)**, 24815-24830 (2020)
19. A. Caplin, M. Ghandehari, C. Lim, P. Glimcher, G. Thurston, Nature communications, **10(1)**, 1-11 (2019)
20. K. Hayes, G. Blashki, J. Wiseman, S. Burke, L. Reifels, International journal of mental health systems, **12(1)**, 1-12 (2018)
21. F. Castelli, Journal of travel medicine, **25(1)**, tay040 (2018)
22. K.F. Davis, A. Bhattachan, P. D'Odorico, S. Suweis, Environmental Research Letters, **13(6)**, 064030 (2018)
23. S. Wilmer, *Performing statelessness in Europe* (2018)
24. D. J. Kaczan, J. Orgill-Meyer, Climatic Change, **158(3)**, 281-300 (2020)
25. S. Behrman, A. Kent (eds.), *Climate refugees: beyond the legal impasse?* (2018)

E3S Web of Conferences **250**, 05004 (2021) https://doi.org/10.1051/e3sconf/202125005004
TRESP 2021

Strategizing sustainable development: the case of the Ural industrial regions

Irina Rakhmeeva[1]*, *Ivan* Antipin[1]

[1] Ural State University of Economics, 8 Marta str. 62/45, 620144 Yekaterinburg, Russian Federation

Abstract. The remit of the article is regional development strategizing. Regional development strategy as part of regional regulatory environment could ensure the task of sustainable development due to determining priorities for resource conservation, development of a "green" economy, and restoration of natural resources. This is especially important for the extractive economies of the Russian regions. On the contrary, the research revealed the preservation of extractive policies in most of the Ural regions. The manuscript contains the description of the ecological and economic position of the regions of the Ural macroregion, that have a high anthropogenic impact on the environment because of their nature of economy. We analysed the content of the development strategies of the regions of the Ural macroregion and the issues of sustainable development. The Chelyabinsk region and the Republic of Bashkortostan didn't include sustainable development in the top priorities. It leads them to stick their place in the tail of the National Environmental Rating of Russian Regions. Orenburg region's strategy conclude the most quality mechanisms for ensuring environmental safety in regional strategic documents. The author highlighted the comprehension of the deep interrelationships of ecological well-being with all spheres of human life in the development strategy of the Sverdlovsk region. The regional development strategy is an important mechanism for ensuring a balance between industrial growth and environmental conservation. Therefore, the authority of traditional industrial regions of the Urals should strengthen the component of sustainable development in their strategies.

1 Introduction

The set of characteristics of the modern Russian Federation determines the relevance of the research. First, the importance of the strategic approach in management is increasing.

Second, the objects of strategic planning are becoming more complex and the new objects (macroregion, parts of the subject of the Russian Federation, etc.) require specific approaches.

Third, there are no unified methodological approaches to strategizing the socio-economic development of a territory in Russia.

Fourth, the authorities do not sufficiently involve various groups of stakeholders in the processes of strategizing the socio-economic development of a region.

*Corresponding author: smartreg66@gmail.com

Economists, sociologists, regionalists, political scientists, and other researchers have developed many theories, models and approaches to analyse the process of strategizing the socio-economic development of a territory of various hierarchical levels, their individual territories, industries, etc.

The issues of studying the existing shortcomings of strategic planning for the Russian Federation are relevant despite the successes of recent decades in improving these processes [1; 2, p. 2].

Human economic activity related to extraction natural resources and transformation the biosphere since ancient times. The growth of human needs and productivity determined the increasing environmental load. A society and authority intervened purposeful in industrial activity when the biosphere's recover ability ceased to correspond to the damage caused by human. L.J. Ramirez Lopez and A.I. Grijalba Castro indicated the need for environmental planning in the space of the territory [3, p. 181].

The United Nations Conference on Environment and Development in 1992 triggered a boom of institutional adjustment process for shift to the new paradigm of sustainable development [4].

The impact of industry on the environment is local. This determines the relevance of local level measures and the focus of regionalists 'attention on this problem.

We emphasized significance of regulation in industrial regions with high ecological load [5], and in extractive economies regions with insufficient recovery measures [6, 7]. Therefore, the object of the research is the Ural macroregion. It is the main industrial centre of the country. 7 industrial regions are part of it.

The aim of the manuscript is to analyse the degree of inclusion of the sustainable development priority in the regional strategic goals in the Ural macroregion. We done 3 challenges. At first, to clarify the role of the regional development strategy in ensuring the sustainable development. Second task is to describe the ecological and economic state of the regions of the Ural macroregion. Third issue is to analyse the content of the development strategies of the regions of the Ural macroregion and to examine the degree of elaboration of the sustainable development priority in these documents.

2 Materials and methods

The methodology is based on regional and institutional economics.

The concept of high causality of the level and the sustainability of regional development by the quality of regional regulatory environment [8, p. 289] is the core of the research. The regional strategy is the heart of the regional regulatory environment. It forms the main development goals and determines the directions (tools) for achieving them, and coordinates development priorities [9, p. 1022]. "The stability of ecological and socio-economic environment depends on the level of economic activity measured by the capacity of territorial ecosystem" [10, p. 1029].

According to the normative framework approach, sustainable development is the provision of "a balanced solution to socio-economic problems and problems of preserving a favourable environment and natural resource potential in order to the needs of present and future generations" [Decree of the President of the Russian Federation No. 440 of 01.04.1996 "Concepts of the Transition of the Russian Federation to Sustainable Development"].

The regional development strategy is a key tool for the formation of a regional policy that can ensure the sustainable development of the territory. It should include a description of the current environmental situation and key problems, mechanisms for solving them, and indicators for the target state. Problems and contradictions in strategic planning did not arise spontaneously. There are no uniform rules for the formation (development and updating) of socio-economic development strategies, including uniform requirements for the composition of participants of the strategizing, in the Russian Federation and in the world community at present. This leads to a significant differentiation in the content and the quality of regional development strategies. The papers of Russian researchers [11-13] revealed methodological approaches and methods for the strategizing the socio-economic development of various territories (cities, regions, macroregions).

Regional authorities should involve all beneficiaries in the designing of strategies. The attraction of scientific and educational organizations will create a theoretical basis for the strategy [14, p. 102].

The key research method is the structural-logical monographic analysis of strategic documents. This method together with the regional analysis of the structure of economic and ecological conditions allows us to predict the prospects for the sustainable development of the territory.

The empirical basis of the study is the official regional development strategies.

The authors rely on the National Ecological Rating of Russian regions to comparing the ecological state of the regions of the Ural macroregion. The rating is based on the principles of sustainable development of the Declaration of The United Nations Conference on Environment and Development in 1992. The rating includes environmental protection, industrial-ecological and socio-ecological indexes.

The authors suggest characteristics of the ecological and economic system of the Ural macroregion.

The regions of the Ural macroregion are industrial. Manufacturing industries (metallurgy, metalworking, machine industry) are the key industries of the regions of the Ural macroregion. The gas and petrochemical industries are also developed in the Orenburg Region, Perm Krai, and the Republics of Bashkortostan and Udmurtia.

According to the study of the sustainability of the ecological and socio-economic environment of the Ural Federal District, "The Chelyabinsk region is the least stable, which is associated with both natural conditions and the specifics of the economic structure" [10, p. 1034].

The regions of the Ural macroregion have many mineral deposits. This determined the rapid growth of the mining industry and the extractive nature of the economy of these regions at the dawn of their formation in the historical context.

The industrial production were declining in the second decade of the XXI century in Russia as a whole and in the regions of the Ural macroregion [15, p. 29]. This leads to a decrease in the environmental load without targeted actions of local authorities.

The high level of industry requires large volumes of production and distribution of electricity, gas, and water. "Significant material flows in the electric power industry lead to a significant share of greenhouse gas emissions" in the Ural regions) [10, p. 1034].

Table 1. The position of the regions of the Ural macroregion in the National Environmental Rating.

Region	Place Composite index: environmental protection, industrial-ecological and socio-ecological indexes	
	2020	2015
Chelyabinsk region	83 46: 27, 47, 63	84 36: 20, 39, 46
Sverdlovsk region	81 47: 36, 46, 59	85 36: 35, 32, 39
Orenburg region	72 50: 31, 41, 70	75 40: 27, 29, 62
Kurgan region	60 55: 46, 37, 71	72 41: 37, 26, 58

Republic of Bashkortostan	50 58: 43, 53, 73	49 44: 36, 39, 56
Perm Krai	25 63: 50, 60, 73	28 48: 43, 47, 53
Republic of Udmurtia	15 65: 53, 52, 78	36 46: 40, 32, 63

The analysis of the National Environmental Rating of Russian regions gives the following conclusions. The composite index and its components increased in all regions of the Ural macroregion over the past 5 years. The Republic of Udmurtia and the Kurgan Region were able to improve their positions on all directions in the rating significantly. Perm Krai, Sverdlovsk and Orenburg regions were able to rise to 3 places. The Perm Krai received the result due to successful work in all 3 areas of the rating. The Sverdlovsk and Orenburg regions increased the rating mainly due to the socio-ecological index, slightly due to the industrial-ecological index. The situation with nature protection has not changed in these two regions at the same time. The Chelyabinsk Region and the Republic of Bashkortostan remained roughly in the same positions.

3 Results and discussions

This section presented the results of analysis of the implementation of the priority of sustainable development in the strategies of the regions of the Ural macroregion.

The development strategy of the Chelyabinsk region until 2035 calls the unfavourable environmental situation in the region, the high environmental load from the organizations of ferrous and non-ferrous metallurgy, the high degree of accumulated damage to the environment as a weak side of the economy of the Chelyabinsk region. The key tools of the target scenario for the development of the Chelyabinsk region are the creation of favourable conditions for investment, the modernization of the economy, the formation of new high-tech industries and the growth of productivity. Creating a network of competitive economic centres (cities and urban agglomerations) for the economic and geological exploration of the territory of the Chelyabinsk region in the long term is a strategic priority. The development strategy of the Chelyabinsk region pays extraordinarily little attention to the task of ensuring environmental safety in comparison with other issues. A significant part of the ecological activities is focused on strengthening government control and monitoring.

The primary tasks of the Sverdlovsk region development are to improve the quality and standards of life of the population, to create favourable conditions for a long, safe, healthy, and prosperous life of citizens. The Government of the Sverdlovsk region considers the high level of environmental impact of technogenic factors as one of the most important problems. The balanced development is among the 3 development priorities. The direction of the development strategy of the Sverdlovsk region until 2030 "Balanced development" includes the purpose of sustainable environmental development of the territory. The "Clean Environment" project would solve this problem.

Improving the ecological situation is a strategic task of the priority of a comfortable living environment in the Orenburg region. A special chapter of the development strategy of the Orenburg region until 2030 covered environmental policy. It includes specific mechanisms and target indicators for improving the quality of the nature and living conditions of the population, reducing the ecological load on the natural complexes, and activating economic mechanisms for rational use of natural resources.

The development strategy of the Kurgan region until 2030 includes the priority of balanced spatial development. Ensuring environmental well-being is part of this priority. It includes the rational use and reproduction of mineral resources, the preservation of a favourable environment, water, forests, and biological diversity. The strategy describes the objectives and targets for each area. The authors consider the measures of the strengthen the requirements for the processing of extracted resources, and the involvement of waste in economic turnover in the strategy of the Kurgan region as essential.

The development strategy of the Republic of Bashkortostan until 2030 does not contain any identification of environmental problems in the region as significant for overcoming. The description of the regional ecological situation in the strategy highlights the more prosperous status of the Republic of Bashkortostan in comparison with other industrial leaders of the country. The development of a "green" economy, resource- and energy-saving technologies is a factor of the transition to a new model of economic growth in the strategy. The development strategy of the Republic of Bashkortostan focuses on measures to reduce the amount of greenhouse gases. It is also noted that the processing of secondary material resources is not developed in the republic and this issue needs to be worked out.

The environmental management section of the development strategy of Perm Krai until 2026 indicated the following problems: the increasing amount of household waste, the threat of environmental pollution and the low efficiency of the use of natural resources. Ensuring the safety of the ecological environment is one of the key tasks. However, the authors of the strategy didn't disclose the content of this task.

The basic and target development scenario of the Udmurt Republic until 2025 involves the active exploitation of available natural resources and does not contain a key priority of environmental safety. Ensuring the environmental sustainability of the territory and energy conservation are the sub-tasks of the direction "Reducing the infrastructure constraints of development".

We would like to say the following for discuss the problem.

The reflection of the priority of sustainability in the reginal development strategy determines the overall environmental policy of the regional authorities and the degree of further elaboration of this issue.

The regions of the Ural macroregion are highly industrial and have a significant anthropogenic impact on the environment. Therefore, the problem of environmental safety is presented in all regional development strategies. We noted the depth description of the relationship between environmental well-being and all spheres of human life in the development strategy of the Sverdlovsk region. The strategy of the Orenburg region includes the most quality mechanisms for implementing environmental policy.

The lack of environmental priority in the top strategic goals fixed the Chelyabinsk region in the tail of the National Environmental Rating and led to the loss of the one position by the Republic of Bashkortostan.

Regions with extractive economies (developed oil and gas production complex) continue to focus on the extensive exploitation of minerals in their strategies. This is contrary to the conditions of sustainable development.

4 Conclusions

The region's development strategy is an important mechanism for ensuring sustainable development and a balance between industrial growth and environmental safety. Traditionally, the industrial regions of the Urals should strengthen the sustainable development component in their strategies.

L.A. Sverdrup showed on the example of Norway that eenvironmental considerations must be institutionalised into the existing political system [16].

The declaration of specific sustainable development goals in the regional development strategy contributes to their further inclusion in the program documents and allocation of budget funding. The development strategy of the region as the central link of the regional regulatory environment determines the contours of innovative and technological development and the formation of a "green" economy, lean manufacturing [17], impact on corporate strategy [18].

Digital tools can improve the efficiency of the resources exploitation [19], the organization of monitoring and control [20].

The authors formulated several recommendations aimed at improving the quality and viability of socio-economic development strategies, and increasing the sustainability of territorial development based on the results of the research of the processes of strategic planning and strategic management in the Russian Federation [21], and the development strategies of the regions of the Ural macroregion.

First, regional authorities should pay serious attention to the analysis of the environment of the object of strategic planning. The quality and depth of trend analysis are the most important conditions for the optimal determination of the directions of strategizing and the viability the regional socio-economic development strategies.

Also, the stage of environmental analysis requires a clear choice and application of strategic analysis methods. A large portion of the socio-economic development strategies of municipalities and regions of the Russian Federation contain SWOT-analysis, but it is performed correctly rarely. The postulates of the "classical" SWOT-analysis of the Harvard School of strategic Planning are fundamental. It is also possible to use significance coefficients for each of the SWOT-analysis factors. We recommend to apply other methods (PEST-analysis, strategic synthesis, etc.) in addition to SWOT-analysis.

Second, the goal-setting stage should include formulation a mission and decomposition of goals clearly. The general (key, top, major) strategic goal is decomposed into lower-level goals. As a rule, the goals of the first level are the goals of strategic directions, the goals of the second level are the goals of strategic programs, the goals of the third level are the goals of strategic projects. Compliance with such a hierarchical structure of goals ensures the relationship between strategic and operational (tactical) tasks and activities.

Third, regional authorities should elaborate the list of strategic directions of the socio-economic development in details, including a system of performance indicators. The achievement of indicators characterizes the degree of implementation of the strategy, the results of achieving of strategic goals and completing strategic tasks annually. We recommend to include performance indicators for each strategic direction. The principles of sustainable development dictate the need to include economic, social, and environmental development among the strategic directions. Regional authorities should actively engage in the development of environmental education of the population and the ecological culture of entrepreneurs. These tasks can be implemented within the framework of the state information policy [22].

Fourth, the regional socio-economic development strategy should provide all the processes in areas of strategic planning object (municipality, region of the Russian Federation, the country as a whole, etc.). The Russian local authorities consider only legally defined powers and competencies often.

Fifth, we recommend specifying the subjects, mechanisms, principles, and stages of monitoring and controlling the implementation of the socio-economic development strategy. A scheme for the organization of the strategic management corresponds to the strategic directions, priorities of strategic development and the actual structure of the authorities. Responsible executors and co-executors of the strategy are not only the local authorities but also absolutely all stakeholders.

The study of spatial development of the entire country and its territories has been popular in recent years in the Russian Federation. Accordingly, the sixth recommendation is coordination of socio-economic and territorial development in the socio-economic

https://doi.org/10.1051/e3sconf/202125005004

development strategies. It means that each strategic direction should contain guidelines for spatial development.

The authors concluded that the processes of strategic planning and strategic management play a crucial role in the system of long-term planning. The strategizing could generate positive impulses in the socio-economic development of the Russian Federation and its regions and require further improvement. The success of the socio-economic development of the Russian regions and country depends on the optimality (efficiency) of the strategic planning and strategic management among other things.

Acknowledgements

The reported study was funded by RFBR, project number 20-010-00824 "Incremental approach to the formation and implementation of strategies for the socio-economic development of regions of different hierarchical levels of the Russian Federation: Uniform rules for strategizing".

References

1. E.B. Lenchuk, Innovations, **2 (256),** 24 (2020)
2. Ya.P. Silin, Ye.B. Dvoryadkina, I.A. Antipin, Upravlenets – The Manager, **9(6),** 2 (2018)
3. L.J. Ramirez Lopez, A.I. Grijalba Castro, Sustainability, **13(1)**, 181 (2021)
4. T.G. Ribeiro, V.J. Rodrigues, Environmental Politics, **6(1)**, (1997)
5. A.S. Nikitina, A.V. Ruchkin, N.N. Startseva, O.M. Trofimova, N.K. Shemetova, N.A. Yurchenko, International Transaction Journal of Engineering, Management and Applied Sciences and Technologies, **10(14)**, 10A14O (2019)
6. F. Ekardt, J. Stubenrauch, B. Garske, B. Jacobs, Land, **9(3),** 83 (2020)
7. P.A. Gurianov, Journal of Mining Institute, **208**, 18 (2014)
8. I.I. Rakhmeeva, R-economy, **6(4),** 280 (2020)
9. Yu. G. Lavrikova, V.V. Akberdina, A.V. Suvorova, Economy of Region, **15(4),** 1022 (2019)
10. A.Y. Davankov, D.Y. Dvinin, Y.A. Postnikov, Economy of Region, **12(4),** 1029 (2016)
11. N.Yu. Vlasova, Administrative Consulting, **3 (15)**, 18 (1999)
12. A.L. Gaponenko, Spatial Economics, **4**, 40 (2005)
13. V.V. Kafidov. Moscow: Delo ANE (2015)
14. D. Gallardo-Vázquez, J.A. Folgado-Fernández, Land, **9(4)**, 102 (2020)
15. I.A. Antipin, N.Yu. Vlasova, O.Yu. Ivanova, Upravlenets – The Manager, **11(6),** 28 (2020)
16. L.A. Sverdrup, Environmental Politics, **6(1)**, 54 (1997)
17. N.N. Trofimova, European Proceedings of Social and Behavioural Sciences EpSBS, 1316 (2020)
18. Zaytsev, D. Rodionov, N. Dmitriev, R. Faisullin, Academy of Strategic Management Journal, **19(4)**, 12 (2020)
19. I.A. Magomedov, A.M. Bagov, A.L. Zolkin, European Proceedings of Social and Behavioural Sciences EpSBS, 553 (2020)
20. Agumbayeva, E. Chmyshenko, N. Pulyaev, D. Bunkovsky, K. Kolesov, E. Amirova, Journal of Advanced Research in Law and Economics, **10-6(44),** 1861 (2019)
21. I. Antipin, N. Vlasova, Journal of New Economy, **21(3),** 73 (2020)
22. E. Ustinovich, Comparative Politics Russia, **4(2)**, 4 (2013)

Digital transformation of energy sector companies

Victoria Akberdina[1] *, *Ainur* Osmonova[2]

[1] Department of Regional Industrial Policy and Economic Security and Department of Public Sector Economics and Regional Development, Institute of Economics of the Urals Branch of the Russian Academy of Sciences, Moskovskaya st., 29, Yekaterinburg, Russian Federation
[2] Department of Accounting, Analysis and Auditing, Kyrgyz Russian Slavic University, 6 ave. Chui, 720022 Bishkek, Kyrgyzstan

Abstract. Digital transformation is an ongoing process that is driven by the recent advances in digitalization as well as the development of information and communication technologies (ICT) that penetrate all socio-economic fields of everyday life and business. In this paper, we describe the digital transformation of energy companies. We show that successful transformation is based on skills, expertise and knowledge of the employees that need to be created and maintained. In addition, we show that digital competences become a key element in building capacities that are required for the digital transformation. This is of a particular importance for the energy companies that are experiencing major changes on the path of transition toward low-carbon economy and renewable energy.

1 Introduction

The utility's digital transformation of energy sector would provide all the consumers with access to a more efficient, reliable, and cost-effective energy system. This digital transformation of the energy sector is an ongoing process that involves all countries in the world [1, 2]. Many countries (such as Germany or the United Kingdom) are more successful in this than the others. In summer 2020, United Kingdom managed to generate its electricity without using its coal plants and there are plans to reach about 80% of carbon-free transport and heating by 2030 [3, 4]. In the United States, the plans to shift towards the green and fossil-free economy with zero greenhouse gas emissions are estimated at 2050 [5]. Of course, this process will take a long time, since the major changes that are needed are not in the industry itself but in the minds of the people. Therefore, it is important to assist individuals and companies in finding solutions to accelerate the transformation of the energy sector into a renewable energy dominated sector, while electrifying the use of sectors to achieve deeper decarbonisation [6]. The digital transformation in the renewable energy sector allows it to automate the development of new technologies such as smart grids, energy storage and smart grid systems. There should be solutions for individuals, cities and businesses, with a focus on energy efficiency and security, drawing on the expertise of energy sector experts in its

* Corresponding author: akberdina.vv@uiec.ru

member countries. These solutions should allow the business companies to develop personalized services that are tailored to the expectations of customers. Digitalization is driving major changes in the energy sector, driving innovation and influencing business models. Future drivers of the energy markets would create a digital backbone that supports the transition from fossil fuels to renewable energy and creates an energy system that is connected to our personal and local ecosystems [7, 8].

In addition to that, recent changes in the energy industry in connection with COVID-19 and the pandemic are leading companies to think about how innovative technologies can affect their future business. The use and application of the latest technologies to existing operations will enable more efficient and efficient use of energy resources in the future which will create pioneers of digital transformation with trusted partners worldwide [9, 10].

Digital technologies are already playing a crucial role in accelerating the decarbonization effort. However, in order for digitalisation to realise its full CO_2 potential, rigorous analysis is needed to guide good policy. The energy sector which is driven by advanced technologies, declining costs and the emergence of new markets, is on the cusp of a new digital age, with far-reaching implications for climate change and energy security. The trend towards greater digitisation of energy is made possible by advances in data analysis and networking. This includes the falling costs of sensors, the increasing amount of data from sensors and data storage and the emergence of new markets. Major industries and sectors have increasingly adopted these technologies in recent years, and the energy sector is no exception [11, 12].

The energy and utilities sectors are the main participants in the digital transformation of the energy sector. Electricity is generated by transmitting it through transmission and distribution networks and transmitting it to final consumers in the form of generation, transmission, distribution and storage. Digitalisation has accelerated the transition from a multi-directional to a decentralised energy system, in which all sources of demand are actively involved in balancing supply on a scale. Highly connected systems are emerging that blur the distinction between traditional providers and consumers and open up opportunities for both [13]. Digitalization has also transformed sectors such as road transport, where connectivity, automation, and further electrification of mobility could dramatically transform the sector. Automation and data also influence the approach to production, increasing the efficiency of factory environments. Having seen the effectiveness of innovative technologies in other industries, the energy sector nowadays does not hesitate to adapt to emerging technological trends, including the use of smart grids and, in all clarity, technologies such as smart meters [14]. Although they have the potential to transform the energy sector equally drastically, they have yet to penetrate it and make their mark and significantly penetrate it. These technologies are reinventing the way energy and utilities do business, engage, and interact with their customers and the masses. The digital transformation solutions the energy sector needs to be innovative and better equipped for the future are becoming very important. IoT, data analysis and machine learning are becoming the main elements in supporting the digital transformation of the business models and processes of the established energy companies.

2 Digital technology for producing and distributing energy

Decentralised energy resources are being used in an increasingly decentralised energy system. Driven by the falling cost of renewable energy sources such as wind, solar and decentralised renewables, fossil-fuel generators have been the main beneficiaries in recent years. These changes present significant challenges to the future of power generation and distribution in the United States, China and the rest of the world [15].

At the same time, new resources and technologies offer opportunities: the cost of electricity infrastructure can be reduced, enabling the energy sector to achieve ambitious decarbonisation targets. As utilities integrate information and communication technologies to modernize the power system, there is an opportunity to be reliable and cost-effective - and to effectively increase decentralized generation as an alternative to centralized power plants and the provision of renewable energy [16]. Existing low-cost decentralised power generation methods such as wind and solar can be used to generate electricity through a combination of solar panels, wind turbines and other renewable sources. Decentralised generation also benefits the environment, as its use reduces the amount of electricity that needs to be generated in central power plants. This in turn could reduce the environmental impact of centralised electricity generation, in particular in the case of wind, solar and geothermal generation. Smart Grid technology is an opportunity to modernise energy distribution and ensure that the network can ensure communication between utilities and consumers. Smart grids also offer the potential for decentralised energy, and this can become a reality if every house becomes a miniature power station [17]. Since each house produces its own electricity, the electricity can be diverted from the plants, even from the house itself, if the plants fail. Rooftop solar panels and smart grids go hand in hand, but that will become a reality in the near future, when homes become miniature power plants.

Figure 1 that follows outlines the model of digital transformation that is typically achieved through capability and competence. This is completed using strategic decisions and investments and yields knowledge, skills and expertise that are required for the successful digital transformation of energy companies.

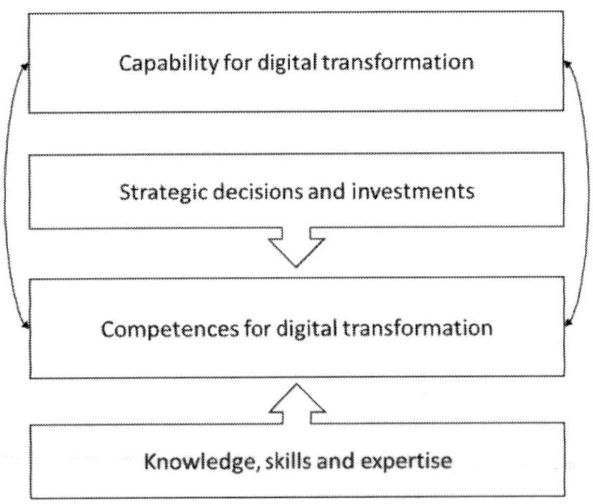

Fig. 1. Digital transformation through capability and competence

The most important of the digital transformation in the energy sector is that we get it right as the electricity grid develops and cities commit to buying electricity from renewable-energy developers. By continuously collecting and synthesizing data from millions of smart sensors, future power grids will depend on coordination through automated smart meters, smart grids and smart energy management systems. Almost $25 trillion will be invested in the modernized smart energy infrastructure in the United States and other countries over the next 10 years [18]. It becomes clear that we need to invest in edge analytics and computing as we strive for a smarter, more efficient and more cost-effective energy system. When people first hear about smart grid initiatives, they often connect smart meters to smart grids. But smart meters are a development that began decades ago, leading to the creation of the United States Energy Information Administration (EIA) and the National Renewable Energy Laboratory. Smart grids make electrical systems more reliable and efficient, helping utilities to reduce power losses, identify and resolve problems faster. Micro grids can play an important role in building a low-carbon future by bringing resilience to the main grid, optimising energy costs, enabling renewable energy hosting, improving the integration of electric vehicles and improving energy availability. Smart grids can help consumers manage energy consumption in times when demand is significantly higher or when reduced energy requirements are needed to support system reliability. Smart appliances in homes, offices and factories can inform consumers about their energy management systems at a time when appliances are consuming relatively much more expensive electricity [19]. These alarm systems help consumers and their intelligent systems to optimally adapt settings which, if supported, can reduce consumers' energy costs. The topic also includes strategies on electricity and electronic management in the context of the use of renewable energy and the conversion of renewable energy. It is about the use of smart grids, smart energy management systems and smart devices to improve energy efficiency and efficiency. Planning and operation of energy systems related to renewable energies and the use of intelligent energy management systems in the context of electricity generation and distribution systems.

However, investment in smart homes and buildings could provide enough energy to generate more electricity than Italy and Australia combined. Buildings consume the most electricity, and the digitalization could reduce energy consumption by about 10% by improving operating efficiency in addition to the efficiency gains from LEDs. By using decentralised energy sources such as solar, wind, water, geothermal and biomass, digitalisation can blur the distinction between supply and demand and provide consumers with opportunities to interact directly with demand or supply in real time [20]. This type of embedded demand - lateral flexibility - is also a key element of the vision for the future of renewable energy in the energy sector.

3 Self-generation of renewable energy

Solar and wind turbines are a great way to reduce energy costs, increase property values and benefit the environment. In order to promote the self-generation of renewable energy, many energy companies have a tariff for the net energy meter service, which covers the cost of using electricity from solar panels, wind turbines, photovoltaic systems and other renewable energy sources [21]. Solar modules or wind turbines have improved significantly in the last decade, and incentives and financing from the federal, state and local governments have made it easier than ever to install in people's own household systems. These programs enable private and business customers to promote renewable energy without having to make costly investments. Customers generators and plants are designed to operate at a lower cost than conventional fossil fuel generators such as coal, natural gas and nuclear power. All of this allows private and small-business owners to generate electricity from renewable sources to sell the surplus power immediately. Moreover, this provides incentives to promote the development of existing and emerging decentralised energy resources such as solar, wind and geothermal energy. Such programs boost the efforts to promote energy efficiency, increase renewable energy, and deliver cost-effective, reliable, and clean energy to the households and businesses. Incentives for the qualification of decentralised energy systems for electricity meters, which are to be installed on the customer's side, are also poplar [22]. Qualified technologies include solar, wind, geothermal, hydro, biomass, waste heat, biofuel and wind power plants. Generators must use micro-turbines or fuel cells and be powered by renewable energy sources such as wind or solar energy or biomass or biofuel. The generator must be located on the customer's premises behind the meter and must cover part or all of the customer's electricity needs. Customers can support green electricity by contributing a monthly contribution to their electricity bill. This can save them money on their electricity bills and reduce their electricity bills. Self-producing energy means that a user or group of users saves energy and generates locally by using renewable energy sources [23]. Then the participants of Green Switch can claim that the electricity comes from renewable sources. The self-production of energy from renewable sources such as sun, wind and water are crucial for a sustainable economy and society. Decarbonisation of the energy sector is feasible, but supply is not nearly sufficient to meet demand. In Europe, an estimated 8 billion EUR has already been allocated to the use of renewable energy in the energy sector. This explains why fossil fuels need to be replaced by renewable energy, according to a report by the International Energy Agency. In 2000, renewables accounted for nearly 9% of the share, and wind power provided more than half of all the electricity produced in the United State, with 41 states and two territories in operation. Wind energy includes both large wind farms (large turbines that can generate electricity) and small turbines on commercial sites. Looking at the wind map for the United States, one can see where power is generated, and looking at the wind maps for each state and area more information can be obtained [24].

The "do not need" segment has been growing for years as prices for renewable energy fall. Battery-powered energy storage is needed if someone wants to accumulate energy for later use when there is less sunlight. Electronic devices convert the direct current generated by photovoltaic systems into alternating current. Mini - wind energy uses the kinetic energy of the wind as a generator for a small generator. The electricity generated by small wind turbines can be used to meet domestic needs, making it an ideal solution for self-consumption of electricity. Solar panels, also known as photovoltaics (PV) and wind turbines have been used in local renewable energy projects. The cost of solar panels has fallen by almost 70% in the last decade, contributing to a large number of installations across the country. With the introduction of renewable energy sources, customers now have an option for renewable energy for the first time in their lives [25].

4 Microgrids and local energy solutions

Microgrid is a localised network that separates itself from the traditional network, operates autonomously and functions as if the main network had failed. Microgrids can strengthen grid resilience, help alleviate grid disruptions and support a flexible and efficient electricity grid by allowing for the increasing use of decentralised energy resources such as wind, solar, hydro, geothermal, biomass, wind and hydro [26]. A microgrid can be self-sufficient in generating, transmitting, distributing and storing electricity and can operate independently in the event of a major power failure or temporary shutdown of the primary grid. This technology thus gives local energy producers greater independence and helps to keep costs down [27].

Sustainable microgrids could also empower the poor by expanding access to Internet and electricity in rural areas by providing such services without central grids and gasoline-powered generators [28]. Certain developments have also shown that micro-grids can turn renewable energy into a source of electricity for households in some rural areas that generate more than enough electricity, with surplus energy being sold back to the central grid. Microgrid is an on-site energy system that integrates decentralised energy resources for electricity generation. Generally, a microgrid represents a decentralised energy resource load that can be operated, locally controlled or operated independently of the main grid. Microgrids operate independently of the power grid and can be operated for up to 30 days without a central power grid. Microgrids are an on-site energy system with a large number of distributed energy resources and loads, which are operated, landed or locally operated and controlled independently [29]. Microgrids can provide temporary local power generation to avoid peak loads and store excess solar and wind energy, allowing them to send electricity when needed or work around the clock. Together, these price reductions will enable variable clean energy storage to be integrated into the renewable micro-grids that support the main electricity grid and can operate independently. Today, sophisticated energy management systems enable project developers to develop microgrids that can incorporate decentralized energy sources such as solar, wind, biomass, heat and power plants, geothermal, hydroelectric and wind turbines. The programme aims to use microgrids to power the mission - critical facilities in the event of a loss of supply or power. Most importantly, they can also keep power plants and other critical infrastructure such as hospitals, schools, hospitals and hospitals running if the main grid fails.

With the infrastructure and resources, microgrids can support many of the operational goals and challenges that distribution networks currently face. Utilities are exploring the potential of new sources of revenue to make microgrids a viable alternative to traditional power generation and distribution systems. Currently, the United States electricity grid is centralized, with a handful of large utilities handling the bulk of electricity generation [30].

While local consumers can be serviced by the utility in the event of a power failure, new, more decentralized systems can also be used to address systemic risks such as large-scale blackouts. A micro grid is a system of generation, energy storage and load that normally operates without connection to a traditional supply network. One can say that a microgrid is a local energy grid with control capabilities, which means it can be disconnected from the traditional grid and operated autonomously. Different from traditional models, microgrids are defined as electricity production at the place of consumption, which is independent of other generators. There can be no need to be disconnected from a traditional power supply network, which is what micro grids need to operate without power in the event of a power failure. The microgrid can be isolated from the conventional grid and continue to supply electricity. In the event of a power failure or other power failure, it can operate autonomously. A precursor to microgrids is a simple system that consists largely of a central power plant that supplies a single building or campus with diesel generators as a backup. There are several semi-autonomous nodes that operate in parallel and each have its own power supply and distribution system. Microgrids may never be abolished, but bringing more power generation, management, and consumption under local control will make everyone less dependent on them.

All of the above offers a new approach to aggregating small, decentralized energy resources that does not include physical switches that can link them to a grid. Virtual aggregations can combine the results and treat them as a single entity participating in the network services market.

5 Conclusions

Overall, it becomes clear that successful digital transformation is based on skills, expertise and knowledge that need to be developed through investments and public and private support and guidance. Surely, this represents a long-range process that can be helped by the advancements in digitalization and information and communication technologies in all socio-economic fields of business and society. However, this process has already started and is likely to continue in the face of the commitments to fighting the effects of global warming and climate change that are in turn marked by the decarbonization of the energy sector and the massive deployment of renewable energy sources such as solar batteries or wind turbines.

Thence, one would agree that digital transformation of energy companies is a challenge since these companies are faced with numerous profound changes as they are getting involved into the ongoing transition to the low-carbon economy and (self-generated) renewable energy sources.

Acknowledgements

The article was prepared with the support of RFBR grant No. 20-010-00719 "Simulation of cross-industrial networking processes in the industrial complex based on hybrid technologies".

References

1. E. Abad-Segura, M. González-Zamar, J. Infante-Moro, G. Ruipérez García, Sustainability, **12(5)**, 2107 (2020)
2. T. Kurbatova, Marketing and management of innovations, **4**, 374-383 (2017)
3. H. Cockburn, *https://www.independent.co.uk/climate-change/news/coal-free-power-uk-record-time-2020-how-long-renewable-energy-a9570891.html* (2020)
4. D. Newbery, M. Pollitt, R. Ritz, W. Strielkowski, Renewable and Sustainable Energy Reviews, **91**, 695-707 (2018)
5. E. Larson, C. Greig, J. Jenkins, E. Mayfield, A. Pascale, C. Zhang, J. Drossman, R. Williams, S. Pacala, R. Socolow, EJ Baik, R. Birdsey, R. Duke, R. Jones, B. Haley, E. Leslie, K. Paustian, A. Swan, *Net-Zero America: Potential Pathways, Infrastructure, and Impacts*, Princeton University, https://acee.princeton.edu/rapidswitch/projects/net-zero-america-project (2020)
6. V. Arabzadeh, J. Mikkola, J. Jasiunas, P. Lund, Journal of Environmental Management, **260**, 110090 (2020)
7. K. Hojckova, H. Ahlborg, G. Morrison, B. Sanden, Research Policy, **49(8)**, 104046 (2020)
8. M. Jayaratne, G. Sullivan Mort, C. D'Souza, Sustainability, **11(24)**, 7076 (2019)
9. M. Fu, H. Shen, Energy Research Letters, **1(1)**, 12967 (2020)
10. R. Elavarasan, G. Shafiullah, R. Kannadasan, V. Mudgal, M. Arif, T. Jamal, S. Subramanian, V. Balaguru, K. Reddy, U. Subramaniam, Applied Energy, **279**, 115739 (2020)
11. N. Kittner, F. Lill, D. Kammen, Nature Energy, **2(9)**, 1-6 (2017)
12. S. Tagliapietra, G. Zachmann, O. Edenhofer, J. Glachant, P. Linares, A. Loeschel, Energy Policy, 132, 950-954 (2019)
13. N. Hossein Motlagh, M. Mohammadrezaei, J. Hunt, B. Zakeri, Energies, **13(2)**, 494 (2020)
14. G. Rausser, W. Strielkowski, D. Streimikiene, Energy & Environment, **29(1)**, 131-146 (2018)
15. V. Madonna, P. Giangrande, M. Galea, IEEE Transactions on Transportation Electrification, **4(3)**, 646-659 (2018)
16. T. Napp, S. Few, A. Sood, D. Bernie, A. Hawkes, A. Gambhir, Applied Energy, **238**, 351-367 (2019)
17. G. Dileep, Renewable Energy, **146**, 2589-2625 (2020)
18. U.S. Department of Energy, *https://www.energy.gov* (2020)
19. I. Abubakar, S. Khalid, M. Mustafa, H. Shareef, M. Mustapha, Renewable and Sustainable Energy Reviews, **67**, 235-245 (2017)
20. S. Hu, D. Yan, E. Azar, F. Guo, Building and Environment, **175**, 106807 (2020)
21. C. Ines, P. Guilherme, M. Esther, G. Swantje, H. Stephen, H. Lars, Energy Policy, **138**, 111212 (2020)
22. R. Fakhry, The Electricity Journal, **30(6)**, 31-41 (2017)
23. A. Papadopoulos, Renewable Energy, **149**, 103-114 (2020)
24. L. Bird, M. Bolinger, T. Gagliano, R. Wiser, M. Brown, B. Parsons, Energy Policy, **33(11)**, 1397-1407 (2005)

25. D. Gielen, F. Boshell, D. Saygin, M. Bazilian, N. Wagner, R. Gorini, Energy Strategy Reviews, **24**, 38-50 (2019)
26. B. Chen, J. Wang, X. Lu, C. Chen, S. Zhao, IEEE Transactions on Smart Grid, **12(1)**, 18-32 (2020)
27. A. Hirsch, Y. Parag, J. Guerrero, Renewable and sustainable Energy reviews, **90**, 402-411 (2018)
28. M. Khodayar, The Electricity Journal, **30(4)**, 68-74 (2017)
29. S. Rangu, P. Lolla, K. Dhenuvakonda, A. Singh, International Journal of Energy Research, **44(13)**, 9889-9911 (2020)
30. A. Cheng, B. Tarroja, B. Shaffer, S. Samuelsen, Journal of Power Sources, **401**, 175-185 (2018)

From sustainable habits to sustainable energy use: social sustainability in the context of digitization

Maria Skivko[1*], *Elena* Korneeva[2,3], *Olga* Korableva[3,4]

[1]Samara National Research University, 34, Moskovskoye shosse, Samara, 443086, Russian Federation
[2]Financial University under the Government of the Russian Federation, 49, Leningradsky Prospekt, Moscow, 125993, Russian Federation
[3]Togliatti State University, 14, Belorusskaya St., Togliatti, 445667, Russia
[4] Volga Region State University of Service, 4, Gagarina str., Togliatti, 445017, Russia
Togliatti State University, 14, Belorusskaya St., Togliatti, 445667, Russia

Abstract. This paper focuses on the social pillar of sustainability and explores the interconnection between social habits and conscious consumer choice in digital time. In particular, the paper analyses the digital culture's involvement in all areas of today's human life, and its possibility to define, transform, and interpret various social values and attitudes. At the examples of "green" apps use and "green" blogs following, this paper demonstrates how digital culture offers instruments and information for the development and diffusion of sustainable habits, in particular for dealing with sustainable energy use. Moreover, some considerations regarding the benefits of digital culture for business strategies and sustainable community development through digitization are proposed.

1 Introduction

Nowadays, the concept of sustainability affects many areas of human activities. Furthermore, it evolves rapidly, provokes discussion at the national and international levels, and inspires innovations and innovative approaches in various industries and business sectors. These innovations and the evolution of sustainable discussion at the level of an average consumer transform consumer preferences and result in some transformations in social values and social habits. The development of digital media and digital communication strengthens the sustainable discussion in the digital space providing access to information about sustainable practices to everyone. This article highlights the potential of green blogs and green apps to implement sustainable habits into the everyday routine and motivate to further follow them by using digital devices. In particular, the article explains the possibility of applying the practice of domestic sustainable energy consumption with the help of green bloggers and green mobile apps.

* Corresponding author: maria.skivko@gmail.com

Although the concept of sustainability has many interpretations and applications, there is still no universal definition of its pillars, especially of the notion of social sustainability [1]. Generally, it may concern social values and social standards, social institutions and participation, equal opportunities, or democratic rights. By connecting environmental protection with social justice through political involvement, social sustainability is responsible for harmonic interaction between people and the environment. Moreover, social sustainability defines, articulates, and regulates those social values and social attitudes that ensure social welfare for future generations.

Many scholars emphasize that social sustainability has not been explored enough, and no clear definition has been given yet. Thence, some researchers pay more attention to explaining socially oriented practices that influence the construction of conceptual framework [2, 3]. Others use the list of indicators that articulate social sustainability's main tasks in the modern world [4, 5]. Littig and Griessler define social sustainability as a "quality of societies" that is expressed through the interaction of people and nature and expresses the satisfaction of basic and extended human needs without harming natural sources over time and notions of social justice, social participation, and human dignity are taken into account [6]. As Boström argues, social aspects of sustainability deserve more attention from the theoretical perspective and empirical investigation due to its ability to be incorporated into the sustainable practices and facilitate changes from the individual perspective [7].

As far as sustainable development generally aims to fulfil the needs of the present generation, not affecting future generations' needs, it is possible to suggest that the social sustainability concept embraces mainly the area of human demands. Talking about basic needs (e.g., food, access to water, housing, health care, sexuality) and additional needs (e.g., education, social relationship, leisure) concerns the relationship between human behaviour and nature. This is the expression of social sustainability articulated in lifestyle, habits, and social practices.

This article considers social sustainability under the scope of interaction between social habits and social well-being. Social habits impact individuals' conditions and quality of life; the switch to sustainable social habits can improve living conditions and lower the negative environmental effect by using sustainable alternatives and practices.

2 Social habits, digitization, and conscious consumption

Social habits as a way of behaviour towards others can be determined and influenced by the social environment, cultural background, education, or particular lifestyle. Moreover, social habits as regular practices are deeply connected to the consumer behaviour patterns: individuals follow the same social attitudes, consumer standards, and consumer preferences by regular or particular shopping occasions.

Nowadays, it is possible to suggest that digitization, technological progress and the possibilities of digital communication also determine social habits and consumer preferences. Digital technologies participate in many everyday activities of individuals aiming to simplify routine tasks and improve the quality of life (to remind, control, calculate, measure). In the digital space today, it is possible both to do business and to entertain oneself: doing shopping, studying, working, and communicating became possible due to the technologies and digitization, almost without borders and limitations.

Furthermore, access to the specific types of information has dramatically changed in the last years. The development of social media simplified digital communication and defined new opinion leaders. Those new authorities prefer to use digital technologies and digital media (microblogs, video channels, social networks accounts, blogs) for sharing their personal experience, expertise, or any specific information [8, 9]. Plenty of bloggers occupy people's minds today and due to the access to information and the internet attract followers from all over the world.

Therefore, these digital authorities, bloggers, transmit not only their opinion but their habits, daily routine, social and cultural values, and individual interests. Besides, they determine consumer trends in various areas. They share the details from private life and motivate their followers to repeat the patterns. For instance, fashion bloggers, mostly women, demonstrate ideal fashioned bodies, fancy outfits, and perfect makeup, promoting the standards of beauty and creating an ideal visual image. As another example, food-bloggers can share their secrets of cooking, promote any nutritional diet, or create food experiments and motivate followers to increase their interest in cooking, gastronomic secrets, or nutrition.

Bloggers today represent the model of consumption created in the digital space that interacts with followers and influences their social habits. Some bloggers support the idea of extensive consumption, for example, and emphasize the importance of following fashion trends – in the same way regarding clothes, accessories, beauty products, or electronic devices and cars. Others, those who share and promote sustainable habits for everyday life, can have the same power to influence and determine the social habits of followers under the sustainable perspective [10].

With climate change, the risk of global warming, and some hazardous consequences for the world economy, today's sustainable discussion develops not only at the world politics level. At the regular consumer level, it becomes essential to be aware of the outcomes of climate change, sustainable use of resources, and sustainable alternatives for the everyday routine. Digitization and digital culture expanded the ways to find the relevant information, talk about the environment and sustainability, and spread the environmental news around the globe.

In these circumstances, bloggers can become this source of necessary information and expertise to define, promote, and explain common sustainable habits. Conscious consumption that is connected to sustainable habits can change its role from the pattern of a small group of adepts of sustainable practices to the main pattern for the majority. Using different forms and types of context (personal stories, visual data, results of individual experiments), bloggers sometimes articulate these topics that are not often covered by the conventional media. Moreover, bloggers can attract attention to quite complicated topics and provoke relevant discussion in digital media. This discussion shapes the representation of sustainability in the digital space.

3 Green blogs for sustainability

Blogs today are considered as a strong marketing instrument to affect consumer behaviour [11]. The more sustainable discussion spread to the world, the more it intervenes in the digital space and personal blogs. Among the variety of blogs, it is relevant to mention those that: i) talk professionally about sustainability (e.g., climate change, greenhouse effect, carbon footprint); ii) promote a sustainable lifestyle and everyday practices (e.g., zero-waste lifestyle, minimalism, principles of reuse, reduce, recycle); iii) explain different topics under the sustainable perspective (e.g., sustainable fashion, sustainable architecture, sustainable art).

In the first case, confirmed field experts provide the approved knowledge on specific topics. Work professionals, academics, coaches, popular speakers for sustainability talk about individual responsibility in consumption and advocate for changes through arguments and scientific data. Expert knowledge is based on facts and less present the daily routine cases, so these blogs' potential audience should have a true interest in the promoted topics. Special attention should be given to big corporations' green blogs (e.g., Google, Microsoft) that also join the sustainable discussion and explain the sustainable agenda from the corporate perspective and their particular input [12, 13, 14].

In the second case, non-experts and non-celebrities express their interest and involvement in sustainability and share with follower's specific knowledge based on practice. These sustainable, or green, bloggers can, at the examples from individual experience and personal daily routine, explain and articulate several norms, regulations, and ideas of sustainability and provide useful advice and tips for sustainable living [15]. They explain how to organize garbage and recycling at home, test new organic skincare or cleaning products, or calculate how much money it is possible to save just by switching off the house's light. Those bloggers transform the ideas of individual responsibility, social values, social involvement, readiness for sustainable development changes into an easy-going discussion, using typical instruments of blogging and marketing (e.g., story-telling, visualization, experiments, tests). This type of content is quite common and popular in the digital space; the audience follows and tracks the everyday sustainable patterns and tends to repeat them at home. Some of these bloggers are well-known globally for their innovative interpretation of sustainable ideas, such as living zero-waste as Bea Johnson [16] or possessing a minimum of stuff and still staying happy as the Minimalists [17].

The third case can be presented by various bloggers from different topics: fashion and beauty industries, architecture, arts, health care, nutrition and diets. As far as the sustainable discussion is growing and concerns more and more areas of human life, even for bloggers, there is much potential to produce new digital content by uncovering the possibilities of sustainable perspective in different subjects. Consumption and fashion are deeply interrelated, and, for instance, fashion bloggers today are those influencers who define many consumer preferences and consumer behaviour models, describe current trends, promote designers, and give inspiration. So, sustainable fashion bloggers talk about all these things but under the sustainable perspective (e.g., fair conditions of clothes production, raw materials extraction, eco-friendly technologies, labour rights, recycling of items) [18]. These bloggers can promote as sustainable goods and services in the fashion industry as well as practices of sustainable consumption (including reuse, reduce, recycling principles).

What is important to mention is the power of social media influencers. On the one hand, they can help regular consumers change their everyday habits into sustainable ones. On the other hand, any business can benefit from collaborating with bloggers to promote their goods and services more sustainably.

Green bloggers writing about sustainable energy use explain how to make a home a bit more sustainable: it can involve advice on planning and budgeting expenses for household, advantages and disadvantages of using smart meters or solar panels, specifics of low carbon heating, or energy-saving and eco-friendly practices. Performing this information in the readable for everyone form, green bloggers promote how to have a comfortable and healthy home with affordable energy-efficient consumption. The energy efficiency topic can be for someone a complicated issue, especially regarding the ways how to implement it in the household; with the help of green bloggers, it seems to become possible to promote the sustainable practices of energy consumption and efficient business solutions (e.g., use of renewable energy, smart meters, solar grids) for this task.

4 Green apps and sustainable energy use

Particular attention should be given to green, or sustainable apps. With the development of sustainable discussion that challenges many industries to become greener and produce less negative environmental impact, the green IT concept started to develop and attracts the attention of business, mainly start-ups, and conscious consumers [19, 20]. Green IT can be understood as practices of environmentally sustainable computing and IT that aim to reduce the use of hazardous materials, maximize energy efficiency, and lower negative environmental impact (e.g., using green hosting provider, image optimization, removing unused features from apps). Such instruments as environmental internet portals, professional or personal blogs, wikis, or interactive simulations of the environmental impact of different consumer practices represent the ways of environmental involvement and the strategies to deal with it at the global and local levels.

Green apps aim to help users join the global movement of sustainable development ideas at the local level of everyday practices and consumer behaviour. In other words, the development of green apps can foster progress in sustainable development. Green apps can be used in the private household or at work; in both cases, they can give suitable information about certain sustainable habits, the necessity to change consumer behaviour, or the advantages of using those digital technologies in everyday life. Moreover, they can establish and support the motivation to follow a sustainable lifestyle, in some cases by compensating or rewarding.

For example, some apps use gamification principles to attract users by the habitual format; by setting the final goal and game rules, these apps support users in this game to create a sustainable habit [21]. Feedback and app support settle productive results in creating a habit and increasing the users' motivation to continue those practices.

Regarding the energy use apps, it is possible to denote the ways how they perform the information, for instance, as an educational tool (lectures, video materials, illustration), entertainment (games), source of various information (useful tips, guides, check-lists, recommendations, indicators), transformation tool (to control, measure, plan, analyze the consumption) [22].

By talking about energy use apps, it is important to mention that one of the efficient ways to attract followers' attention is to motivate them through financial benefits. Not every consumer is ready to change everyday habits just for the benefits of the environment; but the most would instead do that if they get the advantages for the household budget provided. So, many energy use apps primarily emphasize those tools implemented that help calculate the costs, track, and control consumer use of energy to save money or get some extra bonuses.

Thence, financial motivation for sustainable habits can be considered one of the most influential instruments in communication with consumers. Green apps help to define which home appliances consume the most energy, energy usage and energy consumption per day, per week, or per month; some of them provide additional services as, for instance, real-time tracking, remote control, save the history of energy consumption, or just easy tips for saving energy and rational energy consumption.

The easier is the app to install and configure, the more chances it has to stay in the household as a helpful tool. Smartphone, tablets, or wearables with different sensors and hardware modules allows to learn new information and apply it at the practice immediately; ubiquitous access to information through digital technologies shapes the priorities of new generations.

Any business that applies green IT in its production process can expect the brand's image benefits. Moreover, it can foster investments or collaboration for the sustainable advocacy, and decrease certain expenses by using innovative sustainable technologies.

5 Conclusions

Generally, our article presented the perspective on green blogs and green apps as powerful instruments to influence consumers and create sustainable social habits. As the social pillar of sustainability, social habits represent the intention for social well-being and conscious consumption. In the world of digital technologies and digital communication, it becomes easier to introduce, control, monitor, and check various social habits with the help of digital devices and the expert opinion of digital influencers. Additionally, from the marketing perspective, these digital tools and authorities can be considered quite significant instruments for business development, especially in the case of start-ups developing mobile apps.

It is true that the information from bloggers can be considered dually, as a source of unconventional truth and experience or as a source of low-quality information that is difficult to check. It also makes sense to differentiate professional green blogs and personal green blogs that produce and deliver different types of information and the opinion of a different quality. However, the fact that environmental discussion in the digital space is growing every year means the implementation of sustainable practices and habits into everyday reality.

Green apps that help to monitor domestic energy consumption are considered as a technological innovation that can be applied at every household with the help of digital devices. Additionally, green apps facilitate the spread of environmental knowledge among individual consumers and motivate them to join the community of conscious consumers and benefit from this lifestyle.

Furthermore, the business sector can also get some benefits and profits from collaborating with green bloggers and supporting the development of green apps. Such collaboration strengthens the advocacy for sustainable discussion changes and raises questions about individual responsibility and individual choice for a sustainable lifestyle and social practices.

References

1. R. H.W. Boyer, N.D. Peterson, P. Arora, K. Caldwell, Sustainability, **8(9)**, 878 (2016)
2. J. Lu, L. Ren, J. Qiao, S. Yao, W. Strielkowski, J. Streimikis, Sustainability, **11(15)**, 4128 (2019)
3. N. Dempsey, G. Bramley, S. Power, C. Brown, Sustainable Development Bradford, **19(5)**, 289 (2011)
4. S. McKenzie, Hawke Research Institute working paper series. University of South Australia 2004. **27**: https://tmuk.pw/socialsustainability.pdf
5. M. Missimer, K.-H. Robèrt, G. Broman, Journal of Cleaner Production, **140**, 32 (2017)
6. B. Littig, E. Griessler, International Journal of Sustainable Development, **8**, 1 (2005)
7. M. Boström, Sustainability: Science, Practice and Policy, **8(1)**, 3 (2012)
8. M. Pedroni, Film, Fashion & Consumption, **5(1)**, 103 (2016)
9. D. Jiménez-Castillo, R. Sánchez-Fernández, International Journal of Information Management, **49**, 366 (2019)
10. J. Haider, Journal of Documentation, **68(5)**, 639 (2012)
11. S.-C. Chen, C.-P. Lin, Technological Forecasting and Social Change, **96**, 40 (2015)
12. Google Sustainability blog: https://blog.google/outreach-initiatives/sustainability/
13. Microsoft Green Blog: https://blogs.microsoft.com/green/
14. Panasonic Green Living: https://na.panasonic.com/us/green-living
15. S. Joosse, T. Brydges, Environmental Communication, **12(5)**, 686 (2018)
16. Zero Waste Home: https://zerowastehome.com/blog
17. The Minimalists: https://www.theminimalists.com
18. M.L. LeHew, S.U. Patwary, Paper presented at the Third International Conference of the Sustainable Consumption Research and Action Initiative (SCORAI), Copenhagen, Denmark, 27-30 June 2018
19. S. Murugesan, IT Professional, **10(1)**, 24 (2008)
20. F. Loeser, Paper presented at 19th Americas Conference on Information Systems (AMCIS)At: Chicago, IL, August 2013
21. G. Rausser, W. Strielkowski, D. Štreimikienė, Energy & Environment, **29(1)**, 131-146 (2018)

https://doi.org/10.1051/e3sconf/202125006002

22. B. Brauer, C. Ebermann, B. Hildebrandt, G. Remane, L.M. Kolbe, Paper presented at the Conference: Proceedings of 20th Pacific Asia Conference on Information Systems (PACIS 2016), Chiayi, Taiwan (2016)

Corporate sustainability and social responsibility

Zeynegul Samaibekova[1]*, *Gulzhamal* Choyubekova[2], *Kerezkan* Isabaeva[3], and *Asel* Samaibekova[4]

[1]Institute of Economics and Management, Kyrgyz State University of Construction, Transport and Architecture named after N. Isanov, Maldybaev str. 34b, Bishkek, Kyrgyz Republic
[2]Department of Public Health and Health care, Kyrgyz State Medical Academy named after I.K.Ahunbayev, I.Akhunbaeva str. 92, Bishkek, Kyrgyz Republic
[3]Department of Economics and Enterprise Management, Kyrgyz State University of Construction, Transport and Architecture named after N. Isanov, Maldybaev str. 34b, Bishkek, Kyrgyz Republic
[4]Department of Economics and Management, Kyrgyz State University named after Arabaeva, Razzakov str. 49, Bishkek, Kyrgyz Republic

Abstract. Our paper focuses on the links between corporate sustainability and social responsibility. Corporate Social Responsibility (CSR) emerged as a tool for linking the priorities of business companies (making money and achieving profit) with the priorities of citizens and society. Bringing together the many different parts of a complex CSR programme into a single central system is crucial. Moreover, we discuss the role of corporate structures in the development of social organisations and their impact on society, as well as on corporate social responsibility and the impact of the social entrepreneurship model on the economy. It appears that companies can bring important benefits to society if they are responsible for the quality of the goods and services they produce and develop new goods or services that generate economic growth. The long-term benefits for investors therefore allow companies to invest in product innovation, thereby delivering high-quality products that improve people's standard of living. Business companies thus meet the needs of society and offer important benefits to society in the form of new jobs and economic opportunities for those in society who depend on the company's good services. While companies seek new economic opportunities and regain public confidence, the creation of shared values and the pursuit of financial success is becoming increasingly important for companies in a way to support sustainable development and fighting global warming and climate change.

1 Introduction

Corporate Social Responsibility (CSR) is a comprehensive concept that can take many forms depending on the company and industry [1]. Commitment to CSR means that a company works in a way that promotes society and the environment, rather than contributing negatively [2]. Technology can help implement a CSR program that includes reducing any

* Corresponding author: samaibekova@mail.ru

company's carbon footprint for the benefit of the planet. Companies can not only strengthen their brand, but also benefit from society in other ways, such as in the form of education, employment and the environment. Corporate sustainability and social responsibility (CSR) have become a strategic imperative for companies, as they are an integral part of the optimal long-term strategy [3, 4]. Therefore, companies should make decisions and develop processes to achieve greater financial success where consumers are an important stakeholder and where consumers are an important and important stakeholder for the long-term success of the company. But as important as CSR is to the community, CSR is valuable to businesses. Businesses worldwide are developing strategies to become better corporate citizens and communicating their activities to internal and external stakeholders, who can involve a range of groups [5].

While organizations want to promote employee engagement, build brand values and, above all, achieve sustainable change in the local community, improving CSR management around sustainability has become a central pillar of their cause. Modern CSR managers behave in a promising way and help them and their initiatives to achieve high organisational goals [6].

In addition, human resources professionals can understand their influence and respect for CSR, and these insights can be shared with the rest of the organization to integrate into their business model. Human resources management can play an important role in developing a corporate social responsibility strategy to ensure that it is what it is doing here [7]. They can reward and incentivize good decisions and initiatives and encourage collaboration and collaboration between employees, families of employees and other stakeholders. In addition, human resources departments play a crucial role in ensuring that companies implement CSR responsibility programmes. Thy can increase the awareness of the employer brand, increase employee engagement, increase shareholder value and increase the value of their shareholders. HR can monitor the introduction of CSR and manage its implementation while documenting and celebrating success across the company, tracking and reporting on success.

2 Strengthening business connection with the society

It takes a lot of time and energy to build good, strong and lasting business relationships today, and people do not seem to want to participate, but it is an integral and necessary part of anyone's success. In society, people function in relationships and they are in business [8]. But lasting business relationships do not develop or develop without dedicated and consistent work. Although there are many strong relationships that have been established in our community, some are not the most effective, and sometimes it is not what one knows, but who she or he knows. People learn the skills to build and maintain relationships over time.

In order to attract and retain new and existing customers, one needs to devote resources and invest in building strong, long-term business relationships that help people recognize the value and achieve their goals with a specific product or service. Simply put, to be successful, a company must build a positive relationship with its customers. This is because building relationships with new and potential customers enables companies to offer a more personalised and enticing customer experience [9]. It is more than just responding to e-mails on time, sending e-mail marketing campaigns with discounts, or being polite on the phone. One should think of improving customer contact points with the sales, marketing and support teams. In most cases, one can handle communication with loyal customers willing to support one's business through mouth marketing. Ultimately, this becomes a measure of a business company's ability to build a positive relationship with a company's customers [10].

To build a positive relationship with their customers, companies often rely on technology and create a central command centre to manage, monitor and communicate with customers. Some companies have managed to build a stable business that has strengthened over the years, while others have had huge volatility and eventually closed. While many factors lead to success, one essential step is to establish a strong business relationship with a customer who returns with time and sends future business as a recommendation. A strong customer relationship requires active communication, and that means that the customer must feel comfortable approaching the business company and its managers. The customers should know that the business is always available by phone, social media, email or in person [11].

One of the biggest dangers of communication is that a customer cannot find out if the other person understands the message she or he is trying to convey. Poor workplace communication can lead to a culture of back-biting and blame-pointing, which can affect stress levels when the customers do not understand or feel deceived. If a business company works well and motivates people to come to work and do good work, it has a positive effect on morale. This is a fundamental attitude that can bring endless benefits to any business. The simple act of appreciation can increase productivity, improving the quality of work and achieving better returns. This can take the form of better customer service, not to mention a happier environment, which generally creates a better working relationship between a business company and its employees as well as its customers.

3 Business company and its contribution to social development

In general, there is a strong link between social media and social networks such as Facebook and Twitter. Running corporate responsibility campaigns as a quick marketing program can backfire if the company does not hold out. One can adopt socially responsible business practices by applying a one-off action rather than a long-term strategy [12].

Employees and consumers react positively to companies that take on long-term social responsibility but not always positively. The apparent acceptance of CSR in the economy has been taken up by many economists, but not Milton Friedman who claimed that the only social responsibility of companies was to maximise their profits by doing business in open and free competition without fraud or deception. Friedman pointed out that shareholders, customers and employees could spend their own money on social activities if they wanted to. However, he also pointed out that there were many circumstances in which company managers could take measures that served the long-term interests of the company owner but indirectly had positive social effects [13].

In the search for the best categories, the only categories in which a significant proportion of companies were in the category "goods". In the area of human rights, all categories of business companies behave similarly, with some firms showing relative strength. This is understandable, because companies in this category face strong opposition from activists who are concerned about the unskilled labour that is normally employed in developing countries during the production process.

Modern theoretical and empirical analyses show that companies can engage in strategically socially responsible activities and at the same time increase their private profits. Becoming a socially responsible company can be easy, even if there are a few caveats. Any company should avoid participating in charitable efforts that are not related to its core business or that in any way violate its ethical standards. Instead of blindly sending money to completely independent organizations, you can find non-profit organizations your company believes in, projects and communities one works with. Corporate citizenship is one of the most important aspects of corporate social responsibility. Corporate citizenship is important to business and thence we can ensure that organisations are socially responsible. Businesses are interchangeable, but the latter is only one facet of the former [14]. Companies can exercise social responsibility by donating money, products or services to social causes and non-profit organisations. Charitable donations include donations and grants to non-profit organizations. Large companies tend to have ample resources to benefit charities, local communities, and programs; but for small businesses, such efforts can make a big difference. One of the greatest values a company should promote is the ability to provide knowledge. If there is a particular charity or programme you have in mind, contact the organisation and ask them what organisation would help you best and why. This is not the only way to actively participate, but it is crucial that we share what we know, and only then can we do our best together. If one is considering a sustainable activity that is not yet legally mandatory, it is not good to wait until it is legal. If one adopts socially responsible standards early on, she or he can set the bar and refine the process over time and refine the process further in the future [15].

Their impact measures can make a real difference in the world and appeal to socially conscious consumers and employees. Corporate social responsibility was introduced as an approach to international development. It enables certain stakeholders to redefine the meaning of international development success. The interests of companies are not adequately aligned with those of the poor, and explains why CSR does little to change this. There is a historical relationship between business and development and compares how expectations of companies that live by CSR practices differ from those of the past. There should be a new societal approach to CSR, while the new challenge for companies is the lack of a clear understanding of their role in social development. The focus will be on the role of corporate responsibility in the development of social and environmental sustainability [16].

4 Social and environmental programmes and CSR

All of the above being said, one can recall that corporate social responsibility refers to the growing practice of for-profit organizations aligned to relevant purposes and good social programs [17]. A recent employees survey found that more than half of respondents are willing to pay more for products and services from companies that are committed to positive social and environmental impacts, and two-thirds would prefer to work for these companies [18]. Companies that hire people who are strongly committed to social justice, environmental sustainability, and human rights also benefit from the public interest in making the world a better place, according to a new study. It is important to recognize and celebrate socially responsible companies, to promote their work and to show other organizations how they can successfully integrate social welfare into their mission.

Furthermore, a commitment to CSR means that a company works in a way that promotes society and the environment, rather than contributing negatively to it [19]. In this way, companies can benefit society and at the same time increase their brand and thus their profit and reputation. Corporate social responsibility enables large and small companies to implement positive changes. As important as CSR is to the community, it is valuable to businesses and to society as a whole. Consumers feel that they are playing their part in using products and services from socially responsible companies. When companies choose to do what is in the best interests of their customers and the community, they benefit financially by building trust with consumers. The more socially responsible a company is, the more supportive communities and consumers become. The more visible and successful a company is, the more responsibility it must assume for its employees, the families of its employees and the community as a whole. Although it is not as good - and is promoted by large corporations - small and medium-sized enterprises are creating social responsibility programs as part of their business model.

Companies that implement CSR benefit in many ways. For example, Starbucks has long been known as one of the world's largest companies, strongly committed to social and environmental responsibility. Starbucks says it has achieved many of its CSR milestones since it opened. With the increasing use of corporate responsibility, it is becoming increasingly important to have a socially conscious image [20]. Consumers, employees and stakeholders prioritize CSR when choosing a brand or company, and they make companies responsible for driving social change through their business beliefs, practices and profits. To illustrate just how critical social responsibility becomes, one recent research found that more than 60% of Americans hope that, without government regulation, companies will drive social and environmental change, underscoring the importance of corporate responsibility as a key element of their business model [21].

One of the ways global communications companies can achieve social impact is through private corporate social initiatives which support promising start-ups focused on environmental, social, health, human rights, and social justice issues. Such instantiates support for-profit start-ups by guiding companies through the early stages of business, providing access to mentorship, and facilitating pitch placement for potential investors. Many companies have national sustainability programs and invest millions in sustainability and conservation projects. By combining philanthropy and the corporate sector through CSR, stakeholders can bring about environmental change while encouraging other companies to follow in their footsteps and be environmentally friendly. Businesses have a strong influence on the economy and have more resources than non-profit organizations to change their policies, which allows them to act as change agents.

All in all, corporate social responsibility is an opportunity for companies to make a difference in the community. With available corporate profits, CSR programs can not only invest in non-profit sectors, but also promote positive change in communities. There is an increasing pressure on American companies to make meaningful contributions to the greater social good by consumers, workers and investors. Today, many consumers state they would be more likely to buy a company's products and services if the company were to address social, economic and environmental issues. A 2019 study of millennials by Deloitte found that climate change and environmental protection are at the top of millennials' concerns. CSR is a fast-growing field but most large companies have not yet adopted a CSR programme which is both sad and surprising [22].

5 Conclusions

Businesses are highly profit-oriented hybrid social organizations, which sometimes find it difficult to focus on economically sustainable models rather than on solving social problems. Companies that seek profit are often presented as organizations with charitable goals, assuming that a profit-oriented institutional environment performs better than a charity-oriented one. For profit is also encouraged to donate to charity, to establish affiliated foundations and to pursue social goals in addition to maximizing profits through state stakeholder laws. These include companies with profit structures that integrate social welfare into their objectives, as well as those without profit structures that integrate social goods into their objectives. If a for-profit company can tap a large pool of investment capital, a social mission could fit well into it. But if it is not part of the legal structure, then a profit that brings social benefits must have a founder or board of directors who firmly believes in the cause and is willing to invest money in it, even if it conflicts with the company's profit objectives.

Companies that go further than traditional CSR are often referred to as purpose-oriented organisations. In addition, social welfare is not only part of the corporate structure, but also the business model and business strategy of the company. We often hear in the business world that we are focusing on entrepreneurial practices that lead to higher monetary profitability. We give people who are involved in the organisation a sense of importance, draw on their support, explain how they make a difference and give something back. Since these two types of entrepreneurs are driven by different motivations, we must bear in mind that their businesses will function a little differently. Social enterprise means using market-based business strategies to advance the social and environmental impact of an organization. Social enterprises, on the other hand, focus on achieving social, environmental and economic profitability. In other words, social enterprise constitutes a market-driven approach for businesses that is used effectively to solve social problems. Social entrepreneurs face serious challenges in finding innovative and practical solutions. Companies need steady capital flow, innovation requires experimentation, and there is little incentive to pay for unproven approaches. As a result, those who finance social entrepreneurship projects focus on short-term gains, not long-term social impact. Social entrepreneur projects tend to offer investors lower returns than other opportunities. Relations between social entrepreneurs and investors can be strained by conflicting objectives and conflicts of interest, but also by the fact that innovation requires experimentation. All in all, corporate social responsibility helps to establish corporate sustainability and helps businesses to grow and to evolve in accordance with the environmental and green agenda of the 21st century.

References

1. B. Xia, A. Olanipekun, Q. Chen, L. Xie, Y. Liu, Journal of Cleaner Production, **195**, 340-353 (2018)
2. A. Dwekat, E. Seguí-Mas, G. Tormo-Carbó, P. Carmona, Corporate Social Responsibility and Environmental Management, **27(6)**, 2879-2892 (2020)
3. J. Lu, L. Ren, J. Qiao, S. Yao, W. Strielkowski, J. Streimikis, Sustainability, **11(15)**, 4128 (2019)
4. M. Ashrafi, M. Adams, T. R. Walker, G. Magnan, International Journal of Sustainable Development & World Ecology, **25(8)**, 672-682 (2018)
5. L.M. Edinger-Schons, L. Lengler-Graiff, S. Scheidler, G. Mende, J. Wieseke, Business Ethics: A European Review, **29(3)**, 510-527 (2020)
6. R. Cannas, G. Argiolas, F. Cabiddu, Journal of Sustainable Tourism, **27(1)**, 139-155 (2019)
7. M.A.L. Agudelo, L. Jóhannsdóttir, B. Davídsdóttir, International Journal of Corporate Social Responsibility, **4(1)**, 1-23 (2019)
8. Y. Doz, Human Resource Management Review, **30(1)**, 100693 (2020)
9. W. Wikhamn, International Journal of Hospitality Management, **76**, 102-110 (2019)
10. M. Preziosi, P. Tourais, A. Acampora, N. Videira, R. Merli, Journal of Cleaner Production, **237**, 117659 (2019)
11. F.S. Foltean, S.M. Trif, D.L. Tuleu, Journal of Business Research, **104**, 563-575 (2019)
12. V. Kumar, A. Lahiri, O.B. Dogan, Industrial Marketing Management, **69**, 147-160 (2018)
13. T.L. Carson, Business and Society Review, **123(2)**, 217-242 (2018)
14. P. Wan, X. Chen, Y. Ke, Journal of Cleaner Production, **259**, 120877 (2020)
15. J. Lu, L. Ren, S. Yao, J. Qiao, W. Strielkowski, J. Streimikis, Energies, **12(18)**, 3417 (2019)
16. G.K. Stahl, C.J. Brewster, D.G. Collings, A. Hajro, Human Resource Management Review, **30(3)**, 100708 (2020)
17. P. Gazzola, M. Ratti, S. Amelio, Management Dynamics in the Knowledge Economy, **5(3)**, 355-376 (2017)
18. L. Kong, M.S. Sial, N. Ahmad, M. Sehleanu, Z. Li, M. Zia-Ud-Din, D. Badulescu, Sustainability, **13(3)**, 1499 (2021)
19. D. Smith, E. Rhiney, International Journal of Corporate Social Responsibility, **5(1)**, 1-12 (2020)
20. E. Gozdan, A. Sudolska, Journal of Corporate Responsibility and Leadership, **6(2)**, 29-47 (2019)
21. P. Shrivastava, M.S. Smith, K. O'Brien, L. Zsolnai, One Earth, **2(4)**, 329-340 (2020)

22. Deloitte, *https://www2.deloitte.com/content/dam/Deloitte/global/Documents/About-Deloitte/deloitte-2019-millennial-survey.pdf* (2019)

https://doi.org/10.1051/e3sconf/202125006004

Competitive advantage of the sustainable digital economy

Oksana Nurova[1]*, *Tatiana* Freze[1]

[1]Institute of Engineering and Environmental Security, Togliatti State University, Belorusskaya str. 14, 445020 Togliatti, Russian Federation

Abstract. This research focuses on factors that influence the competitive advantage of the sustainable digital economy. Digital economy helped to speed up the exchange of information flows which, in turn, impacted on the creation the new forms of business and socio-economic interactions. Such sectors as banking and finance or hospitality and services are all influenced by the advancements of the digital economy that include sharing economy or Blockchain technology and its related applications. In this article, we show that digital economy is more sustainable due to the fact that it requires less energy sources and can itself contribute to the formation of the green economy and the acceptance of the renewable energy. All these features constitute its competitive advantage that needs to be further supported and fostered. In addition, we discuss key strategies and the development of skills to adapt to a digital economy and provide an overview of high-quality training for high-tech workers. It was assumed that in future, people would be able to easily acquire new skills through vocational training and vocational training. However, the spread of digital technologies has changed the way we work and has increased the need for high-quality training for workers in a wide range of professions. To use such technologies effectively, workers in an increasing number of sectors and professions need both general and advanced ICT skills.

1 Introduction

The main feature of the digital economy is that geographical location is no longer a competitive advantage and that the main role is played by the use of digital technologies such as cloud computing, mobile devices and smart contracts [1]. The digitalisation of the economy is considered a key factor in developing sustainable competitive advantages for the economy in general [2]. Traditional industrial structures and business models are being destroyed, and the boundaries between business and technology functions are blurring. Technology has become a catalyst for corporate strategy and transformation, expectations of IT have shifted, prompting many organizations to rethink the role of technology and rethink traditional operating models and organizational structures [3].

In general, there is a new approach to help business and technology functions collaborate more effectively, innovate and create new sources of value. Technology has become a crucial

* Corresponding author: oksana.nurova@gmail.com

part of the global economy and a sustainable competitive advantage that is growing exponentially. The merging of the physical and digital worlds confuses the value chain, blurs the boundaries of the industry, and disrupts traditional value-added models. We expect that more than 90 percent of global production, retail, finance, healthcare, energy, and transportation will be connected to Industry 4.0, where digital platforms will be coupled and connected to their physical counterparts [4]. Therefore, the digital economy presents a challenge to these companies in terms of compliance requirements.

Digital innovation also brings new challenges in dealing with the integration of the physical and digital worlds, as well as a number of new supply chain challenges. It is our duty to force these companies to adapt their organisational and production patterns in order to remain competitive [5]. What is new is the exploration of a selective approach that focuses on start-ups with better potential, creating a specific business environment consistent with their ability to promote and sustain their growth. Another challenge is that we want to define a development path for developing a sustainable competitive advantage in the digital economy and a sustainable competitive advantage. There is the need to align entrepreneurship with regional economic development. Moreover, there is a dynamic importance of an ecosystem useful for scale development and for business development and highlights the benefits that this ecosystem offers. In addition to the challenges associated with digital technology, there are also a number of challenges associated with digital technologies, such as access to capital, the use of technology and developing a competitive advantage in the digital economy. For example, insurance companies use digital networks to stay ahead of development, and collaboration is critical for companies to harness the power of the Internet of Things (IoT). Today's technology start-ups compete for digital innovation by developing solutions that leverage and create synergies in industries such as home automation and healthcare. Innovators enable reusable solutions to be created that allow local teams to design their products and services and gain a competitive advantage [6].

In the environmental field, our aim is to become a global role model in the digital economy. For example, the European Union has become a pioneer and needs to review its objectives in this area. The broader strategy for the future of Europe includes a strategy for the digital economy. This is not only the goal for Europe, but for all EU member states and the European Union [7].

As the digital economy continues to evolve, it may seem more difficult than ever to maintain our market leadership. Even if one keeps pace with the pace of innovation in other sectors of the economy, such as technology, this can cause serious problems. Add to that the fact that technologies are converging exponentially, and countless technologies are changing business models overnight, and it may seem that a sustainable advantage is impossible. Today, the only way for most companies to maintain their competitive advantages is to adapt to continuous digital transformation, to meet new pressures, keep pace with competition and take advantage of new technologies. By bringing advantages to already more developed economies, first-movers can assume a dominant position in national economies that will be difficult for latecomers to challenge. Thence, the question is how the majority of developing countries can maximize the potential opportunities of the digital economy. Global competitors will not have an easy time outdoing themselves, so developing countries should be able to create sizeable niches that allow rapid growth, as Singapore and South Korea have done. This means looking at building local strengths, recognising what local businesses can do and taking advantage of the specific local circumstances. For example, some countries set up a regulatory laboratory to anticipate and develop future legislation on the use and application of new technologies. Business hubs to promote healthy competition in the telecommunications sector and a regulatory environment conducive to the diffusion of new technology. To achieve these goals, policymakers could introduce testing and learning approaches to developing new technologies and their solutions [8]. Through testing and learning, they can encourage companies to innovate and test new digital products.

2 Technology platforms and digital sustainable economy

Globalisation and the digital economy have brought unprecedented growth to all segments of the public and private sectors and opened up a globally accessible market. The world has become a marketplace for companies of all sizes to provide cross-border services to a global customer base, and geographical reach is growing [9]. All of these provide a reflection on the role of technology in the future. One needs to stress the need for governments and the private sector to work together to create a new digital ecosystem. Some countries such as China or India are well on its way to becoming a digitally advanced [10].

Similarly, emerging technologies such as cloud computing are reducing the cost of data and turning fixed-asset investment into operating costs into a pay-as-consumer model. A connected nation is transforming into a digital economy with a more efficient and efficient public and private sector. Driven by falling costs and increasing data availability, India is digitalizing faster than many developed economies and emerging economies, and is already on the fastest-growing base in the world in terms of the number of mobile phones, tablets, and smartphones. Indians spend more time on social media, download more than twice as much data as their counterparts in other countries, and now account for a share of GDP compared to the last decade. More than half of all mobile phone purchases in India (52.5%) are now made on mobile devices, up from 45% in 2010. The Global E-Commerce Framework sets out five principles that guide the development of the digital economy and its continued high growth potential [11, 12].

New technologies reduce costs, enable market share and offer new products and services that were not previously needed. Technology platforms include, but are not limited to, cloud computing, mobile devices, social media, e-commerce platforms and mobile applications. This, combined with the development of new business models and the use of technology in the digital economy, contributes to the development of new activities in both the private and public sectors. New technologies offer lower barriers to entry, lower operating costs and lower prices, thereby reducing costs. The digital economy, individually and collectively, is using technology to overhaul traditional exchanges and enable new ones. Many entrepreneurs have taken advantage of technology - and driven digital economies to create new businesses and new business models that would not have existed without the size and scale they have today or in previous generations [13].

There are various concepts for business platforms that can help to structure the technological framework that allows the digital platforms to go live. This can bring the business and technology solution back to completion and make it available to the customers and partners in a variety of business models. The customers should identify the business opportunities that arise from the introduction of innovative digital solutions. Digital transformation transactions with the customers are managed in a variety of business models, from small businesses to large corporations and large corporations [14].

All in all, it becomes clear that leading digital platforms have an impact on individuals, markets and society. It has redefined the world economy by creating a new economic space and changing the way analogous activities are conducted. It has had a significant economic impact on people and businesses, and the first is to refocus organizations to make them more responsive to the needs of their customers and customers. Secondly, we need to use digital technology to change the nature of businesses and the way the economy is changing. We can think of it as a digital transformation or a digital transformation in two parts. One is its use and its destination, the other is actually changing the way we make this journey. However, considering the potential that one sees in science fiction films, where everything is connected with a snap of a finger and everything happens in a flash, it will take more than digital infrastructure and technology to realize this potential of the digital economy.

3 Stages of development of the sustainable digital economy

Today, sustainable digital technology accounts for a share of GDP compared to the last decade and represent a significant part of the overall growth of the global economy in recent years. New technologies reduce costs, provide market access, offer new products and services that were not needed before, and provide access to new markets and new opportunities for new businesses [15]. Together with the rapid growth of the digital economy and the development of new business models, this has contributed to the development of new activities in both the private and public sectors. The Global E-Commerce Framework sets out five principles that guide the implementation and development of digital economies in order to maintain their high growth potential [16].

The artificial intelligence (AI) is considered a production factor based on its self-acquired skills, which have developed noticeably over time, and its use in machine-based technologies. The potential benefits could be even greater if governments invest more in innovative technological advances and the adoption of artificial intelligence that enhances the business sector. AI is seen as adding new productivity factors to existing factors. Therefore, the question is how the majority of developing countries can maximize the potential opportunities of the digital economy [17].

Some developing countries should be able to create sizeable niches that allow rapid growth, as Singapore and South Korea have done. By bringing advantages to already more developed economies, first-movers can establish a dominant position in national economies that are difficult for laggards to challenge [18]. This means building on local strengths, recognising what local companies can do to make it less likely for global competitors to compete with them, and taking advantage of the specific local circumstances.

The role of digital transformation is developing with the accelerated introduction of the third platform [19]. More technologies are added to the third platform and more opportunities to experience the additional layers of innovation, innovation and innovation in the digital economy. Through cost optimization, innovation, and the creation of competitive differentiation, digital transformations are becoming the cornerstone of progress. But we must not be under any illusions that, despite these technologies, there is still a human dimension, which is strengthened by the process of information and the development of the third platform. Digital transformation efforts and strategies are often the result of a combination of government, private, and civil-society initiatives. In countries like Japan, digital change is even aimed at influencing all aspects of life through the Society5 initiative, which focuses on education, health, employment, education, and social mobility [20]. It is a question of harnessing the power of technology to accelerate the impact on society in such a way that priority is given to social and economic development and the development of the third platform. The information-driven ecosystem of the digital economy in the United States and other countries is the result of a combination of government, private, and civil-society initiatives, led by the United States Department of Commerce, the Federal Reserve Bank of New York, and the National Institutes of Health. Stagnant countries, such as Australia and the United States, have strong growth histories, but are slowing down and at risk of falling behind. Some countries including New Zealand and the UK have shown high levels of digital development, continuing to lead in innovation and new growth. Stagnant countries (including Australia and the United States) have low growth rates and high innovation rates, but have strong growth histories [21]. The first call for project proposals focuses on forward-looking areas, with a focus on the development of new technologies and new business models. The strategy prioritises innovation and innovation - centred projects in areas such as data analysis, data science and data management. It focuses on new technologies, innovation, technological development, the development of the digital economy and technological innovation.

4 Science and education in the digital economy

The role of teachers is changing in the digital economy and the changing value orientation of consumers in society is creating a new type of student within the education system. The education of the future is currently going through a timeframe for educational institutions that are taking advantage of the unique opportunities of networks and digital technologies to involve both direct and indirect actors in education processes [22]. This is the so-called "network interaction" of students, teachers, parents and other educational actors. This allows the traditional framework of education to be extended to ensure continuous improvement of educational pathways. At the same time, the psychological and educational characteristics of the modern generation should be taken into account when designing a digital educational environment [23].

The modern transformation of our education system shows that reforms aimed at training future staff lead to people who need to develop and retrain in accordance with the demands of the labour market. The global education market is an important contribution to the growth of the global labour market and to the global economic growth of the future. In these circumstances, the improved education model has become the basis for the development of the global labour market and the growth of the digital economy. The construction of the technology centres in some cities leads to the development of a high-quality research and development environment for the future of technology. Much work is being done to train highly qualified engineers and future scientists and to further improve the quality of the research and development environment for the future of technology in the country. Scientific knowledge describes the world through technology and the transformation of the modern world. It plays a crucial role in the development of new technologies such as artificial intelligence, robotics and machine learning, among many others.

A post-industrial society needs an economy based on knowledge, new technologies and human capital. There is a huge gap between what is required for a person and what it required of everyone. This has come to the fore in recent years with the rise of the Internet of Things and the development of smart cities. This era as we know it is about to change fundamentally the nature of human capital and the role of education in our society [24].

In dozens of countries in Asia, Europe and America the development of human capital and education as the basis for the digital economy are given particular importance. These activities aim to build human potential consistently through a combination of science, technology, engineering and mathematics (STEM) education and the use of technology. Large companies are trying to create public-private partnerships in higher education [25]. By strengthening integrated management and research projects to address the key problems facing the economic sector, we will further strengthen the role of science and education as a basis for the development of human capital and the digital economy. The large educational complex is based on joint action between universities and industry, with a focus on research and development and the development of new technologies and services. There are services based on big data processing technologies that use AI methods such as artificial intelligence, machine learning and big data analysis. In some sectors, such technologies face critical regulatory obstacles, which is why global development institutions that promote the global digitization agenda have developed a set of regulatory frameworks for the use of big data in developing countries. These are the legal obstacles to working with citizens' personal data.

5 Conclusions

Overall, digital innovations are clearly changing the demand for skills, creating a huge skills gap that can hold back economic growth. Apprenticeships have the capacity to close this skills gap by equipping people with the skills they need to adapt to the emerging digital economy and equipping them with skills for jobs that may become obsolete during the transition. Although the emergence of digital economies is creating new jobs all the time, many of these jobs could become obsolete in the future. Training courses should not be limited to traditional craft professions, as their scope to provide an effective and efficient training model is also increasingly recognised in the technological sector. There are good jobs that do not require a bachelor's degree to secure them and they can lead to the development of high-quality apprenticeships for these growing sectors of the digital economy by providing access to high-quality training in digital skills such as web development, data analysis and machine learning. The improvement of labour systems can be supported through improved access, better labour market analysis and the promotion of cooperation between industry and employers. Digital degrees and apprenticeships will be designed by leading employers in the digital sector. One should aim at creating high quality training programmes that meet the standards defined by the employer with a purpose to produce top-notch graduates in the fields of technology, including computer science, computer engineering, information technology, data science and data analysis.

Today, many employers coordinate work-based learning and work closely with learning platforms such as Google, Microsoft, IBM and others. Employers coordinate work-based learning and work closely with their digital partners and governments through the learning platforms. The strength of this approach is that it gives employers a clear understanding of the skills they need to build their business. Organizations must define precisely the skills that help workers move forward and show them the new career paths that can result from employees' time and effort. As technology and influence continue to transform today's labour market, developing new basic skills has become imperative, both at the personal and corporate levels.

At the same time, the new technology has demonstrated the ability to perform tasks that workers did not perform as well as they did in the past. While this trend continues, workers need to upgrade their existing skills and acquire new skills throughout their careers. This new environment requires an overhaul of the outdated model of education in which people could expect to acquire education and skills at a young age that will last their entire careers. Employers must also be given additional tools to acquire new skills and learning opportunities during their careers. Furthermore, employers play a unique and critical role in training workers, but most education efforts are not relevant to current and future needs. Moreover, employers, individuals, and universities must work together to modernize their education programs and find timely programs that are relevant and worthwhile. Although many, if not all, industries have already been destroyed, the good news is that researchers expect these jobs to increase overall employment opportunities. Infosys and the Milken Institute conducted a survey to find out what employees want from training programs and what employers need to stay competitive. At-the-job training is limited as an option for prospective workers. For workers already in higher-skilled positions, vocational training is more than twice as common, making those in more automated positions even more vulnerable to displacement. Digital technologies are key to reducing this gap, as tools can increase the reach and accessibility of training and skills tools and are at the top of the demand. Many of today's workers have not been trained since the 1970s, when the ubiquitous technologies of today did not exist. Nowadays, digital skills are the top priority to succeed in today's workplace for most of the young employees. Any college degree is good enough to give virtually anyone the best career path. As the global economy continues to undergo sustainable digital transformation, high-tech skills will become a passport to job and personnel security.

References

1. T.J. Sturgeon, Global Strategy Journal, **11(1)**, 34-57 (2021)
2. S. Bag, G. Yadav, P. Dhamija, K.K. Kataria, Journal of Cleaner Production, **281**, 125233 (2021)
3. J. J. Yun, X. Zhao, J. Wu, J.C. Yi, K. Park, W. Jung, Sustainability, **12(5)**, 1883 (2020)
4. S. Shan, Y. Luo, Y. Zhou, Y. Wei, Technology Analysis & Strategic Management, **31(4)**, 406-420 (2019)
5. H. Viswanathan, P.E. Mogensen, IEEE Access, **8**, 57063-57074 (2020)
6. H. Legenvre, M. Henke, H. Ruile, Journal of Purchasing and Supply Management, **26(1)**, 100596 (2020)
7. D.J. Teece, Research Policy, **47(8)**, 1367-1387 (2018)
8. W. Strielkowski, T. Veinbender, M. Tvaronavičienė, N. Lace, Economic Research-Ekonomska Istraživanja, **33(1)**, 788-803 (2020)
9. S. Dhir, R. Rajan, V. Ongsakul, R.A. Owusu, Z.U. Ahmed, Thunderbird International Business Review, **63(1)**, 43-61 (2021)
10. A.K. Awasthi, J. Li, Renewable and Sustainable Energy Reviews, **76**, 434-447 (2017)
11. M. Fromhold-Eisebith, G. Eisebith, World Development, **123**, 104614 (2019)
12. A. Borthakur, M. Govind, Journal of Environmental Planning and Management, **62(4)**, 717-740 (2019)
13. N. Abbas, Y. Zhang, A. Taherkordi, T. Skeie, Internet of Things Journal, **5(1)**, 450-465 (2017)
14. M. Kotarba, Foundations of Management, **10(1)**, 123-142 (2018)
15. D. Trabucchi, T. Buganza, E. Pellizzoni, Research-Technology Management, **60(2)**, 43-52 (2017)
16. D. Tolstoy, E.R. Nordman, S.M. Hånell, N. Özbek, Journal of World Business, 101165 (2020)
17. M. Zhang, M. Chen, *China's digital economy: Opportunities and risks* (2019)
18. D.N.Y. Mah, D.M.W. Cheung, Sustainability, **12(12)**, 4818 (2020).
19. K.S. Warner, M. Wäger, Long Range Planning, **52(3)**, 326-349 (2019)
20. A.V. Mavrodieva, R. Shaw, Sustainability, **12(5)**, 1893 (2020)
21. K. Rijswijk, L. Klerkx, J.A. Turner, NJAS-Wageningen Journal of Life Sciences, **90**, 100313 (2019)
22. N.N. Pokrovskaia, M.Y. Ababkova, D.A. Fedorov, Education Sciences, **9(3)**, 183 (2019)
23. V. März, G. Kelchtermans, K. Vermeir, Journal of Educational Change, **18(4)**, 439-464 (2017)
24. C. Zheng, J. Yuan, L. Zhu, Y. Zhang, Q. Shao, Journal of Cleaner Production, **258**, 120689 (2020)
25. C.S. Chai, The Asia-Pacific Education Researcher, **28(1)**, 5-13 (2019)

Soft monetary policy in a sustainable economy as a rigid game

Inna Čábelková[1,2]*

[1]Department of Trade and Finance, Faculty of Economics and Management, Czech University of Life Sciences Prague, Kamýcká 129, 165 00 Prague, Czech Republic
[2] Faculty of Humanities, Charles University in Prague. Pátkova 2137/5, 182 00 Prague, Czech Republic

Abstract. Over the past decade, central banks in a majority of developed countries injected astonishing amounts of money into national and international economies in the hope of helping real sectors and with worries of high inflation. Neither of these came true. This paper describes the reasons for such unusual dynamics using a case of the sustainable economy. The three factors considered include the increased propensity to save, the decreased money multiplier, and substantial growth in the financial markets. The mathematical model studies the effect of the new money created on the real sector via the effect on real consumption depending on the share of the new money received by the less and more affluent part of the population. The results suggest, that the higher is the proportion of new money allocated to the poorer part of the society, the higher is the effect of overall money on the real sector if the propensity to consume in this part of the population is held constant.

1 Introduction

After the financial crisis in 2007-2009, central banks of almost all developed countries practiced the policy of monetary easing and injected astonishing amounts of money into the relevant economies. In the 2020 year alone, the central banks of developed economies "printed" an astonishing $14 trillion on top of the trillions of dollars injected into the economies after, and in response to the crisis of 2007-2009 [1]. This enormous inflow of new money is traditionally associated with higher inflation [2, 3]. However, no significant inflation was observed. Moreover, most of the countries, the Czech Republic included, were balancing on the border of deflation [4]. Allen and Geels [5] in this respect explained the decoupling of monetary policy and inflation by an increased propensity to save and increased labor supply caused by globalization. Similar results were received by Van Doorslaer and Vermeiren [6].

This paper further analyses the association between inflation and money supply from the perspective of the effects of printing new money on the real sector consumption depending on the share of the new money given to less affluent parts of the population. Methodologically the paper relies on deductive argumentation and mathematical modeling.

* Corresponding author: cabelkova@pef.czu.cz

The results suggest that new money may be viewed as "perpetuum mobile" which, after the initial emission, are created and destroyed on financial markets, making little harm to the real economy in a form of inflation, and little good in a form of increased consumption. This Perpetuum mobile is however temporary. If most of the new money is directed to more affluent parts of the population, they are more likely to increase inflation in the financial markets while the positive effect on the real economy can be minimal.

2 Materials and methods

This paper relies on mathematical modeling and deductive argumentation to analyze the effects of money creation real consumption depending on the initial distribution of this money to "rich" and "poor" groups of the population that are differentiated according to their propensity to save and consume.

2.1 The inflationary effect of newly-emitted money: What has changed?

There are major factors that differentiate the current situation in developed countries from the one 20 years ago. First, in times of recession, endangered by losing their jobs or facing lowering incomes consumers postpone major purchasing of durable goods and reduce the expenses on necessities. The lowering aggregate demand pushes the prices down and creates the tendency for deflation. There was no such pronounced tendency in the countries where the printing money led to inflation or hyperinflation.

Second, the economic crisis increased the propensity to save [7, 8] at the expense of propensity to consume, caused by the unpredictable future and lower expectation for future incomes. Simply speaking, people and firms tried to return the debts if possible, save more money, and not take that many loans. On the other hand, the banks, faced with a higher risk of default on credits, had to make their criteria on loans stricter and choose more specifically to whom they will give the loan. These tendencies of both of the sides – the banks on one side and the people and firms on the other – led to the credit crunch, which reduced the money multiplier and effectively reduced the total amount of money in circulation. This money needed to be substituted by the money created by the central banks. In the times and countries with no efficient fractional reserve banking with extended reliance on loans, no money can be destroyed by not taking loans or returning the old ones.

The third reason why unprecedented money injections by the central banks did not lead to significant inflation in the developed economies over the last decade was the quick development of financial markets [9; 10]. Boosted by technological innovations, namely the invention of computers, the development of financial markets was unprecedented over the last 30 years. Not mentioning other roles, like enabling corporations to gain access to large amounts of capital without the necessity to apply for a loan in a commercial bank, enabling people to multiply their savings via investments in the financial markets, providing the financial institutions with a new source of income, etc., the financial markets offered new types of assets, on which people and institutions could spend their money, thus diverting this money from the real sector of the economy.

The idea is both simple and ingenious. The extra liquidity created by the central (and commercial) banks instead of going to the markets of goods and services, thus increasing demand and creating pressures for high inflation there, were redirected to the financial markets, where, they create the price bubbles, but the real sector, which most of the people care for more, stays unaffected. Till the time the financial bubble bursts, as happened in 1930, 2007 and several times in between. Yet, before that happens, financial markets can absorb vast amounts of liquidity.

3 Empirical model

As is stated above, this role of financial markets to absorb excessive liquidity might be viewed positively to avoid inflation in the real sector. However, it can be viewed negatively from the point of view of boosting demand, the lowering of which was caused by the inequality of distributions of incomes between imaginary employers and employees presented in a simple model above. From the point of view of insufficient real-sector demand, the tendency of newly printed money to escape to the financial markets significantly reduces the effectiveness of this monetary mechanism to regulate the real economy and exposes the world economy to the supposedly coming long-term recession from which there will be no cure. Similarly, the unavoidable burst of the bubbles on financial markets will eventually endanger the whole banking sector, which may have unpredictable negative effects on the real economy. The question arises on how to avoid these unfavorable outcomes. One, and maybe the only, way which was (again temporarily) effective in the past is again to bailout the problematic banks with the money of a) the government, thus creating even more debt of taxpayers and reducing their purchasing ability in the future even more or b) the central bank, thus fueling another portion of new money to the economy, which eventually will arrive at the financial markets and the whole process will repeat itself. The new Perpetuum mobile would be created following the stages presented in the following process (see diagram 1).

Given that the whole process implies both creation and destruction of the new money in financial markets and the process is likely to be repeated, the question arises whether after the ith iteration of the whole cycle the amount of money will increase, decrease or stay the same. If by this cycle the financial markets can destroy more money than they created, the whole system represents an unprecedented mechanism to effectively neutralize and eventually destroy the new inflationary money. This mechanism did not exist before financial markets were developed. If not, the whole system will operate with increasing amounts of money in each subsequent iteration. In any case, within the iteration, the mechanism presented in diagram one enables effectively divert inflationary money from the real economy.

An interesting view on the described cycle would be from the point of view of insider trading and overall strategies of operating and accumulation of capital on financial markets. Here, the big experienced players (or possibly the players with insider information) would have more probability to survive the burst of the financial bubbles either because of adequate investment strategies, because of more information, or because of the high probability to be bailed out by the government in the case of companies which are "too big to fail". On the other hand, small, inexperienced investors with limited or no bargaining power to be bailed out are likely to lose. Thus, from the point of view of the creation of inequality, the whole cycle is likely to prefer "rich" at the expense of "poor", thus perpetuating and reinforcing the inequality. The few exceptions would be the "poor" investors organized to the investment funds or pension funds, who are either experienced enough to foresee the burst of the bubble or big enough to reach for government support in the case of failure. The recent experience from the USA, however, makes the author be skeptical that the government will bail out the pension schemes at least in this country. The situation in Europe might be different.

Stage 1. New money (M) is created by the central bank due to the lack of demand and slowdown in the real sector economy.

Stage 2. This money will partially go to financial markets (M_f) with the proportion β, then:

$$M_f = \beta * M$$
$$\beta \leq 1$$

the rest will go to the real sector:

$$M_r = (1-\beta) * M$$

Stage 3. If the real economy does not correspond to the rise of the prices of assets on financial markets and does not provide sufficient footage for the growth of the asset prices the bubble on financial markets accumulates. The money created by the central bank that went to the financial markets is multiplied by the banking sector and via leverage trading. The resulting newly created amount of money that is operating on the financial markets equals to:

$$M_f = m * \beta * M$$

Where m is the real money multiplier and

$$1 \leq m \leq 1/RR$$

Where RR is the reserve ratio of commercial banks
Alternatively, if the currency drain ratio is taken into account

$$1 \leq m \leq (1 + CurrencyDrainRatio)/(CurrencyDrainRation + DesiredReserveRatio)$$

Stage 4. The bubble on financial markets bursts, significant wealth in the forms of assets is extinguished, but also a significant amount of money is deleted. To the knowledge of the author, there is no study quantifying whether the amount of money deleted was lower or greater than the initial amount of money created by the central bank and multiplied by the commercial financial institutions.

Stage 5. The central bank injects a new portion of the money into the economy to bail out the banks that were damaged on stage 4, and start the economy that is recovering from the last financial breakdown and the process starts anew.

Fig. 1. Money creation and destruction in the financial sector

Two things can be said about Figure 1: First of all, the author realizes, that the relationship of the outputs on the markets of goods and services and the prices of the assets on financial markets is not easy to describe and, in some time periods the financial markets seem to be completely detached from the markets of goods and services, which, according to the ideas of the author may correspond to the period of bubble creation, the author still has a hope to hypothesize, that if the real sector earning keep up with the earnings of the financial sector, the whole financial markets may be long term stable and not producing extended bubbles. This hypothesis, however is yet to be tested. Second, another assumption of this diagram is that the transfer of the liquidity from the market with real goods and services to the financial markets and back after the initial money injection of the central bank is zero.

The unresolved question from the last exercise that remained to explain is what determines the parameter β, namely, which proportion of the newly created money will stay in the real economy and will be invested in the real sector production, and which proportion will be invested (most likely with appropriate leverage) on the financial markets.

Arguably the capital, if not regulated, goes to the sector with the highest possible returns. So, if in the real sector the distribution of income is highly unequal, the aggregate demand for real sector products and services falls (due to the negative relation of propensity to consume an income, as was explained above), which, if no significant technological innovation reducing costs happen, will lead to lowering returns to capital in the real economy. On the other hand, the returns to capital in financial markets are highly dependent on the flow of liquidity that is invested there according to the principle of a financial bubble. More money available lead to higher demand, which, if the supply of the financial instruments stays the same or does not grow that quickly, will drive the prices of financial instruments up and bring higher returns for the holders of the financial instruments if they sell them (the returns from dividends are not following this logic as they are more dependent on the real outcomes of the enterprises). The bigger is the return to financial capital the more new money is invested, which increases the demand for financial instruments even more and the whole process repeats itself. The side effect is also that the capitalizations of the enterprises increase, meaning they can take even more credits and invest them into the financial markets again, which creates another smaller upward-directed spiral. Obviously, there will come a point when no new money is available, or when, for some reason, investors lose their trust in financial instruments. In this case the bubble bursts.

To describe the processes above more formally, imagine the following model:

Let M be new money injected into the economy possibly because of a new financial stimulus of the central bank. This money by some economic mechanisms, which we will not describe here will eventually arrive to two groups of people. The first, for the sake of the model, will be called "poor" and will get α*M of new money. The second, denoted as "rich", will get $(1-\alpha)$*M. So, α is the proportion of the whole money that arrives to poorer parts of the population and can be partly regulated by the government in various types of government policies. For example, so-called "helicopter money" that is highly discussed nowadays will more likely be targeted to the "poor" parts of the society or evenly for everybody. Similarly, all the ideas related to unconditional income or increase are also likely mostly to rich the "poor". An increase in state-financed retirement benefits will supposedly rich the whole population, but more effect on increased consumption will have in the poorer parts. The other possibilities are to increase the transfers for families with small children or other socially relevant policies. On the other hand, bailing out problematic commercial banks and other globalized firms will more likely help the "rich". It is necessary to note, that both of the types of policies increase budget deficit which will have to be repaid in the future by either increasing taxes, inflating the currency, currency devaluation (if the debt is in local currency but bought by foreign companies), or will be erased by the default on government debt as happened in Russia at the end of 90th.

Both groups (rich and poor) have two options what to do with the money – they can either consume or save (and eventually invest). The propensity to consume of the poor group of people (let us denote it βp) is likely to be much higher than that of the rich group of people (denoted as βr, for more discussion on income and propensity to consume see [11]).

To be even more precise, it is important to note that some people that we would put to the category of rich can spend all the fortunes they have and are left with significant debts. These stories are frequently discussed in the mass media where the main heroes are singers, prize or lottery winners, and other types of people that got their fortune quickly. However, arguably these people will not stay rich for long, and though their consumption may be enormous in individual merits, it may be viewed as minor in the merits of countries or nations.

For others, the proportions of new money, savings, and consumptions are listed in the following table (see Table 1).

Table 1. New money, consumption, and savings of "poor" and "rich" parts of the population

	"poor"	"rich"
New money	$\alpha*M$ (1)	$(1-\alpha)*M$ (4)
Consumption	$C_p= \beta_p*\alpha*M$ (2)	$C_r= \beta_r*(1-\alpha)*M$ (5)
Savings	$S_p= (1-\beta_p)*\alpha*M$ (3)	$S_r= (1-\beta_r)*(1-\alpha)*M$ (6)

In this case:

$$Consumption\ total = C_p + C_r = \beta_p*\alpha*M + \beta_r*(1-\alpha)*M = M*(\alpha*(\beta_p - \beta_r) + \beta_r) \quad (7)$$

And, as it was stated above:

$$\beta_p \geq \beta_r \quad (8)$$

From the formulas, 7 and 8 it follows that the new consumption generated by the new money is a positive (in our case linear) function of α (which is the share of new money that is received by the poor part of the population), in case if the propensity to consume of the poor group of people (β_p) greater than the propensity to consume if rich (β_r). This latter is the case (formula 8). Thus, the higher is the income inequality, represented by α, the lower is total consumption. This holds even in the case new money is printed.

4 Discussion

Let us be reminded, that lower consumption implies lower income of real sector firms, which, if not offset by the new technology that reduces costs of production, leads to lower returns to capital in the real sector and greater motivation to speculate on financial markets. Here it is also important to clarify one simplification that was made in the model and argumentation above. We did implicitly assume that most, or all the speculations on the financial markets do not affect the real economy, which is not always the case. One could imagine a situation when through financial markets a new high technology firm will get the funds for further development, which leads to more productive technology in the real world, which reduces the costs of production and increase returns to capital in the real economy. This was the former real meaning of financial markets – to design a mechanism to raise capital for the large projects in the real economy [12]. However, the development of the last 20 years and financial crises before (the famous crisis of the thirties of the last century and the smaller crises in between then and now) prove that financial markets are subject to the creation of significant bubbles, which divert astonishing amounts of liquidity from the real sector.

The other limitation of the argumentation above is that we do not consider the multiplication effects of the commercial banks which can multiply the money initially injected by the Central bank differently in case of consumption and investment. In fact, if the additional money is used for direct consumption, the immediate multiplication effect of the commercial banks caused by the fractional reserve system may be zero. The secondary multiplication effect may have its rationale in the idea, that the money, obtained by the seller of the good bought for the new money if stored in a bank, will increase its liquidity and possibility to issue more loans. The latter is the essence of a money multiplier. In the case of saving and investment, the money multiplication is more direct and obvious. New money may be included in the reserves which again will allow the banks to lend more.

The last limitation of the model and argumentation above is that we do not take into account the multiplication of the consumption after the first good bought for a newly created money is bought (and consumed). The seller of this goodwill gets the extra money which may increase his consumption as well, and so on. However, all the multiplication effects do not go against the main argumentation and just magnify the effects described above.

5 Conclusions

The emission of new money, or the strategies of central banks of quantitative easing bear hope of more economic growth, which, however, is often accompanied by higher inflation. The unprecedented quantitative easing and "printing" new money in most of the developed economies over the last 10 years did not bring either. This paper discussed the reasons why the economic dynamics in response to soft monetary policy over the last 10 years were so different from the one before. We suggested, that firstly, frightened by the last financial crisis, consumers postponed their purchases of goods, that were not considered necessary, thus diminishing the upward pressure on prices. Secondly, the increased propensity to save and return the loans taken before decreased the money multiplier which partly compensated the increase in the money base. Thirdly, the extensively growing financial markets absorbed excessive liquidity, again, diminishing the inflationary pressured on the market of real goods and services. The adverse side of this process is the limited effect of the new money on the real sector of the economy (as opposed to the financial one).

This paper developed a model, that shows, that the distribution of the effects of new money on the financial and real sectors depend on the share of this money received by more and less affluent parts of the public. The results suggest, that the effect of "printing" new money on the sustainable economy is highly contingent upon the share of the new money that reaches more and less affluent parts of the population. This paper may be interesting for both researchers and policymakers as well as sustainable economy specialsts.

References

1. A. Murphy, *Economics-2020: What Happens When Everything Shuts Down Except the 'Money Printing Presses'* (2020).
2. E. Dean, J. Elardo, M. Green, B. Wilson, S. Berger, *Monetary Policy and Economic Outcomes. Principles of Economics: Scarcity and Social Provisioning* (2020).
3. F. Collard, H. Dellas, Journal of Money, Credit and Banking, **39(2-3),** 713-731 (2007).
4. J. H. Christensen, J. M. Gamble, J, S. Zhu, FRBSF Economic Letter, **11**, 1-5. (2020).
5. E. M. Allen, H. J. Geels, *https://papers.ssrn.com/sol3/papers.cfm?abstract_id=3763499* (2021)
6. H. Van Doorslaer, M. Vermeiren, New Political Economy, 1-20 (2020)
7. I. J. Dos Santos Felipe, *Individual Behaviors and Technologies for Financial Innovations* (2019)
8. O. Primierova, I. Abakumova. Naukovi zapiski NaUKMA: Econimichni Nauki, **159**, 72-78 (2014)
9. Y. Li, S. Wang, Q. Zhao, Finance Research Letters, 101642 (2020)
10. D. Schell, M. Wang, T. L. D. Huynh, Journal of Behavioral and Experimental Finance, **27**, 100349 (2020)

https://doi.org/10.1051/e3sconf/202125006005

11. R. E. Hall, J. B. Taylor, *Consumption and Income. Macroeconomics: Theory, Performance, and Policy* (1986)
12. K. Pilbeam, *Finance and Financial Markets* (2010)

Digital business models and company growth opportunities in the energy market

Boris Kovalenko[1*], *Elena* Kovalenko[2], *Tamara* Yakovleva[1]

[1]Institute of Economics and Management, Herzen State Pedagogical University, Moika str. 48, 191186 Saint Petersburg, Russian Federation
[2]"ComplexSnab" Co Ltd., Vedeneeva str. 12, 195427 Saint Petersburg, Russian Federation

Abstract. We are currently witnessing significant changes in the global energy market. The energy industry is entering the stage of the 4th energy transition, which is characterized by an increasingly large-scale increase in the use of renewable energy sources and a decrease in the share of fossil fuels. The energy sector is also strongly influenced by the trend of digital transformation and the use of digital technologies. The aim of the work is to study the possibility and conditions of using digital business models by companies in the energy sector to improve competitiveness and market growth. To achieve the goal of the study, the following tasks were set: to analyze the main trends and the state of digitalization of the energy industry; analyze digital business models and assess the possibilities of their use by energy companies; to formulate approaches to transforming the activities of energy companies in the transition to a digital business model. The authors hypothesized that the use of digital business models will allow energy companies to remain competitive and gain access to new markets by introducing new technologies. To conduct the study, the methods of microeconomic and industry analysis, systemic and comparative analysis, and analysis of the organizational behavior of the company were used. The results of the analysis showed that the global energy market is characterized by growth in dynamics and volatility. In order to adapt to changing conditions and maintain competitiveness, energy companies need to take advantage of digitalization and shape a digital strategy. One of the basic elements of an organization's digital strategy is a successful digital business model. The article discusses the types and features of digital business models, and also formulates approaches to transforming the activities of energy companies in the transition to a digital business model.

1 Introduction

Currently, digital technologies (social, mobile, analytical, cloud, Internet of things, etc.) are penetrating all spheres of society [1]. Artificial intelligence and robots are changing the look of industries [2, 3] and the way organizations operate. The digitalization processes are

*Corresponding author: kovalenkoboris2015@gmail.com

actively developing in the energy industry. Digitization enables companies to expand technical capabilities to improve customer engagement and satisfaction, streamline business processes, and increase employee engagement [4- 6].

The results of the activities of companies in different industries show that for business differentiation, it is necessary to provide customers not only with better quality or lower prices, but to a greater extent offer a better business model. Thus, in order to succeed in the competitive struggle, companies radically change their business models [7]. The purpose of the article is to analyze digital business models that companies use to increase competitiveness in the market and identify approaches to transforming the activities of energy companies in the transition to a digital business model.

2 Digital technologies and the energy market

The energy market is increasingly influenced by factors such as the decarbonization of the world economy, such as the response to climate change and global warming; distributed generation; implementation of projects aimed at energy saving and energy efficiency growth; an increase in the number of electric vehicles; use of energy storage systems (batteries and fuel cells) [8-12].

Thus, according to Navigation Research, Bloomberg Energy Finance [4], more than 200 MW of new generating capacities were commissioned in the world in 2017. If we consider the structure, then almost equal amounts of centralized generation and distributed generation capacities were introduced. Distributed generation is expected to reach more than 250 MW by 2025, while centralized generation will be even lower than the 2017 level (less than 100 MW).

The energy sector is also strongly influenced by the trend associated with digital transformation and the use of digital technologies (cloud, big data, mobile, Internet of things, artificial intelligence, etc.) in the economy. Energy companies are already actively using digital technologies to automate all stages of the value chain (Table 1).

Table 1. Digital solutions used in energy companies to automate the value chain [5]

Digital solutions	Use in energy companies
Drones and industrial robots	Southern Company, Swissgrid, Snam, RTE
AI and predictive analytics	DUNE ENERGY, PG&E, Exelon, TATA POWER
Using chatbots for customer service	Exelon, E.ON, Göteborg Energi
Customer profiling based on BigData	Innogy, HydroOttawa, LondonHydro

The total volume of the global market for digital technologies in the energy sector in 2017 amounted to $ 52 billion. The market is expected to grow to $ 64 billion by 2025. Areas of automation include maintenance of power plants, installation and operation of "smart" meters, automation of distribution networks, operation of home energy management systems, ensuring the stability of the operation of energy facilities, etc. [4].

The analysis shows that digital technologies are already successfully used in the energy sector to automate tasks for managing the current operating activities of enterprises: asset management, maintenance and repair, procurement, billing and other basic business processes [6].

In addition, digitalization contributes to the development of new energy technologies, such as distributed integration, and makes it possible to coordinate the work of complex energy systems.

According to Kearney [13], there is a trend towards decentralizing the value chain of energy enterprises and expanding the list of services that they can offer to consumers. Personalization of customer service and provision of services together with partners from other industries is considered as a strategic goal.

To successfully adapt to the ongoing changes, as well as to use the opportunities provided by digitalization, it is advisable for companies to form a corporate strategy that allows them to offer new customer value that meets or anticipates customer expectations. In this case, it becomes necessary to adjust the current business model and transform it into a digital business model.

3 Features of digital business models

There is a plethora of definitions of the concept of "business model". We will use the definition formulated by Osterwalder and Pigne [14] that a business model serves to describe the basic principles of the creation, development and successful operation of an organization.

The results of the analysis show that business models generated by digital technologies can be classified according to different criteria. Consider some of these criteria and their associated business models. There is an approach that involves two possible types of business models: a customer acquisition model or a digital decision model [7].

The business model of attraction involves a comprehensive, individualized approach that ensures customer loyalty. This business model focuses on building trust and customer loyalty in order to achieve high brand loyalty. The implementation of this model involves multichannel interaction with customers, a deep understanding of their needs and rapid response to changes in these needs (Kaiser Permanente). The business model of digital solutions is since customers are provided with products or services with the addition of information, thereby creating a new value.

This business model means the integrated nature of the products and services that are offered to customers in order to meet their needs. Moreover, the strategy is aimed at gradually moving away from the sale of individual products to more complex products with additional value (Shindler Group, Apple). The successful implementation of such a business model requires a digital operating platform that combines all corporate functions in a single unit. The goal is to ensure access for all corporate units to a single database of customers in order to increase their satisfaction and loyalty. In addition, the presence of a corporate culture loyal to digital transformation is important.

Following a dynamic approach, when any kind of business model can be the starting position for the company, and the other kind can be a guideline, we can consider the following typologies of business models. For example, Weill and Woerner [15] offer an approach based on changing the business model in two directions: i) movement in the direction from control in the value chain to participation in the ecosystem; and ii) movement from low customer awareness to full knowledge (purchase history, goals, etc.). As a result of the intersection of these areas, four digital business models can be distinguished:

1. Supplier (P&G, Vanguard, Chubb Group):
- Implementation of the product through intermediaries;
- Low impact on the end user;

- Corporate competencies are based on cost reduction and gradual innovation.

2. Modular manufacturer (PayPal, Apple Pay, Square):

- Products / services ready for use;
- The ability to connect to any platform;
- Corporate competencies require continuous innovation for market leadership.

3.Omnichannelity (Nordstrom, Walmart, Citibank):

- Wide multi-product and multi-channel interaction with customers in accordance with their life cycles;
- Close relationship with the end user;
- Integrated value chains.

4. Ecosystem driver (Amazon, Microsoft, WeChat, Fidelity):

- The destination point of the ecosystem;
- Search for suppliers and competing products to satisfy customers;
- Use of customer data from all points of contact;
- Provide an excellent customer experience;
- Implement a rental approach.

According to the approach proposed by Linz et al. [16], two megatrends are currently influencing the activities of organizations - digitalization and service orientation – a services-oriented. Digitalization enhances the technical capabilities of firms to develop, produce, deliver their proposals and manage large-scale interactions with customers, the services-oriented approach represents a fundamental change in the field of value proposition for customers towards the joint creation of value and individualization.

The influence of the aforementioned mega-trends guides the decision-making process on choosing a business model in two dimensions: transaction coverage and proposal customization.

Moving along measurement data shows the capabilities that a company can use to develop or modify a business model. Following along the axis - transaction coverage shows the movement of the company from single and independent transactions (single products) to complex and interconnected transactions (related products) with repeated transactions.

Proposal customization is the movement from standardized, batch and automated offers to individualized, created jointly by the company and the client.

Table 2. Types of Digital Business Models [16]

Transaction Coverage - High	**Platform Oriented Business Model**	**Solution Oriented Business Model**
Transaction Coverage - Low	**Product Oriented Business Model**	**Project Oriented Business Model**
	Standardized level of offers customization	Individualized level of offers customization

As noted by Loucks et al. [17], there is an approach that involves the classification of business models depending on the type of value for the client that they provide: i) the value of the value; ii) the value of the interaction; iii) the value of the platform. Let us consider the features of these business models.

The value of the value: The most intense competition between companies occurs within the framework of this strategy. Companies greatly reduce the cost of the product for customers. Great opportunities for lowering prices are provided by product virtualization. In the framework of this approach, there are such business models as:

- Price transparency;
- Free / very low price;
- Reverse auction;
- Buyers association;
- Consumption-based pricing.

The value of interaction.

This approach is based on the fact that customers are offered a more convenient way to obtain a product or service. As digitalization intensifies, the value of interaction increases. This is due to the fact that now in the value chain you can select only those elements that attract the buyer and transfer them to digital devices.

The most successful digital companies are disaggregating the products of companies from different industries. They provide an opportunity for customers to select (and pay) only those products and services that are valuable to them, excluding other elements included in the chain and leading to an increase in prices.

Digitalization enables new companies to personalize their products and services and provide them at lower prices than traditional companies.

This approach complicates the activities of established companies that are accustomed to competing either through differentiation or brand. Examples of business models based on the value of interaction are:

- Empowerment of customers;
- Individualization;
- Receiving results in real time;
- Reduction of disagreements;
- Digital technology.

The value of the platform.

The value of the platform is a new form of competition, which arose precisely in the context of digital transformation. The business platform is non-linear and is determined by network effects.

Network effects are determined by Metcalfe's law, according to which the value of the network grows in proportion to the square of the number of users. These effects arise due to various forms of interaction between network participants:

- Dependence on each other;
- Equal relations with each other;
- Interconnection in the process of game models;
- The presence of feedback between the participants

Due to the existence of such forms, the overall effect of network activity exceeds the results of interaction within the framework of individual areas of its functioning. Thus, the presence of the platform allows the client to get additional value associated with connecting him to the network. For companies, such a business model is very attractive.

In the event that the platform is effective, then it has a very high degree of competitiveness. The result of such an activity is the approach when "the winner gets everything." The owners of the platform earn the bulk of the profit in individual markets. Examples of such platforms are Apple, iTunes, Facebook, and other business models. Within this framework, the following business models are distinguished:

- Crowdsourcing;
- Ecosystem;
- Communities;

- Digital market;
- Data orchestrator.

The analysis shows that at present, companies use a variety of business models to maintain and enhance their competitiveness. For the energy companies, the problem arises of choosing the most relevant digital business model and developing a strategy for the transition from the current business model to a digital one.

4 Approaches to transformation the activities of energy companies in the transition to a digital business model

The choice of an acceptable digital model should be determined by the real capabilities of the company and how it wants to compete. In what directions do the leaders and organizations of the energy industry need to decide in order to move to the transformation of the business model in the direction of its digitalization? The analysis shows that, first of all, it is necessary to assess organizational competencies in the following areas, the characteristics of which are presented in Table 3 that follows.

Table 3. Directions for assessing the organizational competencies of energy companies for the transition to a digital business model

Organizational competencies	Directions for assessing
The impact of digitalization on company products /services	- A service or product is provided using the Internet; Whether there is an influence of companies in other industries on the product / service; - Is there a possibility of replacing the product / service with another digital product; Possible financial losses on the company's products / services from digitalization
Competitiveness	The quality and relevance of the information that the company places for customers; The possibility of cross-selling and organizing multi-channel services for customers; - The quality of the internal platform of the company; The ability to connect external suppliers, partners, customers to the platform
Current business model	- Linear model (single transactions, omnichannel, etc.);
Organizational effectiveness	- The possibility of multi-channel customer experience; - Analytics about customers during their life cycle; - The possibility of expanding and developing partnerships; The presence of a brand that allows you to dominate in case the client needs a product / service
Corporate Culture and Leadership	The deep competencies of the Board of Directors and the General Director in the field of digitalization; The impact of the director of information technology on the activities of the Board of Directors; - Organizational flexibility and adaptability; - The propensity of organizational culture to digitalization

The assessment will allow leaders to understand which business model is current and what level of digital culture the company has in order to begin the transformation towards a digital business model.

Once the assessment has been completed, a vision for a new digital business model, a digital strategy and an implementation plan must be formulated. An analysis of the experience of companies that have carried out the transformation of business models shows that a digital strategy is not just measures to automate individual functions. Digital transformation must be part of the corporate strategy and requires the active involvement of the company's top management.

When developing a digital strategy, it is necessary to determine the main directions of change and the necessary resources (financial, technical, human, etc.) for the transition to a new business model. It is important to form a cross-functional team of specialists who will launch change projects. In addition, it is necessary to develop the digital skills and competencies of all employees in the company.

Many companies introduce additional levels of hierarchy for technicians: at one level, there are specialists who perform standard tasks, and at a higher level, those who perform more complex creative tasks.

For the implementation of projects related to the expansion of the list of services of energy enterprises, it is advisable to interact with technology companies and start-ups. These joint ventures enable energy companies to leverage digital skills and know-how that they lack.

5 Conclusions

Overall, it appears to us that the success of energy companies in the digital environment presupposes not a simple use of digital technologies to automate certain functional areas, but a change in production and management business processes, value chains, and ways of doing business.

It becomes apparent that digitalization provides energy companies and their leaders with opportunities to qualitatively change their business and increase its competitiveness through the introduction of digital business models. These models improve customer service and reduce costs through standardization and streamlining of operations.

Finally, we can conclude by stating that digital business models allow in the short term to improve the efficiency of current operating activities, and in the long term - to offer customers new customer value.

References

1. K. Shvab, N. Dehvis, *Tekhnologii chetvertoy promyshlennoy revolyutsii* (2018)
2. A. Ross, *Industrii buduschhego* (2016)
3. N. Bostrom, *Iskusstvennyi intellekt. Etapy. Ugrozy. Strategii* (2016)
4. Ministerstvo tsifrovogo razvitiya, svyazi i massovykh kommunikatsii RF, *https://digital.gov.ru/uploaded/files/tsifrovaya-energetika16x915.pdf* (2019)
5. Assotsiatsiya Tsifrovaya Energetika, *https://www.digital-energy.ru/wp-content/uploads/2020/04/strategiya-tsifrovoy-transformatsii-elektroenergetiki.pdf* (2020)
6. TADVISER, *https://www.tadviser.ru/index.php* (2020)
7. MITSloan Management Review, T*sifrovizatsiya: Prakticheskie rekomendatsii po perevodu biznesa na tsifrovye tekhnologii* (2019)
8. V. Smil, *Energy Transitions: Global and National Perspectives* (2017)

9. D. Ergin, *Novaya karta mira: Energeticheskie resursy, menyayuschchiisya klimat i stolknovenie natsi* (2021)
10. D. Ergin, *V poiskakh energii: Resursnye voiny, novye tekhnologii i buduschee energetiki* (2017)
11. V. Smil, *Energetika: mify i real'nost'* (2012)
12. V. Smil, *Energika i tsivilizatsiya* (2020)
13. RBK, *https://trends.rbc.ru/trends/innovation/5d6796719a7947b5b36a5972* (2020)
14. A. Osterval'der, I. Pin'e, *Postroenie biznes-modelei. Nastol'naya kniga stratega i novatora* (2012)
15. P. Vail, S. Vorner, *Tsifrovaya transformatsiya biznesa: Izmenenie biznes-modeli dlya organizatsii novogo pokoleniya* (2019)
16. K. Lints, G. Myuller-Stivens, A. Tsimmerman, *Radikal'noe izmenenie biznes-modeli: Adaptatsiya i vyzhivanie v konkurentnoi srede* (2019)
17. D. Lauks, D. Makolei, E.H. Noronkha, M. Uehid, *Kak pobezhdat' didzhital-novatorov ikh zhe oruzhiem* (2016)

Small business and its place in promoting sustainable development

Elena Korneeva[1,2*], *Natalia* Skornichenko[3*], *Tatiana* Oruch[3*]

[1] Financial University under the Government of the Russian Federation, 49, Leningradsky Prospekt, Moscow, 125993, Russian Federation
[2] Togliatti State University, 14, Belorusskaya St., Togliatti, 445667, Russian Federation
[3] Volga Region State University of Service, 4, Gagarina str., Togliatti, 445017, Russian Federation

Abstract. Sustainable development is becoming a very important issue in the 21st century. Facing global changes such as the global warming, global climate change, as well as other pressing issue, all spheres of economy and social life need to take part in mitigating them and preventing disasters from happening. Our article studies the role of the small business in the above processes and the place of the small business in promoting sustainable development through its actions, public and social awareness and responsibility. We show how even small and medium enterprises can become a decisive power in tackling the climate change and promoting green thinking and sustainable awareness. This can be achieved through enhancing social responsibility of business companies which can greatly contribute to supporting the Sustainable Development Goals (SDG) and promoting sustainable economic growth.

1 Introduction

In general terms, the objective models of development chosen by individuals, communities and societies for themselves and their communities concern living together on one planet and coexisting in one space and time [1, 2]. Each community or society has set itself the objective of achieving the objectives of the development model it chooses, taking into account the needs of its citizens and the social and economic conditions of that community, community and society. At the operational level, social sustainability results from actions in key thematic areas that cover the social sphere of individuals and society, from capacity building and capacity development to environmental and spatial inequality. Indeed, the term "social sustainability" and its contributions include a shared enthusiasm for sustainable development and a commitment to social and economic sustainability [3, 4]. New issues of sustainability, such as climate change, environmental degradation and human rights violations, as well as economic and social inequality and poverty are becoming more pressing with each year. As Nobel laureate Amartya Sen reminded us, "social sustainability" is about taking into account the human aspects of development in everything, including the economy, as part of a holistic approach to economic and social development, and about not only

* Corresponding author: korneeva1207@yandex.ru

economic sustainability, but also the need to take into account the humanitarian aspects of development [5, 6].

The concept of "social sustainability" can be called unfair because it focuses on economic sustainability, but not on social justice, which relates to the fairness of all people. There are, in fact, other aspects of economic and social development which, significantly, also belong to the area of social sustainability. Sustainable development requires that we are involved in decisions that affect our communities, not only in economic development, but also in social development. Promoting social justice requires progress by countries, organizations and communities whose social well-being can be sustained in the long term. Social sustainability is the ability to achieve lasting good social improvements for all [7]. In addition, social sustainability ensures that the most disadvantaged communities are better off and can maintain their economic and social rights in the long term, regardless of their socio-economic status. Real sustainability and a true circular economy can only be achieved if the economic, social and environmental aspects of sustainable development - economic growth, economic development and social justice are balanced [8].

Sustainable development is economic development that takes place without the depletion of natural resources. Sustainable development can be promoted including the tripartite partnership with such international powerful organizations as the World Bank, the International Monetary Fund, and, for example, the European Union, which in turn include tripartite partnerships with governments, NGOs, the private sector, and civil society. Such partnerships establish the relationship between economic and social development in the context of sustainable development. Sustainable development is called a "sustainable development society" and is linked to economic, social, environmental and cultural development, as well as the development of a sustainable economy and society [9]. The agreements can be translated into concrete actions with tangible results and we are making progress in implementing these goals in line with the United Nations Sustainable Development Goals. The objectives include economic, social, environmental, health and social development, as well as economic development. Innovative technologies and creative ideas are brought into play to capture an expanded range of values and disciplines. It is essentially considered development that meets the needs of the present without affecting the capabilities of future generations. Sustainable development requires improving the quality of life for all by improving access to health, education, employment, housing, health care and other basic services [10, 11].

Today's interconnected global economic system requires an integrated approach to promoting responsible, long-term growth, to ensure that no nation or community is left behind. To preserve our environmental heritage and natural resources for future generations and to preserve natural habitats, economically viable solutions need to be developed to reduce resource consumption, stop pollution and preserve natural habitats in line with the objectives of sustainable development. People worldwide need access to clean water, clean air, clean drinking water and healthy food. This includes the use of renewable energy sources such as wind, solar and hydro power, as well as the development of sustainable agriculture. Each of these pieces is part of the puzzle of a sustainable society, as they are fundamental components of daily life. Sustainable development is the process of thinking, planning and implementing ways to achieve development that is economically, socially and ecologically sustainable. This represents the need to develop a number of ways in which development can be achieved through the use of renewable energy, sustainable agriculture and sustainable water and energy sources. A second, more recent approach suggests that the area of sustainability is social in nature, including environmental, economic, political, and cultural sustainability. While the ecological sphere defines the embedding of humans in the environment, the areas of social sustainability depend on relationships between the social and the natural. Social sustainability is achieved by actively supporting the development of healthy communities that are worth living in, and that create a sustainable environment for human health, well-being and economic prosperity.

Our article aims at assessing the place and the role of the small business in fostering and promoting sustainable development. We show that small business companies might be as effective as large corporations when practicing corporate social responsibility and playing by the rules of sustainable economy.

2 Sustainable development and small business

Sustainable development truly penetrates all spheres of economy and social life. For example, inclusive companies can appeal to low-income markets while generating reasonable profits and having a tangible impact on development. The question is how can small businesses contribute to these ambitious development goals and remain financially sustainable? [12]. The engagement can be transformative and put the world on a path to inclusive, sustainable, and resilient development. Sustainable supply chains ensure that business companies are offered greater opportunities to innovate in new markets and open them up to gain a competitive advantage in international trade. We need to be able to bridge the gap between the product demand, which meets the needs of people in poverty, and the supply chain. New markets open up new business opportunities as we are ultimately the key to the business growth and success. Connecting to these goals would save many small and medium enterprises (SMEs) time and money, improve their brand image, build resilience to uncertainty and save time, money and time for all SMEs involved. It is more than just the responsibility of an SME owner to build a business to shape the economic, social and environmental sustainability that prevails in the world, and it is now her or his own social responsibility as a citizen and an entrepreneur [13].

Half of small businesses consider it important to lead the company into a sustainable future without even thinking about it. For example, Nordic companies with 0-19% employees (in fact, SMEs) have difficulty understanding the governance part of ESG, especially when they come across governance parts of it [14]. This is due to the fact that in the Nordic countries sustainable development became a part of the primary school curricula and is accompanying people all their lives. Small businesses employ up to 90% of workers in developing countries, so it is not difficult to see how important it is for growing companies to participate in the agenda [15]. Employees of small and medium-sized enterprises often feel strongly about the sustainability, and some say that sustainability is a moral imperative. If one is a leader who wants to improve her or his environment, friendly efforts or an employee who wants to motivate her or his team, there are steps to help to create a sustainability plan for any SME. This can make a difference when it comes to helping our planet. One can start by creating a "sustainability plan" that guides her or his activities and outlines the goals and future plans.

The good news is that there is growing evidence, from the organizations such as the United States Environmental Protection Agency (EPA), that provide certified embedded systems, that reducing one's SME's carbon footprint can also help improve the outcome [16]. It is often said that in the 21st century, SME customers are increasingly committed to sustainability. Thence, SMEs that apply sustainable business practices have increased the number of employees with at least two years of experience in sustainability and have reduced costs [17]. The benefits of integrating sustainability into a company's core business strategy and reporting on it can more than offset the reduced risk of not being able to achieve any company's sustainability goals in the next five to ten years. Sustainable Business and Sustainability Development in Small Business (SMP) provides university research on the impact of sustainability on small businesses and sustainability development. It also becomes apparent that SMEs can also work together to remove obstacles to achieving environmental and economic sustainability by learning and exchanging ideas for sustainable businesses [18]. Development of practical business strategies to promote the environmental, economic and sustainability of existing and new companies and organisations, as well as development of tools for success in the emerging sustainable economy. Several recent studies on corporate citizenship found that over two-thirds of millennials believe that the private sector has a very important role to play in achieving global goals to which companies of all sizes can contribute [19]. The Sustainable Development Goals (SDGs) are very important in the above context. It is a set of 17 "global goals" led by the United Nations to change the world around a common agenda for sustainable development. These objectives provide a framework for formulating a global vision for the future of the global economy and the development of sustainable economies.

3 Social responsibility for sustainable development

As it has been already noted above, sustainable development and corporate social responsibility are closely interlinked business concepts that have greatly influenced corporate governance in the early 21st century [20]. Socially responsible companies conserve the ecological resources that are crucial for future generations. Corporate Social Responsibility (CSR) means that corporate and environmental responsibility must be reconciled with the return to communities that support business and long-term business success [21, 22]. The development of corporate social responsibility has been driven in part by increasing consumer social awareness and accountability expectations. Each of the four pillars of corporate sustainability has a direct impact on the long-term sustainability of a company's business model and operations. In recent years, there has been a shift away from traditional growth, profit, and maximization models to a more sustainable business model. The term "paradigm" is used because corporate sustainability is a combination of the four pillars of corporate social responsibility (CSR) and sustainability management. This is an evolving concept that managers have embraced as an alternative to the traditional model of "growth" and "profit maximization," according to a recent report by the Centre for Sustainable Business. Sustainable development is a comprehensive dialectical concept that balances the need for economic growth with environmental protection and social justice. A glance at the literature suggests that the concept of sustainability of companies takes up elements of established concepts. Awareness of the importance of business growth and profitability requires companies to adopt sustainable development strategies, especially those related to environmental sustainability, human rights, social justice, and economic development [23].

In recent years, big business has joined the ranks of development agencies and actors involved in promoting sustainable development. The term was popularized in a report by the World Commission on Environment and Development (WCEC). The situation that is developing is in stark contrast to the past, when big business was less concerned about the environment and more concerned about the welfare of its employees. The growing attachment of senior citizens and the emergence of a new generation of leaders suggest that all divisions are beginning to reshape their relationships with both the environment and their diverse stakeholders. A closer look at the reasons why these sectors and companies are changing shows that the answer has less to do with new ethical concerns that have been found than with changing the nature of their business model. These include so-called win-win opportunities for improved competitive advantage, such as the way production and marketing are organized globally, and new business models.

Although such drivers could encourage companies to focus more on environmental and social concerns, we argue that this process of change is likely to remain fairly fragmented and unevenly distributed across companies, countries, and sectors. As millennials and Gen-Z move into the investor space occupied by Gen-X and boomers, corporate social responsibility for securing investment is likely to become even more important. Companies need CSR plans that are focused on longevity, flexibility and responsiveness. There is a clear need to strengthen trends in corporate environmental or social responsibility so that companies can make more meaningful contributions to sustainable development. When different business units have different priorities and objectives that may be aligned, companies may have difficulty getting everyone on the same page. It is vital that companies form a united front in their positions on political, environmental and social issues. Social responsibility can be as simple as applying soft legal principles to promote universal ethical standards in the relationship between private and public enterprises. A unified approach to corporate social responsibility for sustainable development can help ensure the long-term sustainability of the global economy and the future of our planet.

This is also called corporate social responsibility, and refers to a form of self-regulation that integrates various disciplines, including business ethics, human rights, social justice, environmental sustainability and corporate governance. You can donate money, strive to reduce your environmental footprint, implement fair hiring practices, and only work with suppliers with similar values. One can see that the term "Corporate Social Responsibility" is very broad and complex and has different meanings. According to UNIDO, CSR is a form of self-regulation that integrates social and environmental concerns and is seen as a way for companies to achieve a balance between economic, environmental and social needs. Companies take on activities with the aim of the welfare of society, but they have a different meaning. The economic, legal and ethical responsibility of a company is the value it creates for society and for various stakeholders, including its human resources. Many commentators have argued that a CSR agenda has the potential to bring businesses to businesses. In addition, it might be that achieving leadership in business unit is becoming crucial for the success of the Corporate Social Responsibility agenda for sustainable development.

4 Conclusions

Therefore, it is apparent that sustainable development is required to achieve a prosperous, inclusive and environmentally sustainable society and to reach economic, social, cultural, environmental and social justice and economic growth. The global economy has created great wealth and technological miracles, but it has also created a society deeply divided between the haves and have-nots - an economy that has destroyed the natural environment and even threatened humanity in order to spur economic growth. The basic idea of sustainable development is that economic growth is not enough for human well-being. It makes no sense to promote an economy in which the poor are harmed by the exploitation of the rich and which threatens the environment, human health and the livelihoods of millions of people, especially in poor countries.

The sustainability of an economy depends on the introduction of appropriate production, distribution, and consumption to promote environmental sustainability, as well as on economic growth. This is why small business enterprises that often constitute the backbone of any economy (in fact making up to two-thirds of all enterprises in some economies, such as, for example, in Southern European countries) should adapt this concept and embed it into their daily functioning.

In addition, everyone has to understand that sustainable development is not only understood from the point of view of economic development, but also in the context of social and environmental development. This has become the core of mainstream sustainability thinking and a key element of the Brundtland report's analysis. The importance of integrating the social dimension has been recognised, but what is its role in defining sustainable development? It builds on indicators developed since 1996, which focus primarily on economic and environmental interactions. This work reflects a broader catalogue of literature that takes into account sustainable development indicators that use the three pillars that seem to be building over time. The hypothetical case mentioned above illustrates the importance of the integration of three interrelated areas of sustainability in the concept of sustainable development and their role in the development of sustainable development agenda. This example may be overly simplistic, but it contextualizes how economic, social, and environmental sustainability substrates can interrelate and promote sustainable business and balanced economic growth.

References

1. B. Singh, M. Keitsch, M. Shrestha, Sustainable Development, **27(2)**, 237-244 (2019)
2. E. Tarkhanova, Terra Economicus, **16(2)**, 75-82
3. W. Strielkowski, E. Lisin, E. Astachova, Entrepreneurship and Sustainability Issues, **4(4)**, 591-600 (2017)
4. V. Amrutha, S. Geetha, Journal of Cleaner Production, **247**, 119131 (2020)
5. W. Holford, Futures, **105**, 143-154 (2019)
6. V. Ivlev, M. Ivleva, International Conference on Contemporary Education, Social Sciences and Ecological Studies, 869-873 (2018)
7. J. Fleetwood, Sustainability, **12(12)**, 5027 (2020)
8. V. Panchenko, Y. Harust, Y. Us, O. Korobets, V. Pavlyk, Marketing and Management of Innovations, **1**, 256-264 (2020)
9. H. Kopnina, Education Sciences, **10(10)**, 261 (2020)
10. W. Leal Filho, S. Tripathi, J. Andrade Guerra, R. Giné-Garriga, V. Orlovic Lovren, J. Willats, International Journal of Sustainable Development & World Ecology, **26(2)**, 179-190 (2019)
11. E. Lisin, D. Shuvalova, I. Volkova, W. Strielkowski, Sustainability, **10(4)**, 1111 (2018)
12. J. Rendtorff, Local Economy, **34(6)**, 510-524 (2019)
13. D. Ortiz-Avram, J. Domnanovich, C. Kronenberg, M. Scholz, Journal of Cleaner Production, **201**, 254-271 (2018)
14. S. Gillan, A. Koch, L. Starks, Journal of Corporate Finance, 101889 (2021)
15. S. Raza, M. Minai, A. Zain, T. Tariq, F. Khuwaja, International Journal of Entrepreneurship, **22(4)**, 1-13 (2018)
16. G. Wu, Corporate Social Responsibility and Environmental Management, **24(6)**, 661-675 (2017)
17. M. Kimanzi, V. Gamede, International Journal of Economics and Finance, **12(2)**, 453-468 (2020)
18. S. Kraus, J. Burtscher, T. Niemand, N. Roig-Tierno, P. Syrjä, Sustainability, **9(10)**, 1828 (2017)
19. S. Ghosh, J. Rajan, International Journal of Sustainable Development & World Ecology, **26(4)**, 344-353 (2019)
20. A. Scherer, C. Voegtlin, Academy of Management Perspectives, **34(2)**, 182-208 (2020)
21. J. Lu, L. Ren, J. Qiao, S. Yao, W. Strielkowski, J. Streimikis, Sustainability, **11(15)**, 4128 (2019)
22. G. Barauskaite, D. Streimikiene, Corporate Social Responsibility and Environmental Management, **28(1)**, 278-287 (2021)
23. A. Androniceanu, Amfiteatru Economic, **21(52)**, 503-519 (2019)

Sustainability of micro-enterprises in the digital economy

Oksana Mukhoryanova[1]*, *Larisa* Kuleshova[2], *Nina* Rusakova[3], and *Olga* Mirgorodskaya[4]

[1] North-Caucasus Federal University, Pushkin str., 1, Stavropol, 355009, Russian Federation
[2] Stavropol State Agrarian University, Zootechnical line, Stavropol, 355017, Russian Federation
[3] Rostov State Transport University, Rostov Rifle Regiment of the People's Militia Square, house 2, Rostov-on-Don, 344038, Rostov Region, Russian Federation
[4] Stavropol Institute of Cooperation (branch) of BUKEP, Stavropol, Goleneva str., 36, Stavropol, 355035, Russian Federation

Abstract. This paper aims at investigating the predisposition leading to the sustainability of micro-enterprises in the digital economy, especially the sharing economy. This area represents a new field since the research of the impact of the sharing economy on small enterprises is still in its infancy. We study the role of the entrepreneurial approach and entrepreneurial philosophy of the small business with regard to the digitalization and the sustainable development and growth using examples from the European Union and the United States. Some common features and trends are derived and the outcomes are discussed. Our results point at the fact that by creating an economy for micro-entrepreneurs, the sharing economy thrives on traditional industry disrupted by technology. Since micro-enterprises constitute a backbone of the economy in many developed and developing countries, more research is required to shed the light of the sustainable development of these types of enterprises in the globalized and digitalized world.

1 Introduction

Generally, micro and small entrepreneurs still do not see environmental management as a strategy for innovation that can help increase the profitability of the organisation and thus ensure its survival in the market [1, 2]. Small business has no incentive to meet environmental standards despite the growing market demand for sustainable business practices. A culture of innovation plays a crucial role in maintaining a company's competitiveness [3, 4]. For micro and small enterprises, this role is perceived as much more important, as it contributes to a longer life of the company by helping to reduce its fragility. Small and medium-sized enterprises have one major advantage over larger organisations when it comes to solving sustainability issues [5-7]. This means that they can react very quickly to changes in the business environment. While some enterprises are leaders in sustainable practices, others lack a strategic attitude that sees changes in the business environment as an opportunity rather

* Corresponding author: omukhorianova@ncfu.ru

than an additional burden on the task that requires further change. There are market changes that create sustainability opportunities for innovation and inspiration for employees.

Generally, the innovation process in this environment has emerged as a factor that can be crucial for the survival of micro and small enterprises. Environmental certification cane become an opportunity for small and medium-sized enterprises to prove their ability to enter new markets and to be competitive and recognized as environmentally friendly enterprises [8, 9]. What motivates those seeking official recognition is the legitimacy of what the competitive divide seeks to represent. Innovation in this environment is an instrument that can help small businesses to adhere to environmentally friendly production processes, but without significantly increasing their production and marketing costs. In the current market scenario, micro and small enterprises have adapted to the constant change and invested in innovation. The innovation process has proved to be a factor that has helped companies to hold their own and gain ground in increasingly competitive markets. On the contrary, the goal is to increase profits through sustainable production, and changes in corporate behaviour have taken place to not only increase profitability, but also ensure the survival of the market [10].

Management and control mechanisms influence the way companies produce and enable them to purchase new machinery that helps reduce pollutant emissions and enable them to produce products through environmentally sound extraction of raw materials. In this scenario, it is crucial to include micro and small enterprises in the products and services offered to consumers. As companies invest in environmental certification, competition differences become a growing demand for their products and a competitive gap becomes an incentive for companies to invest in environmental certification. Therefore, it is required to define the role of micro and small enterprises for the sustainability of the global economy and to bring their relevance to this scenario. Stakeholders are promoting the need to educate micro and small entrepreneurs about ways to increase profitability and protect our environment. The Rio de Janeiro Conference back in 1992 discussed the need to link our society's attitude to environmental issues and to create companies charged with ecologically sound production and marketing. With the growing trend of recent years, the organization can increase the sustainability of small businesses in the market by developing a sustainable business model [11, 12]. Environmental management practice has become a contribution to the transformation of the production process, which aims to ensure the sustainability of the environment and the health and well-being of people. These standards are based on the requirements of the United States Environmental Protection Agency (EPA) and the United Nations Environment Programme (UNEP). This change in position affects the strategic role that environmental management plays in our society, even if it is seen as a means of profit-making for capitalist systems [13, 14].

As mentioned above, we believe that it is crucial for public policy to encourage micro-enterprises to comply with environmental standards and to introduce environmental certificates, as these measures contribute to the survival of businesses and the preservation of the environment. In short, there is an urgent need to create a global movement for sustainability and allied innovation in the production process. If environmental certification is introduced but companies do not want to be certified, certification can be a factor that excludes them from the market.

2 Uberization of economy and economic systems

Currently, many researchers write about the "uberization" of the economy [15, 16]. It is symptomatic that the process of regulating joint enterprise activities is subject to constant review, based on empirical evidence that is exemplary in urban policies on Uber, a taxi-hailing app. The complexity and differentiation of sharing practices has gripped the entire urban economy, extending beyond selected sectors to activities that are not limited to traditional modes of transportation, but also include activities such as sharing scooters, originally designed for last-mile commuting.

The sharing economy can be used as an attribute of the city, where innovation thrives and leads to the creation of a common city and a new form of economic and social norms. Together with increased user protection, the economic benefits of sharing, combined with increased user protection, will contribute to the growth of the community and the sharing of these services, leading to faster growth and adoption of values that make sharing a social norm [17, 18]. One can emphasize the new position of the consumer in the sharing economy by defining him as a consumer who grants physical assets to each other, possibly in exchange for money. The benefits and access for consumers are that the sharing economy uses digital platforms to provide customers with access to a wide range of services such as transport, food, clothing and other goods and services. The European Union uses the term "collective consumption" as a reference to the changes that are coming with the emergence of a sharing paradigm [19]. To capitalise on the untapped capacity for trust and verification, the specifications of this new market must ensure that buyers and sellers are aligned. The digitally managed peer-to-peer nature of the sharing economy does not fit well with current laws and regulations.

Sharing economy services do not have to comply with certain rules, which, on the other hand, helps to make the services provided more affordable. This business model has had many growing problems and many ordinary employees have tried to stop the growth of business models. Many are not ready to switch to a model where others already use more than four platforms in their daily lives. Taxi services now have apps that allow consumers to call a car and pay for it, as with Uber and Lyft [20]. Sharing economy platforms can also popularise the use of transit and accommodation services and drive existing companies out of the market. For example, there were protests in Spain in the late 1990s and early 2000s, which turned quite violent at times. Spain has been more or less paralysed by the collapse of the taxi industry and the rise of private taxi companies such as Uber. This is because Uber offers a platform on which other companies can build, such as a fleet of pizza delivery vehicles. It seems that the rise of Uber has changed that, and that Uber is increasing the pie rather than cutting a thin slice of it. In addition to the algorithmic structure that Uber and other sharing economy platforms offer, companies operating on these platforms should also be considered as separate "Uber companies." The collection, processing and use of personal information is the closest study of consumers by digital and physical companies such as Uber and other companies [21].

Digital platforms facilitated by the Internet, such as social media, social networks, mobile apps, cloud computing, data analysis, etc., are restructuring more and more parts of the economy. But as we will examine in the next part, what we share may be more important than what is shared, and there is a reason for that. Just as people who raise privacy concerns about Uber and Airbnb (a property renting app) do so for the same reasons as those who raise privacy concerns about Facebook and Google, so do people who raise privacy concerns about Airbnb [22].

The promise of the sharing economy, rhetorically based on the idea of social reciprocity, continues today. Airbnb, for example, claims to be a community wherever a marketplace is created, where hosts auction off their vacant rooms to tourists. Uber helps underserved communities provide an alternative to more efficient transit access. The sharing economy is a bit difficult to define, but one definition can be defined roughly as one sees it here: you pay to access or rent an object, such as a condo, on Airbnb through an online marketplace.

All in all, trust in the economic community enables trade, while at the same time responding to the needs of local communities and even addressing historical injustices. The sharing economy enables the gig economy, in which people make money by doing a series of small jobs or "gigs." At first, performing meant paid work, such as musical engagements, which could range from a single musical performance of just one night to much more.

3 Sharing economy and micro-enterprises

Broadly speaking, sharing economy literally produces a vast quantity of micro-entrepreneurs. These entrepreneurs often start with a few hundred dollars or less in capital and are often in the early stages of their business. Many of the investors agree with the moral vision of a sharing economy [23].

Then, it is about drawing up a business plan to start in the shared economy, which is very different from traditional business models such as traditional taxi business, with very little traditional financial support, let alone taxes. The micro-entrepreneurs who make the most money on sharing platforms are also those with assets that are often rented out by similarly wealthy people. As for the cold hard net worth, the sharing economy has done little to bridge the wealth gap. It has raised working-class living standards by making it affordable for them to earn an additional income. Today, it is the second largest source of income in the United States, and is expected to be worth $335 billion by 2025 [24].

Sharing economy not only encourages micro-entrepreneurs, but also facilitates the creation of new jobs where there were none before. Faced with the ripple effects of the recession, more consumers are turning to the sharing economy to test the entrepreneurial potential. There is no doubt that certain technology companies are beginning to develop tools to democratize their software and applications to help the shared economy thrive in an ecosystem that could potentially be anyone's game. For consumers, entrepreneurs and businesses alike, the introduction of democratized access and control could potentially disrupt an already disruptive economy.

The network, which we call the "sharing economy," allows people to exchange material and immaterial goods on a large scale. Exchanging relationships undermines traditional retail employment by reducing transactional friction. The sharing economy lacks a common definition, but it is probably more accurate to divide it into several coherent but distinct areas. One can share the same office space with dozens of different companies, get unsecured personal loans directly from her colleagues, and stay with a stranger instead of a hotel when she travels outside the city. At least some variation in the sharing economy's themes is integrated into daily life, from the growing army of people who regularly use Uber to the proliferation of social networks like Facebook and Twitter.

When viewed from the perspective of those providing services, it is often argued that collaborative trading platforms have opened up the possibility for individuals to monetize previously untapped assets. Likewise, there are millions of people who use the services they buy and consume, such as food, clothing and other goods and services. While it is surely true that an increasing number of people are making decent money by renting spare rooms to travellers or offering inner-city commuters parking, cynics argue that the financial benefits for most sharing economy providers are limited.

In addition, there is also a considerable social and environmental impact of the sharing economy, showing that the sharing economy goes much further - and is more far-reaching than ever. Sharing and its effects are now a global phenomenon and could make an important contribution to combating climate change [25]. There are traditional defenders of the economy, but ultimately the battle is between asset owners and consumers. In short, sharing economy is very beneficial for micro-companies and sole traders since it reduces the transaction and incumbent costs allowing them to concentrate on their sustainable agenda and get rid off the redundant administration and red tape.

4 Entrepreneurial behaviour models and Internet platforms

Currently, very little is known about the entrepreneurial behaviour of individuals who start new businesses. To understand the factors that drive individuals to start a new business, this study attempts to assess these locations. In addition, it is interesting to follow the introduction on the entrepreneurship's intentions as assessed using concepts and theories contained in the IT deployment literature. Based on an analysis of the impact of the use of social media as a business platform on business activities and the factors influencing business intentions, a number of factors considered relevant to their impact on business intentions can be singled out. The perceived desirability is the factor that leads a person to start a business, whether that factor is intrapersonal, extra - personal or both [26, 27].

In recent decades, the development of the Internet the way people communicate and do business. The perceived feasibility indicates the degree to which the individual feels comfortable when starting his individual project, and the attitude to the expected returns is taken into account. As such sites have reached their peak popularity, their trends have evolved and flourished in recent years. Companies that focus on social media, social networks, web design, business development, and marketing include venture capital, venture capital and risk management using forecasting models. Harnessing AI's automation and forecasting capabilities can facilitate the process of venture capital creation by accelerating market studies and product testing, automating data collection through social media, and improving market targeting and positioning. A concrete example is a biotechnology company that improves the performance of research and development by using predictive models to find candidate molecules with optimal properties and then test large samples of these molecules. In this case, artificial intelligence has significantly reduced the company's research and development costs and also accelerated the time to market for its new product [28].

All these problems encourage marketers to find new ways to integrate digital technologies into market development. Since the 2010s, marketing automation has developed in response to a constantly changing marketing climate. Consumers have become increasingly aware of the importance of social media, social networks and other digital platforms for marketing and advertising. In addition, consumers are expected to search social media sites such as Facebook, Twitter, YouTube, Pinterest, Instagram and other social networks for product information. Video is also a powerful tool, but it is also increasingly being used in conjunction with social media by B2B companies looking to increase leads and conversions. While video is a great way to share product information, using video during the buyer's journey can also help companies further qualify B2B sales management capabilities [29].

All in all, social media, blogs and forums are all tools that companies can use to communicate effectively with customers. Businesses really benefit from building a strong customer network, so might be really important for companies to improve the way they identify customer problems and offer solutions that build trust and honest relationships with customers. Communication and networking also help to implement business-to-business communication and customer relationship management strategies for companies [30].

Social media has changed the way we communicate with some of the most commonly used business applications on social media such as recruiting, building employee engagement and communication, and expanding employee learning opportunities. Another important application of social media for employers is the use of the employee level as a platform for knowledge exchange. This can be presented in the form of a social network, with a focus on business-to-business communication, customer relationship management and networking. Social media can be turned into a collaborative tool to improve the workflow of work products, with an emphasis on collaboration, communication, collaboration and information exchange.

5 Conclusions

The impact of a strong economy on small businesses is twofold: as business activity increases, so does the ability of small businesses to keep pace with demand by hiring additional staff, expanding sales space, and adding new product lines. This may be seen as a positive, but the downside is that many smaller companies feel overwhelmed when the economy falters, which can lead to mass layoffs and company bankruptcies. Consumers worry about job stability and are likely to be wary of spending that will reduce income for small business owners. Slow profits can make it difficult for small businesses to repay creditors, which can have a negative impact on their long-term viability.

In general terms, the impact of small businesses has broader economic implications for at least two reasons. On the one hand, small businesses are an important route to economic mobility; they enable entrepreneurial individuals to become economically independent. This environment could make the economy more stable and less vulnerable to economic downturns such as a recession. Although the role that start-ups and young technology companies play in job creationism well documented, their contribution to overall productivity has not been widely discussed or even considered. Between the two, small businesses employ nearly half of the U.S. workforce and account for more than one-third of all new jobs in the United States. Translating ideas into the new products and services that people need is the source of prosperity in the developed world.

In order to better understand how innovation contributes to economic development, we have unpacked some examples below. Workers lose their jobs at any given time, but small firms bring more jobs into the economy than they are taken away, resulting in a net increase in the number of workers. If a small business has a low survival or expansion rate, the growth rate of a given year, or even of the economy as a whole, depends on a number of factors. The economy is at the top of the list and is the most important factor for small businesses. Consumer confidence is an economic indicator that measures consumers' overall optimism about the state of the economy. Confident consumers tend to spend and are more likely to be in business, meaning that a business is more likely to thrive when consumer confidence is high. Economic factors that often influence companies include the number of jobs, the size of their workforce, and the level of investment in their business. While periods of low confidence can force a company to cut costs to maintain its profits, a period of high consumer confidence offers an opportunity for new companies to enter the market. There is a continuous positive relationship between consumer confidence and the number of small businesses in the United States. In recent years, entrepreneurship has contributed significantly to the economic growth and stability of developing countries. Financial inclusion and access to finance can be crucial to enhancing entrepreneurship's capacity to promote economic growth and accelerate the achievement of the SDGs in developing countries. Entrepreneurship, when properly exploited, can trigger economic development in emerging economies such as Africa, Asia, and the Middle East. The economic environment plays a key role in determining the success or failure rate of entrepreneurial efforts. In addition to the economic conditions of a country's economy, the economic environment can also determine the success or failure rates of entrepreneurship. Governance is an important factor in the decline in small business success rates, which further supports the findings of this study. If this decline is not reversed, it could have a disturbing impact on economic dynamism and growth. Recent studies show a strong correlation between governance and the success or failure rate of entrepreneurship in a country.

Although small businesses do not have a significant impact on employment growth, new businesses account for more than half of all jobs all around the world. Over the past three decades, companies that are less than a year old have created more jobs than any other type of business, such as hotels, restaurants, and retail stores. Only a small share of all new start-ups in the data make an important contribution to the economy. Large companies, on the other hand, have the resources to finance growth in difficult times and, of course, have a greater capacity to create new jobs and job opportunities at national level. Financial transaction data show that, despite the commonly held view that companies that grow with external financing make the biggest contribution to the economy, organic growth companies also make a significant contribution. Thence, governments should start creating an economy for micro-entrepreneurs that would employ the modern trends and tools such as the sharing economy that are powered by novel technologies and approaches. All of these would lead to the creation of the new class of the sustainable and digital micro-enterprises and would results in the increasing energy efficiency and the transition to the green economy.

References

1. A. Kasim, Y. Ekinci, L. Altinay, K. Hussain, Journal of Hospitality Marketing & Management, **27(7)**, 855-875 (2018)
2. J. Tidd, J. Bessant, *Managing innovation: integrating technological, market and organizational change* (2020)
3. W. Strielkowski, M. Krejcí, I. Čabelková, Business: Theory and Practice **16(3)**, 304-315 (2015)
4. W. Ashton, S. Russell, E. Futch, Journal of environmental planning and management, **60(12)**, 2133-2149 (2017)
5. D. Rita, F. Ferreira, I. Meidutė-Kavaliauskienė, K. Govindan, J. Ferreira, Journal of Cleaner Production, **196**, 985-996 (2018)
6. X. Liang, X. Zhao, M. Wang, Z. Li, Sustainability, **10(11)**, 4242 (2018)
7. L. Johnstone, P. Hallberg, Journal of environmental management, **266**, 110592 (2020)
8. G. Santoro, R. Quaglia, A. Pellicelli, P. De Bernardi, Technological Forecasting and Social Change, **151**, 119820 (2020)
9. E. Kraemer-Mbula, E. Lorenz, L. Takala-Greenish, O. Jegede, T. Garba, M. Mutambala, T. Esemu, International Journal of Technological Learning, Innovation and Development, **11(1)**, 1-30 (2019)
10. S. Chege, D. Wang, Technology in Society, **60**, 101210 (2020)
11. M. Bengtsson, E. Alfredsson, M. Cohen, S. Lorek, P. Schroeder, P. Sustainability Science, **13(6)**, 1533-1547 (2018)
12. B. Todeschini, M. Cortimiglia, D. Callegaro-de-Menezes, A. Ghezzi, Business Horizons, **60(6)**, 759-770 (2017)
13. J. Lu, L. Ren, S. Yao, J. Qiao, W. Strielkowski, J. Streimikis, Energies, **12(18)**, 3417 (2019)
14. C. Haibo, E. Ayamba, A. Agyemang, S. Afriyie, A. Anaba, Environmental Science and Pollution Research, **26(7)**, 7228-7242 (2019)
15. P. Fleming, Organization Studies, **38(5)**, 691-709 (2017)
16. A. Gandini, Human Relations, **72(6)**, 1039-1056 (2019)
17. M. Ritter, H. Schanz, Journal of Cleaner Production, 213, 320-331 (2019)
18. M. Hossain, Sharing economy: A comprehensive literature review. International Journal of Hospitality Management, **87**, 102470 (2020)
19. A. Davies, B. Donald, M. Gray, J. Knox-Hayes, Cambridge Journal of Regions, Economy and Society, **10(2)**, 209-230 (2017)
20. M. Pouri, L. Hilty, Environmental Innovation and Societal Transitions, **38**, 127-139 (2021)
21. P. Diamandis, S. Kotler, *The future is faster than you think: how converging technologies are transforming business, industries, and our lives* (2020)
22. S. Nambisan, M. Wright, M. Feldman, Research Policy, **48(8)**, 103773 (2019)

23. S. Chai, M. Scully, Journal of Business Ethics, **159(4)**, 943-960 (2019)
24. S. Ganapati, C. Reddick, Government Information Quarterly, **35(1)**, 77-87 (2018)
25. S. Gössling, C. Michael Hall, Journal of Sustainable Tourism, **27(1)**, 74-96 (2019)
26. H. Zaheer, Y. Breyer, J. Dumay, Technological Forecasting and Social Change, **148**, 119735 (2019)
27. V. Ahuja, A. Akhtar, O. Wali, Journal of Global Entrepreneurship Research, **9(1)**, 1-27 (2019)
28. A. Oliveira, Biotechnology journal, **14(8)**, 1800613 (2019)
29. J. Paschen, M. Wilson, J. Ferreira, Business Horizons, **63(3)**, 403-414 (2020)
30. W. Wereda, J. Woźniak, Social Sciences, **8(6)**, 177 (2019)

New trends of the economy: a case of digital energy

Aysel Guliyeva[1]*, *Adalat* Bayramov[2], *Sabir* Taghizade[2]

[1]International Economy Department, Azerbaijan State University of Economics (UNEC), 6, Istiglaliyyat str., AZ1001, Baku, Azerbaijan
[2] SABAH CENTER, Azerbaijan State University of Economics (UNEC), 6, Istiglaliyyat str., AZ1001, Baku, Azerbaijan

Abstract. This paper studies the concept of "digital energy", its own unique content, or maybe is it just a fashionable designation for what has been done in the industry since the massive spread of computer technology. Why use the neologism "digitalization" instead of the well-known and well-defined term "automation"? Some sources insist on having its own unique content in terms of "digital energy" and "digitalization". This post sets out our working hypothesis – five distinguishing features by which one can understand whether we are working with the concept of "digital energy" or with something else. In digital energy as a part of the digital economy, the main concept is "economy". The essence of digital energy is the reassembly and development of a set of industrial and economic relations in the industry based on digital approaches and means. Consequently, the essence of digital energy is the reassembly and development of a set of industrial and economic relations in the industry based on digital approaches and means. So, in the phrase "digital economy" (and hence "digital energy"), the defining word is "economy", and the adjective "digital" only indicates the means to achieve the goal.

1 Introduction

The main task of digital energy is to eliminate the sharply growing costs of integrating distributed energy and market transactions [1]. The emergence of a digital platform in any industry significantly reduces transaction costs. And what, in fact, is the goal? According to leading experts, the essence of the modern era is the replacement by machines of ineffective transactions requiring routine human participation from economic and social life. The emergence of a digital platform in any industry leads to a significant reduction in transaction costs and an acceleration of the operational cycles of its participants [2, 3]. Energy is no exception. Digital transformation (digitalization) in the energy sector is, first of all, the creation of new business models, services and markets based on the capabilities of the digital economy [4]. A simple example from another industry: creating an automated dispatch system for a taxi fleet is automation, but Uber, which essentially creates a new business

* Corresponding author: aysel.guliyeva10@gmail.com

model for the same service, without being a taxi fleet and not owning a single car, and makes it cheaper, more convenient and safer is digitalization. The same is true in digital energy: it is important to define a new business model whose potential is unlocked by pervasive communications, machine-to-machine communications and digital modeling. Many digital business models have been developed: demand aggregators, virtual power plants, virtual distributed energy storage, energy hedging, etc. [5, 6].

A digital energy project always involves the use of digital models of fragments of the real world. A characteristic feature of the digital economy (energy) is the creation of new business models that combine the physical and digital worlds. And such unification is possible when smart machines begin to form and use digital models of the physical world. This is what ensures the independence of machine decision making in near real time. Sometimes it seems that for new business models of digital energy, the emergence of means of informing a person who will still make decisions for himself is enough. But experimental projects have shown that after a while people cease to be interested in new patterns of behavior that impose too high demands on them [7, 8]. Cyber-physical systems need to take over the initiative, which should act as a miraculous laxative from one old advertisement "without disturbing human sleep." And for this, machines must be guided not by standards and rigid algorithms, but by goals set by people and digital models of fragments of the real world. "Digitalization" is the liberation of a person from machine functions and the emergence of new, more creative forms of employment [9, 10].

In high-tech industries, there is a high additional load in adjacent areas of activity. It is still poorly described how the digital economy (energy) affects people and their employments in socially useful work. Often people record their fears of being unemployed and unemployed. To this, experts in the digital economy say that high-tech industries create a high additional load in adjacent areas of activity.

But there are deeper answers to questions about the changing role of humans in the digital world. Back in the discussions of the 1960s, it was discussed that the meaning of the cybernetic/digital transition is not to create a machine that would be smarter, stronger and more perfect than a person, but to make the living person again smarter and stronger than anything else [11]. The world of machines he created, which got out of his control and enslaved him. The task is to transform a person from a raw material and a means of technical progress, from a detail of "production for the sake of production" into the highest goal of this production, into an end in itself. This means that a project in the field of digital energy is always the emergence of a new truly human labor with a large share of scientific, technical, artistic and social creativity [12].

2 Materials and methods

Digital energy is the transaction machines of new markets, models and services. The digital transition is often referred to as a new technological revolution [13]. But it would be strange to call a revolution a phenomenon and processes that have long become commonplace. Automation of production, technological and management processes is the greatest achievement of the second half of the 20th century. Process control systems for power plants, dispatch control systems, automation of active energy devices, accounting and office automation systems - all these are important and relevant areas of the evolutionary development of the industry [14]. But this is not a new technological revolution; it is not a digital economy or digital energy. It seems that it is precisely in the change in the ways of organizing economic relations, leading to the effective involvement of millions of new subjects and the smart machines behind them, that a new qualitative leap should take place. First of all, digitalization will not be subject to technical systems and internal business processes, but the relationship between people, companies and institutions. Digital energy is, first of all, the transaction machines of new markets, business models and services. The New World changes its energy circuits and there are several reasons for this. And the first is the active relocation of people to cities.

In 2012, the share of the world's urban population exceeded 50%. By 2020, 60% of the population is expected to live in cities in China alone [15]. The growth of old and the emergence of new cities creates a demand for the transition to urban energy of a new generation: with a high concentration of capacities, a margin of safety, occupation of a small area, an acceptable price for connecting to the infrastructure and differentiated cost of electricity for different groups of consumers.

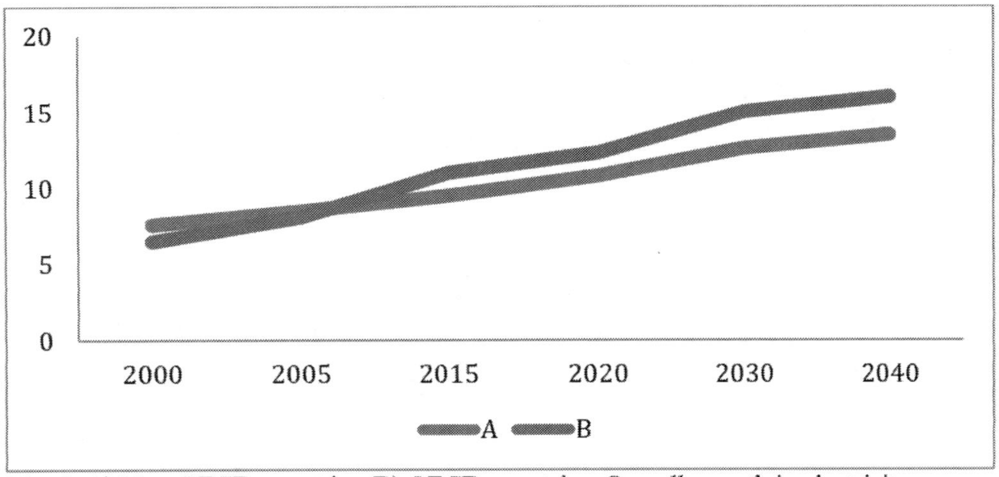

Fig. 1. A) Non-OECD countries, B) OECD countries. Overall growth in electricity production, 2000-2040 [16]

In addition, energy solutions must be environmentally friendly. According to the International Energy Agency, the electricity industry is the source of 42% of anthropogenic greenhouse gas emissions and a significant source of pollutant emissions [17]. Cities with a high concentration of gas and especially coal power generation are already today hazardous to health and cannot grow.

3 Empirical model

The development in modern cities of digital systems, additive and high-precision industries, which are highly sensitive to the reliability of power supply and the quality of electricity, determine the qualitative change in demand. According to forecasts, by 2030, of the customers in the developed countries will consume up to 30% of electricity [18].

The growing requirements for the quality of energy and the environmental friendliness of its production will be accompanied by the need for a sharp increase in its volumes. According to calculations, by 2035 additional electricity consumers will be: 1.3 billion people who currently do not have access to electricity; 2.7 billion people who cook with wood; another 1.6 billion people due to population growth in the world [19, 20]. This will lead to the fact that the consumption of electricity on the planet will grow one and a half times, and its geography will change greatly.

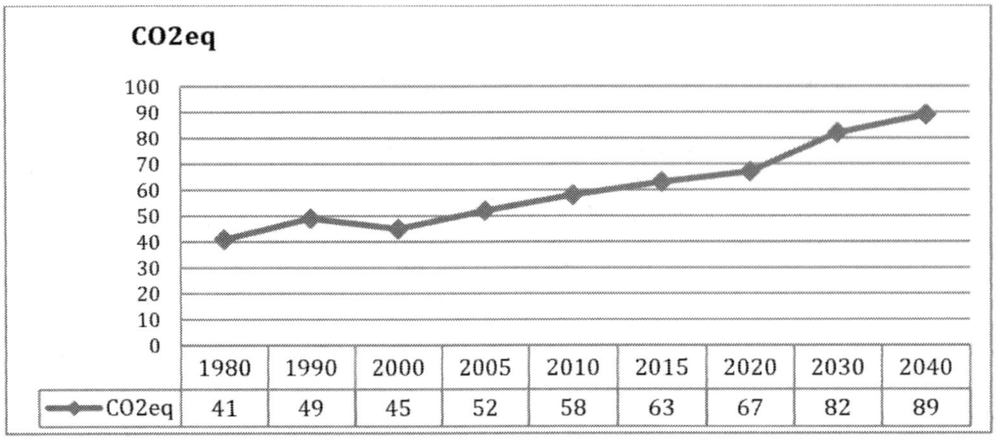

Fig. 2. Greenhouse gas emissions and target reduction levels to keep global warming within

The question arises - where to find finances to solve such a large-scale task? Traditional industrial-type energy systems were built in a situation of high savings and strong government investment, as well as investments from private sources. Today, this is not the case: the level of savings has fallen over the past 30 years, and countries are overwhelmed by the costs of servicing financial deficits and social obligations. Bank capital, on the other hand, due to changes in the model of energy markets and the presence of more attractive investment objects, does not show interest in such projects.

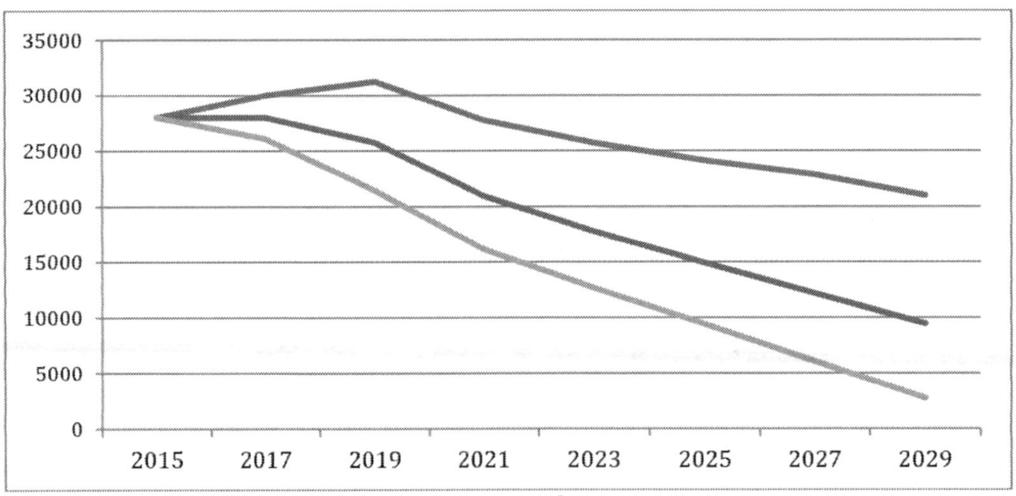

Fig. 3. Forecast balance of demand for power supply

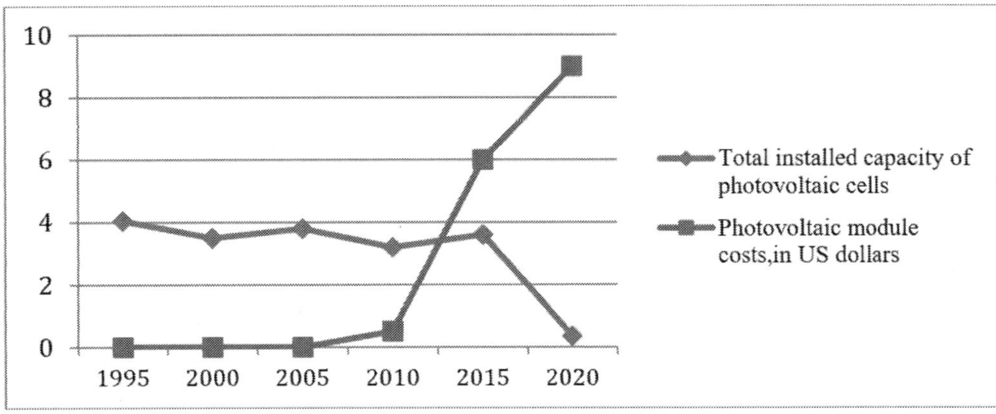

Fig. 4. The speed of development of photovoltaic technologies: the ratio of the installed capacity and the cost of modules.

4 Conclusions

Thence, the question arises: what to do about it all? How to meet the growing demand, including for high-quality and environmentally friendly power industry in the absence of funding and long project implementation terms? For this, a strategy was developed, the so-called "energy transition", which became possible due to the following technological and market trends.

First, there is a reduction in the cost of new technologies for the use of renewable energy sources (RES). For example, the cost of wind turbines from 1980 to 2013 decreased by 10 times, and photovoltaic modules in 2014 fell in price by 75% compared to the 2009 level.

Second, the share of distributed energy is growing, associated with the speed and ease of installation and maintenance of equipment, and the comparable cost with connection to the public network. In the case of further improvement of storage systems (accumulators) of electricity, one can expect a sharp increase in this trend.

Third, energy saving technologies and practices are being disseminated. According to the IEA, the energy intensity of the GDP of the organization's member countries is decreasing by 2% annually.

Fourth, the proliferation of digital networks and intelligent control systems, as a result of which the network infrastructure is being transformed into a new cyber-physical platform for flexible and efficient energy supply to consumers.

Fifth, consumers themselves are turning into suppliers of electricity and are in conflict with the norms of traditional regulation of the electricity market, and even demand “Energy Democracy”. For example, in Germany, half of the renewable energy market is controlled by households and farmers.

Sixth, the emergence of new technologies in the financial sector (block chain, Smart Contract, Decentralized Autonomous Organizations), which creates potential opportunities for attracting new private investments in the energy sector and the formation of various energy exchange practices.

All this makes it possible to start the energy transition today. Moreover, not only in developed countries, but also in developing countries, which simply cannot build a traditional power system due to its high cost (for example, the cost of the US power system is $ 1 trillion), the need for complex competencies and a large number of skilled labor for management and maintenance.

What will the energy transition look like in general terms? First, technically, this is the creation of a more flexible architecture of energy systems due to distributed generation, the development of "smart" grids and the energy storage market, as well as due to the emergence of active ("smart") consumers. Secondly, it is the formation of a new business model for the electric power industry: instead of the traditional value added chain "generation - trading - transmission - distribution", the model of the "Internet of energy" and the provision of services in a new interactive environment. Thirdly, this is a change in state regulation - a gradual transition from support for renewable energy sources and competition in the electricity market to consumer support, integration of local solutions; from energy supply to "interconnection of capacities" within the framework of the "Internet of Energy", the transition to a "flexible market".

References

1. J. Niu, Z. Tian, J. Zhu, L. Yue, Energy Conversion and Management, **208**, 112575 (2020)
2. J. Mendling, G. Decker, R. Hull, H. Reijers, I. Weber, Communications of the Association for Information Systems, **43(1)**, 19 (2018)
3. W. Strielkowski, *Social Impacts of Smart Grids: The Future of Smart Grids and Energy Market Design* (2019)
4. A. Alekseev, S. Lobova, A. Bogoviz, Y. Ragulina, International Journal of Energy Economics and Policy, **9(5)**, 274-280 (2019)
5. V. Munyon, W. Bowen, J. Holcombe, Energy Research & Social Science, **38**, 19-27 (2018)
6. J. Yun, X. Zhao, J. Wu, J. Yi, K. Park, W. Jung, Sustainability, **12(5)**, 1883 (2020)
7. L. Groeger, K. Bruce, I. Rolfe, Industrial Marketing Management, **77**, 102-115 (2019)
8. E. Clemons, *New Patterns of Power and Profit: A Strategist's Guide to Competitive Advantage in the Age of Digital Transformation* (2018)
9. J. Zhou, P. Li, Y. Zhou, B. Wang, J. Zang, L. Meng, Engineering, **4(1)**, 11-20 (2018)
10. F. Miele, I. Tirabeni, Sociology Compass, **14(6)**, e12795 (2020)
11. J. Carrasco-Sáez, M. Careaga Butter, M. Badilla-Quintana, Sustainability, **9(12)**, 2258 (2017)
12. W. Strielkowski, E. Volkova, L. Pushkareva, D. Streimikiene, Energies, **12(7)**, 1392 (2019)
13. M. Loock, Energy research & social science, **69**, 101740 (2020)
14. D. Horváth, R. Szabó, Technological Forecasting and Social Change, **146**, 119-132 (2019)
15. Y. Li, X. Kong, Z. Zhu, Sustainable Cities and Society, 61, 102326 (2020)
16. OECD, *https://www.oecd.org* (2020)
17. IEA, *https://www.iea.org/reports/global-energy-co2-status-report-2019* (2019)
18. Energypost, *https://energypost.eu/end-energiewende* (2019)
19. Y. Wei, H. Liao, *Global Energy Development and Energy Poverty* (2018)
20. M. Rajaeifar, H. Ghanavati, B. Dashti, R. Heijungs, M. Aghbashlo, M. Tabatabaei, Renewable and Sustainable Energy Reviews, **79**, 414-439 (2017)

Energy research and social sciences: thinking outside the box

Wadim Strielkowski[1,2]*

[1] Prague Institute for qualification enhancement, Trnkovo nám. 1112/2, 15200 Prague, Czech Republic
[2] Centre for Energy Studies, Prague Business School, Werichova str. 1145/29, 15200 Prague, Czech Republic

Abstract. This paper aims at explaining the universality and broadness of the research in energy studies. Specifically, it wants to show that the energy research is not a solely engineering or natural sciences field and how it can be done in social sciences. The paper draws some relevant examples including energy research in literature and poetry, history, religion, art, as well in other social sciences and humanities. In general, it becomes apparent that energy research can boast vast depths and angles that are worth exploring for any social scientist. Given the key importance of energy research in the third decade of the 21st century and the worldwide focus on the renewable energy sources, electrification of transport and heating in the face of the threatening global warming and climate change, it seems relevant to focus on researching the perspectives and paradigms for the traditional and renewable energy sources in the 21st century using the toolbox of the social sciences.

1 Introduction

Energy and sustainable development issues have now become an essential part of our daily lives. We constantly read about sustainability and green energy and see them in the news, follow the discussions on oil and gas prices, read articles and blogs with reasoning about the safety of nuclear energy, as well as being bombarded with the accounts of the necessity to shift to the renewable energy sources (RES) or the electric vehicles that are going to help the humanity to embrace its carbon-free future [1, 2].

Replacing traditional energy production with RES is a multi-factor issue, the pros and cons of which are viewed differently by governments, technology companies, markets, and the general public [3, 4]. There are significant public doubts about the advisability of investing in renewable energy sources. Research on energy and sustainable development in sociology and economics can focus on factors related to public opinion about renewable and traditional sources of energy [5, 6].

One way or another, energy-related topics have found their way into social sciences and humanities. Energy research is no longer the domain of electrical engineers and technicians

* Corresponding author: strielkowski@prizk.cz

but a broad and multidisciplinary area that attracts economists, sociologists, anthropologists, historians, or artists, just to name a few [7-9].

However, quite often we are confronted with the fact that social scientists cannot understand the full potential of energy research in social sciences and are reluctant to engage into pursuing the energy-related topics considering them too "technical" and "mathematical" [10, 11]. This is far from truth – energy research is abundant in economic and social issues related to the energy production and consumption. For example, one can think of the interaction of various energy-related activities and processes with social structures that shape the interaction of people with the energy system. This is due to the fact that energy production, distribution and consumption have both technical and human components. Energy-related research features many social issues such as the relationship between energy consumption and energy security, energy efficiency and renewable energies, as well as energy use and its effects on human health and their well-being.

This short but insightful paper is intended to explains the universality and broadness of the research in social sciences and humanities in the relation to the energy studies. The paper attempts to demonstrate how the energy research can be done and provide inspiration for the future research in this field. This work draws from many examples including energy research in literature and poetry, history, religion, art, as well in other social sciences and humanities.

2 Materials and methods

Energy production, distribution and consumption have both technical and human components. Indeed, there are a number of social and technical issues related to energy production and consumption and energy-related humanities [12, 13].

One can see that energy research in social sciences covers a wide range of topics that revolve around societal processes and influences, including the impact of energy on the health of the people that are affected by it as well as the impact of climate change [14].

The study of the social structures related to energy systems and the way in which people interact with them. The latter, in particular, resembles the interaction between energy and social systems, such as the relationship between the human body and the environment [15]. With the increasing interest in how this understanding could be directly translated into policy interventions, energy-related social sciences have been studied in a number of different ways. There are major differences between disciplines and sub-disciplines, which relate primarily to theorization, definition and research problems, but it should not be forgotten that social sciences in general and energy research in particular share many of the same principles and scientific approaches [16].

In addition, one should not forget about the interdisciplinary research that overlaps academic disciplines, sectors and policies and is increasingly expected to develop new insights and help to address various situations. In recent years, it was expected that the academic discipline of the social sciences and humanities would contribute to this new knowledge and touch upon the research in energy economics and policy, energy and religion, energy and philosophy, energy and arts and literature, just to name a few [17].

In the section that follows, this paper will present a series of examples of interdisciplinary energy research applied from a social science and humanistic perspective. All of them constitute the interdisciplinary cooperation between the social sciences and humanities in the energy sector. This approach would help to revert the situation which we often witness nowadays - researchers and policymakers continue to focus on only one side of the energy dilemma, and social, humanities, and arts sciences are marginalized in energy research. Surely, this has to change very soon. The one-sided emphasis on science raises many questions and leads to frustration.

3 Energy@Everywhere

Let us show some practical examples from the energy research in social sciences that involve top-notch research and publications in prestigious academic journals and books. One can start with Economics and proceed towards other social sciences.

3.1 Energy economics and sociology

It is clear that Economics might be a science about the allocation of scarce resources but it is also an exciting field of research that studies energy as a commodity. Take electricity for example. In fact, electricity is a special commodity that has to be consumed as soon as it is produced and which cannot be effectively and sufficiently stored for long periods of time [18].

And there are also energy prices, tariffs, quotas and other elements that are deeply rooted in Economics. Quite often, these economic tools are not designed in a best way and customers suffer as a result of that. For example, Simshauser [19] investigated a case of Southeast Queensland in Australia which features one of the highest penetration rates of domestic solar photovoltaic (PV) panels in the world and found that the massive increase in solar PV resulted in a huge transfer of wealth and costs between customer groups due to unfair tariffs [20].

There are also topics in energy that can found its use in sociology. Public and social acceptance of energy project and initiatives, attitudes to the energy-related issues and threats are all at the crossroads of sociology and economics.

Moreover, economists and sociologists can study such issues as willingness-to-pay for various energy-related projects, for example the industrial CO_2 capture and storage products and technologies [21]. Other good examples might be the public acceptance of renewable energy sources [22], behavioural changes thanks to the deployment of smart meters that help to save energy in households [23], the new approaches to the Corporate Social Responsibility (CSR) with regard to the development of energy projects [24, 25], or the perceived threats to urban energy networks [26].

3.2 Energy and philosophy

The topics of energy and sustainable development are also relevant in philosophy [27]. Why does a person need energy and how much? Should this energy be free or should one pay for it? Does a person know how to use it correctly? Will the person destroy our planet in pursuit of new sources of energy? Why are mineral deposits unevenly distributed across countries and continents? The mission of the philosophy of energy is to explore and clarify the space in which the so-called energy transition to sustainable energy development (also called "green innovation") and renewable energy is taking place [28].

3.3 Energy and religion

Energy is also a constituent part of the religion and religious studies. For example, there is an Energy@Cambridge's Grand Challenge called, 'In Search of "Good" Energy Policy' at the University of Cambridge that has an aim to identify optimal practices and principles for designing and implementing the best possible energy policies. A part of this Grand Challenge is a serious of workshops on the topic, some of which are dedicated to finding the links between energy and religion.

For example, one of these workshops established that Pope Francis's 2015 encyclical "Laudato Si': On Care for our Common Home" apparently makes a strong appeal about the necessity of integrating energy policies and environmental policies with social concerns and grasping that ecological debt and other imbalances represents the issues of social and moral justice [29].

3.4 Energy in literature and arts

Lots of attention is paid to energy projects and various endeavours related to the conquest of nature for the extraction of new deposits or forms of energy in the literature and poetry. These topics were especially popular during the grandiose construction projects and "Communist" projects of the Soviet era. One can recall poems like the following one:

"Once the man river Dnieper told:
"I will lock you up with a wall!
You will be jumping from the top,
Moving the machines along"

Samuil Marshak "The war with Dnieper" (1930) [30]. Own translation

Energy and sustainability issues are also widely reflected in art and film [31]. Suffice it to recall the posters or mosaics of the Soviet era, which extolled industrialization, electrification, the use of the "peaceful atom" and so on to the skies (see Figure 1).

Fig. 1. Soviet era poster glorifying energy projects

Prior to the collapse of the Soviet Union, many films and TV series were made, in which the heroes conquered the virgin lands, performed the five-year plans in three years, or struggled with nature and its elements in the forests, steppes, mountains and seas of the endless Soviet Motherland.

Many of the energy-related stories of the Soviet era are reflected in contemporary cinema, both Russian and Western. Suffice it to recall the recently released HBO series Chernobyl (2019) that gained wide popularity among the viewers.

https://doi.org/10.1051/e3sconf/202125007001

3.5 Energy and history

The history of energy is an interesting topic in itself. Human history resembles a cycle: the new is always a well-forgotten past. Until the 18th century and the Industrial Revolution, all energy produced by mankind for its own needs was renewable. Humans used horsepower to travel on land, wind power to travel across seas and oceans, burn wood in stoves and fireplaces to heat their homes, and harness the power of streams and rivers to run mills and other simple equipment. From today's perspective, all of these energy sources were renewable and sustainable.

One of the very interesting topics in the history of energy is the development of the electric power industry in the Soviet Union after the October Revolution of 1917 [32, 33]. The Soviet plan for the electrification of GOELRO was once admired by Herbert Wells, the famous science fiction writer and author of the bestseller "Time Machine" (1895).

4 Conclusions

There are various reasons why social scientist want and need to focus on the research in energy-related topics. First of all, it might be due to their multidisciplinary nature or the search for new exciting and unexplored topics for research. Second, it might be an attempt to publish their research in academic journal with high impact factor (the highest impact factor (IF) measured by the prestigious Web of Science database among the journals in the social sciences barely exceeds 5-6 with a median around 1-2, while in case of energy-related journals it can easily exceed 10 or be even higher). But most important of all is that social scientists need to learn how to think outside the box.

Overall, it becomes clear that energy research can boast vast depths and angles that are worth exploring for any social scientist. Given the key importance of energy research in the third decade of the 21st century and the worldwide turn to the renewable energy sources, electrification of transport and heating in the face of the threatening global warming and climate change, it seems relevant to focus on researching the perspectives and paradigms for the traditional and renewable energy sources in the 21st century.

The research in energy and energy-related issues must go beyond the dimensions of technology and economics and include social and human elements, as well as addressing issues such as the relationship between energy production, distribution and consumption and their impact on human health and well-being.

References

1. A. Pegels, G. Vidican-Auktor, W. Lütkenhorst, T. Altenburg, Journal of Environment & Development, **27(1)**, 26-45 (2018)
2. W. Strielkowski, Economics and Sociology, **10(1)**, 310-318 (2017)
3. M. Wang, Y. Li, Z. Cheng, C. Zhong, W. Ma, Journal of Cleaner Production, **278**, 123944 (2021)
4. I. Čábelková, V. Blaginin, W. Strielkowski, A. Platitsyn, E3S Web of Conferences, **208**, 09001 (2020)
5. S. Ntanos, G. Kyriakopoulos, M. Chalikias, G. Arabatzis, M. Skordoulis, Sustainability, **10(3)**, 687 (2018)
6. I. Čábelková, W. Strielkowski, D. Streimikiene, F. Cavallaro, J. Streimikis, Technological Forecasting and Social Change, **163**, 120477 (2021)
7. I. Ingeborgrud, S. Heidenreich, M. Ryghaug, T. Skjølsvold, C. Foulds, R. Robison, R. Buchman, R. Mourik, Energy Research & Social Science, **63**, 101398 (2020)

8. M. Biresselioglu, M. Demir, B. Solak, A. Kayacan, S. Altinci, Science of The Total Environment, **729**, 139027 (2020)
9. T. Dunlop, Energy Research & Social Science, **56**, 101216 (2019)
10. J. Markard, Nature Energy, **3(8)**, 628-633 (2018)
11. W. Strielkowski, *Social Impacts of Smart Grids: The Future of Smart Grids and Energy Market Design* (2019)
12. G. Sonetti, O. Arrobbio, P. Lombardi, L. Lami, S. Monaci, Energy Research & Social Science, **61,** 101342 (2020)
13. B. Van Veelen, A. Pinker, M. Tingey, G. Aiken, W. Eadson, Energy Research & Social Science, **57**, 101240 (2019)
14. S. Hallegatte, J. Rozenberg, Nature Climate Change, **7(4)**, 250-256 (2017)
15. C. Fontes, F. Freires, Renewable and Sustainable Energy Reviews, **82**, 247-259 (2018)
16. H. Mohajan, Journal of Economic Development, Environment and People, **7(1)**, 23-48 (2018)
17. J. Koehrsen, S. Dickel, T. Pfister, S. Rödder, S. Böschen, B. Wendt, K. Block, A. Henkel, Current Sociology, **68(6)**, 738-760 (2020)
18. B. Dunn, H. Kamath, J. Tarascon, Science, **334(6058)**, 928-935 (2011)
19. P. Simshauser, Australian Economic Review, **47(4)**, 540-562 (2014)
20. W. Strielkowski, D. Štreimikienė, Y. Bilan, Renewable and Sustainable Energy Reviews, **77**, 461-473 (2017)
21. V. Kashintseva, W. Strielkowski, J. Streimikis, T. Veynbender, Energies, **11(10)**, 2787 (2018)
22. I. Čábelková, W. Strielkowski, I. Firsova, M. Korovushkina, Energies, **13(7)**, 1742 (2020)
23. G. Rausser, W. Strielkowski, D. Štreimikienė, Energy & Environment, **29(1)**, 131-146 (2018)
24. J. Lu, L. Ren, S. Yao, J. Qiao, W. Strielkowski, J. Streimikis, Energies, 12(18), 3417 (2019)
25. J. Lu, L. Ren, J. Qiao, S. Yao, W. Strielkowski, J. Streimikis, Sustainability, **11(15)**, 4128 (2019)
26. I. Čábelková, W. Strielkowski, F. Wende, R. Krayneva, Energies, **13(21)**, 5659 (2020)
27. J. Rendtorff, *Philosophy of management and sustainability: Rethinking business ethics and social responsibility in sustainable development* (2019)
28. R. Geerts, B. Gremmen, J. Jacobs, G. Ruivenkamp, Towards a philosophy of energy, Scientiae Studia, **12(SPE)**, 105-127 (2014)
29. J. Chaplin, *From Inspiration to Implementation: Laudato Si', Public Theology and the Demands of Energy Policy*. In: M. Ozawa, J. Chaplin, M. Pollitt, D. Reiner, P. Warde, (eds.) *In Search of Good Energy Policy*, 257-270 (2019)
30. Culture.ru, *https://www.culture.ru/poems/42772/voina-s-dneprom* (2021)
31. I. Samorukova, Vestnik of Samara University: History, pedagogics, philology, **23(4)**, 59-68 (2017)
32. A. Gibadullin, Statistika i Ekonomika, **15(1)**, 17-24 (2018)
33. R. Davies, R. Davies, M. Harrison, S. Wheatcroft (eds.), *The economic transformation of the Soviet Union, 1913-1945* (1994)

Building major competencies for enhancing higher education for sustainable development

Alexander Ivanov[1]*, *Svetlana* Tolstikova[1], *Larisa* Ovcharenko[1], and *Tatiana* Morozova[1]

[1]Department of General and Practical Psychology, Moscow City University, 4 Vtoroy Selskohoziajstvenny proezd, Moscow, 129226, Russian Federation

Abstract. This paper focuses on the specifics and preconditions for building major competencies that are envisaged for enhancing higher education for sustainable development (ESD). It looks into the professional development of specialists in higher education, factors affecting individual stages of professional development of these specialists, as well as into the role of the environment in their professional development. Furthermore, it studies the formal and informal learning as the important setting leading to the creation of the major competencies for linking globalization, complexity and environmental responsibility and channelling them to students through the teaching process in higher education.

1 Introduction

Higher education institutions (HEI) assign key competences for sustainable development that enable professionals to cope with the challenges in different socio-economic environments. Universities are becoming increasingly competent - based on the development of skilled workers who must be able to face the demands of organisational competition and the labour market [1-3]. Therefore, universities integrate their curricula with a wide range of skills, such as knowledge, skills and skills, as well as skills in the field of sustainable development.

This educational movement is supported by the need to rethink the purpose of education and to change the curriculum and educational practices. Universities are making important changes to their programmes to boost students' professional skills, while others are planning to improve their programme's learning outcomes. This underlines the significant progress and potential of education for sustainable development (ESD), although more efforts will be needed to move towards a sustainable future. Higher education institutions play a crucial role in implementing and promoting sustainable development initiatives [4, 5]. One of the key questions to be answered is whether university leaders, lecturers and students are doing a good job in implementing sustainable development in higher education in their institutions. This reinforces the need to strengthen education at all levels, build on the success of the ESD and translate it into political and institutional contexts [6, 7].

According to a United Nations report, justice means understanding a student's learning needs in order to reduce the barriers to his or her academic success. With increasing demands

* Corresponding author: aivanov7@rambler.ru

on society and the environment due to increasing urbanization, migration and human migration, it is clear that global action is needed to build a more sustainable future [8]. Building a sustainable future means more than just creating green campuses and implementing recycling efforts and initiatives for global civic engagement.

In recent years, the concept of education and sustainable development has become a central educational initiative, helping to address many of the problems associated with human development. Given the importance of higher education for the development of a more sustainable future, it can serve as a powerful tool for creating this education, according to a new study. As the world becomes more globalised and interdependent, the role of higher education in building a sustainable future has become more important. There is increasing interest in contributing to the United Nations Sustainable Development Goals by educating pupils about the importance of sustainable development and its role in human development. This the way how leading higher education institutions can contribute to the necessary societal change beyond their classrooms [9, 10].

There is a theory of economic growth, stressing that higher education was an important factor in improving the regional economy and human capital. This theory confirms the positive role of education in the development of sustainable economy.

2 Teaching environmental responsibility in higher education

Given the increasing demands on society and the environment resulting from increasing urbanization, increasing human migration, and increasing energy demand, it is clear that global action is needed to create a more sustainable future. In recent years, the concept of education and sustainable development has become a central educational initiative, helping to address many of the problems associated with human development [11]. As the world becomes increasingly globalized and interdependent, efforts to build a sustainable future will become more important. Given the importance of higher education for the development of a more sustainable world, it can serve as a powerful means of creating this education. What inspires us to inspire environmentally-oriented higher education courses are the need to understand the development of sustainable environmental literacy.

By promoting environmental knowledge, dispositions, skills and behaviour relevant to the sustainability of the environment and human health and well-being, especially designed courses can improve students' ability to influence changes in higher education and promote a better understanding of environmental issues and their impact on human development [12]. Future research will be necessary for any study aiming at developing a significant new understanding of the impact of environmental issues and issues in their application to education in general and their role in environmental education. In particular, we could focus on the interaction between different aspects of environmental competence [13]. This information will allow a better understanding of the efforts and results of higher education in terms of environmental literacy, as well as specific measures to assess the impact of education on students' ability to influence changes in education in general and environmental issues in particular.

In general, measuring changes in environmental literacy in the context of developing a sustainable environmental education system has two advantages. First, we need to measure where environmental education is located in the sustainability movement at university locations. Second, it seeks to identify the role environmental education could play in promoting the necessary movement outside the classroom [14]. The specific aim is to educate students about the need for sustainable development by integrating sustainable development issues into all aspects of teaching, research and services. Research has shown that academics are particularly interested in filling their curriculum with sustainability concepts. This means reorienting the education system to help people think and act in ways that promote a more sustainable planet. In practice, this means equipping students with the knowledge, skills, attitudes and values needed to build a sustainable future [15]. Colleges and universities must produce students who are capable of dealing responsibly with the environment. To this end, students should develop a value system that emphasizes responsibility for themselves, others and the planet, conducts authentic interdisciplinary learning activities, and cultivates critical and creative thinking skills. The urgency that fuels this response suggests that higher education must continue to take a holistic approach to environmental education, not just one - outside the classroom [16]. By promoting the skills necessary to cultivate the knowledge, skills, attitudes and values necessary to create a sustainable future for our planet, our children and grandchildren, and the future of the world, it will be ready to provoke and sustain the necessary environmental responses.

Thence, higher education must help to develop and promote the ability of students to promote environmental improvement measures. Higher education institutions play a crucial role in implementing and promoting sustainable development initiatives. One of the most important questions to be answered is what university leaders, faculty and students should do to implement sustainable development. Justice means understanding students' learning needs while reducing the barriers to academic success. Building a sustainable future means more than just creating green campuses and implementing recycling efforts and global citizens' initiatives [17]. Areas of great opportunity include expanding formal environmental education and aligning students, faculty and staff with environmental programs and goals on campus. Together, these institutions are perhaps the most responsible for educating science students on environmental issues, staffing and implementing recycling programs, and establishing sustainable development initiatives such as green spaces.

One of the most encouraging findings is that there is strong support for greening and management systems at institutions of all types and sizes, including public and private colleges and universities, as well as private universities. While there are signs that colleges or universities are working to teach students about environmental issues, trends in higher education curricula are relatively weak compared to those in "greening" and "farm management" systems. Across regions and countries, there is work being done to teach and demonstrate sustainability as a practice. In fact, many universities worldwide offer programs large or small in environmental studies, and these programs promote students' interest in sustainability and environmental issues such as environmental protection and management systems. But only a small share of universities requires all students on campus to take environmental studies courses, regardless of their focus. Even if they are studying biology or environmental sciences, students at many institutions can complete their studies without acquiring basic environmental knowledge. Clearly, this is a problem that needs to be solved in the nearest future.

3 Role of the environment and climate change in education

There is no doubt that climate change has consequences for the earth and human life, and many students want to play an active role in combating it, but are not sure how to do so in the classroom [18]. Although the majority of scientists worldwide believe that climate change is occurring and man-made, there are no government standards requiring schools to teach it in social sciences, and educators have the ability to incorporate elements of it into students' lessons to ensure that students have all the knowledge they need to address the problem in any capacity that is their own. Universities should help students in their efforts to take an active role in learning about climate change and perhaps address it as an issue in the future. Climate change is no longer just of interest to scientists, but also to educators [19]. Teaching in social studies remains in the hands of individual teachers and is decided by individual school districts. As students learn about climate change in science classes, they will be able to make informed decisions about policy institutions that are responsible for policies that affect greenhouse gas emissions. Teachers say they need improved curricula and policies, and a recent finding demonstrate that teachers feel they have the knowledge and skills to teach climate change. Students have to learn and talk about climate change to develop critical thinking skills, solve problems, and work with others to find solutions to the complex challenges of today and the future. Almost half of students are aware that climate change is man-made, but do not believe that human action and mitigation will be effective. The new revision of science standards emphasizes climate change, offers some advice to science teachers, and encourages them to include the topic in the curriculum. The use of science as an appropriate educational context in which to discuss global warming and climate change should be adequately supported [20].

Institutions of higher education need to engage in a discussion on how to teach the climate in a way that does not contradict the scientific consensus on the causes and effects of global warming and the role of human activities. Important climate-related scientific concepts are already being incorporated in primary school, but climate change is explicitly addressed only in middle school. Many of the committee members felt that they needed to answer questions about the effects of climate change and the eco-fears that it causes in younger students, and that this was not an issue that could only be addressed by science teachers. Environmental principles and concepts should be a part of the strategy which provides guidelines for teachers to implement new scientific standards. Some schools, however, still struggle to include these studies in their curriculum. Public schools need to educate young people about current and future environmental challenges, and create a group of environmental leaders that will achieve that goal [21]. A scholarship program can be established to support climate change education programs for young people and provide teachers with training and professional development related to improving climate literacy among students [22].

4 Formal and informal learning in sustainable higher education

Formal learning should be structured and give pupils, teachers and institutions the opportunity to ensure that learning objectives are achieved. Informal learning, on the other hand, takes place and continues to take place in the absence of formal learning, which is highly structured and institutionally promoted and takes place outside the traditional classroom, such as in a classroom or informal environment [23, 24]. A quick look at the typology shows that formal learning involves an educational process whose primary task is the creation of new knowledge. Learning is described by various classifications that are linked to the environment and circumstances in which learning is most likely. Specific pedagogy and materials that facilitate the educational process are generally provided in a formal learning context. A non-formal learning environment is found where the primary mission is not necessarily education. Learning that takes place in a specific learning environment the result of an activity of interest to an individual or group and cannot be recognised as learning in the traditional sense of the word (e.g. formal or informal). Learning that is planned and structured as a series of short courses organized in response to identified interests and needs. Learning that is not formally organized by the employer into a program or curriculum. Learning that takes place in a work context that relates to an activity of interest to the individual or group (e.g. vocational training, on-the-job training, etc.) [25]. Informal education is closely linked to informal learning, which takes place in relationships between members of society. It can be seen that different parties are involved and it may or may not be explicitly promoted. Informal education is created when concepts are adapted to the individual needs of each student [26]. What students learn in higher education is a function of both formal and informal experience. Formal learning takes place in the classroom and related activities structured by teachers and others to help students achieve certain cognitive and other goals. Informal learning includes a wide range of activities such as lectures, quizzes, workshops and other forms of learning. The basic argument here is the obvious point to make, that much of our learning takes place outside the formal confines of the classroom. One reason is that learning cannot be replaced by education, as we see in lifelong learning. Learning is a process that takes place all the time and requires intention and commitment [27, 28]. The focus on learning is important, but if it is all we think about and what we do, we are doing everyone a disservice.

While universities have historically been characterized by the emphasis on traditional teaching methods required in the classroom, the future of universities is determined by the response of universities to informal learning. Since informal learning and formal learning overlap, it is not easy to separate purely student social activities from those related to learning, as they often take place outside the traditional classroom. The balance between formal and informal frameworks must change to demand greater autonomy from students and to better support them in their learning.

This has been around for several years, but traditionally it has placed a high value on formal learning processes over informal learning as it is learned in the workplace. Moreover, anthropologists have found that complex learning can take place in indigenous communities that do not have formal educational institutions, as in the case of indigenous peoples in South America and Africa. Non-formal learning includes some of the same aspects of formal learning, but typically manifests itself as short-term, voluntary programs outside the framework of traditional educational institutions [29]. The need for this category is evident in programmes designed to learn outside the framework of educational institutions. However, since learners participate in these programmes in an educational environment, researchers have considered them formalising in order to be involved in the area of informal learning. Common ground between the two often includes the need for students to follow a planned curriculum and to use formal and non-formal learning methods.

5 Professional development of specialists in higher education

Specialists may have a variety of educational qualifications, but most have behaved well and require relevant professional experience. Some education and development specialists need a Bachelor's degree, others a training, and most do not need one. Others may have behaved well and need work experience, but most do not, and some may behave well but not well enough for the job [30]. Many positions require work experience, and employers may prefer to hire employees from companies that work in the industry or use technology-based tools.

Consultancy and employment management firms are hiring more specialists to train and develop their staff, as more companies seek to hire them to cut costs and cope with the high costs of recruiting and training staff in higher education. Many employers prefer a Master's degree because most apprenticeship or development jobs require a Bachelor's degree and a degree can offer a competitive advantage [31, 32]. If a company wants to recruit and retain the best people for a particular field, it can rely on the expertise of a graduate program in the professional development of specialists in higher education. Graduates with the certification will find good employment opportunities, and those with a Master's degree in the professional development of higher education specialists will do even better. Business cycles and economic conditions influence the demand for employees in the education and training of skilled workers.

Nowadays, the number of disciplines in higher education and the degree of professional development of skilled workers vary according to the area of responsibility. Graduates who obtain certificates will enjoy the best opportunities for advancement. Interpersonal skills must be imbued with skills such as communication, interpersonal skills, communication skills and communication skills. Over the next decade, above-average growth in employment opportunities is expected [33].

Employment in the training and development of skilled workers is expected to increase as companies develop new media and technologies and introduce them into their training programmes. Workers in many professions will have to attend training and qualification courses throughout their careers, creating a demand for worker-led training activities [34]. For example, organizations are increasingly using video, audio, video and other forms of higher education and online courses in their training programs. In addition to the available certification programs, there are many online courses that can help with personal and professional development [35]. To lead effective communication and faster progress there are courses that help learners achieve excellence in their careers. These include courses entitled around the lines like "Becoming a Successful Leader" and "Leading Effective Communication," as well as a number of other courses run by leading non-profit organizations. For specific expertise required in the financial world, professional training courses, including courses in finance, accounting, financial management and financial planning, as well as a range of other courses are offered. In general, the professional development of specialists in higher education nowadays is a serious and complex process that requires lots of time and effort.

6 Conclusions

Overall, it becomes clear that we need to offer teachers and young people fair access to rich, relevant learning and develop leadership capacity at all levels. Modern universities are forging links with industry to establish research laboratories and scientific centres. Competitive higher education depends on the development of a strong relationship between universities and industry and the integration of research and education. For more students and more institutions to benefit, universities, research institutions and other institutions need to take sustainability issues and challenges more into account and allocate resources. Universities develop a long-term commitment to the sustainability of their research, education and research institutions. Senior administrators at universities and colleges who are leaders in sustainability could help inform and train key government agencies and policymakers. Higher education is facing rising costs and challenges such as climate change and the effects of global warming. If one says that an educational institution committed to building a sustainable and humane future and wanting to learn to build it would be an exciting and challenging place, he or he should place it in the context of a competitive higher education environment.

The Sustainable Development Goals clearly show that higher education is part of these goals, as it is part of high-quality education and can offer opportunities for lifelong learning. Furthermore, universities can help to promote human sustainability, and therefore the need for them to develop a sustainable perspective is the main focus. Specifically, SDG Goal 4.3 aims to promote equal access to higher education regardless of gender and socio-economic status. This can be reflected in the SDG's Goal 4.7, which aims to equip learners with the knowledge and skills for sustainable development. Specifically, the United Nations proposes to foster the development of higher education as a key component of the sustainable development goals. All of these makes it very relevant to invest into boosting the major competencies for enhancing higher education for sustainable development.

References

1. E. Abad-Segura, M.D. González-Zamar, Journal of Cleaner Production, **294**, 126133 (2021)
2. E. Fleacă, B. Fleacă, S. Maiduc, Sustainability, **10(4)**, 1032 (2018)
3. F. Findler, N. Schönherr, R. Lozano, D. Reider, A. Martinuzzi, International Journal of Sustainability in Higher Education, **20(1)**, 23-38 (2019)
4. A.V. Agbedahin, Sustainable Development, **27(4)**, 669-680 (2019)
5. M. Singer-Brodowski, A. Brock, N. Etzkorn, I. Otte, Environmental education research, **25(4)**, 492-507 (2019)
6. J.H.P.P. Eustachio, A.C.F. Caldana, M. Will, A.L. Salvia, I.S. Rampasso, R. Anholon, M. Kovaleva, Sustainability, **12(9)**, 3761 (2020)
7. W. Leal Filho, M. Will, A.L. Salvia, M. Adomssent, A. Grahl, F. Spira, Journal of Cleaner Production, **232**, 1394-1401 (2019)
8. E. Boeren, International Review of Education, **65(2)**, 277-294 (2019)
9. V. Grebennikova, T. Bonkalo, N. Nikitina, Z. Gardanova, O. Grebennikov, Espacios, **40(44)**, 25 (2019)
10. J. Mensah, S.R. Casadevall, Cogent Social Sciences, **5(1)**, 1653531
11. J. O'Flaherty, M. Liddy, Environmental Education Research, **24(7)**, 1031-1049 (2018)
12. J. Tai, R. Ajjawi, D. Boud, P. Dawson, E. Panadero, Higher Education, **76(3)**, 467-481 (2018)
13. I. Molderez, E. Fonseca, Journal of Cleaner Production, **172**, 4397-4410 (2018)
14. C.R. Clark, J.E. Heimlich, N.M. Ardoin, J. Braus, Environmental Education Research, **26(3)**, 381-399 (2020)
15. N. Nikitina, E. Romanova, E. Komarova, S. Tolstikova, V. Grebennikova, Review of European Studies, **7(3)**, 66–79 (2015)
16. C. Marouli, Sustainability, **13(5)**, 2901 (2021)
17. S. Yáñez, Á. Uruburu, A. Moreno, J. Lumbreras, Journal of Cleaner Production, **207**, 57-66 (2019)
18. S.R. Stapleton, Environmental Education Research, **25(5)**, 732-750 (2019)
19. C.A. Jones, A. Davison, Geoforum, **118**, 190-200 (2021)
20. M.C. Monroe, R.R. Plate, A. Oxarart, A. Bowers, W.A. Chaves, Environmental Education Research, **25(6)**, 791-812 (2019)
21. M.A. Windschitl, D. Stroupe, Journal of Teacher Education, **68(3)**, 251-261 (2017)
22. K.C. Busch, J.A. Henderson, K.T. Stevenson, Environmental Education Research, **25(6)**, 955-971 (2019)
23. L. Janssens, K. Smet, P. Onghena, E. Kyndt, International Journal of Training and Development, **21(2)**, 92-112 (2017)
24. A.A. Holland, Computers & Education, **128**, 214-226 (2019)
25. A. Lantz-Andersson, M. Lundin, N. Selwyn, Teaching and Teacher Education, **75**, 302-315 (2018)
26. S. Colognesi, C. Van Nieuwenhoven, S. Beausaert, European Journal of Teacher Education, **43(2)**, 258-276 (2020)
27. P. Jarvis, *Adult education and lifelong learning: Theory and practice* (2004)
28. M.L. Bernacki, J.A. Greene, H. Crompton, Contemporary Educational Psychology, **60**, 101827 (2020)
29. N. Denzin, Y. Lincoln, L. Smith (eds.), *Handbook of critical and indigenous methodologies* (2008)
30. G.M. Alam, A.R. Forhad, I.A. Ismail, Technological Forecasting and Social Change, **159**, 120184 (2020)

31. D.O. Jewell, S.F. Jewell, B.E. Kaufman, Human Resource Management Review, 100749 (2020)
32. S. Wilkins, S. Neri, Journal of Studies in International Education, **23(4)**, 451-472 (2019)
33. E. Smith, P. White, Journal of Science Education and Technology, **28(1)**, 26-40 (2019)
34. H. Schroth, California Management Review, **61(3)**, 5-18 (2019)
35. A. Stathopoulou, N.T. Siamagka, G. Christodoulides, European Management Journal, **37(4)**, 421-431 (2019)

Using interactive technologies and distance learning in sustainable education

Anna Rybakova[1*], *Aleksandra* Shcheglova[2], *Denis* Bogatov[2], and *Liudmila* Alieva[3]

[1]Center for Additional Professional Education Moscow Institute of Physics and Technology, Kerchenskaya str., 1A, bldg. 1, Moscow, Russian Federation
[2]Russian State Social University, Wilhelm Pik str. 4 p. 1, 129226 Moscow, Russian Federation
[3]Department of Russian history, Pskov State University, 2 Lenin Square, 180000 Pskov, Russian Federation

Abstract. This paper focuses on the use of interactive technologies and distance learning in sustainable education. It discusses how remote learning technologies can positively influence students' learning and entry in sustainable education. The paper looks at the use of distance learning in higher education as a means to help students in the built environment and its use within the education system. It studies and expands the theoretical research on the benefits of distance learning, where the study is remote and there is no personal contact with staff or students, and examines the impact of distance learning on the student's learning experience. It also proposes and evaluates potential solutions to overcome the barriers to learning in the built environment and create successful virtual learning communities, recognising that such improvements must be reconciled with the primary benefits identified. The paper provides an overview of sustainable distance learning within higher education and discusses the differences between learning outside the structural environment of a profession, what it means for the student's learning experience and the potential to overcome barriers to distance learning. This is a very timely topic in the times of COVID-10 pandemic. Lockdowns of the economy and social life impacted all spheres of education with schools and universities closed for long periods of time and all teaching moved to online and distance mode. However, coronavirus pandemic also brought the digital surge in the system of education, including the sustainable education. All these innovations might stay after the pandemic and help the education to evolve and to embrace more novel trends and technologies.

1 Introduction

Since distance learning is continuing to expand, educators must be careful to examine the problems caused by this paradigm shift. Distance learning is becoming a common practice, as evidenced by the number of universities offering distance learning programmes, companies offering distance learning programmes and the number of distance learning courses - K-12 learning projects that have already been set up or are currently in use [1, 2].

* Corresponding author: anya_rybakova@list.ru

It was also noted that more and more students do not want and need traditional and flexible working hours. Thus, remote study became an ever more popular method, in order to obtain higher education which becomes clear by the number of the correspondence courses, which are offered at the university [3]. Distance learning has the potential to reach many more people through the use of online courses and other forms of learning such as social media and online learning. In literature, the term "distance learning" is used both for online-based programs and for programs that also include personal contacts. One has to distinguish the distance learning from a situation when courses are held at remote universities with local staff and students and staff never meet in person. Distance learning benefits students when and where they can devote themselves to learning, and it benefits all students, regardless of when, where or how they do it [4]. Information and communication technologies (ICT) can create distance learning institutions - learning platforms that enable many students who are far from the centre of learning to form themselves. Distance learning can support their families and be given the opportunity to work while raising their family or working full-time. In addition, the use of e-learning systems can offer a number of other benefits, such as helping to teach students in low-income and countries with limited access to educational resources, and providing students from other countries with the opportunity to learn globally and access higher education.

There are a number of other benefits of distance learning technologies, such as helping higher education institutions to combat the high cost of education and lack of access to resources for students in low-income countries. Institutions use distance learning technologies and often expect student outcomes to soar. Only by combining the right distance learning technologies with trained teachers and the focus on sustainable education, the educational institutions can offer more personalised teaching to positively impact students' learning outcomes in higher education [5, 6].

Distance learning technologies can also help rural and remote schools and universities to overcome the disadvantages of geographical isolation by expanding course offerings and learning opportunities, and connecting teachers and students with a broader range of resources and materials. Given the high cost of new technologies, more rural districts may be able to take advantage of distance learning [7]. Surveys of rural school districts show strong support for distance learning as a way to address teacher shortages and meet the needs of their students. Half of those surveyed supported distance learning as an alternative to traditional teaching and learning methods, and half favoured the use of online courses [8].

Universities and schools need to enhance their competences in distance learning. In doing so they might learn from each other. For example, the University of North Dakota offers a 34-hour virtual education program that focuses on the use of technology in education, communication and social media in the workplace. The curriculum combines training in psychology and communication and prepares graduates to use the latest technologies for educational purposes. It includes the leading curriculum for learning in virtual education (LCLE) and the university's first ever virtual learning course, Virtual Learning. The 100% online format can be completed in just two years and allows for a flexible schedule for professionals [9]. This can be a challenge for some lecturers and they may need to be trained professionally. It is believed that doctoral students will prefer online courses due to their better access to online resources and more flexible schedules. It is believed that distance learning at a higher education institution can be an important part of a student's development as an ICT specialist. If pupils' ICT skills can be improved and if pupils' attitudes towards online learning are positively influenced, distance learning in higher education could be used as a tool to increase the number of pupils who could have taken part and the quality of education.

2 Effective learning environments for sustainable education

Teaching and learning methods tend to humanize and represent a new form of educational quality that suggests that they could potentially stimulate adequate action for the environment and sustainability, both at individual and collective levels. Projects of change appear to have been an important step in the development of a quality education that is more responsive to the needs of students and the environment, as well as to their own needs. During the implementation of the change projects, the participants show a strong interest in learning contexts by using and promoting teaching and learning methods that emphasize critical, cooperative and experiential values [10, 11].

However, the need to rethink education at all levels, including teacher training, cannot be separated from the current debate on the quality of education. Drawing on the results of a regional research programme involving ten countries in Southern Africa, the researchers argue that quality education not only focuses on learning for a sustainable future, but also on the processes that create learning links, such as learning in connection with social, cultural and economic development, as such processes tend to deepen understanding of context, concepts and practices, while improving inclusiveness and knowledge-based access [12]. All of these seem to offer a unique opportunity for learners to develop a deeper understanding and appreciation of the social and cultural context of their learning environment, both inside and outside the classroom.

It is therefore crucial that the knowledge, attitudes and behaviour of teachers are supported in the design of teaching and collaboration in the classroom. Over the past two decades, tech companies invested a lot of time, effort, and money in research into effective teaching and learning methods, and has decided that one of the best investments, governments can make to improve student learning is in teacher professional development [13]. Unless there is a strong focus on teacher training as a key component of a sustainable education system, this will only lead to disappointing results. There is a guidance design paradigm known as ADDIE, an acronym that stands for analysis, design, development, implementation and evaluation. It is the basis for designing effective learning environments for sustainable education systems. It notes that this paradigm, while generally accepted, does not take into account the importance of design for sustainability [14].

Generally. there are some key concepts that enable students to discuss together their importance for the design of innovative and sustainable educational approaches. It intends to guide its students in taking into account the learning environments used to teach students from underdeveloped regions of the world such as Africa, Asia, Latin America and the Middle East. Learning through design - based education and thus embedding the principles of self-determination - regulated learning, sustainable learning and sustainable teaching and learning practices. With regard to the above, some researchers note that even if a learning environment is designed to promote self-regulated learning [15]. Students can perceive the environment in two different ways. On the one hand, they might find this environment motivating and stimulating, on the other hand, threatening [16]. These phases should be coordinated when designing learning environments for sustainable learning and sustainable teaching and learning practices.

Lecturers can encourage students to consider their campus, whether it is in the city or in the countryside, as a classroom for sustainability. This can help students to understand and discuss sustainability issues and initiatives in the community and to present them to faculty and staff. Using an ecological footprint calculator is a useful tool for educators who want to engage students in wide-ranging topics [17]. Higher education offers a wide range of learning services to teachers who are considering including policies, research and community engagement in their courses. Raising awareness of the impact of environmental issues on students, faculty and staff is crucial to bring about positive change. Education must be a big part of the solution, and we should be grateful to schools and other educational organizations that are pioneering and inspiring others to do the same. It offers the opportunity to involve young people in complex sustainability issues and help to build a bridge between science and the humanities. Efforts are also being made to integrate curricula in these areas with project-based learning opportunities, while ensuring a high degree of sustainability in the curriculum, such as the use of renewable energy and education for sustainability. There are also some educational institutions that are recognized as the ones that save energy, reduce costs, provide environmentally sustainable learning environments, protect health, and provide environmental education to promote academic achievement and community engagement. This approach protects health by reducing the chemical environmental hazards that negatively affect children's growth and development [18]. Students, teachers and school staff are trained in student-led green teams to create a healthier school. A better learning environment promotes health and education for all by reducing the risk of environmental chemicals and hazards that negatively affect childhood and young adulthood growth and development.

3 Interactive educational process

In the long run, educators, school districts, and states can use the information technologies worldwide to support continuous learning improvement and innovation. Families can be more confident about how they and their child learn throughout the school day. The method of "thinking, mating and sharing" is a great way for teachers to hear from students, even those who are quiet in class. When students feel comfortable in a class to exchange ideas, they can learn through collaboration and engagement. Students are more involved in class discussion and participation and make cumulative assessments of their progress. Properly implemented, this saves teachers time, helps students get involved, keeps students prepared for class, and keeps teachers informed throughout the learning process [19].

Teachers and trainers can also be supported by observing students and seeing if they understand the material under discussion. By creating these conditions, the activities of the teacher can be replaced by activities for the students aimed at developing the ability to communicate effectively, analyse alternative opinions, and to build intellectual viability, confidence and ability, which also constitutes the ability to make informed decisions. The role of teachers is also changing, as they do not provide ready-made answers to the questions discussed, as in traditional teaching, but rather encourage students to seek solutions independently and in an informed manner [20].

It is generally accepted that knowledge can be acquired by listening to lectures and reading texts, as well as other forms of learning such as reading, writing and writing. In some cases, it is even the information that is transferred from one form of learning process to another, such as from a lecture to a book, or even from the classroom to the office. In summary, teaching and targeted guidance in the learning process are complex and can take many forms. One of the goals is to create comfortable learning conditions in which students feel comfortable themselves, which makes the learning process productive [21]. Consider measures of students' desired learning, including overlapping goals that are taught as tested goals, and control of student behaviour and classroom activities. Under the best conditions, it takes many years of experience for most teachers to fulfil the ideal teaching practices that they have set themselves as protections for undergraduate students, but it is critical that classroom activities are taken into account. Interactive learning is an organization of the educational process, whose essence is the interaction of students, teachers and other students in the classroom in many ways. Interactive activities and lessons include the organisation, development, dialogue and communication that lead to the development of students' knowledge, understanding and understanding and the learning process [22]. Interactive learning takes into account the needs of the students, the students analyse their own activities, their personal experiences are included, targeted adjustments of knowledge are made and optimal results are achieved. Interactive learning expands the educational offer, helps to establish emotional contact with the students, teaches teamwork, takes into account the needs of each student, expands and expands, expands and expands. Educational institutions worldwide use active and interactive forms of study to develop the professional skills of students.

4 Using interactive technologies for multidisciplinary sustainable education

Multidisciplinary university usually involve a wide range of sub-fields, including digital media, streaming media and social media. Through education, digital technologies affect our lives and our societies. There is a number of interesting aspects, some of which are interesting for both practising managers and scientists [23]. Today's students and pupils must be entrepreneurs who monitor market trends, exploit innovation, create intellectual property, and apply aesthetic solutions to enhance the value of products, services, and experiences. The design, development and evaluation of technologies should be carried out in a participatory design process involving teachers and students. A rigorous scientific approach should not exclude the possibility of being inspired by the humanities, and especially the arts. Start has developed a growing awareness of how art and science are strongly interlinked when it comes to the design process and the development of innovative products and services. In the case of multisensory technologies in education, we believe that the exceptional ability of the arts to communicate content through sound, music and visual media represents a significant added value to the educational perspective. This is an opportunity to design and develop technologies based on sound scientific knowledge and tailored to the real needs of teachers and students. We believe that this becomes a key element in effectively supporting an embodied, enactive educational approach that uses the best sensory modalities to teach concepts in schools. We already know that a multidisciplinary approach is needed to understand how these technologies work, what benefits they can have, how people interact while working on the project, and what is learned. Based on our experience with technologically advanced learning projects, we suggest that we consider it important to "fully seize" this opportunity before taking advantage of it [24].

On the other hand, there is little research on the benefits of working on interdisciplinary technology-enhanced learning (TEL) research. Scanlon et al. reported on an analysis of prototype TEL projects funded by the National Science Foundation and the Office of Science and Technology Policy at the U.S. Department of Energy [25, 26].

Other fields of interest demonstrated the creation of knowledge, but there was relatively little in comparison to the interest in studying it. On the one hand, there are many examples of projects that work in teams, such as in the fields of computer science and technology, but also in other areas of science and in human-computer interaction. In general, their research objectives are to promote multidisciplinary scientific research and interdisciplinary cooperation in the field of computer science and technology. These goals require the collection of recognized scientific credibility and cutting-edge expertise in a wide range of disciplines, including engineering, mathematics, physics, chemistry, biology and physics. There is a need to use the resources available to maintain interdisciplinary cooperation in order to advance research and development of TEL technologies. Although some aspects such as data collection have improved, the results of some studies suggest that the use of ICT itself is not sufficient to change team dynamics and improve the delivery of knowledge and the quality of education. There are also human and organizational factors that support the successful use of technology in education process [27, 28]. Specialists who understand how to use and interpret technology and information systems, experts who design clinical databases for teams, design interactive systems and analyse clinical records during meetings are of particular importance. ICT alone is a tool that can be used to improve the learning process, but there is also a need for human organisational factors to support the successful use of these technologies in health care. They have a range of tools to capture and describe user needs and design interactive systems, including conceptual structural design. They also have an access to ethnographic and other qualitative research, which seeks to understand the needs and practices of people in the use and potential use of interactive technologies and in their use. In fact, identifying these requirements and developing designs is an active research topic that calls for the use of existing and novel techniques.

The use of ICT tools would allow children and young adults to have the same teaching methods thus naturally breaking down barriers between peers and facilitating social interaction. We believe that they are essential to effectively support educational approaches that use the best techniques that are appropriate for communicating concepts. They allow for the development of different strategies that teachers can choose more flexibly. Some newer technology can be designed to introduce children to the geometric concepts and forms as well as to explain them complex concepts such as global warming, climate change, or sustainable development.

5 Conclusions

The results of this paper can lead to the development of a new way of structuring learning so that students can learn more effectively and efficiently than ever before. In this respect, vocational school teachers and university lecturers alike face the challenge of developing and implementing teaching methods aimed at activating the interests and abilities of pupils and their learning ability.

One of the ways to organise the educational process is to use interactive learning methods such as seminars and training elements for better explaining concepts such as sustainable growth or global environmental changes. Professors and teachers are considering the possibility of designing an active and interactive form of work for students, as well as introducing seminar and training elements into practical teaching, which is an innovative approach to educational processes.

All in all, the novel technologies could and should be used in distance learning and education in general and for sustainable education in particular. Recent COVID-19 pandemic that helped to digitalize the education at all levels became a powerful trigger that would help to accomplish the transition to digital and online education that has the benefits of all powerful tools, like augmented reality or virtual reality, for explaining and visualizing the complex issues. Sustainable education in schools and universities would only benefit from this transition.

References

1. K. Lee, H. Choi, Y. Cho, The Internet and Higher Education, **41**, 25-33 (2019)
2. U. Kaden, Education Sciences, **10(6)**, 165 (2020)
3. Z. Gardanova, V. Ponkratov, N. Kuznetsov, N. Nikitina, O. Dudnik, E. Latypova, S. Shcherbatykh, Journal of Open Innovation: Technology, Market, and Complexity, **6(4)**, 1–21 (2020)
4. L. Dos Santos, International Journal of Education and Practice, **8(1)**, 134-148 (2020)
5. N. Colvard, C. Watson, H. Park, International Journal of Teaching and Learning in Higher Education, **30(2)**, 262-276 (2018)

6. E. Agormedah, E. Henaku, D. Ayite, E. Ansah, Journal of Educational Technology and Online Learning, **3(3)**, 183-210 (2020)
7. D. Zinger, J. Sandholtz, C. Ringstaff, The Rural Educator, 41(2), 14-30 (2020)
8. A. Picciano, J. Seaman, I. Allen, Journal of Asynchronous Learning Networks, **14(4)**, 17-35 (2010)
9. R. Ferdig, E. Baumgartner, R. Hartshorne, R. Kaplan-Rakowski, C. Mouza (eds), *teaching, technology, and teacher education during the COVID-19 pandemic: stories from the field* (2020)
10. P. Payne, Journal of Environmental Education, **51(2)**, 113-143 (2020)
11. W. Penuel, R. Riedy, M. Barber, D. Peurach, W. LeBouef, T. Clark, Review of Educational Research, **90(5)**, 627-674 (2020)
12. H. Lotz-Sisitka, E. Rosenberg, P. Ramsarup, Environmental Education Research, 1-29 (2020)
13. M. Goralski, T. Tan, The International Journal of Management Education, 18(1), 100330 (2020)
14. C. Budoya, M. Kissaka, J. Mtebe, International Journal of Education and Development using ICT, **15(1)**, 1-20 (2019)
15. J. Broadbent, M. Fuller-Tyszkiewicz, Educational technology research and development, **66(6)**, 1435-1455 (2018)
16. G. Goloshumova, O. Ershova, V. Salakhova, A. Kidinov, S. Nalichaeva, V. Yanysheva, Eurasian Journal of Biosciences, 13(2), 1867-1874 (2019)
17. W. Villegas-Ch, X. Palacios-Pacheco, S. Luján-Mora, Sustainability, **11(10)**, 2857 (2019)
18. N. Anwar, N. Mahmood, M. Yusliza, T. Ramayah, J. Faezah, W. Khalid, Journal of Cleaner Production, **256**, 120401 (2020)
19. E. Tokareva, Y. Smirnova, L. Orchakova, Education and Information Technologies, 24(5), 3219-3234 (2019)
20. Y. Waghid, N. Davids, *Teachers Matter: Educational Philosophy and Authentic Learning* (2020)
21. C. Rapanta, L. Botturi, P. Goodyear, L. Guàrdia, M. Koole, Postdigital Science and Education, **2(3)**, 923-945 (2020)
22. I. Backfisch, A. Lachner, K. Stürmer, K. Scheiter, Computers & Education, 104159 (2021)
23. A. Lytos, T. Lagkas, P. Sarigiannidis, K. Bontcheva, Information Processing & Management, **56(6)**, 102055 (2019)
24. S. Seufert, J. Guggemos, M. Sailer, Computers in Human Behavior, 115, 106552 (2021)
25. L. Daniela, A. Visvizi, C. Gutiérrez-Braojos, M. Lytras, Sustainability, **10(11)**, 3883 (2018)
26. E. Scanlon, S. Anastopoulou, G. Conole, A. Twiner, Frontiers in Education, **4**, 134 (2019)
27. S. Kim, M. Raza, E. Seidman, Research in Comparative and International Education, **14(1)**, 99-117 (2019)
28. R. Malik, Journal of Sustainable Development Education and Research, **2(1)**, 9-20 (2018)

Regional and global implications for the sustainable development and education

Natalya Nikitina[1]*, *Raisa* Krayneva[2], *Alexandr* Platitsyn[3]

[1]Togliatti Academy of Management, 5, Yagodinsky forestry, Stavropol district, Samara region, 445144, Russian Federation
[2]Financial University under the Government of the Russian Federation, 49, Leningradsky Prospekt, Moscow, 125993, Russian Federation
[3]Togliatti State University, 14, Belorusskaya St., Togliatti, 445667, Russian Federation

Abstract. Sustainable development and education are linked into a holistic and complex system that might help to transform the society by imposing the sustainable development goals and objectives. Moreover, it might contribute to the creation of the new type of global citizens who would act at regional and global levels with a focus on preserving the environment and promoting the sustainability agenda. Global civic education has been supported by the United Nations Development Programme (UNDP), the World Health Organization (WHO) and other international organizations as a means of promoting international development but the results might vary across regions and countries. We show that all of the above leads to a global approach to the development of the education system, not only for students but for society as a whole. The results of our research are likely to feed into the development of the initiatives which develop environmentally friendly attitude and promote community environmental management by creating participatory networks for the nature restoration and protection

1 Introduction

Sustainable development became a mantra for the 21st century that is repeated in mass media and in the public discussions. This is not a surprise since many serious reports, studies and initiative, such as Agenda 21 or the Paris Agreement of 2015 demonstrated how alarming the situation with the global climate change might be and how hard the whole humanity needs to work together in order to prevent global environmental disasters threatening our sole existence from happening [1, 2].

One of the most powerful tools for promoting sustainable development is sustainable education that includes a number of tools and approaches targeted at showing the people that the environmental threats are very real but that some well-balanced and well-managed strategies and approaches might preserve the planet we live on for the future generations to come [3, 4].

* Corresponding author: nikitinanv_doc@mail.ru

One of these approaches is the social ecology. Social ecology is considered to be a traditional ecological study in which students study how many factors need to interact with nature to create the world's ecosystems [5]. Social ecology aims to investigate how individuals interact and react with their environment and how these interactions affect society and our environment as a whole. Social ecologists examine how individuals and collective institutions interact and interact, looking at the overall picture of the system as a whole, facilitating greater communication and social cohesion, and solving society's collective problems. This facilitates the development of a more holistic view of society and its interactions with nature, which has a positive impact on the resilience of the community by enabling the collective synthesis and use of ecological knowledge. The research also shows how periods of abrupt environmental change can bring coastal communities together, creating opportunities to play a more active role in managing their natural resources [6]. This might provide new insights into the role of social ecology in various ecosystems and offer new insights that are important for the sustainability of decisions on the management of natural resources [7].

In this paper we propose to draw on the work of ecosystem scientists to develop a framework that considers environmental and sustainable education as part of an ongoing social-ecological process that includes and promotes a positive feedback loop between environmental responsibility and social-ecological processes and has regional and global implications. This view has practical implications; for example, such views may indicate that environmental and educational programs help reinforce positive feedback loops represented by existing administrations and community practices, rather than acting independently of the surrounding social and environmental systems. The perspective presented in this paper is intended to stimulate discussion on how existing approaches to environmentally friendly learning and education can be extended. We draw on the research of social-ecological systems and try to integrate environmental education into a conceptual model that describes hierarchical ecology and feedback resistance. In this paper we build on previous work on systems thinking as an integral part of ecological competence. We look at learning through the lens of interaction, participation, management and action, but in this view, we may neglect other important processes that take place within the same civic-ecological context discussed in the work. We continue to refer to socio-cultural learning theory and propose that the educational programmes in which individuals participate can be viewed from an environmental perspective.

It is apparent that much of the current thinking on sustainability can be rooted in the ability of ecological systems to explain how they can recover on different time and spatial scales in response to changes in social and environmental conditions [8]. We find that we can reconnect with past experience if we try to explain the ecology of environmental education. A number of attributes characterize systems that are able to maintain important functions in the face of human - caused or other disturbances and rebuild after a threshold of change. The first type of resilience that emphasizes adaptation and adaptability would be that of an urban community that maintains a highway built into its core. Given that many practices of civil ecology are being taken up by urban minorities and immigrants who might not otherwise be involved in the management of natural resources, they represent an opportunity to broaden the range of parties involved in environmental policy and the promotion of sustainability.

Although such practices often involve the involvement of non-governmental organizations (NGOs) and private sector stakeholders, they can also serve as a bridge organization that integrates different perspectives and knowledge into resource management. Resilience, linked to different forms of governance, such as the ability to bring different perspectives on environmental issues, helps foster change among participants in local ecosystems. For example, in some courses pupils or students might explore how to connect and interact with the managed and built environment, including public health, education, health care, and the environment as a whole. Students can develop their worldview by exploring concepts of diversity, social justice and consumerism in relation to the promotion and maintenance of health and well-being [9]. They might look at how connections and interactions between these systems play a key role in promoting, supporting and maintaining the well-being of individuals, communities and environments as a whole, now and in the future. Through exercise and experience, pupils and students might have the opportunity to develop links with the environment and understand the interrelations between health, human beings and the environment [10].

2 Globalization and sustainability

Generally, without the globalization, there would be no ties of economic exchange and political agreement that bind nations and societies, but rather a shared awareness of being part of a global system [11]. Globalisation has brought about an unstoppable change in awareness, and schools are playing an increasingly important role in this. Consciousness gives a deeper understanding of the world and a greater awareness of our place in it and our role in it. The role of education in globalisation is being put very much in the spotlight, but a multitude of other concepts are being used under the umbrella of internationalisation and globalisation with a specific focus on sustainable education [12, 13]. The different terminology refers to those who take a strong approach to internationalisation and also to those who do not. The world of globalisation is increasingly being permeated by the globalisation of education, the globalisation of technology and the internationalisation of culture and society in general.

Currently, the characteristics that lead people to believe that we are in the process of globalization, including the role of the state and the importance of nations, are not as deeply rooted as they may appear [14]. Simply put, in a world of transnational corporations and global trade, the roles of states and their importance to nations are greater. Increasingly, the world shares problems and challenges that do not obey nation states - state borders like climate change, environmental degradation, and human rights violations. The transition into the 21st century is marked by striking differences in the expanding globalisation and globalisation of modern life and culture.

The term globalization, which is called a political - economic and socio-cultural - process, had two positive and negative effects. One consequence is a culture of emerging economies, held together by a common rhetoric focused on the "globalization of life" and "internationalization" of the world economy and society in general, as well as on international relations [15]. The level of economic globalisation has increased, which has a negative impact on the status quo in many countries and in some countries with negative effects on social and economic development. On the other hand, critics of globalisation will point out the negative impact it has on certain nations and sectors that may be exposed to increased competition from international companies. Globalization is important because it is one of the most powerful forces influencing the modern world, to the extent that it can be difficult to make sense of our world without understanding it. It can also have a positive impact on social and economic development, human rights and the environment.

Similarly, one can see that globalization has made it possible for the world's various national cultures to be diverse and inherently different. A truly multinational organization is an office or a supply chain that extends across the world. In this sense, it can be emphasized that the world is experiencing "Americanization" rather than globalization. American hegemony is reflected in the dominance of the Internet: 85% of all websites are in the United States, and American companies control more than 80% of software sales and more than 90% of the world wide web (WWW) [16]. Indeed, there seems to be a strong correlation between the rise of multinationals and the growth of globalization. It is very important that these powerful multinational companies also practice environmental-friendly approach and practice what is called "Corporate Social Responsibility", an approach when businesses take the responsibility for their actions related to the global climate change and attempt not only to act responsibly but also to embed the principles of the sustainable development into their corporate culture and everyday business [17, 18]. This can be achieved through schooling, promotion, as well as educating the company's management at all levels.

3 Global and regional dimensions of sustainable education

The UNESCO Transition Project examines three learning dimensions that give priority to engagement with education for sustainable development (ESD) and Global Citizenship Education (GSE) learning [19]. One can examine the three dimensions of learning reflected in the GCE identified, as well as the global and regional dimensions, in order to identify the impact of these dimensions on the development of education for sustainable development. Figure 1 below shows the development of the Education Index (a part of the Human Development Index (HDI)) in the selected countries [20].

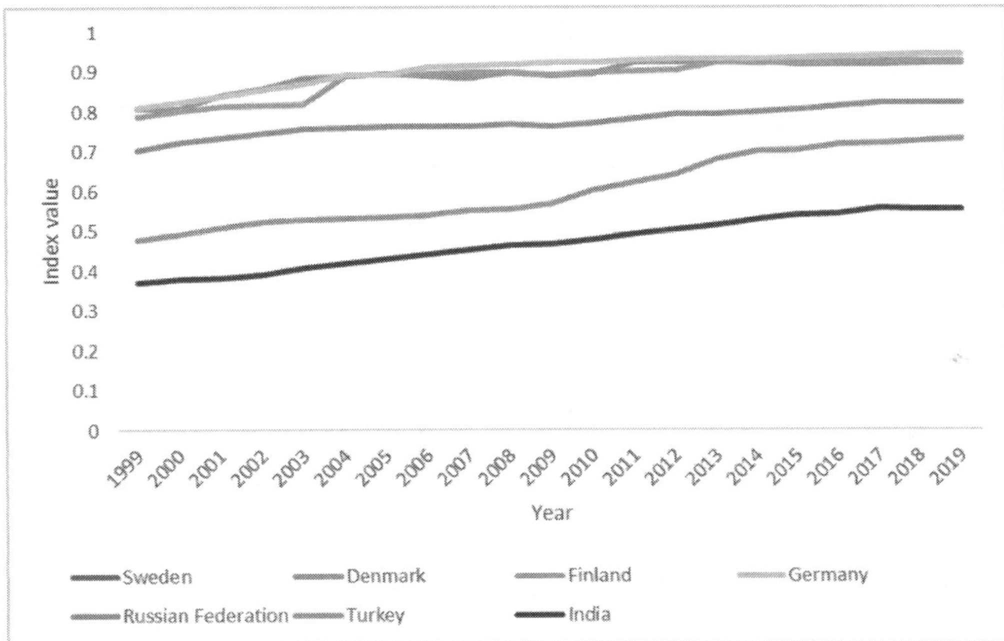

Fig. 1. Education Index in the selected countries (1990-2019)

One can see the comparison between the countries with the highest level of the total progress towards achieving all SDGs (represented here by Sweden, Denmark, or Finland) and the Russian Federation, India and Turkey. Russian Federation that ranks 57th is closer to Sweden, Denmark and Finland that rank the first three than to Turkey (70th) or India (117th). The level of sustainable education and the attention paid to this issue at all levels of educational institutions might and also seems to correspond to this chart.

In general terms, given that many other actors are now contributing to global cooperation, the Sustainable Development Agenda cannot be driven by the international organizations alone, given the importance of the global and its role in developing education for sustainability. The integrated framework for the Sustainable Development Goals (SDG) presented by the United Nations requires the participation of the multitude of actors that now populate the global health landscape. Indeed, an approach to the SDGs could provide an opportunity to build a network of diverse actors by bringing together those working in the health sector. This approach can be brought together with communities, national and regional organisations to provide a framework for the development and implementation of sustainable education for sustainable health and development. Indeed, one of the most important impetus for education for sustainable development comes from the World Health Organization (WHO), and in particular from the United Nations Development Programme (UNDP). In the course of the discussion and formulation of the concept of sustainable development, it became clear that education is the key to sustainability.

In many countries, the education for sustainable development is still shaped by the educational community and the environment plays a central role, as it continues to attach great importance to participatory and inclusive education for young people, which facilitates their own decision-making. The European Commission and UNESCO attach great importance to examining how our actions at local, national and international level are connected to the world. The aim of both organisations is to raise awareness of the importance of education for sustainable development and the role of young people in the development of a sustainable future. Our work as an organisation focuses on promoting participatory and inclusive education as a key component of sustainability education. Stop Climate Chaos, a campaign by the European Commission and UNESCO that brings together local, national, regional and international organisations and individuals to fight together for the reduction of greenhouse gases. The groups involved in this campaign may come from different sectors, may have different reasons for membership, but they are working together for a common good. Climate change campaigns could serve as an opportunity for educators in the fields of environment, development and sustainable development to work towards common goals. We must promote and develop the capacity of civil society groups to work for a sustainable world.

4 Conclusions

All in all, it is clear that recent years are marked by the changes in society and the world economy which are likely to dramatically increase the importance of globalisation for the development of society, culture and economy as a whole. The current (and more recently) epoch of globalization is dominated by the economic, social, cultural and political developments of the 21st century. All these developments are also subjected to the growing importance to understand the concept of globalization and to communicate it to the wider general public. We show how sustainable development began to play an important role in all spheres of human activities and how crucial it might be to provide the education for sustainable development to pupils and students at all levels of the educational process. All of these might create the solid basis for the sustainable growth and development of the society and economy including the strategies for regional development and business agenda of large and small companies.

Our results show that the key to overcoming the problem of the global warming and climate change lies in the development of citizens' initiatives in the areas of public health, education, health care and environmental protection.

Global citizenship which is emerging nowadays is an increasingly common idea: companies use the term to underscore their commitment to corporate social responsibility and sustainability around the world, universities use it to promote global awareness and international education, and there is also a movement that uses online activism and social media around the world to work toward ending global poverty. One might hear the terms' world citizenship 'and' world citizen 'in connection with the global movement for sustainable development and the fight against poverty and climate change. In short, it becomes clear that sustainable education is the tool that might help to promote it with a purpose to increase environmental awareness and positive attitude to the environment.

References

1. C. Hepburn, B. O'Callaghan, N. Stern, J. Stiglitz, D. Zenghelis, Oxford Review of Economic Policy, **36(1)**, S359-S381 (2020)
2. W. Strielkowski, E. Lisin, I. Gryshova, Romanian Journal of European Affairs, **16(4)**, 68-77 (2016)
3. F. Annan-Diab, C. Molinari, The International Journal of Management Education, **15(2)**, 73-83 (2017)
4. A. Aleixo, S. Leal, U. Azeiteiro, Journal of Cleaner Production, **172**, 1664-1673 (2018)
5. D. Hummel, T. Jahn, F. Keil, S. Liehr, I. Stieß, Sustainability, **9(7)**, 1050 (2017)
6. G. Kennedy, M. Raimonet, M. Berman, N. Gaye, J. Huctin, T. Kaleekal, J. Vanderlinden, Global Environment, **11(2)**, 405-433 (2018)
7. G. Cumming, C. Allen, Ecological Applications, **27(6)**, 1709-1717 (2017)
8. G. Fedele, C. Donatti, C. Harvey, L. Hannah, D. Hole, Environmental Science & Policy, **101**, 116-125 (2019)
9. F. Piasentin, L. Roberts, Environmental Education Research, **24(5)**, 694-715 (2018)
10. J. Pretty, M. Rogerson, J. Barton, International Journal of Environmental Research and Public Health, **14(7)**, 706 (2017)
11. S. Babones, J. Aberg, International Theory, **11(3)**, 293-317 (2019)
12. I. Čábelková, V. Blaginin, W. Strielkowski, A. Platitsyn, E3S Web of Conferences, **208**, 09001 (2020)
13. S. Son-Turan, International Journal of Sustainability in Higher Education, **22(1)**, 100-119 (2020)
14. O. Petricevic, D. Teece, Journal of International Business Studies, **50(9)**, 1487-1512 (2019)
15. E. Makarova, E. Makarova, T. Korsakova, Journal of History Culture and Art Research, **8(1)**, 111-127 (2019)
16. M. Cartwright, Internet Policy Review, **9(3)**, 1-18 (2020)
17. A. Scherer, C. Voegtlin, Academy of Management Perspectives, **34(2)**, 182-208 (2020)
18. J. Lu, L. Ren, S. Yao, J. Qiao, W. Strielkowski, J. Streimikis, Energies, **12(18)**, 3417 (2019)
19. M. Gómez-Zermeño, Journal of Sustainable Development of Energy, Water and Environment Systems, **8(3)**, 577-589 (2020)
20. United Nations, Human Development Reports. http://hdr.undp.org Accessed on 15 February 2021 (2021)

E3S Web of Conferences **250**, 07005 (2021) https://doi.org/10.1051/e3sconf/202125007005
TRESP 2021

A visualization of the energy-saving problems

Svitlana Pryshchenko[1]*, *Yevgen* Antonovych[1], *Andryi* Petrushevskyi[1]

[1]Department of Information Technologies and Design, State University of Infrastructure and Technologies, Kyrilivska str. 9, Kyiv 04071, Ukraine

Abstract. This paper focuses on the visualization of eco problems and the latest green technologies, including energy-saving. Today, there are four color-graphic means in advertising: photography, graphics (drawing or computer graphics), font compositions, and often a combination of these. Our study considers the imagery and stylistics of eco poster as a type of public advertising, and ascertains that in the subject of energy-saving a light bulb as the visual stereotype prevails. The paper indicates the importance of system design thinking, the use of creative approaches in the educational process of designers (metaphor, metonymy, hyperbole, association, allegory also) in creating social appeals for a wide audience – from young children to the older generation. Moreover, it emphasized the social value of design, and the aesthetic, moral as well as communicative aspects of visual information of environmental orientation within the context of this study.

1 Introduction

Today, special attention should be paid to advertising and information activities in the social sphere, in particular environmental, because of its important role in solving complex socio-cultural problems and educating society. The world community contributes to the active development of eco design and public advertising, greening Europe for social change and innovation at the global, glocal and local levels [1]. However, this support mainly concerns the financing of specific actions for the protection of the biosphere and the implementation of industrial development in everyday life in developed countries. The global task of today and the near future is to update effective energy-saving programs, which can`t be solved without shifting moral emphasis, guidelines, social priorities and practices in favor of harmonious human development and understanding of nature as the highest value. Critical content analysis of the aesthetic level of public advertising, especially in post-Soviet countries, revealed the use of outdated images, visual stamps and the lack of original ideas. Therefore, our research focuses on the relationship between graphic means and sustainability.

Design as a discipline has long been positioned from the context of developing and catering to the private good, focusing on individual and smaller sections of society – largely leaving its contributions to the common good underexplored. Within this context, the designers act as the important players in the formation and amelioration of any society [2].

Nowadays, the massive anthropogenic impact on the environment is apparent and clear. This puts an emphasis on the creation of the programs aiming at managing on the natural resources and conducting effective environmental protection. The authors consider this topic quite relevant. These are the development of social responsibility, eco consciousness,

[1] *Corresponding author: akademiki@ukr.net

creative self-expression and aesthetic education of pupils and students in the context of smart consumption strategy, popularization of eco-trends and eco design, promotion of clean cities, the formation of a new lifestyle in urban and rural environment, analysis of urban ecology, waste disposal, bioethics, studying the experience of European countries, improving the image of Ukraine in these matters.

And in a previous study, we noted that scientific articles in the "Nature" journal, publications of international organizations Greenpeace (USA) and World Wildlife Fund (Switzerland) are the most authoritative among foreign studies of environmental issues and the promotion of their solutions. In addition to publications and conferences, these organizations actively organize environmental seminars, exhibitions, fairs of eco products and eco design, as well as disseminate information about the environment through the Media. Although we still see an insufficient number of environmental social challenges in the media space – outdoor advertisements and single articles on news portals are the most common [3].

Noting the insufficient number of environmental social appeals in the modern media space (sometimes in Ukraine we can find outdoor advertising and some articles on news portals), we present an impressive fact – according to the monitoring of the Institute of Mass Media in 2016, only 1% of materials in Ukrainian online media, covering mainly foreign experience [4]. Over the next 5 years, the situation has not changed.

2 Materials and methods

Socio-humanitarian knowledge of the 20th and 21st centuries characterized by the emergence of new interdisciplinary areas, among which should be singled out "visual researches". The system-structural method chosen for this study allows the analysis of Design as a complex field of activity at the junction of science, art and technology: both at the level of analysis of individual factors and at the level of their synthesis in the disclosure and understanding of functional, technological, marketing and cultural aspects of visual information. environment: urban, subject, virtual, etc. In Design, a system is understood as a set of necessary and sufficient elements that are in an established relationship and form a single whole. The set of components covers socio-economic issues, problems of synthesis of utilitarian and aesthetic aspects, environmental friendliness, rational use of materials. Each structural element of the form in the design performs certain functions that meet the needs of the system as a whole. Thus, Design can be understood as a creative method, process and result of art and design activities to meet the utilitarian, social and aesthetic needs of the consumer, and the poster in this case is a tool that can be used productively to harmonize human life.

The article "Green gold" highlights that the round-up on last year's global investment in clean energy shows that renewables continued their rapid ascendancy, bringing multiple benefits for all [5]. In recent years, many authors analyze not only the problems of energy-saving as such, but also related educational and behavioral aspects, as well as development strategies in Future [6-16].

Sociologist Robin Cohen believes that posters, placards, advertising appeals are examples of creolization, as they combine the verbal part with the visual, which belongs to a different sign system than language. According to him, creolization occurs when designers choose certain elements of cultures, give them new meanings and creatively combine to create new images that become products of mass culture [17].

Advertising is a manifestation of semantic concepts, and the global trend is the direction of advertising graphics in the direction of transferring the functions of influence from its verbal part to the visual. The visual series no longer performs an illustrative-decorative function, but is formed as a visual-verbal model, the language of which is based on the techniques and means of fine arts. The use of visual means in advertising should be due to the focus on the target audience, taking into account actual topics.

The meaning of the information should be perceived by the person "in motion", and this will determine the visual shape of the poster – size, conciseness of the image, easy-to-read font, bright colors. In advertising posters often uses photography, symbolic elements, associative and emotional methods of influencing potential consumers. Eco poster belongs to non-commercial mass media. The ability to see and reflect an idea in a poster, and poster thinking – the main feature of a poster artist. The poster made at a high artistic level can be an important factor in positive aesthetic action. For more than a hundred years, "green" advertising has been educating environmentally conscious people, becoming brighter and more expressive.

The history of eco poster began in 1906, when the American Public Association organized the protection of Niagara Falls from the damage caused to it by energy companies. New trends in Art (constructivism, futurism, cubism, symbolism) stimulated the further development of the poster. Since 1961, the World Wildlife Fund has been actively covering environmental issues, and nowadays (Fig. 1).

The famous Swiss graphic designer Mueller-Brockmann believed that the poster is the most effective means of visual communication, it should stimulate, explain, provoke, persuade. Its form should be easy and accessible, however, to achieve this is much more difficult than to express oneself the artist in the painting [18].

Fig. 1. WWF, the World Wildlife Fund energy-saving poster, 2016 [19]

It should be noted that in the Soviet Union in the second half of the 20th century, a large number of propaganda posters in various graphic techniques were created on the use of atomic energy for peaceful purposes, directed against the nuclear arms race. The technique of work execution – watercolors, gouache, colored pencils. Such posters at exhibitions, in schools and industrial premises raised patriotic feelings through acquaintance with historical events in the life of the country (Fig. 2), and the whole World. The main role in raising the poster was played by competitions and exhibitions held by the Union of Artists and publishing houses in Moscow, Kyiv and Kharkiv.

Fig. 2. Soviet posters, 1980s: 1. Don't Be Like That! Save Electricity!
2.Atom to work for Peace, for Communism! 3.We give warmth
4.Nuclear energy for the cause of Peace!

Fig. 3. Visual stereotypes for energy-saving

In Ukraine, the Kharkiv State Academy of Design and Arts became the founder of the International Triennial of eco posters "4th Block". The triennial owes its name to Unit 4 of the Chernobyl. The "4th Block" exhibition was first held in 1991 as a tribute to the heroes who took part in the aftermath of the Chernobyl explosion. The exhibition has become an important event in the tandem of graphic design and ecology. The triennial collection includes several thousand works. The main purpose of this exhibitions is to draw the attention of the world community to environmental problems, the consequences of social, man-made and natural disasters (Fig. 4). The Ukrainian Association of Graphic Designers aims to unite efforts to help solve environmental problems through poster art, education of environmental awareness. Due to this, the association has achieved wide international recognition. The "4th Block" is not an elite creative forum closed to self-sufficient formal searches. It has a special social orientation, close and understandable to everyone and attracts more and more participants from different countries and continents.

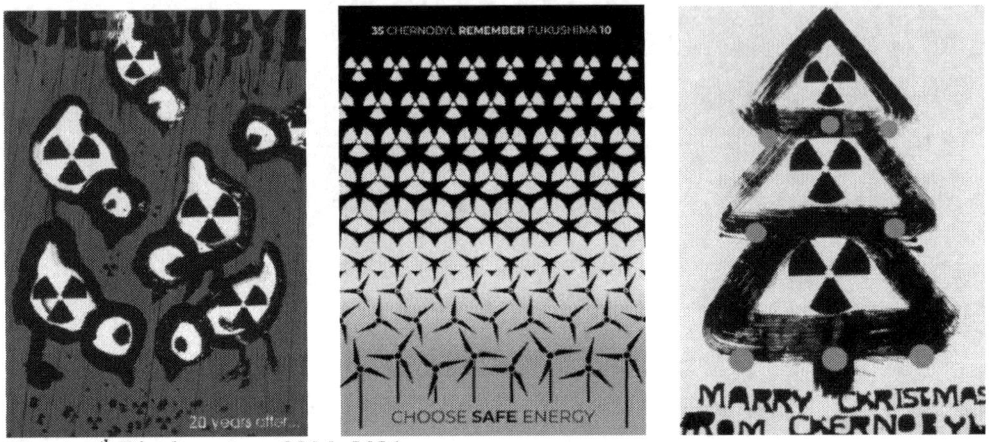

Fig. 4. 4th Block posters, 2006–2021

3 Results and discussions

Designers, who working in the media, based on the perception of color drawings and photographs, must develop strategies for visualizing the invisible – advertising ideas: this is a fundamental paradox of advertising. Compared to verbal language, visual elements are perceived faster, easier, more accurately. The poster must conform to the important principle of semantic integrity, which consists of physical, psychological and symbolic totality and strong internal connections. All color elements: illustrations, headings, text blocks, corporate constants, are combined into a common image, interact closely and determine its effectiveness, not only commercial but also social. Given the largest coverage of the population, the most effective means for eco poster is outdoor: billboards, city lights, advertising on transport and in the subway [20].

In the comparative study, the authors identified significant shortcomings of modern advertising graphics: the predominance of stereotypes, primitiveness, vulgarity and pop art, kitsch, eclecticism, as become cultural dominants. There is no visual culture of color, its conciseness and restraint. We distinguish three main visual means – photography, graphics (drawing or computer graphics), font compositions, and 4th, the most common tool – a combination of them: photos, drawings, or ornaments with background geometric screensavers. As a rule, modern posters are a combination of different visual means in one way or another. But the contemporary poster needs interesting images, non-standard, emotion, humor or satire, so the use of creative technologies in creating an advertising appeals – metaphors, metonymy, allegory, association, hyperbole – is becoming increasingly important. Among the common stylistic trends are minimalism, constructive style, postmodernism (eclecticism in a combination of various visual means: drawn characters, graphic elements, text blocks, photographs, works of art).

The use of visual *metaphors* in the poster indicates that they have significant potential and persuasiveness, perhaps on a subconscious level, and, as a result, have a result. Metaphor as a creative technique and means of communication becomes a basic characteristic of modern visual culture, a tool for integrating abstract and concrete, interpretation of complex aspects of worldview, ideas and combining images with emotions, a means of creating and emphasizing new meanings, which is so important in advertising.

Hyperbole is the second most frequently used technology in advertising. Exaggerating the positive characteristics of a product or service enhances the expressiveness and emphasizes the originality of the message.

When creating public advertising, designers should choose a few basic *associations,* rather than trying to gather all the feelings and associations. It is clear that this is difficult and requires appropriate visualization. By developing one (or more) stable positive associations, the customer and the designer get the opportunity to increase the social value of the selected object.

Allegory – dissent, creative advertising technology that avoids banality, cliches, presenting objects.

Metonymy ranks last among advertising technologies in terms of frequency of use, as it is not as simple a technique as it seems at first glance. The transition from form to form, renaming is more complicated than metaphor – the foreign language of metonymy is based on the definition of one object through another in close connection, comparison, irrational in its nature.

British designer Tolly defines "new simplicity" with visual trends of the present and the near future: clear structure of advertising information, conciseness, minimalism in the use of graphics, pronounced contrast, and limited colors. Tolly claims that decorative design, posters, branding, packaging, printing, and WEB design have saturated the market over the last decade. Minimalism is not so much a trend as a "barometer of social change". In the 1950s, graphic minimalism was a way to overcome the dominance of advertising. Modern minimalism is a form of restraint that comes to the fore as a reaction to clearly expressive graphics, as a response to the visual excess of the 1990s – early 2000s. There used to be a lot of interest in the culture of postmodernism, and everyone who showed itself reflected it in the poster. It was very decorative and very "loud". According to him, we are now in a period of austerity, and this ideologically visualizes minimalism [21].

The current environmental situation gave rise to the organizing a joint German-Ukrainian project "ECO-culture" on the base of Munich Ludwig Maximilian University (LMU) for holding a contest of environmental drawings for children and youth in the framework of the Munich–Kyiv partnership (September – December 2015; the concept and presiding over by Pryshchenko). The aim of the project was a development of eco thinking, eco education and long-term collaboration of the civic communities of Germany and Ukraine, popularization of eco trends, forming a new lifestyle with the means of poster art, and promoting the concept of clean cities, consumer culture and energy-saving. It was the first step. In eco poster competition was attended by more than 70 children from different countries living in Munich: Ukraine, Italy, Germany, Poland, Bulgaria, Greece, Latvia, Hungary, France, Serbia, India, Kosovo, Croatia, Romania, and Iraq. In 2016, she conducted a thematic workshop "Environmental public advertising as a visual-verbal model in the structure of modern media" for LMU students.

The next stage was organizing the exhibition based on the results of eco poster competition among students of the National Academy of Culture and Art Management, whose posters were presented in May 2017 at the "Pecherskaya" gallery (Kyiv). A total of 90 posters were presented in the exhibition. All contestants were awarded an ecological kit (eco notebook, eco pen and eco package) from recycled packaging from the company "Tetra Pak", and the winners – diplomas, design books, and art materials.

Thus, as a further development of the project, in early 2021 the State University of Infrastructure and Technologies (Kyiv) began the International virtual eco poster competition among students from European countries: Poland, Germany, Ukraine, Czech Republic, Lithuania, Latvia, Belarus, Moldova, Georgia, Croatia, Greece, Azerbaijan, Armenia also. The main goal of the competition is to draw public attention to modern environmental threats, to identify the creative potential of young people in caring for natural resources, understanding significant biodiversity loss, climate change, biosphere protection, energy control, and possible ways to solving these eco problems.

International socio-cultural designing in this direction is quite interesting and promising, the main focus of which will be ecology and in-depth study of regional cultures and their manifestations in the modern media space, holding exhibitions of eco posters within the integrated Carpathian summer and winter schools in Ukraine.

4 Conclusions

Overall, we can conclude that energy-saving problems can often be neglected in today's globalized and sophisticated world if compared with waste management or global warming and climate change issues. The system-structural analysis of the poster's visual means revealed tendencies of straightforwardness and stereotypes, if these images are created without design education and a deep understanding of the subject.

Prospects for the use of research results are as follows. The obtained scientific results will be important at the theoretical, practical and educational levels and will generally contribute to the development of eco poster. Some of the statements are debatable and require further research in the field of visual culture, digital media in order to implement modern design concepts of energy-saving. Design and advertising face the growing imbalance of resources and consumption in the future, and must be based on new socio-cultural principles of harmonious coexistence of the individual, society, state, economy, culture, etc.

References

1. European Green Week, *https://www.eugreenweek.eu* (2020)
2. Swissdesignnetwork, *https://swissdesignnetwork.ch/symposia/design-as-common-good* (2020)
3. O. Novikova, S. Pryshchenko, Herald of the National Academy of Culture and Art Management, **3**, 57-61 (2016)
4. Detector Media, *www.detector.media/monitoring/article/119349/2016-10-04-ukrainski-internet-zmi-zgaduyut-pro-ekologichnu-temu-lishe-v-1-svoikh-materialiv-imi* (2016)
5. Green gold. Nature energy, 2016. **1,** 16011; https://doi.org/10.1038/nenergy.2016.11
6. U. Ablayeva, N. Normatova Problems of Architecture and Construction, **2,** 1, 7 (2019)
7. R. Bull, J. Romanowicz, N. Jennings, M. Laskari, G. Stuart, D. Everitt, International Journal of Sustainability in Higher Education, **19(7),** 1220-1238 (2018)
8. P. Fraternali, F. Cellina, S.L.H. Gonzales, M. Melenhorst, J. Novak, C. Pasini, C. Rottondi, A. Emilio Rizzoli. Visualizing and gamifying consumption data for resource saving: challenges, lessons learnt and a research agenda for the future. Energy Informatics, **2,** 22 (2019)
9. T.G. Holmes, Proceedings of the 6th ACM SIGCHI Conference on Creativity & Cognition, 153-162 (2007)
10. M. Iordache Platis, J. Romanowicz, Sustainability, **12(22),** 9626 (2020)
11. K. Koroleva, M. Melenhorst, J. Novak, S.L.H. Gonzales, P. Fraternali, A. Rizzoli, Energy Informatics, **2(1),** 1-20 (2019)
12. Y. Liu, J. Liang, X. Wang, Y. Ouyang, Sustainable Development, **3,** 116-122 (2013)

13. S. Pahl, J. Goodhew, C. Boomsma, S.R. Sheppard, Frontiers in Psychology, **7,** 92 (2016)
14. I. Čábelková, W. Strielkowski, I. Firsova, M. Korovushkina, Energies, **13(7),** 1742 (2020)
15. Y. Tange, S. Kiryu, T. Matsui, Y. Fukuyama, IEEJ Transactions on Electronics, Information and Systems, **134(1),** 78-84 (2014)
16. E. Van der Werff, D. Taufik, L. Venhoeven, Energy Research & Social Science, **54,** 26-33 (2019)
17. R. Cohen, Globalizations, **4 (3),** 369-384 (2007)
18. J. Müller-Brockmann, *History of the Poster* (2004)
19. WWF, *https://wwf.panda.org/wwf_offices/switzerland* (2016)
20. S. Pryshchenko, *Design and Advertising* (2020)
21. S. Tolley, *The new simplicity in Graphic Design* (2016)

Family leadership and environmental sustainability: a case of the single fathers in Russia

Olga Bezrukova[1*], *Valentina* Samoylova[1], *Maria* Yashina[1]

[1]Department of Sociology, Saint-Petersburg State University, Universitetskaya Emb. 7-9, 199034 Saint-Petersburg, Russian Federation

Abstract. Nowadays, the traditional perception of the family is changing. However, understanding children preferences and shaping their views of the world still remain the key prerequisites for the environmental sustainability. The purpose of the article is to analyze models of single fatherhood, to study the motivation and structure of factors that determine the involved fatherhood making, the specifics of mother's and parent family's influence on the paternal practices implementation. Our research testifies to the fact that single fathers tend to become family leaders and undertake responsibility related to childcare in the context of transforming marital and family relations as well as facing global environmental issues. The results of the study show that single fatherhood is usually a forced situation caused by death or severe illness of the mother, her deviant behavior and leaving the family, deprivation of parental rights, divorce consequences, long-term separation of spouses, use of modern reproductive technologies of surrogacy. It is concluded that the scenarios of the single fatherhood becoming – planned or casual – are associated with the cause of the child appearance in the family. The significant differences are found in the social and cultural capital of the single fathers which might have different impacts on the level of environmental education they can pass on to their children.

1 Introduction

The number of the single fathers grow throughout the world, and the same trend is observed in Russia. According to the 1989 Population Census, the share of the paternal monoparenthood was only 1.0% of total number of the families [1].The results of the 2010 All-Russian Population Census demonstrated increase of the paternal families share in total number of the family cells with minor children under 18 up to 3.7% [2] and the 2015 Microcensus gave us 3.2% [3]. These phenomenon is caused by the recent changes in our society such as globalization, precarization, but also global climate change and all its adverse effects.

The research interest shown by the Russian authors in the problems of the paternal families can be observed in the analysis of the causes and typology of the male

*Corresponding author: o.bezrukova@spbu.ru

monoparenthood, characteristics of a social portrait of the single father with children, experienced difficulties, educational practices [4-10], contradictory generation of "caring" masculinity and involved fatherhood [11, 12] as well as in the study of protection of rights, technologies, and measures enabling to support the single fathers as new clients of the social services [6, 8, 13]. Deficient researches and poor studies of the single fathers as a social group give rise to its polar assessments: from positive heroic images of forced single fatherhood due to severe illness, death, or deviance of the mother [14] to negative ones relating to the fathers estranged or underperforming their parental responsibilities [8, 15].

A separate category of the Russian publications is devoted to description of practices of working with fathers in various governmental and non-governmental organizations. In the presented studies, the main factors of the increasing paternal monoparenthood are family instability and transformation of family values; decrease in parental role asymmetry due to breaking traditional gender order; fatherhood development as a parental substitute, appearance of a "new" masculinity and alternative models of paternal behavior, growth of paternal movements, prevalence of labor precarization, educational and professional mobility [4, 12]. The foreign researchers state that although, as before, the majority of single parents are mothers, the number of single fathers who are the main guardians of their children is gradually increasing [16]. Some of the few studies devoted to such families review the content and specifics of the parental functions of the fathers compared to the mothers. The analysis of four categories identified for child care activities (routine care, games, management, and training) shows that the single fathers spend far less time for child care than the single mothers do (average time spent by the fathers is 54 minutes and the same by the mothers is 1 hour 39 minutes) in all categories of the child care activities except for the games [16]. These data confirm the earlier studies [17, 18] and fit into a gender paradigm according to which women and men behave in accordance with the social expectations and standards regarding their gender. As a result, the gender contrasts that are considered to be natural and justified ones are acknowledged, and it is reasonably to expect in the future that single fathers will be less involved in most child care activities than the mothers creating certain risks to the well-being of their children. At the same time, it is noted that the structural characteristics of the immediate environment affect the balance of the parental functions: the gender differences in the total time of child care are mitigated if female family members are near the father and child or, on the contrary, time shortage of communication with the child increases if the family includes other male family members [16].

In any case, the father bears the primary responsibility for the children upbringing and his participation and involvement become necessary conditions for transfer of social capital resources to the child. According to the study of [19], the teenagers from the families with the single fathers report of the higher level of participation of their fathers than the teenagers from the families with married birth-fathers or stepfathers; however, the levels of involvement, supervision, monitoring, and closeness with the children of the single guardian fathers is lower than the same of the guardian mothers. The families with the single fathers which do not differ from the families with both parents in terms of the fathers' education are less vulnerable in terms of socio-economic resources than the families with the single mothers [19]. The economical benefits allow children from these families to receive more resources for intellectual development that is confirmed by the higher test results [17]. At the same time, in terms of financial and material resources, such families are still inferior to two-parent families [20]. More educated fathers are often more bound with their children, and the presence of the uneducated single father limits the children ability to generate human capital [21] and reduces possibility of their successful social adaptation during the transition to adulthood.

The studies of the family social environment show that the single guardian fathers less often use reputing parenting practices and more often adhere to the strategies of permissiveness and non-intervention. The facts suggests that the teenagers from such families have a higher level of isolation than the children from the two-parent families [20]. The above studies allow to conclude that the single fathers can have poor resources including social support to fully cope with the parental functions and ensure the well-being of their children, and the most noticeable shortage concerns their educational competence.

2 Methodology and methods

The research methodology includes consideration of the fatherhood as a social institution, a system of rights, obligations, social expectations, environmental and natural concerns, as well as the requirements applied to the man as the parent and rooted in the regulatory system of culture and in the family structure (fatherhood); as an individual and reflective social practice (fathering), multipurpose and multidirectional event, complex phenomenon consisting of many structural components and exposed to various influencing factors [11, 22-25]. In addition, we examined the fatherhood in terms of social changes in manliness, generation of non-dominant masculinities, construction of "caring" masculinity by the men and society, "new culture" of fatherhood [26, 27]. The responsible fatherhood is an integral part of the concept of "responsible parenthood" which is built based on the adoption of a new masculinity model and includes such indicators as emotional closeness with the children, involvement in direct care, communication and games with the child, child minding, responsibility for their physical and personal development [23, 28].

The purpose of the article is to analyze models of the single fatherhood, to study the motivation and structure of factors that determine the involved fatherhood making, the specifics of mother's and parent family's influence on the paternal practices implementation. The main tasks of the study include examination of the typical scenarios of the man becoming the single father, their social problems and practices, development of recommendations regarding single fathers' support.

Empirically, the article is based on the data of the study carried out in 2019 in Saint-Petersburg. The targeted sample group included 17 interviewees as the single fathers. They were from 37 to 50 years old. They had experience of living with the children aged 9 months and older. The children age varied from 2 to 15 years. 9 persons had higher education and 8 persons had vocational secondary education. Among them, 7 fathers were widowers, 7 fathers were divorced with the child living with the father, 1 father was married, 1 father was in the unregistered marriage, and 1 case was with the mother was deprived of the parental rights. 11 persons had one child, 5 fathers had two children, and 1 father had three children. The financial situation was assessed as good by 6 persons and as satisfactory by 11 persons. An in-depth interview was used as the study method. The interview duration ranges from 1 to 3 hours.

3 Results/findings and discussion

The common feature of the fathers in this target group is the fact that they bring up their children alone without the wife but still to give their offspring the best knowlede and the education of the surrounding world [29]. The reason for this situation is the basis for their classification and determines, to a great extent, the patterns of their paternal behavior. Based on the life circumstances, the fathers become single parents as a result of the wife's death or family leaving by the child mother including the divorce. The family leaving by the mother can be caused by her asocial behavior entailing sometimes the deprivation of her parental rights. The mother can have no asocial behavior, but due to difficult relations between the spouses, she voluntarily refuses parenthood and agrees with the court decision that the child will live with the father after the divorce. In addition to the formal grounds for classification, the accepted approach to fatherhood understanding as a part of the personality sphere of the man can be taken into account. For men of this group, the main characteristics are implementation of the "responsible fatherhood", active involvement in the caring and raising the children not so much due to the circumstances but as a need to fulfill oneself as a father.

The initial scenarios of the single fatherhood becoming connected with the circumstances of the child appearance in the family – *planned or casual* – can differ. Without regard to this circumstances, the significant differences are found in the social and cultural capital of the single fathers. The fathers with the expanded capital are usually involved into active emotional and educational relations with the child. The children participate in leisure activities of their fathers, adopt their experience in everyday communication; the families actively travel together, visit the cultural and leisure events. The fathers exercise the initiative in the event of any psychological difficulties of the child, visit medical and other professionals. They build trust-based relations with the child without any violence, establish positive interconnections with educational and health institutions that the child attends. They position themselves as independent, accomplished, and parents; they do not resort to financial assistance from the governmental structures.

In case of the poor social and cultural capital of the father, there is a shortage of parent-child relations, low level of trust and mutual interest; the child grows to a large extent as an autonomous subject that does not demonstrate the need (however, this does not mean that the child has no need) in affection and emotional closeness with the father. The fathers prove themselves as passive participants in relations with the children, they mainly focus on the financial component of support. The father fulfills himself primarily as a breadwinner or provider of means of substance that are often quite modest. The child's upbringing style is mainly permissive.

One of the key issues with the increased explanatory and prognostic ability is the fatherhood motivation. What place does the parenthood / fatherhood take in the person's value system and among his other social roles? The answers to these questions show that some of the single fathers are characterized by high emotional involvement in the child's life. For them, birth of the child, especially if this is not the first child, was originally desired. Absence of the child mother creates the conditions for maximum self-fulfillment as the parent. In spite of the objections of the surrounding community, such father has not doubts that the child will remain with him. The children are not perceived as a burden despite the difficulties of combining the parental and professional functions. The child for the single father is something "my own" and it is not possible to give this "his own" to someone. However, fatherhood understanding takes place in different ways and depends on early involvement in caring for the child. When the desired children appear, and they are, as a rule, the second and subsequent children, the perception of oneself as a father arises immediately. For men who are highly focused on work and have the desired children, generation of the paternal feeling occurs when emotional involvement into the child upbringing appears and the need to take care of the child subtilizes the fatherhood feeling.

The factors affecting the "involved" fatherhood making depend on the current life situation. In a widowerhood situation, the close relatives of the father become full second parents. In this case, the gender of the relative does not matter in caring for the child. The role of the wife's relatives is often connected with the socio-economic condition of the father's family. In case of the low economic level, the relations are often negative, and in case of the high economic level, the relations are mutual assistance.

For the father, the professional interests and career go into the background as significant time expenditures are required for the child raising and care. Therefore, the father is forced either to resort to assistance of the relatives or to transfer to the part-time work. In case of divorce, a significant part of the father's life that takes away his physical and emotional resources (which are necessary for communication with his own child) is the struggle with the judicial system for his paternal rights including determination of the child's place of residence. The relations with his wife's relatives are usually negative, the assistance can only be obtained from his family. In highly profitable and highly professional groups, the father makes a conscious decision to transfer to the part-time work (which, nevertheless, allows him to have an acceptable income level) to have more time for his child. The greatest difficulties are observed in professionally incompetent groups with low income when the work is the main source of income, full-time employment is indispensable, and the relatives can offer only physical assistance. In this case, the child remains without necessary attention of the father and the significant resource in the upbringing begins to belong to the educational institution. At the same time, the living conditions and and financial situation as a whole do not play a decisive role in development of the involved fatherhood. The available financial resources facilitate implementation of care for the physical well-being of the child and development of his abilities; however, the main factor of establishing close paternal-child relations is the motivation- and value-based component of the fatherhood.

Implementation of the paternal practices depends on the influence of the surrounding community as well as the upbringing experience in the parental family and cultural inheritance patterns that each person has. Due to the absence of the child mother, the fathers do not implement the parenting practice used by the mother. The upbringing forms and methods used in the parental family of the father or his own approaches are implemented. In case of divorce, the mother's practices are generally not acceptable for the fathers; they are rejected as wrong and adverse for the child.

The fathers demonstrate different scenarios of intergenerational relations: from full adoption and positive attitude as well as transfer of the parental relations to relations with their own children to complete rejection, rethinking and searching for their own models of interaction with the children. The typical feature for the single fathers participated in the study is early family leaving for various reasons: the beginning of occupational training, early marriages, conflicts with their parents. However, the relations with their own mothers are more definite and positive than with their fathers.

The study revealed the specifics of the intergenerational differences in the life purposes, scenarios, and practices of fatherhood. The intergenerational differences are caused by the socio-historical dynamics between the childhood of the fathers and the childhood of the modern children, technical capabilities of today life, requirements of the educational system.

In general, the fathers can be divided into three groups: the first group is characterized by the positive relations with their fathers and the communication experience transmitted to relations with their children. The second group has negative relations with the fathers due to violence applied to the interviewees in childhood or due to insurmountable mindset differences; in this case, the alternative methods of interaction with their own children are searched. The third group includes the interviewees who lost their fathers at an early age; in this case, they search the fatherhood model by themselves. Regardless of belonging to a particular group, it is important to take into account that the fathers' childhood was during the Soviet period, and the "Soviet" fathers living in the families had primarily to actively work and only then they could take part in the children upbringing. Therefore, all interviewed fathers remember more about female (mother, grandmothers) participation in their upbringing compared to the male one.

4 Conclusions

Thus, the poor social and cultural capital of the father results in a shortage of parent-child relations, low level of trust, lack of affection and emotional closeness with the father. However, the fathers who have the expanded cultural capital are active, involved in the upbringing of children, and open for communication with them. The different scenarios of intergenerational relations with own fathers and families are described. Thus, the positive relations with the fathers result in adoption of their model of fatherhood, transfer of the parental relations to relations with their own children; the negative relations cause refusal; and the conflicting or missing relations lead to rethinking and searching for own models of interaction with the children. Our paper examined the typical scenarios of a man becoming a single father, social problems, and paternal practices based on the interviews conducted in Saint-Petersburg with 17 single fathers raising minor children.

The results of our study show that single fatherhood is as a rule a forced situation caused by the death or severe illness of the mother, her deviant behavior and leaving the family, deprivation of parental rights, divorce consequences, long-term separation of spouses, use of modern reproductive technologies of surrogacy. Our research testifies to the fact that single fathers tend to become family leaders and undertake responsibility related to childcare in the context of transforming marital and family relations.

The analysis of the interviews has demonstrated that the single fathers experience significant difficulties in arranging their family's life, raising children, communicating with family members, employers, and teachers. This situation is usually typical for the fathers raising children without any help from their relatives. The support of the single fathers raising the children shall be improved within several main areas. It is important to improve the family law by enhancing the legal protection of the fathers and their right to raise children in case of divorce, deviant behavior of the mother or deprivation of her parental rights. According to the fathers, it is necessary to involve the fathers themselves in examination of draft legislation related to support measures for the single parents and paternal families. Many families with the single fathers has a difficult financial situation which requires an expansion of the financial support for such families at the federal, regional, municipal levels (provision of benefits, subsidies, allowances, payments to the single fathers and children). A hot topic is improvement of training of the social work specialists focused on the single-parent families support. It is important to promote the development of charity, civic initiatives aimed at the single-parent (paternal) families support. There is a need for information support, digitalization of social services including consulting on websites, forums, and chat rooms, broadcasting the problems of the single fathers to the media in order to attract public attention to the paternal monoparenthood. Moreover, there is also a need in helping single fathers to become aware of the environmental issues and to convey this knowledge to their children raising their environmental awareness and helping them to get familiar with the sustainable eduction.

Acknowledgment

The reported study was funded by RFBR, project number 19-011-00543.

References

1. Goskomstat, *https://www.gks.ru/free_doc/new_site/perepis2010/croc/perepis_itogi1612.htm* (1989)
2. Federal State Statistics Service of the Russian Federation, *https://www.gks.ru/free_doc/new_site/perepis2010/croc/perepis_itogi1612.htm* (2010)
3. Federal State Statistics Service of the Russian Federation, *https://www.gks.ru/free_doc/new_site/population/demo/micro-perepis/finish/micro-perepis.html* (2015)
4. G.L. Voronin, A.L. Ianak, Woman in Russian Society, **1(86)**, 53-66 (2018)

5. O.S. Kirillova, L.E. Petrova, Perm University Herald, **3**, 128–136 (2014)
6. N.Y. Egorova, A.L. Yanak, Vestnik of Lobachevsky State University of Nizhni Novgorod, **2(34)**, 42-46 (2014)
7. I.O. Shevchenko, RSUH/RGGU Bulletin, **4(126)**, 63-172 (2014)
8. A.I. Mikhailova, Single-Parent (Father) Families: Sociological and Managerial Aspect, Cand. sociol. sciences thesis, Transbaikal State University, 2014
9. A.V. Ermilova, Woman in Russian Society, **(78)**, 69-79 (2016)
10. N.L. Moskvicheva, A.A. Rean, S.N. Kostromina, N.V. Grishina, E.V. Zinovieva, Psychological Science and Education, **24(3)**, 5–18 (2019)
11. O.N. Bezrukova, Sociological Studies, **11**, 118-130 (2013)
12. O.N. Bezrukova, V.A. Samoylova, Sociological Studies, **7**, 116-125 (2017)
13. O.V. Kuznetsova, Bulletin of Chelyabinsk State University, **17(372)**, 89–94 (2015)
14. O.N. Bezrukova, V.A. Samoylova, Sociological Studies, **7**, 90-101 (2019)
15. O.V. Stasenko, Woman in Russian Society, **3**, 78-82 (2010)
16. Y. Lee, S.L. Hofferth, Journal of Child and Family Studies, **26**, 3439–3451 (2017)
17. M. Dufur, N.C. Howell, D.B. Downey, J.W. Ainsworth, A.J. Lapray, Journal of Marriage and Family, **72**, 1092–1106 (2010).
18. J.L. Hook, S. Chalasani, Journal of Marriage and Family, **70**, 978–990 (2008).
19. S. Demuth, S.L. Brown, Journal of Research in Crime and Delinquency, **41**, 58–81 (2004)
20. J. Bronte-Tinkew, M.E. Scott, E. Lilja, Journal of Marriage and Family, **72(5)**, 1107–1127 (2010)
21. P. Donati, R. Prandini, International Review of Sociology, **17**, 209–223 (2007)
22. R. LaRossa, *The Modernization of Fatherhood: A Social and Political History* (1997)
23. W.J. Doherty, E.F. Kouneski, Journal of Marriage and Family, **60(2)**, 277–292 (1998)
24. S. Coltrane, *Handbook of Contemporary Families: Considering the Past, Contemplating the Future* (2004)
25. I.S. Kon, *Man in a changing world* (2009)
26. D. Lengersdorf, M. Meuser, *Balancing Work and Family in a Changing Society. The Fathers' Perspective* (2016)
27. S. Ruby, S. Scholz, Österreich Z Soziol, **43**, 73–83 (2018)
28. J.A. Levine, E.W. Pit, *New Expectations: Community Strategies for Responsible Fatherhood* (1995)
29. A. Mahidin, S. Maulan, Procedia-Social and Behavioral Sciences, **38**, 324-333 (2012)
30. K. Malone, *Children in the Anthropocene: Rethinking sustainability and child friendliness in cities* (2017)

Development of teachers' professional skills in education for sustainable development

Svetlana Tolstikova[1]*, *Larisa* Osechkina[2], *Elena* Tabolova[2], and *Galina* Travinova[3]

[1]Department of General and Practical Psychology, Moscow City University, 4 Vtoroy Selskohoziajstvenny proezd, Moscow, 129226, Russian Federation
[2] Institute of Advanced Training and Professional Retraining, Moscow State Regional University, Vera Voloshina str., 24, 141014, Moscow region, Mytishchi, Russian Federation
[3] Department of Psychology, Moscow International University, Leningradsky Prospekt, 17, 125040 Moscow, Russian Federation

Abstract. The article considers the skills and features of the personal development of a teacher in education for sustainable development (ESD). It contemplates over the influence of age differences on the formation of the professional and communicative culture of a specialist in higher education. Moreover, it describes the necessary conditions for the professional formation of the professional and communicative culture of a teacher in the development and promotion of her or him as a professional that is capable of raise the environmental awareness in students and help them to nourish their sustainable citizenship. In addition, it discusses the role of teacher training in education for sustainable development and stresses the importance of reorienting teacher training to take sustainability into account in the context of education policy and the implementation of the Sustainable Development Goals (SDGs).

1 Introduction

The educational processes at universities including sustainable education (or the education for sustainable development or ESD) are systematic and aim to make use of the knowledge acquired in professional and everyday life. University education is a crucial point because it is systematic and aims to develop knowledge in the educational field and its application in the real world that is burdened by contemporary problems such as global warming and climate change [1, 2]. It is aimed not only at students but also at professors who aim to promote their students to higher academic achievement and professional success. In contrast to secondary education institutions, students in higher education institutions have more opportunities because they have the opportunity to choose from a variety of educational institutions such as schools, colleges, universities and colleges [3]. Teaching quality becomes a key component of education plans. Professional development, teacher preparation and research also support the importance of teaching qualities and identify content-specific pedagogy as a key component. This is a good example of how we can integrate best practices in professional development and research in preparation for teaching into our education

* Corresponding author: tolstikovasn@mgpu.ru

systems [4, 5]. For instance, government-funded school-supporting reform programs in the United States are focusing primarily on teacher training. Many universities in that country departments offer a wide range of professional and creative training programs for teachers, administrators and administrative staff [6].

In the age of digital media and personalized content, creative professionals need to master the latest media strategies, understand and harness the power of storytelling, align a compelling creative perspective with the brand's needs for communication, master digital marketing and social media strategy, use the action possibilities and master storytelling to use it [7]. These skills can prepare today's students to become a digital media, digital marketing and social media strategy professional. Communication cultures are university students or interdisciplinary researchers with a special knowledge of languages. This creates an environment in which different departments meet on topics relevant to their respective subjects [8].

In the process of vocational and communicative cultural education, the principles of education and individualization require consideration of the psychological and professional characteristics of the students. The professional teacher is responsible for the development of a professional communication culture in design and design culture [9]. In the context of a diverse society, there is a need for teachers in grammar schools and minority schools to master the contents of a diverse pupil education. Various research associations exist on the educational market that offer teachers with different groups of pupils many vocational learning opportunities. Culturally appealing teaching, also known as culturally relevant teaching, is a pedagogy that recognizes the importance of the cultural references of students in all aspects of learning [10].

In general terms, traditional teaching strategies emphasize the dynamics between teacher and student; teachers adhere strictly to a curriculum that supports standardized tests for student knowledge [11]. No one has students with the same background and experience, so the teaching is different. Teachers are experts, but they adhere strictly to the curriculum, which supports standardized testing as the only measure of the knowledge students receive. For teachers, it is not only important to give a high-quality lecture, but also to deal with difficult people. Students generally learn interpersonal skills in school, both as students and as teachers [12]. These skills are crucial for interacting with other people at the individual or group level, and there are many different types of interactions between students, teachers and other students. The concept of immediacy is the physical, emotional and psychological closeness that arises through the interaction between students, teachers and other students in a classroom or group of students and teachers. The pedagogical content of knowledge is a special combination of content and pedagogy, uniquely constructed by teachers, and thus represents the most important component of the teacher-student relationship in the development of interpersonal skills. It consists of an integrated knowledge that represents the wisdom accumulated, the knowledge of other people's experiences and the experiences of the teacher and the students [13, 14]. The development of the pedagogical content of knowledge is influenced by factors related to the personal background of the teacher. This includes the theory learned during the preparation of the teacher, the experience gained through ongoing school activities, and the context in which he or she works.

2 Historical challenges in education climate change issues

Today's challenges in higher education should reflect the key issues of our time. It is important to give students the opportunity to make informed decisions about the future of their own lives and the society in which they live. To give a few examples of classroom activities, it is essential for teachers to design a curriculum that is coherent, reflects continuity, builds on students' ideas and experiences, and is guided by scientific consensus

[15]. The time needed to teach about global warming and climate change, as well as the curricula and educational policies associated with the subject, make this teaching even more difficult. However, we must recognise the impact of climate change and support teaching about it for the betterment of our students and the future of the planet. Studying programme should aim at helping people understand the effects of global cooling today and improve climate literacy, especially among young people. In addition, they should aim to make climate education a central part of today's education system [16, 17].

Although the involvement of countries is crucial to solving these problems, we must call on all people and governments to recognize the seriousness of the threat of global warming and to develop effective strategies to reduce the potential consequences of global climate change [18]. World's superpowers like the United States should lead the developing world in promoting sustainable economic development in poor countries and in building global consensus on the need to mitigate climate change. If climate change undermines educational performance, it could have a worsening effect, exacerbating the direct effects of climate change over time. The measures we are taking today could make a significant difference to our descendants, and they could be measured by the impact on our children and grandchildren, as well as on other generations. Climate change and its effects on human health and human capital are associated with changes in utero and early childhood climate. We have to focus on climatic conditions during childhood and uterus, as they can negatively affect long-term human capital outcomes, including educational attainment and income [19].

The IPCC warns that to avoid a catastrophic global cooling, we must achieve or exceed a minimum temperature increase of 2 degrees Celsius above pre-industrial levels by the end of the century, and this is concentrated in the tropics [20]. The longer we take, the more we will be dependent on costly technologies that could have a detrimental impact on human rights. For this reason, tackling climate change may well be one of the most pressing challenges of our time, even in the face of overwhelming scientific evidence, not only for the future, but also for our future. The planet's climate is changing, and it has become clear that humanity has caused most of the warming of the last century by releasing heat - the separation of gases commonly referred to as greenhouse gases into the force of modern life. However, the current warming phase is occurring faster than previous events [21].

Scepticism is a right-wing ideology that has joined the argument that it should not exist, because climate change requires coordinated collective action - the kind of ideology that is repugnant to that ideology. Carefully organized campaigns have used such scientific uncertainties to create political uncertainty, because those who fund the case against the reality of climate change have a massive interest in the fossil-fuel economy.

More insidiously, scepticism can also lead to the notion that it enables ordinary people to question the claims of the scientific elite. The belief that global warming is caused by human activity, that climate change is beginning to take effect and will soon pose a serious threat, has increased in recent years. Some studies conclude that more than 90 percent of the world's climate scientists actively publish believe that human activities contribute to rising global temperatures [22]. Although many people are sceptical of the arguments and evidence of climate scientists, the overall belief in a global cooling in the twenty-first century has grown steadily in the recent years.

3 Building cultural relationship with the environment

To promote environmental and climate awareness in the classroom, teachers must understand the individual students. This strategy will foster cultural awareness and strengthen students' sense of identity. To do this, the students must be introduced to each other, each individual must be included in the lessons and be taught what makes them unique [23].

To promote students' cultural awareness, we must take into account the various aspects of our culture that can influence the perception, attitudes and behaviours of the students. It is important to recognize the impact skin colour has on broader culture and how it can affect a student's perception of other students. Although ethnicity is sometimes used as a synonym for race, it can also be defined as a culture derived from a nationality. Personal identity is shaped by the lived experiences shared with people of their own culture during the development of an individual, as well as by the people and environment with which they interact throughout their lives. Questions related to identity development, which are important in psychology, sociology and anthropology, are central to our understanding of who we are. When personal identity emerges, our study of the various forms of human behaviour can be improved by identifying the social processes that influence identity formation and the ethical principles that underlie individual action. Given the importance of education for identity development and the role of teachers in this process, students must be aware of their own school experience and their relationship with the environment. The goal of school psychology is to work with parents, teachers and students to promote healthy learning environments that focus on the needs of children. Psychology includes the term "procedural memory" as a synonym for implicit memory [24, 25].

The High School Personality Questionnaire deals with the relationship between the teacher's personality and the behaviour of his students in the classroom. It is important that teachers are aware of the impact of evaluations and increased accountability, which help to create a more progressive educational environment [26]. People with different cultural traditions may have an educational approach that differs from the mainstream approach of schools. Although the education of pupils is the main objective of schools, teachers have different ways of achieving this goal. For example, there are differences between the traditional approach of the Catholic Church and the modern approach, according to which younger children are generally taught by older children and not by adults. This is a different approach to what might be considered when a Polynesian student attends an American school. For this reason, a school's subculture is different from that of the classroom, where colleagues are not allowed to share a teacher's preferences. Culture has a crucial impact on educators in the field of teaching, as student bodies in many developed are going to be overwhelmingly non-white within a couple of decades [27].

Teachers cannot control what students are exposed to outside school, and they cannot control what students see in the media. They can promote effective communication by creating a classroom that fosters a positive relationship between students and their teachers and the environment around them [28]. If we create such an environment, teachers will find that they are less likely to engage in class discipline and to spend more time teaching and learning to students. It is important that teachers first explain to their students what discrimination and insensitive behaviour looks like. By creating a classroom atmosphere that encourages conversations about diversity and enforces a zero-tolerance policy for negative language, teachers can teach their students to respect differences. Next, they must develop, explain, and implement strategies that leave room for insensitivity.

4 Professional development in education for sustainable development

While the participation of principals in workshops based on school materials can theoretically address the problems of teaching sustainable development and environmental awareness, a culture change is needed. Teachers who feel intimidated are vulnerable, especially in environments where leaders do not normally teach, and there is therefore a need for professional development of teachers and leaders in the context of sustainable development [29, 30]. An important message is that we need time to focus on effective teaching and develop a shared responsibility.

School programme represent a response to the government-funded framework which sets out sustainable development opportunities for teachers, school leaders and teacher leaders as part of a wider strategy to preserve development opportunities. It provides opportunities to alleviate the challenges facing the education system in terms of the development of teachers, principals and the leadership of teachers in the education sector. The vocational development programmes and educational materials provide groups, educators and young people with the skills to understand the relationships necessary for sustainability and to find integrated solutions to environmental challenges. The overall objective is to improve the number and quality of teachers in the education sector and their skills [31]. This objective is to be achieved by improving and facilitating school relationships and cooperation between communities, using sustainability as an inclusive force and encouraging schools themselves to become learning communities. Training programs for professionals, including professional development for university teachers and lecturers are needed. This also serves as a resource for organisations, institutions and individuals worldwide that are involved in sustainability education initiatives [32].

As it has been noted, there is a need for a new generation of managers dedicated to training future managers to tackle complex social issues and manage sensitive environmental resources. Since the mid-1990s, a number of researchers and leaders have proposed assessing how environmental education affects teacher training. While the lens of sustainability encourages us to look at innovation within tradition, the learning processes that typically underpin teachers' professional development are paradoxical to the goal we are striving for. This does not mean that there is a lack of knowledge about the development of the teaching profession. It is an opportunity to highlight innovative programmes, as we strive to contribute innovation to teachers' professional learning. Learning can be combined with the transformative power of nature and sustainable learning to make a meaningful contribution to a more sustainable future. There are many teacher trainings programs that provide teachers and school districts with comprehensive support in implementing a sustainable learning approach to teacher development in education for sustainable development [33, 34].

To illustrate the work and transformation approach that is based on the learning process outlined by UNESCO, one can recall the existing programmes that have shown lasting positive changes and results in the form of places of learning and sustainable education. In general, teacher training requires teachers to learn in school with their peers. It also describes the active learning methods that can be used for the professional development of teachers, thus offering teachers opportunities to improve themselves. The native language of the course is also used as a learning material for the professional development of the teacher [35].

Although the time for professional learning is short, not focused and disordered, teachers can receive a great deal of help in developing these activities with sufficient time, which is important for their professional development. To encourage teachers to change their classroom habits, school leaders should allow sufficient time during the school year for teachers to develop their careers.

5 Conclusions

Our paper provided an insight into what works for teachers in the context of environmental education and built upon the research on the impact of environmental education on teacher training in sustainability education. When it comes to incorporating sustainable development into the curriculum, higher education has focused on education, which aims to develop specific knowledge and competence for, for example, the climate crisis and renewable energy, and to develop skills for autonomous, responsible and reflective citizens. Nevertheless, the goal must be to develop skills for sustainable development that combine international and high education in order to move closer to what some scientists call "local citizenship", which actually means that education that is globally engaged and locally integrated. To achieve the Sustainable Development Goal 4.7, we need an understanding of what is needed to achieve it and how we will train and train. It is clear that global citizenship requires ethical, responsible, and open teachers and students to become responsible citizens of their countries and the world as a whole. In the face of the ongoing crisis of climate change and its impact on the global economy, it is essential that education provides young people with a source of hope and that they envision and implement a more sustainable future by developing the values, knowledge and skills to do so.

We can conclude this article by stating that a central objective of the strategy aiming at developing of teachers' professional skills in education for sustainable development is to design an educational system for sustainable development based on the principle of life-long learning. The main themes of the strategy are the importance of education as a means of social and economic development and the promotion of sustainable livelihoods.

References

1. J. Radianti, T.A. Majchrzak, J. Fromm, I. Wohlgenannt, Computers & Education, **147**, 103778 (2020)
2. M. Doolan, B. Piggott, S. Chapman, P. Rycroft, Australian journal of teacher education, **44(6)**, 91-108 (2019)
3. A. Koricich, X. Chen, R.P. Hughes, The Review of Higher Education, **41(2)**, 281-305 (2018)
4. J. Barenthien, E. Oppermann, Y. Anders, M. Steffensky, International Journal of Science Education, **42(5)**, 744-763 (2020)
5. I. Čábelková, V. Blaginin, W. Strielkowski, A. Platitsyn, E3S Web of Conferences, **208**, 09001 (2020)
6. B.A. Jacob, L. Lefgren, Journal of Human Resources, **39(1)**, 50-79 (2004)
7. D. McGruer, *Dynamic digital marketing: master the world of online and social media marketing to grow your business* (2019)
8. M. Odinokaya, T. Krepkaia, O. Sheredekina, M. Bernavskaya, Education Sciences, **9(3)**, 187 (2019)
9. M.G. Sergeeva, N.G. Bondarenko, T.A. Shebzuhova, B.A. Solovyov, D.V. Parinov, L.A. Shvedov, A.P. Ovchinnikov, Amazonia Investiga, **8(18)**, 5-14 (2019)
10. S. Hollie, Teacher Education Quarterly, **46(4)**, 31-52 (2019)
11. M. Şen, C. Öztekin, B. Demirdöğen, Journal of Science Teacher Education, **29(2)**, 102-127 (2018)
12. E.A. Tokareva, Y.V. Smirnova, L.G. Orchakova, Education and Information Technologies, **24(5)**, 3219-3234 (2019)
13. V. Kind, K.K. Chan, International Journal of Science Education, **41(7)**, 964-978 (2019)
14. Z. Deng, Teaching and Teacher Education, **72**, 155-164 (2018)
15. N. Iivari, S. Sharma, L. Ventä-Olkkonen, International Journal of Information Management, **55**, 102183 (2020)
16. D. Rousell, A. Cutter-Mackenzie-Knowles, Children's Geographies, **18(2)**, 191-208 (2020)
17. E.J. González-Gaudiano, P.Á. Meira-Cartea, The Journal of Environmental Education, **50(4-6)**, 386-402 (2019)
18. A. Miles-Novelo, C.A. Anderson, Current Climate Change Reports, **5(1)**, 36-46 (2019)
19. H. Randell, C. Gray, Proceedings of the National Academy of Sciences, **116(18)**, 8840-8845 (2019)

20. IPCC, *https://www.ipcc.ch* (2021)
21. J. Henderson, D. Long, P. Berger, C. Russell, A. Drewes, Educational Studies, **53(4)**, 412-425 (2017)
22. S. Capstick, L. Whitmarsh, W. Poortinga, N. Pidgeon, P. Upham, Wiley Interdisciplinary Reviews: Climate Change, **6(1)**, 35-61 (2015)
23. T.W. Farmer, J.V. Hamm, M. Dawes, K. Barko-Alva, J.R. Cross, Educational Psychologist, **54(4)**, 286-305 (2019)
24. C.L. Bagnall, Y. Skipper, C.L. Fox, British Journal of Educational Psychology, **90(1)**, 206-226 (2020)
25. D.L. Gray, E.C. Hope, J. S. Matthews, Educational Psychologist, **53(2)**, 97-113 (2018)
26. A. Di Fabio, D.H. Saklofske, Personality and Individual Differences, **123**, 140-144 (2018)
27. A.Y. Lee, A.J. Lee, Journal of Curriculum Studies Research, **2(2)**, 40-59 (2020)
28. A.J. Samuels, SRATE Journal, **27(1)**, 22-30 (2018)
29. B. Türkoğlu, Sustainability, **11(18)**, 4925 (2019)
30. R. Mintrop, *Design-based school improvement: A practical guide for education leaders* (2020)
31. A. Mogren, N. Gericke, Sustainability, **11(12)**, 3343 (2019)
32. Q. Iqbal, N.H. Ahmad, Sustainable Development, **29(1)**, 108-119 (2021)
33. A. Oke, F.A.P. Fernandes, Journal of Open Innovation: Technology, Market, and Complexity, **6(2)**, 31 (2020)
34. N. Podgorodnichenko, F. Edgar, I. McAndrew, Human Resource Management Review, **30(3)**, 100685 (2020)
35. M. Rieckmann, *Education for sustainable development goals: Learning objectives* (2017)

Role of education in implementing the sustainable development strategy

Raisa Krayneva[1]*, *Aleksandr* Rudenko[2], *Roman* Motylev[2]

[1]Financial University under the Government of the Russian Federation, Leningradsky Prospekt 49, Moscow, 125993, Russian Federation
[2] Saint Petersburg State University of Architecture and Civil Engineering, Vtoraya Krasnoarmeiskaya str. 4, Saint Petersburg, 190005, Russian Federation

Abstract. Our paper aims at assessing the role of education in implementing the strategy for sustainable development. We show how education might be used to promote the provisions set out by the United Nations Development Programme (UNDP) and the UNESCO Sustainable Development Goals (SDGs). We argue that sustainable education might help to give the young people necessary knowledge and skills that are required to understand what the SDGs represent and to take part in implementing the necessary changes and mitigating the climate change and the global warming as well-informed and concerned citizens. Understanding and embracing the concept of sustainability represents one of the main issues of the society and education plays a key role in that process. Thence, we think that the role of education is sustainable development strategy is crucial which requires additional support for education from the governments and private actors.

1 Introduction

It was the Agenda 21 that became the first international document to identify education as an essential instrument for sustainable development and to highlight areas of action for education [1]. Sustainability in education and training and sustainable development are interchangeable terms that describe the practice and teaching of sustainability [2]. A definition of "education and sustainable development" (ESD) is an interdisciplinary learning methodology covering integrated social, economic and environmental aspects of education as well as social and economic aspects [3]. This academic approach can help graduates to develop their knowledge, talents and experience to play a role in environmental development and become responsible members of society. They are not only aware of their responsibilities, but also gain a deeper understanding of how the world works and how their actions can contribute to sustainable development. We must equip them with the skills, knowledge and values they need to create a fairer and more sustainable world when they enter society.

Education and sustainable development should be a part of formal education in schools and it can articulate and describe what is needed to equip students for sustainable development [4]. It is also a matter of communicating the scientific and theoretical background to these questions. Education and sustainable development enable people to

* Corresponding author: RKKrajneva@fa.ru

acquire the skills and knowledge necessary to build a sustainable future for themselves, their families and society.

In addition, education and sustainable development take into account key issues of sustainable development in the curriculum, such as health, education, health care, environmental protection, social justice, human rights and the environment [5]. Furthermore, education and sustainable development promote the knowledge, skills, understanding, values and measures needed to create a sustainable world, ensure environmental protection and protection, promote social justice and promote economic sustainability. This requires teaching and learning methods that motivate learners to change their behaviour and take measures for sustainable development.

The concept of education and sustainable development was developed largely from environmental education, which aims to develop people who care about their environment. Recognizing this change and the need to take action, the United Nations Development Programme (UNDP) and the UNESCO Sustainable Development Goals (SDGs) were launched in 2015 [6]. The decade that UNESCO has led is supported by the conviction that education is the way to shape the values, skills and knowledge needed to build a sustainable society. This was also reflected in the Paris Agreement that was adapted in the same year and paved the way towards mitigating the climate change [7].

The aim of education and sustainable development is to enable people to make the decisions and take the measures necessary to improve their quality of life without endangering the planet. In short, education and sustainable development enables individuals to contribute to the achievement of the Sustainable Development Goals (SDGs) defined by the United Nations (UN) by equipping them with the knowledge and skills they need to understand what the SDGs are and to participate as informed citizens in implementing the necessary changes [8]. By developing state-of-the-art knowledge, skills and skills in the fields of sustainable development, education and training, it is an essential contribution to the efforts to achieve the SDGs by enabling them to contribute to sustainable development by promoting social, economic and political change and by changing their own behaviour. It can provide the necessary skills to enable individuals to deal with the specific challenges facing the humanity and thus facilitate its achievement. In order to achieve sustainable development, environmental protection must be an integral part of the development process. Competence design refers to the ability to apply knowledge of sustainable development and to identify the problems of unsustainable development.

Moreover, sustainable development is considered a key component of achieving the Sustainable Development Goals and the Universal Declaration of Human Rights [9]. Eliminating poverty and reducing disparities in living standards in different parts of the world are key to achieving sustainable development and meeting the needs of a majority of people. Nations should work together to preserve, protect and restore the health and integrity of the Earth's ecosystems. In addition, nations work together to preserve, protect and restore the health, integrity and earth ecosystems for the benefit of all people, the environment and humanity as a whole. It is generally recognised that sustainable development is fundamentally about achieving a balance between the economy, society, people and the environment. Where the international community recognizes the reciprocal nature of human activities and the environment, developing countries must recognize their responsibility for the health and integrity of their environment, as well as for their own economic development [10].

Nevertheless, it is important to recognize that the definition of sustainable development can often controversial to some people, as it can be considered too vague and therefore prone to misinterpretation. It is also crucial to integrate sustainable development into the context of education. The development of a comprehensive understanding of the relationship between education and the health and integrity of our environment should be a top priority. People around the world need to be educated about how to improve the quality of life without creating problems for the future or unfairly affecting other people's lives [11].

All in all, we think education is an important tool for society to face and address the challenges of sustainable development. It may be a key element to teach people to respect the needs of the environment, but it is not the whole answer to the problem. If we are jointly responsible for the world's survival, lived sustainability should become part of what we teach the next generation. Thence, the role of education in implementing the sustainable development strategy is hard to underestimate. All players involved in this process should therefore concentrate on delivering sustainable education and informing the pupils and students on the pressing issues our society faces nowadays. It is not COVID-19 pandemic but the global and ongoing climate changes that pose the threats to the very existence of the human race and we all need to be aware of that.

2 Understanding the sustainability in the society

One would probably agree that it is important to recognize that people in poor countries currently have more access to food, health care, education, and housing than in rich countries, and as countries move out of poverty, their citizens increase their individual consumption. This might be an issue of convergence – poorer countries tend to grow faster than the richer ones [12]. This goal is very important because the planet cannot sustain the increased demand for resources necessary to sustain a growing world population. Therefore, the development of sustainable development goals in developing countries, especially in the Middle East and North Africa, is also crucial to prevent environmental damage in the coming decades. However, SDG12 remains crucial for the long-term sustainability of the global economy and the environment. This idea has generated much resistance, because it poses a significant threat to the status quo, despite the great social, environmental, and economic opportunities [13]. It is important to note that this specific SDGs includes improving public health while at the same time providing adequate nutrition. Recent concepts of sustainable nutrition are important to promote ecological and economic stability while improving the quality of life. In the 1980s, a" sustainable diet "was proposed, combining dietary recommendations with a healthy environment for consumers (using different terms, perhaps), and revived in the 1990s. Sustainable nutrition protects biodiversity and ecosystems, while protecting and at the same time optimizing natural and human resources. Defining a sustainable diet is important, but more work is needed to understand what constitutes a "sustainable" diet [14].

This is an important issue of environmental sustainability because we have no idea how our natural environment is able to sustainably support human life and development despite rapid population growth and climate change. This global concern has been expressed in the United Nations, the World Bank and the International Monetary Fund (IMF), as well as in other international organisations. Sustainability, as designed and defended by the experts from the international organization and think-tanks, refers to the preservation of the health, resilience and adaptability that sustain the planet in conditions in which life as a whole can thrive. Nowadays, sustainability is usually defined as a way for people to avoid the depletion of natural resources while maintaining an ecological balance that does not allow the quality of life in modern society to decline [15]. It is used in connection with the environmental and social problems that societies around the world are facing. In order to make this possible in the long term, the way in which modern society organises itself should be designed and maintained using best practices of sustainability, such as the use of natural resources and the preservation of the environment. This provides a framework for how a society can interact with an environment in a way that does not risk harming resources for the future. Looking at this approach, sustainable development is a development approach that uses resources in a way that allows the resources themselves to coexist.

This could have important implications for other important SDGs that have not been discussed above, such as reducing inequality. While this assessment is consistent with the "Common Socio-Economic Paths" scenario used in the context of the Intergovernmental Panel on Climate Change, it suggests that the United Nations projections presented by the sensitivities analysed are probably too narrow. This is because they will number between 1.5 and 3.2 billion people by 2050 and between 2.4 and 4.3 billion by 2100. Only based on a narrative of stagnant birth rates the population growth sustained this century. The first place shows the global scale of predictions given by the UN projections, and the first place shows an increase of 1.5 billion people by 2050 and 2.4 billion by 2100 [16].

3 Level of education and damage to the environment

Nowadays, one of the most important goals of the preschool and higher education alike is educating schoolchildren and students about climate change and sustainable development which can help them to shape and sustain future policy-making. Today's policy makers are supported by a broader public and international debate to take decisive action now. In the group of countries covered by the World Values Survey 2005-2008, a person is more likely to express concern for the environment if he or she has a secondary education than if he or she has a tertiary education [17]. The results of some studies show that those with secondary school qualifications preferred the environment when forced to choose between protecting them and boosting the economy. Similarly, a survey of respondents in the United States, Canada, Australia, and New Zealand showed that the percentage of people who did not think that people cared too much about the environment rose from 25% in less than - secondary - education to 46% among people with higher education and from 35% to 52% among people with higher education. Many studies have shown that poverty and lack of education correlate strongly and positively. In other words, the higher the level of education of a country's population and the lower the poverty rate of its citizens, the greater the environmental impact on the economy [18].

Another effect of inadequate education is that poor education often leads to an increased likelihood of unemployment and lower wages. Poverty, however, is one of the main causes of educational inequality in countries with low levels of education. Because poor families cannot send their children to school, these children suffer from significant educational inequalities. Fewer jobs can exacerbate economic hardship and poor health, which are common among people with less education. Public schools are poorly equipped with low property taxes, and these low-resource schools have difficulty in offering attractive teacher salaries and maintaining the building fabric properly. But the higher the level of education, the greater the damage to the economy and health. Poverty is one of the most common causes of poverty in the poorest countries of the world, with drought being the most closely linked threat. Lack of access to insurance and social protection means that people in poverty are often forced to use their already limited assets to cover disaster damage, which drives them further into poverty, according to the study. Poor people are more likely to live in vulnerable areas and less able to invest in risk-reduction measures such as health care and education. In the United States, for example, the country with the highest level of education is the most affected by drought, according to the study. People who start life with negative childhood experiences have greater illnesses and difficulties at school and at work, contributing to the link between socio-economic conditions, education and health. Instability in the home and community can have negative effects on economic security and stability of housing, which can also affect the physical and mental health of adults.

One important way to improve outcomes is to address the root causes that primarily expose children to stress, such as poverty, poverty, and lack of access to education and health care. Results from Canada and Germany suggest that childcare can improve the health and

https://doi.org/10.1051/e3sconf/202125007008

well-being of children from low-income and disadvantaged families [19]. No sane person could conclude that we should continue to subsidise childcare from disadvantaged families. But if the primary consideration is the well-being of the next generation, these findings support universal child care. The results from Denmark and Canada reach similar conclusions regarding the negative effects of comprehensive childcare. Studies that have sought to assess the impact of high education on children's health and well-being have produced disturbing results in terms of childhood obesity, deteriorating educational skills, and higher rates of mental illness. This suggests that policies promoting high-quality programs to enable low-income single parents to work full-time could harm disadvantaged children. One of the best studies we have on the impact of Head Start on children's health and well-being, the Child Health and Development Study in the United States (2009), suggests that Head Start's youngest participants have been significantly harmed by the lack of access to quality child care in their communities [20]. This provides compelling evidence for those who fear that expanding early childhood education would do more harm than good to disadvantaged students. They often have less education and fewer resources to insulate themselves against the effects of stress, and they are spared health - which harms stress from prolonged social and economic hardship. Indeed, stress is reduced in countries with higher levels of education, such as the United States and Canada, according to the World Health Organization (2010). Economic hardship and other stressors can have cumulative health effects over time, making people more sensitive to other stressors. According to the World Health Organization, Americans with higher education live longer than their counterparts in countries with lower education levels and less education.

4 Supporting sustainable development

Supporting sustainable development becomes a must for private and public institutions. Financial support is also needed for rush-off times, which are normally associated with activities related to the built environment and infrastructure development. Sustainable, low-income, collaborative design based on the principle of maximizing available renewable resources often does not incur significant additional investment costs. Moreover, successful community activities often find that their ability to generate a sustainable source of income in return for investment can help maintain their current activities and even enable the development of new activities, as described in the Feasibility, Technology, and Operational Needs section. External assistance will help to generate the internal resources needed since 2000 to improve the living and working conditions of people, including the growing number of unemployed and income groups.

Improvements to the sustainable education should be based on the basis of decision-making - the capacity of the government, not only the capacity of local governments, but also at national and regional levels. As it enters a new phase and changes its strategy, the Global Task Force is building on the success of the Sustainable Development Goals (SDGs) over the past two years. The GTF acts as a catalyst for the development of a global strategy for sustainable development (SDSG) and the creation of an international governance framework. The creation of a Global Governance Framework for Sustainable Growth (GGGF) will bring together the major international networks (LRGs), thereby strengthening the capacity of local and national governments to act. These are closely linked to the achievements of the SDGs and in particular their contribution to the implementation of the SSDGs. LRGs focused on how to support the implementation of the 2030 Agenda through bottom-up action and provide a framework for local development policy. Despite the growing recognition that nuclear energy has a crucial role to play in achieving the United Nations Sustainable Development Goals, there is still a lack of awareness and understanding of the importance of local governance structures in contributing to sustainable development. The use of nuclear energy has several sustainability benefits over available alternatives, such as the growing role of nuclear energy in the global economy and the fact that almost every major study outlines plausible pathways to sustainable energy. The competitive position of nuclear energy is robust and our research shows that the approach adopted in its sector is to pass on a number of benefits to future generations, while minimising environmental impacts and burdens. Moreover, it is clear that developing countries will be able to avoid potentially major negative consequences if they switch from their dependence on fossil fuels to their own renewable energy sources, such as biofuels. This is an important issue that should be addressed quickly by the developing and developed countries, which are at the forefront of global efforts to expand global production. In general, the ability to achieve the long-term sustainability benefits of switching from conventional fuels to biofuels would also be greatly improved by using nuclear energy as an alternative to conventional fuels and other renewable sources. This requires the involvement of all sectors of society, including government, the private sector and civil society, as well as the public sector. The Montreal Process provides its 12 member countries with internationally agreed and locally supported instruments to integrate these issues, such as the Montreal Plan, the Paris Agreement on Sustainable Urban Development, and the International Forest Agreement, which applies to forests. ICLEI is a global coalition of more than 1,000 civil society and private sector organisations committed to sustainable urban development. We shape policies and stimulate action to transform the urban environment worldwide, both locally and globally. Decentralised governance is practised in 12 member countries, including the United States, Canada, Australia, Brazil, China, France, Germany, India, Italy, Japan, Mexico, Peru, South Africa, Spain, Switzerland and Turkey. Today, many of the sectors addressed by the SDGs, such as education, health, education and health care, agriculture, social services and the environment, are increasingly covered. Sustainable development depends on decisions - on power and sufficient resources to include everyone's responsibilities. The growing responsibility of governments, especially in the areas of social and economic development, has the potential to ensure more inclusive and representative decision-making.

5 Conclusions and implications

Overall, one can see that education constitutes a key and a very important element of a sustainable development strategy and an essential part of the development process. In order to achieve inclusive and high-quality education, an inclusive quality of education in all its forms must be integrated into the education system. The world has changed with the global movement towards a fairer, more just and sustainable future for all.

For the first time, sustainable development has been enshrined as an international extension of national policy discussions. Sustainable development is a broad concept and has been a guiding principle of the United Nations Development Programme (UNDP) and the World Bank for some time. Our results demonstrate that there is a risk of further global environmental change, and that the lack of sustainable development goals such as the SDGs and MDGs will undermine these risks in the future.

While the objective of being a universal and indivisible link is aimed at implementation, the implementation goals are largely silent on the linkages and interdependencies between the goals, and the final goal is dedicated exclusively to means of implementation (i.e. the development of sustainable development goals/ As our analysis shows, the SDGs and related goals form a series of interconnected goals with a variety of interdependencies and interdependent goals. Understanding the interactions between these objectives is essential to make comprehensive progress towards these objectives and to ensure that progress towards one particular objective does not hamper the achievement of other objectives and vice versa. Human well-being is ultimately based on global natural resources and biodiversity, and the sustainable use of these resources within ecological boundaries is embedded in the basis of sustainable development.

The sustainable development targets are a set of goals to be met by 2030 and cover a range of social, economic and environmental issues addressing key global challenges, including ending hunger and poverty, protecting life, water and land, promoting sustainable production and consumption, and ensuring prosperity by reducing inequality for all. We have to use all available tools to promote them and to embed them into the collective conscience. Doing this through sustainable education might be a very good and effective way how to achieve that objective.

References

1. S. Yáñez, A. Uruburu, A. Moreno, J. Lumbreras, Journal of Cleaner Production, **207**, 57-66 (2019)
2. I. Čábelková, V. Blaginin, W. Strielkowski, A. Platitsyn, E3S Web of Conferences, **208**, 09001 (2020)
3. P. Sund, N. Gericke, Environmental Education Research, **26(6)**, 772-794 (2020)
4. I. Čábelková, V. Blaginin, W. Strielkowski, A. Platitsyn, E3S Web of Conferences, **208**, 09001 (2020)
5. V. Kioupi, N. Voulvoulis, Sustainability, **11(21)**, 6104 (2019)
6. A. Agbedahin, Sustainable Development, **27(4)**, 669-680 (2019)
7. W. Strielkowski, E. Lisin, I. Gryshova, Romanian Journal of European Affairs, **16(4)**, 68-77 (2016)
8. F. Annan-Diab, C. Molinari, The International Journal of Management Education, **15(2)**, 73-83 (2017)
9. N. Chams, J. García-Blandón, Resources, Conservation and Recycling, **141**, 109-122 (2019)
10. D. Celermajer, D. Schlosberg, L. Rickards, M. Stewart-Harawira, M. Thaler, P. Tschakert, C. Winter, Environmental Politics, 1-22 (2020)
11. S. Lee, S. Trimi, Journal of Innovation & Knowledge, **3(1)**, 1-8 (2018)
12. W. Strielkowski, F. Höschle, Technological and Economic Development of Economy, **22(4)**, 617-630 (2016)
13. E. Gawel, N. Pannicke, N. Hagemann, Sustainability, **11(11)**, 3005 (2019)
14. B. Barone, R. Nogueira, J. Behrens, Food research international, **124**, 206-212 (2019)
15. V. Osipov, Herald of the Russian Academy of Sciences, **89(4)**, 396-404 (2019)
16. J. Hickel, G. Kallis, New political economy, **25(4)**, 469-486 (2020)
17. K. Parboteeah, H. Addae, J. Cullen, Journal of Business Ethics, **105(3)**, 403-413 (2012)
18. P. Serneels, S. Dercon, The Journal of Development Studies, **57(1)**, 163-183 (2021)
19. T. Kaiser, J. Li, M. Pollmann-Schult, A. Song, International Journal of Environmental Research and Public Health, **14(9)**, 981 (2017)
20. J. Cassidy, B. Brett, J. Gross, J. Stern, D. Martin, J. Mohr, S. Woodhouse, Development and Psychopathology, **29(2)**, 651 (2017)

Development of smart grid technologies: organizational and communication aspects

Andrey Vlasov[1]*, *Arina* Adamova[1], *Kirill* Selivanov [1]

[1]Department of Design and Technology of Electronic Devices, Bauman Moscow State Technical University, 2-aj Baumanskaj 5, str. 1, 105005, Moscow, Russian Federation

Abstract. This paper concentrates on the organizational and communication aspects of development of the smart grid technologies. The paper highlights the potential of decentralised electricity generation for generating electricity from less energy-intensive and cost-efficient sources. It shows that renewable and unconventional energy sources may be integrated into decentralised electricity grids – the generation lines that have an intelligent grid. In addition, the paper focuses on the benefits and risks of different smart grid applications and their impact. We show that smart grids have the potential to minimise costs, but the use of smart grid technology also affects the level of risk, so the organizational and communication aspects are of a great importance.

1 Introduction

Integrating renewable energy sources and electricity generation into smart grid systems can be an important step in the transition to a more efficient and sustainable energy system [1]. Smart grids can transform a 20th century electricity grid into a 21st century electricity grid, allowing consumers to have greater control over the distribution of electricity and the production of energy from renewable sources such as wind, solar, geothermal, hydroelectric, biomass and hydroelectric, as well as energy storage and transmission [2]. An intelligent grid system improves the efficiency of electricity distribution, takes into account the need for more reliable, cost-efficient and reliable generation, transmission and distribution of energy, and can also take into account energy security, energy efficiency and energy saving [3]. The primary effects of the smart grid technologies are cost reduction, improved grid reliability and resilience, and greenhouse gas emission reductions [4]. Smart grids can also create new markets for the private sector to develop new technologies such as energy storage, energy efficiency and renewable energy generation.

Traditionally, smart grids have been designed to transport electricity generated by central power plants to consumers. Smart grid technologies such as energy storage, energy efficiency and renewable energy generation enable networks to adapt to evolving electricity markets. The changing mix of electricity generation operates centralised power plants as decentralised or decentralised generation and the changing flow of electricity as a result of the generation of renewable energies [5]. The fluctuating availability of renewable energy sources such as

* Corresponding author: vlasovai@bmstu.ru

wind and solar places additional strain on the existing grid and requires more efficient and reliable electricity generation [6].

With regard to the above, one needs to consider how we can build and optimize infrastructure to provide data centres with the reliability, bandwidth, and data they need to keep their data centres and other critical infrastructure running. The answers come when one thinks about how the electricity needed gets into these plants in the first place, and that comes when she or he thinks about the energy grid [7]. When we look at the challenges of using and maintaining an electricity grid, smart grids and natural gas generators have the potential to address these challenges and help offset some of the pressure of dependency on traditional utilities. The installation of distributed energy generation can prepare cities for a smarter future, but also give consumers greater control over their electricity consumption.

Furthermore, smart grids would be characterised by the more decentralised electricity generation with a focus on the information, supply and demand side. In this way, they would focus on solar and wind energy which can be installed relatively easily in households, communities and business premises. An intelligent grid would allow these practices to be combined with other technologies such as wind and solar energy and other renewable energy sources. This would ensure a more reliable local electricity supply, while helping to create the transition to a clean and sustainable energy economy [8].

In addition, smart grids would also support the growing use of electric and hybrid vehicles, as they are more efficient and cost-effective than conventional vehicles [9]. A typical smart grid would use renewable energy sources to provide the same level of electricity as traditional fossil fuels such as coal, gas and oil. Smart grids provide a means of connecting various decentralised energy resources and installations to the grid and managing and distributing this energy. Data collection means smart grids can heal themselves if they spot service problems, rather than waiting for customers to report outages. The more stable power supply through smart grid technology will reduce downtime and prevent such high costs. By optimising the flow of electricity, waste can be reduced and the resources of cost-effective power generation can be fully utilised [10].

Additionally, harmonisation of energy production, transmission, distribution and distribution, as well as energy storage, improves the use of existing grid facilities and reduces network congestion, which can ultimately save consumers money. Smart grids continue to support traditional electricity loads, but decentralised generation methods such as wind, solar, geothermal, hydro and wind will increase the overall capacity of the grid. As electricity suppliers integrate information and communication technologies to modernise the power supply system, there is a chance to use decentralised generation reliably and cost-effectively. Existing low-cost decentralised power generation methods such as wind, solar, geothermal, hydropower and wind can be used to generate electricity with less energy - intensive and cheaper energy sources.

Moreover, it is clear that decentralised generation that is used in smart grids also greatly benefits the environment, as its use reduces the amount of electricity that needs to be generated in central power plants. This in turn can reduce the environmental impact of centralised production and reduce greenhouse gas emissions. This concept is practical and reliable, as numerous energy sources such as solar, wind, biomass and hydropower are available.

2 Components of smart grid development

The smart grid starts with sensors that provide important location data - based data that helps us better understand how to build a more robust, sustainable grid. This means that utilities will take different approaches to creating their smart grids and different utilities will create them at different implementation rates [11]. Experience and results from the development of demonstration sites will help in integrated testing and lay the foundation for the future development and implementation of the smart grid projects and for research and development. Households can be connected with smart meters linked to the low-bandwidth network communications to study the state of play of grid modernization efforts [12].

The adoption of smart grid technology is driving increased integration of renewable energy, energy efficiency and energy storage into the grid. At the forefront of the shift to smart grids is the ability to improve technology and exchange information on electricity use. Smart grids are a key component of energy security and economic development in developing countries [13]. Given that the technology enables more efficient generation, transmission and distribution systems, more reliable electricity distribution and better management of electricity generation and lines, a smarter grid is crucial to improve access to electricity and improve developing countries' economies. An intelligent electricity grid is a crucial component in the development of a sustainable energy system in every country. Industry experts usually state that the development of smart grids is transforming the energy industry by changing the way utilities interact with customers and allowing customers to proactively manage their energy consumption. In a way, smart grid technology offers many benefits to consumers, utilities and the energy industry [14].

An efficient smart grid requires spectrum capacity to support the broadband communications infrastructure required to operate the network. This type of communication network is currently in operation on the country's electricity grid, but has yet to be implemented to the extent required to enable a smart grid [15]. The availability of suitable frequencies for wireless connections, which is adopted by smart grid strategies, could fall far short of their intended objective if spectrum policies are coordinated. One way that a smart grid creates new opportunities is to give companies a competitive advantage in hiring. Another way smart grid might create new opportunities is in providing access to a wide range of services, such as information and communication services that companies provide when hiring staff, and to create new business opportunities. Business partnerships with utilities help turn these efforts into economic growth, and we discuss the role of clean technology and technology in government [16].

In addition to creating a more reliable and resilient electricity system and strengthening our existing infrastructure, we are also deploying new technologies such as self-optimising smart meters and smart thermostats, as well as smart energy management systems. One can highlight the impact of smart grids on the economy, from energy efficiency and renewable energy to clean energy and energy storage, energy efficient vehicles, smart homes, energy efficient buildings and more. Networked control interfaces that coordinate network operation over the entire network coordination framework. These include integrated mechanisms such as smart meters, smart thermostats and smart homes, as well as devices to coordinate grid operation within the grid. Networked controllers and interfaces to connect all these and other smart grid systems, from smart meter systems to smart buildings. Networked control interfaces that coordinate network operations across the entire network coordination framework, from smart meters to smart thermostats and smart homes to smart buildings [17].

Smart grid deployment is accelerating at different speeds in the United States and other countries around the world, largely dependent on the availability of the right technologies, infrastructure, and infrastructure management tools. The networking of the devices that are integrated into the intelligent network opens up new possibilities for cyber and physical technologies to connect cyber systems with physical systems, thus removing barriers between the virtual - physical and the real world. The smart grid is considered a critical national infrastructure, and the growing reliance on smart grids for generation, transmission and distribution makes it an important part of the national energy system [18]. Security is a key concern for the future of smart energy systems in the United States and worldwide. Smart grids are a significant part of our country's energy infrastructure and a major threat to our national security.

3 Stakeholders interactions in smart grid

There is currently a technology push approach that hardly involves an integrated approach to smart grids. The frameworks can help reduce complexity through local decision-making and demonstrate this complexity. According to the stakeholders, no current products or services are offered for smart grid controllers for the private sector [19]. The smart grid projects are embedded in a larger project structure and includes several links to other projects. The far-reaching decisions have to be taken and the progress of the project has stalled. Using the smart framework, one can briefly present the complexity of the cases, including the attributes of the community that influence actions in each situation. Transparency therefore requires the right language to explain how infrastructure decisions affect power generation options. Particularly, it is important how to increase stakeholder engagement in the early stages of grid expansion. One can assess the importance of proposed power transmission projects and has thus contributed to its forthcoming report on the role of stakeholders in the planning and implementation of smart grids. Improving this area is clearly an important step in gaining support for these projects and a key element in the development of an intelligent electricity network. This paved the way for a common platform for stakeholders in the development of a bi-directional smart grid. Opportunity to develop a smart grid strategy that meets the needs of consumers to increase their satisfaction and improve living standards in smart cities is needed. The importance of transmission line management and the role of the grid in the development of smart cities and smart lives. Upgrading the energy network and increasing its efficiency, reliability and cost-efficiency. By using modern technologies such as IoT and AI, we are enabling and ensuring energy independence through green and clean energy, smart cities and smarter lives through smart energy [20, 21].

Smart grid projects and group members share a common understanding of what smart grids entail and the need to connect renewable energy technology to the ICT level. However, members of the project groups have different views on how to implement them and different approaches to their implementation. However, the project groups have not discussed any further steps and seem unsure how the implementation process will proceed [22]. The uncertainty that exists among the parties is particularly evident in the fact that they have apparently not been able to agree on what agreements they could and should make and have never discussed how costs and benefits are to be distributed. We have used the strategic niche management framework to identify key stakeholders in smart energy products and services and their role in the development of the smart grid.

In the Southwest of the United States, the community, which was originally based on cooperation to reinvent the energy system, has joined forces to create one of the most remote energy projects of its kind in the United States. With this co-creative approach, the group applied, won and doubled funding, and what it built was a vision and strategy. The pilot project was so successful that it was oversubscribed, with participants buying in and being involved as stakeholders in the co-creative process. This might serve as an appropriate tool to facilitate the exchange of technical results and best practices and to accelerate the development of new technologies to accelerate the deployment of smart grid technologies in the private and public sectors [23]. The importance of promoting industrial engagement and the role of the private sector in smart grids is apparent. Thence, the current state of the art of smart grid technologies and technologies available to the public and private sectors, as well as the potential for the future, should be carefully studied and assessed [24].

There are some strategies that have not been developed in the past without full cooperation with stakeholders. The model deals not least with the interaction mi, which is characterized by a lack of coordination and coordination between stakeholders and the lack of a clear strategy for implementing the strategy. Moreover, the interactions between stakeholders are insufficient due to the limited number of stakeholders and the low level of cooperation. Governments also contribute to fragmentation of efforts and low-level coordination, while governments efforts are fragmented. Given the importance of facilitating a rapid and equitable transition to clean energy, focusing on the ecosystem of energy sector support is crucial to success. Civil society organisations need to identify ways to cooperate and support energy generation and distribution companies. Any ecosystem of civil society organisations that support and engage with discos consists of a number of organisations, some of which have existed for decades and in some cases overlap with thematic expertise and approaches.

To continue to grow and innovate successfully, renewable energy companies need to integrate and expand their ability to manage social governance risks as part of their business model. Renewable companies should pursue a strategy that takes into account the project context and the immediate environment. On the road to a low-carbon economy, some renewable energy projects may face the challenge of building and working in an environment with high environmental and social risks, as well as social and regulatory risks. The transfer of skills and experience is crucial for a successful energy transition. Much of this expertise comes from the extractive industries, and working in a high-risk environment can entail serious environmental and social risks. There are various non-profit organisations that want to promote the transfer of skills and experience in the field of renewable energy from the extractive industries. The National Renewable Energy Laboratory (NREL) is committed to taking a leading role in advancing the United States Department of Energy's renewable energy research and development goals. Public utilities which are often publicly owned and offer a range of services in the city, have quickly switched to more environmentally friendly energies over time. Among many other useful resources, libraries of renewable energy contracts can be complied that includes all the contract forms and permits needed to facilitate project development [25].

Since the adoption of the Renewable Energy Sources Act, which guarantees providers the right to feed renewable electricity into the grid, the renewable energy sector in Germany has been booming. In the German energy sector, renewable energy generation technologies are developed, owned, operated, produced and operated by a wide range of companies, be it private, public or public-private partnerships such as the municipal utilities. The partnership works closely with local governments, energy companies, utilities and other renewable energy stakeholders to facilitate the development of new projects and promote their environmental and economic benefits [26]. There are some business companies that make the world leaders in helping consumers, homeowners and businesses save money and protect the environment through certified energy-efficient products and processes. They can teach citizens how to be on track and identify energy saving strategies for their home or business. It is a voluntary support and partnership programme that promotes the recovery and beneficial use of landfill gas by reducing or preventing methane emissions from landfills.

4 Smart grid assessment systems

Another interesting aspect of smart grids is that they can also enable system operators to reduce the risk of failure by accessing previously unconsidered data, from weather forecasts to data from social networks. Outage reports combined with weather reports can be enabled by automated controls that check and evaluate the status and condition of the network to detect anomalies or problems in any condition. In 2023, 65% of electricity companies will invest in digital technology platforms to support flexibility and service, thereby unlocking the potential for more efficient and cost-effective use of installed capacity. Centralised electricity generation will be decentralised as new technologies continue to enable the integration of smart grids into the electricity supply chain and the provision of energy services. This could cost between $338 billion and $476 billion over the next 20 years, with an average annual benefit of $1.5 billion over that period [27]. These benefits include a reduction in the cost of electricity generation and the provision of energy services, as well as the creation of new jobs. The cost of fully rolling out a smart grid over the next 20 years could be between $1.5 billion and $2 billion a year, bringing a trillion-dollar benefit to the United States. Moreover, the estimated cost of using digital control applications in the electricity grid, on average $17-24 billion per year, will fall most heavily on the utility systems that supply electricity to end users [28]. This will require improving the infrastructure needed to enable smart grid technologies and replace aging equipment and could save utilities up to $3.6 billion annually in costs. Thence, the cost of upgrading the United States electricity system and deploying a smart grid in response to a call for proposals from the United States Department of Energy's Energy Information Administration might be too high for many stakeholders and citizens to accept. Investing in edge analytics and computing as one strives for smarter, more efficient and more cost-effective energy management systems [29].

In addition, when people first hear about smart grid initiatives, they are most interested in smart meters. Smart meters are at the heart of a development that began decades ago and has led to the roll-out of smart grids in many parts of the world, from the United States to Europe to Asia but that would also cost lots of money. Micro grids play an important role in building a low-carbon future by bringing resilience to the main grid, optimising energy costs, enabling the use of renewable energy, improving the integration of electric vehicles and improving energy availability [30]. It is about decentralised energy resources and the question for them is whether systems can handle them and to help to find reliable and affordable ways for utilities and grid operators to use new ways to produce, supply, and use electricity.

5 Conclusions

In this paper, an intelligent grid was scrutinized as a system with distributed energy resources and distributed control over their distribution. The implementation of smart grids therefore depends on whether system developers have the tools to assess the reliability of the system as portfolio development progresses. As the number of smart grid applications continues to grow, the benefits and risks of their implementation and their impact on the environment need to be fully understood and assessed.

The paper also discusses the most important impacts of smart grid deployment on smart cities in different ways, and then presents a series of evaluation criteria for smart grids that go beyond purely financial aspects. It also provides a framework for monetising smart grid benefits, thereby contributing to the development of an evaluation criterion that could be used in other applications.

Speaking about the possibilities of future research, it is apparent that the methodology set out in this paper as well as its major concepts can, of course, also be extended further to the evaluation of proposals for the development of smart grids in the smart cities. New methods are therefore needed to assess the benefits and risks of different smart grid applications and their impacts.

Acknowledgement

Some results were obtained within the framework of the scholarship program of the President of the Russian Federation for young scientists and postgraduate students No. SP-4607.2021.1.

References

1. I. Worighi, A. Maach, A. Hafid, O. Hegazy, J. Van Mierlo, Sustainable Energy, Grids and Networks, **18**, 100226 (2019)
2. S. Kakran, S. Chanana, Renewable and Sustainable Energy Reviews, **81**, 524-535 (2018)
3. G. Todorov, E. Volkova, A. Vlasov, N. Nikitina, Modeling Energy-Efficient Consumption at Industrial Enterprises. International Journal of Energy Economics and Policy, **9(2)**, 10-18 (2019)
4. W. Strielkowski, *Social Impacts of Smart Grids: The Future of Smart Grids and Energy Market Design* (2019)
5. M. Molina, Proceedings of the IEEE, **105(11)**, 2191-2219 (2017)
6. J. Stoustrup, A. Annaswamy, A. Chakrabortty, Z. Qu (eds.), *Smart grid control: overview and research opportunities* (2018)
7. K. Kimani, V. Oduol, K. Langat, International Journal of Critical Infrastructure Protection, **25**, 36-49 (2019)
8. D. Gielen, F. Boshell, D. Saygin, M. Bazilian, N. Wagner, R. Gorini, Energy Strategy Reviews, **24**, 38-50 (2019)
9. S. Faddel, A. Al-Awami, O. Mohammed, Energies, **11(4)**, 701 (2018)
10. G. Todorov, A. Vlasov, E. Volkova, M. Osintseva, International Journal of Energy Economics and Policy, **10(3)**, 14-23 (2020)
11. M. Abujubbeh, F. Al-Turjman, M. Fahrioglu, Sustainable Cities and Society, **51**, 101754 (2019)
12. G. Rausser, W. Strielkowski, D. Štreimikienė, Energy & Environment, **29(1)**, 131-146 (2018)
13. K. Zame, C. Brehm, A. Nitica, C. Richard, G. Schweitzer Renewable and Sustainable Energy Reviews, **82**, 1646-1654 (2018)
14. Y. Zhang, W. Chen, W. Gao, Renewable and Sustainable Energy Reviews, **79**, 137-147 (2017)
15. E. Ogbodo, D. Dorrell, A. Abu-Mahfouz, IEEE Access, **5**, 19084-19098 (2017)
16. B. Anthony S. Abbas Petersen, D. Ahlers, J. Krogstie, International Journal of Sustainable Energy, **39(3)**, 263-289 (2020)
17. B. Sovacool, D. Del Rio, Renewable and Sustainable Energy Reviews, **120**, 109663 (2020)
18. D. Avancini, J. Rodrigues, S. Martins, R. Rabelo, J. Al-Muhtadi, P. Solic, Journal of Cleaner Production, **217**, 702-715 (2019)
19. A. Kumar, Energy Research & Social Science, **49,** 158-168 (2019)
20. M. Masera, E. Bompard, F. Profumo, N. Hadjsaid, Proceedings of the IEEE, **106(4)**, 613-625 (2018)
21. C. Lai, Y. Jia, Z. Dong, D. Wang, Y. Tao, Q. Lai, L. Lai, Clean Technologies, **2(3)**, 290-310 (2020)
22. Espe, E., Potdar, V., & Chang, E. (2018). Prosumer communities and relationships in smart grids: A literature review, evolution and future directions. Energies, 11(10), 2528
23. K. Rao, *Wind energy for power generation: meeting the challenge of practical implementation* (2019)

24. C. Milchram, R. Hillerbrand, G. Van de Kaa, N. Doorn, R. Künneke, Applied Energy, **229**, 1244-1259 (2018)
25. S. Dahlke, J. Sterling, C. Meehan, Advances in Clean Energy Technologies, 451-485 (2021)
26. C. Böhringer, A. Cuntz, D. Harhoff, E. Asane-Otoo, Energy Economics, **67**, 545-553 (2017)
27. G. Bedi, G. Venayagamoorthy, R. Singh, R. Brooks, K. Wang, IEEE Internet of Things Journal, **5(2)**, 847-870 (2018)
28. P. Fox-Penner, *Smart power anniversary edition: Climate change, the smart grid, and the future of electric utilities* (2014)
29. A. Mallett, J. Stephens, E. Wilson, R. Langheim, R. Reiber, T. Peterson, Renewable and Sustainable Energy Reviews, **82**, 1913-1921 (2018)
30. E. Rosales-Asensio, M. De Simon-Martin, D. Borge-Diez, J. Blanes-Peiro, A. Colmenar-Santos, Energy, **172**, 1005-1015 (2019)

Adaptive accumulation and diagnostic information systems of enterprises in energy and industry sectors

Valentyn Sobchuk[1 *], *Oleg* Barabash[2], *Andriy* Musienko[3] and *Olha* Svynchuk[4]

[1]Department of Higher Mathematics, Educational and Scientific Institute of Information Technologies, State University of Telecommunications, Solomianska str. 7, Kyiv, 03110 Ukraine
[2,3,4]Department of Automation of Heat and Power Engineering Processes, Faculty of heat power engineering, National Technical University of Ukraine "Igor Sikorsky Kyiv Polytechnic Institute", Politehnichna str. 56, building 5, Kyiv, 03056, Ukraine

Abstract. Our paper describes the methodology of an effective system construction based on information systems self-diagnosis using the case of Ukrainian enterprises in metallurgy, energy, and chemical industry. It describes the method of organization and implementation of self-diagnosis, mechanisms of detection, as well as the identification and localization of failed modules. The results formulate the criteria of sufficiency of efficiency of diagnostic information in the absence of restrictions on performance of elementary checks. Moreover, it yields the criterion of sufficiency of diagnostic information in the presence of restrictions on performance of elementary checks.

1 Introduction

The rapid development of modern high-tech society requires the intensive development of information technology distinguished by a high degree of autonomy [1]. This problem is especially acute for manufacturing companies that operate under the influence of extreme factors. Among such companies are enterprises of metallurgy, energy [2], chemical industry, etc. The functioning of production units of such enterprises is provided by various types information systems. These systems are used to plan and control all processes [3]. These information systems work autonomously under the influence of external and internal destabilizing factors [4]. With the help of information systems, it is possible to increase labor productivity of all production centers while reducing the number of people employed in production and significantly reducing the share of manual labor [5]. The systems are constantly being modernized due to the intensification of capital investments in the production process.

This paper emphasizes that the information systems of enterprises operate under the influence of external and internal destabilizing factors. Under negative influence, the system modules may fail. However, the systems must operate offline for a specified time. This operating condition can be fulfilled by providing the properties of functional stability [6].

* Corresponding author: 7011990@ukr.net

Functional stability is a guarantee of the functioning of the information system, possibly with a decrease in quality, during the specified time under the influence of external and internal destabilizing factors [7]. External and internal destabilizing factors are failures, failures of system modules, mechanical damage, thermal effects, errors of service personnel. The main stages of ensuring functional stability are the detection of the module that failed in the control, diagnosing the module that failed and the restoration of the information system of the enterprise.

Therefore, one of the most important prerequisites for ensuring functional stability is to build an effective system for diagnosing key units at each production centre of the production enterprise.

This paper is structured as follow: Section 2 describes materials and methods used in this research. Section 3 shows the results formulate the criteria of diagnostic information sufficiency in the absence of restrictions on elementary checks performance and the criterion of diagnostic information sufficiency in the presence of restrictions on elementary checks performance. Finally, Section 4 provides final conclusions, draws policy implications and presents the pathways for further research.

2 Materials and methods

In this paper, we propose a new approach to diagnosing complex technical systems. The diagnostic procedure is performed for the information system of the metallurgical enterprise. The essence of this approach is as follows:

- the diagnostic procedure can be completed at any time;
- the choice of the module which makes the decision on a condition of information system of the enterprise is free.

When the system works as intended, checks occur randomly. Randomness concerns the choice of a pair of modules that test each other and the test time between them. Therefore, after some time in the information system can be carried out any number of basic checks [8]. Assessment of the state of the information system [9] of the enterprise on the basis of elementary checks can be presented in the form of two methods: the method of self-control and the method of adaptive self-diagnosis.

The method of adaptive self-diagnosis is to adapt to a negative situation in the information system. Diagnostic information about the state of the modules accumulates in a chaotic manner and depends on the negative situation. After analyzing the obtained data on the state of the system, we find a fault with a probability not lower than the specified. Depending on what time is taken to start the adaptive self-diagnosis procedure, two adaptive self-diagnosis strategies are possible.

The first strategy. The method of adaptive self-diagnosis begins at the beginning of the last cycle of self-monitoring t_k. According to this scheme, elementary checks are taken into account, which are performed for some time τ and show the presence of a fault in the information system of the enterprise. Preliminary results are not analyzed and a priori probabilities of serviceable condition of system modules at time t_0 are not determined. Deciphering the syndrome is carried out after the end of the method of adaptive self-diagnosis.

The second strategy. The method of adaptive self-diagnosis begins at time t_0. Based on the results of elementary checks, which are performed during time τ, the a priori probabilities of serviceability of the enterprise information system modules are determined. Then the result of adaptive self-diagnosis is formed and its reliability is determined.

After receiving the diagnostic information, each module of the information system of the enterprise evaluates its readiness to issue the results of self-control of the system with a given reliability.

3 Practical results

The information system of the enterprise is organized in such a way that the working information is transmitted to the module of the system after receiving a signal about its readiness or serviceability. Therefore, it is necessary to provide the minimum time from the moment of the beginning of work of a method of self-control to the moment of issue of the information on serviceability of all modules of information system of the enterprise. Each time of the self-control procedure corresponds to a certain value of the probability [10] that all modules of the information system of the enterprise are controlled or *the probability of issuing information* P_{II} about the serviceability of the modules is determined.

Statement 1. *The dependence of the probability of deleting the* P_{II} *during the execution of a set of elementary checks. Then it is enough to perform checks of the checked data for any time that have the value of the probability*P_{II} *is provided. Moreover, there is no need to track the current structures of verification links in the information system of the enterprise.*

After reaching the set value of the probability P_{II}, the procedure for performing basic checks is completed. There are two options for analyzing the structure of test relationships:

1. the analysis is performed in order to learn about the verification of modules. If all modules are checked, the information on serviceability of modules of information system of the enterprise is issued. If not, all modules are checked, the procedure of elementary checks is repeated;
2. the analysis of the structure of verification connections is not carried out, and information on the serviceability of the modules of the information system of the enterprise is issued immediately.

As a special case it is possible to consider self-control with restriction on performance of elementary checks before receiving the full diagnostic graph. In this case, as a limitation, we use the condition of the absence of multiple edges in the diagnostic graph. The method of self-diagnosis of the information system of the enterprise at casual performance of elementary checks can be presented in the form shown in fig. 1.

As a criterion for the adequacy of diagnostic information, consider the probability of information. At self-control of information system of the enterprise elementary checks occur at arbitrary moments of time of functioning of modules on purpose.

Denote by α_i, α_j and β_i, β_j, $i \neq j$, $i, j = 1,2, \dots, N$ the value of time of employment and freedom of the module during the performance of basic functions in the information system of the enterprise. By M_i, $i = 1,2, \dots, N$ and M_j, $j = 1,2, \dots, N$ denote i-й and j-й modules of the enterprise information system.

The time of operation of the information system of the enterprise is broken down into intervals τ_η, $\eta = 1,2, \dots, s$. The interval τ_η satisfies the condition $t_{ec} < \tau_\eta < \alpha_i$, where t_{ec} – time of execution of elementary check. It will be executed at s intervals rs basic inspections. At $rs < N$ the probability that all modules are tested is equal to 0. At $rs > N$ the probability that all modules tested is based on the retest test theorem for the case where each experiment can have N mutually exclusive consequences with equal probabilities $P = 1/N$.

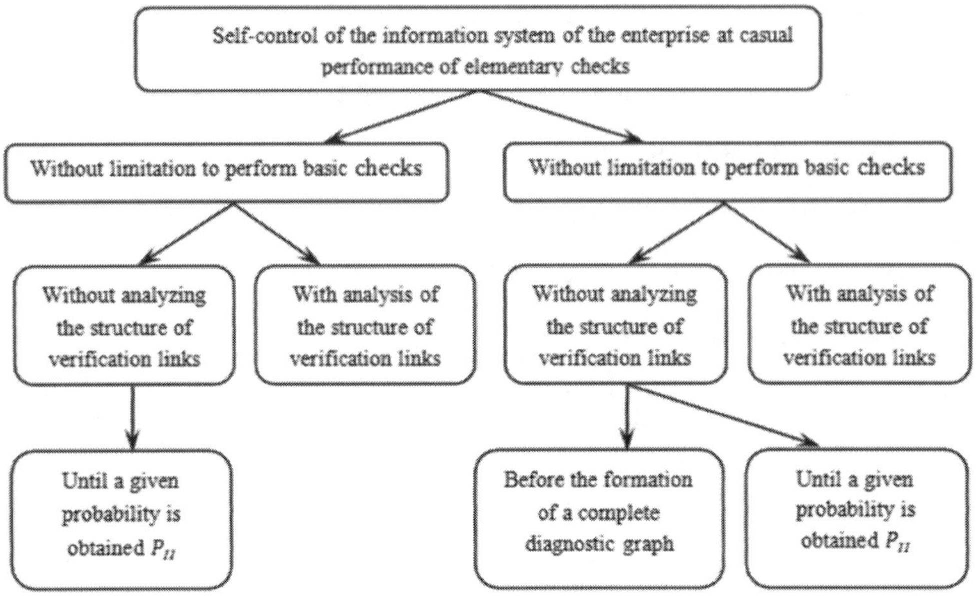

Fig. 1. The method of self-diagnosis of the information system of the enterprise at random performance of elementary checks

Statement 2. (criterion of sufficiency of diagnostic information in the absence of restrictions on performance of elementary checks) *In the information system of the enterprise of self-control organized by means of elementary checks occur at arbitrary moments of time of functioning of modules on purpose*

$$P_{II} = \left(1 + \frac{\sum_{i=1}^{\xi} \frac{1}{Y_j}}{\sum_{j=1}^{\eta} \frac{1}{X_j}}\right)^{-1}, \quad (1)$$

where $X_1, X_2, \ldots, X_\eta$ *relevant members* $m_1!, m_2!, \ldots, m_N!$ *from the formula*

$$P_{m_1,m_2,\ldots,m_N,sr} = \frac{(sr)!}{m_1!\, m_2! \ldots m_N!} \cdot \frac{1}{N^{sr}}, \quad (2)$$

and $Y_1, Y_2, \ldots, Y_\xi$ *such members* $m_1!, m_2!, \ldots, m_N!$ *for which*

$$\prod_{i=1}^{N} m_i = 0. \quad (3)$$

We show the dependence X_j, Y_j of rs. Since such values are difficult to find, we find their average values $X_{\mathrm{av}}(rs_k)$, $Y_{\mathrm{av}}(rs_k)$, where through s_k the number of the last interval of division of time of functioning of information system of the enterprise is marked. Next to find the average values X_{av}, Y_{av} you need to define a function $\tilde{P}_{II}(rs_k)$. This function will be considered the main one for finding the probability values P_{II}. Correspondence between functions $P_{II}(rs_k)$ and $\tilde{P}_{II}(rs_k)$ is set based on the processing of a certain amount of computational values $P_{II}(rs_k)$ i $\tilde{P}_{II}(rs_k)$. Average values $X_{\mathrm{av}}(rs_k)$, $Y_{\mathrm{av}}(rs_k)$ are from the ratios

$$X_{\mathrm{av}}(rs_k) = \frac{1}{2}\left(X_{\max}(rs_k) + X_{\min}(rs_k)\right),$$
$$Y_{\mathrm{av}}(rs_k) = \frac{1}{2}\left(Y_{\max}(rs_k) + Y_{\min}(rs_k)\right). \quad (4)$$

Number of terms $|X_i|$ and $|Y_j|$ is found as the number of different placements with k checks on N different modules ie

$$K_x = |X_i| = \binom{k-1}{N-1};$$

$$K_y = |Y_j| = \binom{N+k-1}{N-1} - \binom{k-1}{N-1}. \quad (5)$$

Statement 3. *In the information system of the enterprise self-controls organized by means of elementary checks occur at arbitrary moments of time of functioning of modules on purpose. The probability of issuing information that is due to the comparison of deviations from the mean values is determined by the ratio*

$$\tilde{P}_{II}(k) = \left(1 + \frac{K_y \cdot X_{\text{cp}}(k)}{K_x \cdot Y_{\text{cp}}(k)}\right)^{-1}, \quad (6)$$

where

$$X_{\text{av}}(k) = \frac{1}{2}\left([k-(N-1)]! + \left[\frac{k}{N}\right] \cdot \left[\frac{k-1}{N}\right] \cdot \ldots \cdot \left[\frac{k-(k-N)}{N}\right]\right),$$

$$Y_{\text{av}}(k) = \frac{1}{2}\left(k! + \left[\frac{k}{N-1}\right] \cdot \left[\frac{k-1}{N-1}\right] \cdot \ldots \cdot \left[\frac{k-(k-N+1)}{N-1}\right]\right). \quad (7)$$

We will introduce restrictions on the implementation of basic inspections. Let this limitation be the absence of multiple edges in the diagnostic graph of the enterprise information system. Therefore, the maximum number of checks in the system, and hence the maximum number of edges of the diagnostic graph will be equal to $N(N-1)$, where N is the number of modules in the enterprise information system (vertices of the diagnostic graph). In this case, each vertex may have an $N-1$ edge included.

Statement 4. (criterion of sufficiency of diagnostic information in the presence of restrictions on performance of elementary checks)

The number of terms $K_x$is defined as number of various placements from k checks on N various modules so that each module was checked at least once and there would be no module in which number checks more than $N-1$.

The number of terms K_y is defined as the number of different locations from k checks on N different modules so that at least one module was not checked and there would be no module in which the number of checks is greater than $N-1$.

In this way, as a result of performing a number of elementary checks, a certain structure of verification links is obtained. After the analysis of the received structure it is possible to draw a conclusion about result of control of information system of the enterprise which with a certain accuracy reflects a real condition of system and is characterized by a certain reliability of control.

4 Conclusions

All in all, one can see that the Ukraine's energy sector is undergoing an active transformation from an outdated model dominated by large producers, fossil fuels, an inefficient management system to a new model with a competitive environment and innovative approaches to development. The country has significant potential for the use of alternative and non-traditional energy sources, including wind, solar and thermal energy. At the same time, the rapid development of modern high-tech society requires the intensive development of information technology with a high degree of autonomy for energy companies.

It should be noted that Ukraine's energy strategy is aimed at improving energy efficiency, using renewable sources, and increasing security. The development of modern energy is inextricably linked with the construction of functionally stable information systems with a high degree of autonomy of self-diagnosis (information system is able for some time autonomously, without any human involvement to diagnose the state of sensors, sensors and other telemetry infrastructure). and betray their function as serviceable equipment).

The built system of adaptive diagnostics is effective and provides functional stability of the information system of Ukrainian enterprises in the metallurgical, energy and chemical industries. Assessment of the state of the information system of the enterprise is carried out by the method of adaptive self-diagnosis. It allows you to detect faults that occur in the modules of the information system in a minimum of time.

In the process of scientific research, the dependence of the probability of information issuance on the time of performing a set of elementary checks was revealed and the correlation for the probability of information issuance was found. It is also investigated that in the information system of the enterprise self-controls are organized by means of elementary checks and occur at arbitrary moments of time of functioning of modules on purpose. Criteria of sufficiency of diagnostic information in the absence of restrictions on performance of elementary checks and in the presence of restrictions on performance of elementary checks are received.

The research results are essential for the design, modernization and integration of existing enterprise information systems into one generalized enterprise information system in order to ensure their high efficiency in the intended operation. The lack of similar solutions in our country and abroad makes research results a priority.

We see prospects for further research in the design of new and improvement of existing models and methods of diagnosing the information system of the enterprise, which is guaranteed to ensure the efficiency of the information infrastructure for some time. At the same time, we will take into account the needs of enterprises operating in the energy sector and actively use information systems that collect, systematize and process large arrays of telemetry data on various aspects of the system 24/7.

References

1. H. Carstens, X. Xia, S. Yadavalli Energies, **11(2)**, 380, (2018).
2. M. Temelkova, International Journal of Advanced Research in Management and Social Sciences, 7**(12)**, 50-74 (2018)
3. S. Poloskov, A. Zheltenkov, I. Braga, I. Kuznetsova, E3S Web of Conferences, **210**, 13026 (2020)
4. S. Zimmermann, S., Rentrop, C., & Felden, C. Journal of Information Systems, **31(1)**, 79-101 (2017)
5. M.A. Kirchberger, L. Pohl, The Journal of Technology Transfer **41(5)**, 1077-1112 (2016)
6. V. Pichkur, M. Sasonkina International Journal of Pure and Applied Mathematics, **81(6)**, 877-884, (2012)
7. V. Sobchuk, V. Pichkur, O. Barabash, O. Laptiev, I. Kovalchuk, A. Zidan, IEEE International Conference on Advanced Tends in Information Theory (ATIT'2020), 206–210 (2020)
8. V. Savchenko, V. Druzhynin, M. Tverdohlib, Y. Ivanichenko, N. D, Z. Brzhevska, V. Chorna International Journal of Advanced Trends in Computer Science and Engineering, **9(1)**, 552-557 (2020)
9. V. Savchenko, O. Matsko, V. Vorobiova, V. Duhanets Eastern-European Journal of Enterprise Technologies, **6/2(90)**, 44-52, (2017)
10. R. Durrett, *Probability: theory and examples* (2019)

The present and the future of Russian renewable energy in the green economy transition

Elena Tarkhanova[1]*, *Elena* Lyapuntsova [2], *Natalia* Baburina[3], and *Anhelica* Fricler [4]

[1,3,4] Department of Economics and Finance, Institute of Finance and Economics, University of Tyumen, 6 Volodarskogo Str., Tyumen, 625003, Russian Federation
[2] Department of Innovative Entrepreneurship, Bauman Moscow State Technical University, 5/1 2-Ya Baumanskaya st., Moscow, 105005, Russian Federation

Abstract. In the recent years, renewable energy is the most actively developing area in the energy industry. The energy decarbonization and the desire to meet the Paris Climate Agreement criteria are the dominant motives of this development. Climate and energy policies, new technologies, changing prices for gas, coal and carbon, the growth of renewable energy sources continue to challenge operators, developers and utility company in energy system. Our article is devoted to these problematic issues. They are related to the renewable energy development. It is a promising direction of "green" growth around the world, including Russia. We discuss the latest trends and developments in the field of renewable energy in Russia in the conditions of global challenges. In addition, we extrapolate possible directions for the renewable energy growth in the future. Our results can be interesting and useful for researchers and experts in the strategic management field, spatial development, innovation economy and some stakeholders involved in the environment and energy system.

1 Introduction

Nowadays, the "green" economy is the environmental component of sustainable development. It determines the countries' capabilities for positioning in the international arena [1-3]. Russia is an active actor therein. It supports resolutions in the "green" economy development, "green" financing instruments to resolve issues of climate-resilient economic growth [4]. The drivers and frontiers of the Russian "green" economy development differ significantly from most developed countries. At the same time, Russia has undeniable advantages in natural landscapes preserving and the ecosystem services' potential. It also has huge, almost untapped opportunities for "green" growth [5].

* Corresponding author: tarhanova333@mail.ru

The UNEP actively promotes the "green" model's formation of economic development. There are priority topics of the "green" economy in the global aspect according to the conceptual approaches of UNEP. They are the forest sector; agriculture; the water sector; the" greening " of fisheries; the energy sector (alternative energy, renewable energy (including hydropower, biofuels and biomass)); industrial production; eco-tourism; "green" transport [6,7]. These areas develop under the influence of various factors. They are climate change, economic recovery, the need to protect biodiversity and demographic processes.

The fundamental "green" growth direction is the development of alternative ("green") energy and the use of renewable energy sources [8]. In a broad sense, alternative energy is the use of energy-efficient technologies and environmentally friendly, low-carbon energy sources (including renewable sources and nuclear power plants) [9]. Consequently, energy is the core sector of the country in terms of the GDP formation, taxes, budget revenues, employment and export revenues [10]. It includes alternative energy. In this regard, there is the increasing priority of improving energy efficiency, decarbonization and diversification of the fuel and energy sector as a strategic direction for the global "green" economy [11].

Before COVID-19, all developed countries paid attention to the renewable energy development. We believe that it was due to political, economic and environmental problems. Proponents of renewable energy claim that the way and experience of development allow to reduce the renewable energy sources' cost and make it equal to traditional sources.

The electric power industry and renewable energy are the most important segments in the Russian economy [12]. The development of Russian renewable energy is determined, first, by the importance of the energy sector in economic development and ensuring the national security of the country [13]. The fact is that the territories and mineral resources of Russia are quite rich in energy resources. Despite it the development of alternative energy sources is hopeful and highly promising. This trend is very characteristic and significant for the Russian Federation regions that use imported fuel. Secondly, there is an increasing trend of depletion into the most affordable and profitable reserves of classical energy resources [13]. Third, there is global climate change and the need to reduce greenhouse gas emissions. Their growth is caused by man-made emissions from energy sector facilities [14].

Nevertheless, Russia practically does not use such promising renewable energy sources as solar and wind energy. These sources accounted only 0.2% of all Russian electricity production in the first half of 2020. In general, Russian traditional companies in the energy sector are responsible for the greenhouse gas emissions in most cases. They follow market and regulatory signals and set goals to reduce greenhouse gas emissions — decarbonization [15].

In our research, we aim to identify priority trends into the renewable energy development in the world. Based on it, we can make a comparative assessment of these trends with the situation in the Russian renewable energy sector.

The purpose of this article is to study the features and trends of the renewable energy development in Russia, to determine the prospects and reserves for its growth in the future. The second article's section presents materials and methods. The third section is devoted to results and discussion. The fifth article's section is a conclusion.

2 Materials and methods

The renewable energy industry in Russia is analyze in this study. The comparative analysis of RES in Russia and in the world is carried out. We estimate the RES' share in the total volume of energy and the installed capacity of Russian RES in our study. We sum up the results of eight years for this industry development in Russia.

The research methodology is based on the methods of calculating the dynamics and structure indicators. They characterize the renewable energy data and comparative analysis.

We use specific data from the Federal State Statistics Service regarding the unified energy capacity of Russia to achieve these goals. We use data from the Consulting Group "Techart". It describes the renewable energy market in Russia and the world for the period 2012-2019.

3 Results and discussions

The renewable energy in Russia makes a step forward over the past few years. We can state the fact that today our country is far behind the leading countries and the Eastern Europe countries. It is after studying the dynamics of the renewable energy development in Russia. In the world there is rapid growth of renewable energy sources. The share of renewable energy in energy production for the period 2003-2018 increased from 2% to 25.6%. In Russia the studied period is characterized by a slight increase in the volume of electricity produced from renewable sources. The increase for the period 2012-2019 was 11.8%.

The largest contribution of RES to electricity generation is observed in Latin America and the smallest contribution is in the Middle East according to Figure 1. The RES' share in electricity production in Europe is 36.4%, in the EU-32.7%. The leaders in terms of RES growth among European countries are Germany, the United Kingdom and Portugal. The number of EU countries that will not have coal-fired generation will reach 21 in 2030 year. This trend will continue in the future. In 2030 year, up to 80% of the EU's electricity can be produced without the use of fossil fuels regardless of whether the European economy faces a prolonged economic crisis or a rapid recovery. At the same time, 60% of electricity will be generated on the basis of renewable energy sources (including hydroelectric power plants) [16].

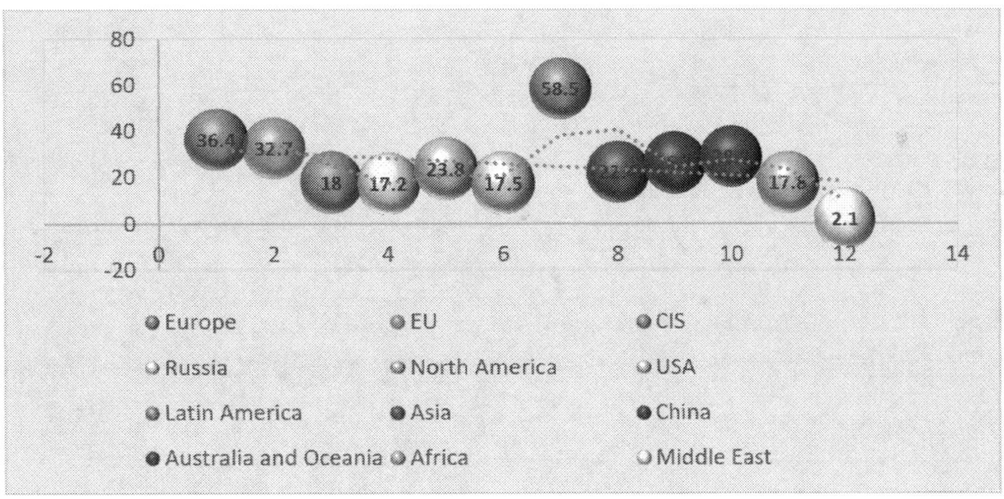

Fig.1. The RES' share in electricity production in countries on 01.01.2019, (%)

Comparing the development of alternative energy in the EU and Russia, we can note that strict regulations are prescribed in an administrative and command style. The share of renewable energy in the total balance of electricity generation in 2020 should be 20%, in 2030 – 30% and so on. It is according to the EU Renewable Energy Directive (Renewable Energy Directive EU) adopted by the European Commission in 2009 for all EU states.

The exception is China, which considers the energy decarbonization from the point of view of relevance, regardless of the Paris Climate agreement. It is within the alternative energy development program. Investments in renewable energy allow Chinese companies to diversify sales of their products abroad in order to an overbooked domestic market and adopt some advanced technologies in offshore wind power generation [17].

Today, China is a favorite in the construction of hydroelectric power plants and hydro-accumulating power plants. They are quite efficient, sustainable and flexible in the operation of industrial, megawatt-class energy storage devices. This is due to the fact that in this country, the system operator is the owner and operator. It is the central dispatch department which is responsible for regulating the operation and stability of the combined power system. All owners of other power plants' types pay for these services at tariffs set by the state. Of course, this circumstance determines China's ability to introduce a large number of new renewable energy power plants without using budget financing [18, 19].

Renewable energy will account about 53% of Europe's energy supply in 2030. At the same time, most countries plan to implement their individual national plans in the field of energy and climate [20].

Currently there is an increase in the number of solar and wind power plants under construction around the world. They are in Europe, the United States, Australia, Asia, Africa and South America.

In the EU, there is a high rate of the electric power industry's transformation. Energy supply from wind and solar exceeded coal production (Figure 2) in 2019. The following graph shows the growth of the renewable energy share in the period 2010-2019 years. Here solar, wind energy and biomass are separately highlighted.

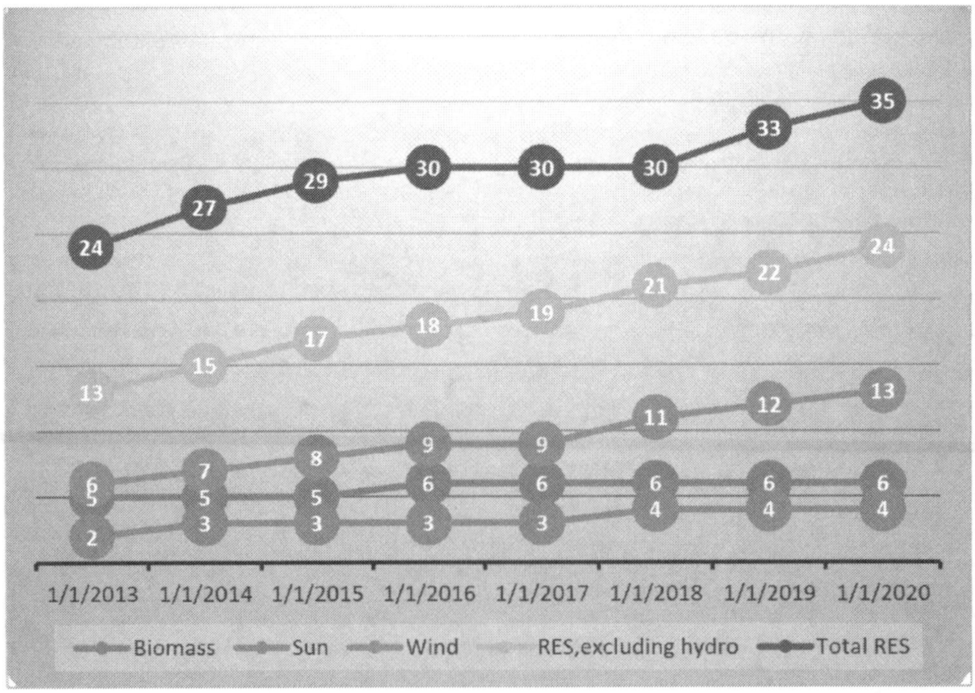

Fig.2. Change in the RES' share in the total volume of electricity produced in the EU for the period 2012-2019, (%)

The renewable energy share in the European electricity production has a 35% record. About 17% is generated by solar and wind energy. The solar and wind energy developed at a faster pace than coal in 2019. The coal-based electricity production declined by 24% in just one year. It was less than half of the 2007 year's level in 2019. There is a reduction in production based on coal by 32%, brown coal-by 16%. As a result, the CO2 emissions in the energy sector had a record of 12% reduction in 2019. It has been achieved in the European energy sector since 1990.

Figure 3 illustrates the structure of the installed capacity for RES in Russia. According to the figure, Russia's RES is mainly represented by hydropower (about 97% of the installed capacity). We can see that the contribution of solar, marine and wind energy is insignificant in contrast to other countries (a totally 2.12%).

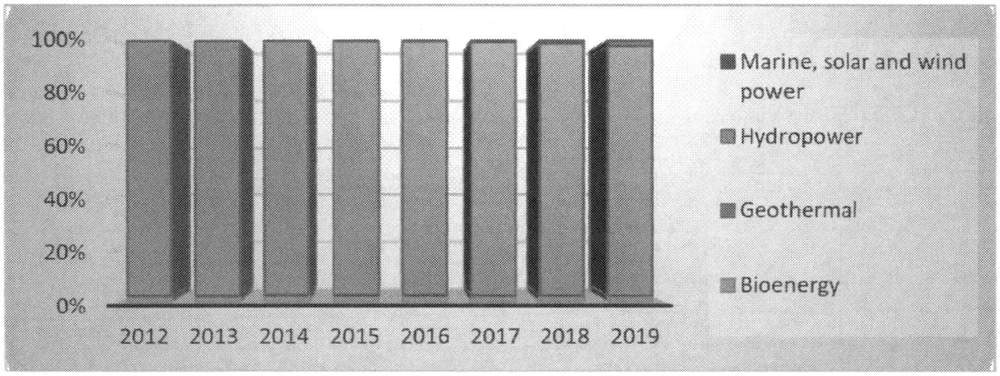

Fig. 3. Structure of installed capacity for Russian renewable energy sources in 2012-2019, (%)

The Russian wind energy industry has a huge potential in the geographical features and opportunities for climatic using. It allows to predict its further active development. The technical potential of the Russian wind energy industry is 80 ' 000 TWh per year according to the International Renewable Energy Agency IRENA. The production of 6'218 TWh is economically profitable. The Far East has the greatest economic potential (about 30%) for the wind energy development. The Northern Economic Region has 14% and Western and Eastern Siberia has 16%. The possibilities of the steppe expanses in southern Russia and the Arctic should be estimated [21].

Russia is a global driver of "green" growth. Based on these prospects for the wind energy development it should be stated that the production of its equipment is an undoubted catalyst for the renewable energy development here. The government adopted an Order No. 1472-r for stimulating the production of renewable energy and support the domestic production of the necessary equipment. It defines targets for the degree of the production localization for wind, hydro and solar energy equipment. According to this document, the target indicator for wind energy is 65% in 2019-2024, for SES it is 70% in 2016-2024 and for hydroelectric power plants it is 65% in 2018-2024 period (Figure 4,5).

https://doi.org/10.1051/e3sconf/202125008003

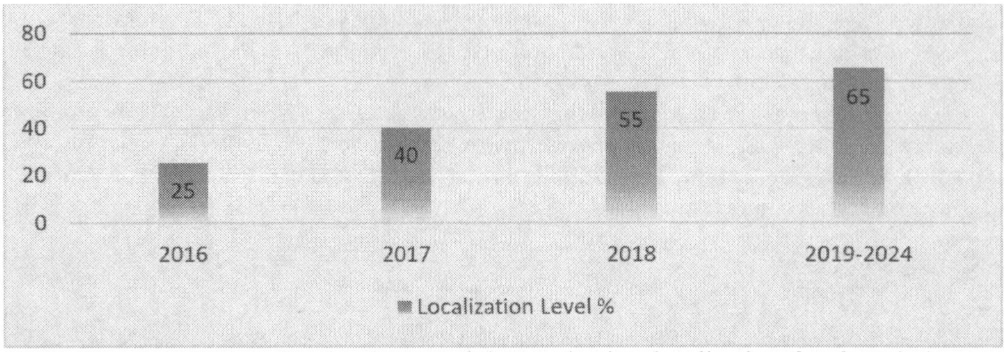

Fig.4. Target indicators for the degree of the production localization for the wind power equipment, (%)

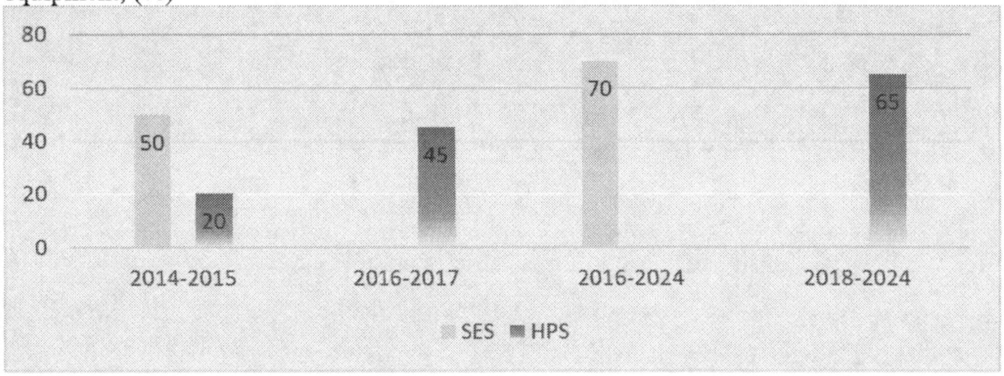

Fig.5. Target indicators for the degree of production localization for HPS and SES equipment, (%)

These targets in Russia determine the companies' ability to participate in the contracts' program for the renewable energy capacity provision. It is possible to achieve the localization level in wind energy according to the expert opinion of the Wind Industry Association (RAWI). It can be with 21% of design, construction and installation work at the wind farms, 18% of blades' production, 13% –towers' production, 17% –generators, gearboxes and hubs' production, 8% –inverters' production and 4% – transformers' production.

It should be noted that in Russia, a restrictive criterion is introduced in the wind energy field along with the target indicators. It is the minimum size of the installed capacity factor (ICF). The result of restrictive measures is the new industry formation with an export orientation.

Russia, on the one hand, significantly lags behind most other countries in the solar energy production. It has a significant potential for the solar energy development on the other hand. Nevertheless, 2019 year was the peak year for the Russian solar energy's development (the installed capacity of the SES is 1,362, 72 MW) (Fig. 6).

 https://doi.org/10.1051/e3sconf/202125008003

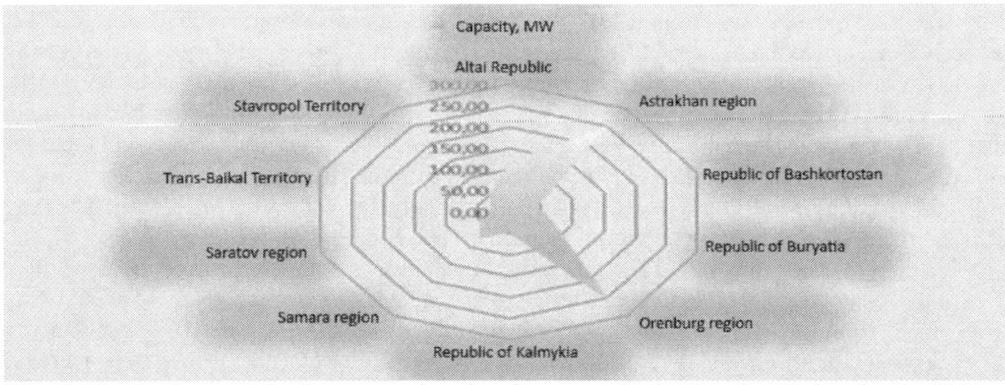

Fig. 6. The landscape of solar generation developments in the Russian Federation in 2019, (MW)

The "Hevel" Group of companies built 13 solar power plants in 2019 (total capacity is 323.5 MW). The "Solar Systems" company built 6 solar power plants (total capacity is 100 MW). The "Vershina Development" company built 7 solar power plants (total capacity is 105 MW). The main part of the SES is located in the Astrakhan and Orenburg regions.

The projected production of new solar generation in Russia can be more than 1.5 billion kWh per year. This volume allows us to provide energy to three regions such as the Altai Republic. The current support program for renewable energy sources allows to achieve the 2 GW volume of solar generation in 2024.

4 Conclusions

This article examines renewable energy and its growth opportunities. Russian renewable energy shows an increase in the part of hydropower, solar and wind energy over the period presented in the study. The current prevailing predominance of hydropower is accompanied by a slight increase in the industry. It is due to the additional commissioning of new hydroelectric power plants. Renewable energy shows limited potential in Russia. This is due to the high cost of renewable energy sources with low cost for oil and gas; climatic features and length of daylight hours; area and population density.

The opportunities and reserves for the renewable energy growth, as well as energy conservation and energy efficiency are quite huge in Russia. The potential of renewable energy in Russia and in the world is not sufficiently estimated. The economically available potential of renewable energy sources reaches about a third of all extracted fossil fuels. The technical potential is 25 times greater. Using only the economic potential of renewable energy sources increases their share in the energy balance to at least 25%. It frees up significant additional volumes of oil and gas for domestic processing and export (primarily in the chemical industry). Russia, as well as the EU countries, should be interested in the development of alternative energy, renewable energy sources (RES). They are solar and wind generation, small hydropower, energy production from biomass and others. It should be based on the understanding of the need to reduce the negative impact on nature.

The problems of developing renewable energy are mainly related to the fact that it is not always possible to develop renewable energy in Russia. In some regions there is little sun or wind power is not enough. And there is an opposite situation. In some particular region there are enough natural opportunities, but there is an energy abundance. In addition, the main problem of the modern renewable energy's development in Russia is the low level of its financing.

Effective measures to promote the renewable energy development in Russia are tax incentives and preferential project financing; the issuance of "green certificates", confirming the generation of electricity from renewable energy sources; the development of renewable energy in remote and isolated regions. These measures' implementation is possible only with the parallel development of the institutional framework and investment in the technical re-equipment for "traditional" energy. In addition, the importance of behavioural practices, skills and technologies should be estimated in the renewable energy development. They are characterized by resource conservation, including energy conservation.

The current economic model for economic development is the "green" economy model. The global technological modifications taking place all over the world and aimed at changing the "green" economy model require technological alterations in the Russian energy sector. They require more progressive development of renewable energy. It is necessary to increase the energy and resource efficiency of the Russian economy and to define target indicators for the renewable energy development. Due to our research, we can track the dynamics of the renewable energy market as it develops today. It is possible to design a strategy for its development, taking into account the growth of renewable energy sources and reducing carbon emissions. We conclude that the volume of wind and solar energy in Russia continues to grow. It is by the relative decarbonization of the Russian Unified Energy System. We can talk about Russia as a major exporter of second-and third-generation biofuels in prediction of the alternative energy development and renewable energy in the long term. The immediate future is the small hydroelectric power plants and installations that use biomass.

All in all, the prospects and reserves for the renewable energy growth have significant potential in Russia. They are subjects for a further study.

References

1. P. Mealy, A. Teytelboym. Research Policy, 103948 (2020)
2. D. D'amato, N. Droste, K. J. Winkler, A. Toppinen. Journal of Cleaner Production, **230,** 460 (2019)
3. E.A. Tarkhanova. Terra Economicus, **16 (2),** 75 (2018)
4. S. Sen, S. Ganguly. Renewable and Sustainable Energy Reviews, **69**, 1170 (2017)
5. J. E. Meyer. The Renewable Energy Transition, **161** (2020)
6. V. Ferreira Gregorio, L. Pié, A. Terceño. Sustainability, **10(11)**, 4232 (2018)
7. M. Kardung, K. Cingiz, O. Costenoble, R. Delahaye, W. Heijman, M. Lovrić, ... & B. X. Zhu. Sustainability, **13(1)**, 413 (2021)
8. L. Stegemann, M. Ossewaarde. Energy research & social science, **43**, 25 (2018)
9. I. G. Zhivotovskaya, T. V. Chernomorova. "Green economy" as a global development strategy in the post-crisis world: Reviews, **188** (2016)
10. M. Ram, A. Aghahosseini, C. Breyer. Technological Forecasting and Social Change, **151**, 119682 (2020)
11. B. V. Porfiriev. World economy and international relations, **60 (9)**, 5 (2016)
12. E. Lisin, D. Shuvalova, I. Volkova, W. Strielkowski. Sustainability, **10(4)**, 1111 (2018)
13. V. Y. Ushakov. Electrical Power Engineering, 13 (2018)
14. V. V. Klimenko, A. V. Klimenko, A. G. Tereshin. Thermal Engineering, **66** (11), 769 (2019)
15. M. Mazzanti & U. Rizzo. Technological Forecasting and Social Change, **115**, 111 (2017)
16. M. A. Aktar, M. M. Alam, A. Q. Al-Amin. Sustainable Production and Consumption (2020)
17. R.A. Epikhina. The Role of Electric Power Sector in China's Global Economic Expansion. Outlines of Global Transformations: Politics, Economics, Law, **12 (6)** (2019)

18. O. Bazaluk, V. Havrysh, V. Nitsenko, T. Baležentis, D. Streimikiene, E.A. Tarkhanova. Energies, **13,** 3113 (2020)
19. M. McPherson, M. Mehos, P. Denholm. Energy Policy, **139,** 111335. (2020)
20. B. Knopf, P. Nahmmacher, E. Schmid. Energy policy, **85**, 50 (2015)
21. G. S. Kust, O. V. Andreeva, D. V. Dobrynin. Arid Ecosystems, **1(1)**, 14 (2011).

Political strategies applied in Romania wind energy market in a medium and long run

Ştefan Ţălu[1]*, *Anton* Nazarov[2]

[1] Technical University of Cluj-Napoca, Cluj-Napoca, 400020, Cluj county, Romania
[2] Ural State University of Economics, 8 Marta str. 62/45, 620144 Yekaterinburg, Russian Federation

Abstract. This paper analyses the main measures to be adopted in the field of wind energy in the medium and long term in Romania and assesses the progress made by Romania in comparison with the European Union and the objectives set by the Community institutions. The paper aims to identify opportunities for the development of the wind energy sector based on existing resources, which will be a positive factor in generating positive economic and social effects. The study indicates that Romania has a high potential for wind resources, which can be harnessed through appropriate policies to support projects related to this type of energy.

1 Introduction

International energy markets are in a complex dynamic with difficult challenges exacerbated by the global financial crisis. These challenges have a decisive impact on energy markets at all levels and bear the imprint of a threat to local energy security. International energy markets are evaluated on several dimensions: technological, climatic, geopolitical, and economic [1,2].

The European Commission (EC) energy policies have linked energy security to the European Union (EU) with sustainable development in a strategy with several objectives, among which we list: - achieving and diversifying a complete internal energy market integrated; - the creation of a functional infrastructure and diversification of external energy supply sources; - quantitative and qualitative increase in energy production in the EU; - increasing the degree of European regional security, as well as protecting the strategic critical energy infrastructure, by coordinating national energy policies, but also by external energy diplomacy. The EU, through the appropriate mechanisms, finances energy projects, primarily those in the field of "clean energy" and the interconnection of European energy markets [1, 3].

Romania in the current geopolitical context must anticipate and position itself flexibly towards the trends on the international markets to the geopolitical resettlements that decisively influence the strategic partnerships in the region. Romania presents a balanced and diversified energy palette, illustrated by energy-producing companies, each focused on the

* Corresponding author: stefan.talu@auto.utcluj.ro

exploitation of a type of primary energy resource, as well as by electricity transmission operators.

Romania currently needs coherent and clean energy policies, as well as pragmatic investments for the growth of the Romanian energy sector. The growth of the Romanian economy from the perspective of the energy sector involves the construction of new energy production capacities; modernization and refurbishment of production capacities; energy transmission and distribution; encouraging the growth of energy-efficient domestic consumption; as well as export [3].

The energy strategy promoted by Romania has clear, well-defined objectives, through which Romania confirms its position as a producer of energy in the region and active factor in national energy management, which will produce substantial changes and will energize the entire energy sector [4,5].

The multiple technological developments have led to the European markets, energy efficiency policies, which have materialized in a slight decrease in energy demand, stimulated by diversification of supply. In Romania, technological development in the energy field has led to relatively high energy prices, largely amortized by subsidies from public budgets, but which have materialized in production increased energy. Through this prism, a decrease in energy production costs was found, in contrast to the accelerated progress of energy consumption management systems [6].

The implementation of IT technologies, through the development and modernization of smart grids with real-time coordination and two-way communication, have led to increased operational capacity for analysis and transmission of large volumes of data, with beneficial implications for optimizing energy consumption. It is found, as with the increase in production decentralized electricity may arise the situation of reorganizing the electricity transmission and distribution system, with the emergence of active consumers and the strengthening of electricity storage capacities.

In the energy sector, climate and environmental policies have a major impact, with the role of reducing pollutant emissions, but also on influencing decisions and social attitudes to choose "clean energy", to model both investment behaviour and consumption patterns. To achieve a sustainable national energy system, by the provisions of the 2015 Paris Agreement, but also of European policies to prevent climate change, it is necessary to implement measures to reduce pollutant emissions in the energy sector, to limit global warming at a maximum of 2 °C from the pre-industrial level. National energy policy planners have several clear objectives for implementing an optimized, economical energy transition process, including increasing energy efficiency and creating a global set of decarbonization indicators.

In Europe's energy supply crises, regional cooperation is an effective solution as it allows the development of energy markets, but also of specific regional energy security mechanisms, which have specified operating rules within the EU. Currently, Romania has an interconnection capacity of 9.5%, especially with the interconnections with Hungary, Bulgaria, and Serbia; but Romania also aims to develop interconnections with neighbouring countries outside the EU (Republic of Moldova, Ukraine).

The main advantages of wind energy are: - it is zero emission of pollutants and greenhouse gases; - no waste is produced; - low costs per unit of energy produced; - low decommissioning costs at the end of the normal operation period, when practically they can be fully recycled. Among the disadvantages of using wind energy we mention: - visual pollution - that is, they have an unpleasant appearance; - noise pollution; - a practical disadvantage is the variation in wind speed.

Being located at the border of the Balkan Peninsula and covering an elliptical surface of 238,391 km^2, Romania is at the eastern limit of the atmospheric circulation generated in the North Atlantic basin, which has a sufficiently high intensity that offers a substantial energy recovery on the ridges of the Carpathian mountains at high altitudes and favourable areas.

Romania has a temperate continental climate of transition, specific to Central Europe, but with a high energy potential, which is highlighted especially in coastal and coastal areas, but also in alpine areas with plateaus and mountain valleys.

On the other hand, the atmospheric circulation generated in the area of the Black Sea and the Russian Plain, in conjunction with the North Atlantic, lead to real possibilities of advantageous energy recovery in the areas of Dobrogea, Bărăgan, and Moldova. Also, in small areas, there are local atmospheric circulations that allow the construction of small wind farms for efficient economic recovery.

Romania has the greatest potential in South-Eastern Europe in the field of wind energy, and the position of South-Eastern Dobrogea is among the first places on the European continent. Wind turbines start from a wind speed of only 3.5 m/s, but in Dobrogea the measurements showed that the speed is 7 m/s at a height of 100 meters.

Table 1. Wind power installed capacity in Romania in MW in period 2009-2019

Year	2010	2011	2012	2013	2014	2015	2016	2017	2018	2019
MW	389	988	1.822	2,773	3,244	3,130	3,025	3,030	3,032	3,040

Hidroelectrica, the main producer in the Romanian energy industry, has proposed in its investment agenda a series of onshore and offshore wind projects, each of 300 MW, to be completed in 2025 and 2026, respectively, through an investment of 26 billion lei, of which almost 7.6 billion in the period 2020-2025. This investment project is the first in Romania, as, at present, there are no offshore wind farms installed in the Romanian Black Sea sector.

The production of energy from offshore wind sources has some important advantages, among which we list: a large number of operating hours with low variability, which implies low forecast errors and low balancing costs compared to onshore wind energy. The offshore wind potential allows investments in production and consumption capacities of "green hydrogen", but also the creation of national industrial entities in the field of offshore wind energy and hydrogen, with a positive impact on the creation of new jobs nationwide.

2 Research methodology

In this study, one question was proposed: What are the appropriate policies that can be applied in Romania, in the wind industry, in the medium and long term to improve the developments in the field of wind energy depending on regional conditions, stimulated by the national regulatory framework?

3 Discussions

The directions of action proposed in wind energy in Romania to be implemented on a medium and long term at a national level are: - increasing the number of auctions and facilitating them to increase the production capacity of electricity from wind energy; - simplification of administrative procedures; - promoting investments in infrastructure and network interconnections. The volatility of energy production in wind farms in Romania requires the entire national energy system, which requires a reassessment of the need for system services and appropriate investments in state-of-the-art power plants, with modern, fast control, and efficient storage systems.

From the analysis of the decisive political factors in Romania regarding wind energy on a medium and long term, several conclusions can be drawn:

- obtaining wind energy is still dependent on public support and folds after major political changes on the political scene;
- there is a lack of a clear, well-defined plan, which due to hasty regulations, without an adequate impact study and by avoiding consulting all actors involved in the wind energy market has led to low distrust of producers in the field, who have not invested capital enough on this segment;
- the existence of high risks, but also of a long-term investment return;
- adoption of advanced technologies in the sector wind energy, by attracting private investment, through supporting scientific research and development strategic partnerships;
- developing cooperation in the field of research scientific studies in the wind energy sector and the transfer of know-how;
- cooperation with the authorities of the partner states for increase infrastructure security [7];
- increasing the management of wind energy companies with state capital in order to increase their value on the medium and long term;
- economic optimization of the assets and investment projects of wind energy companies;
- eliminating bureaucracy through transparency and digitization;
- developing institutional mechanisms by publishing regular reports on public procurement and all sponsorships granted;
- settling and eliminating conflicts of interest between institutions public and state-owned energy companies;
- supporting scientific research, technological development and innovation in the field of wind energy, by participating in international consortia;
- organizing information programs and public debates on major wind energy projects, taking into account the interests of local communities, but also the national interest;
- the development of partnerships with the universities with the wind energy industry, as well as of the research - development - innovation institutes [8-10];
- implementation of physical security measures for critical infrastructure against possible malicious actions; - computer security of energy network control systems by increasing the degree of protection; - attracting European and international funding sources for scientific research in the field.

4 Conclusions

This article proposed a series of local political measures that can be applied in the context of wind energy in Romania to be implemented on a medium and long term at a national level.

The paper identified opportunities for the development of the wind energy sector based on existing resources, which will be a positive factor in generating positive economic and social effects. All in all, our results indicated that Romania has a high potential for wind resources, which can be harnessed through appropriate policies to support projects related to this type of energy.

References

1. Romanian energy strategy 2016-2030, with an outlook to 2050, Ministry of Energy, http://energie.gov.ro/ (2021)
2. Romania wind energy market (2020-2025), https://www.mordorintelligence.com/ (2021)
3. M.I. Aceleanu, A.C. Şerban, D.M. Pociovălişteanu, G.C. Dimian, Energy Sources, Part B: Economics, Planning, and Policy, **12**, 958 (2017)

4. F. Onea, E. Rusu, Energy Explor. Exploit., **34**, 217 (2016)
5. F. Onea, L. Rusu. J. Mar. Sci. Eng., **7**, 142 (2019)
6. N Marinescu, Energies, **13**, 6493 (2020)
7. Ş. Ţălu, Advances in Economics, Business and Management Research (AEBMR), **156**, 522 (2020)
8. Ş. Ţălu, Advances in Economics, Business and Management Research (AEBMR), **105**, 554 (2019)
9. Ş. Ţălu, Advances in Economics, Business and Management Research (AEBMR), **138**, 253 (2020)
10. Ş. Ţălu, A. Nazarov, E3S Web of Conferences, **208**, 09005, 1-6 (2020)

Forming the digital environment for developing the electric power spatial infrastructure system of the region

Elena Shishkina[1] *, *Nadezhda* Surnina[1]

[1]Ural State University of Economics, 8 Marta str. 62/45, 620144 Yekaterinburg, Russian Federation

Abstract. The study was carried out on forming the digital environment for developing the electric power spatial infrastructure system of the region. The system of documents aimed at forming and developing the digital economy and the digitalization of economic activity was analyzed. Target priorities of digitalization of spatial infrastructure systems of the region were determined. The article analyzes the dynamics in the use of information and communication technologies and global information networks by the organizations of the Sverdlovsk region. This dynamic determines the region's high readiness for digitalization. The hierarchical nature of forming the priorities and parameters of digitalization of the region's electric power infrastructure was defined. The analysis was done of the digital environment for developing the electric power spatial infrastructure system of the region based on the national, regional, and corporate level documents, which allowed determining the general targets for digitalization - implementing digital technologies for reliable and accessible electricity supply.

1 Introduction

At the present stage, the socio-economic development of the regions and its level are closely related to the state of development of information technologies [1-4]. In the world ranking of digital competitiveness in 2019, Russia's position was 38th out of 63. In 2019, Russia's domestic expenditures on the development of digital economy from all the sources amounted to 2.1 % as a percentage of GDP and the gross added value of the ICT sector as a percentage of GDP was 2.81 % [5].

The development of digital economy aims both to solve the specific problems of economic activity for increasing the efficiency of the production processes as well as the relationships between the subjects, accelerating the business processes, and to increase the competitiveness of the economy in international markets.

The works of many domestic and foreign scientists are devoted to the study of various aspects of the digital economy formation and digitalization. Within the study, special interest was shown to the works, which analyze the impact of digitalization on the development of the regional economy and its components in the aspect of reindustrialization and sustainable

* Corresponding author: shishea@usue.ru

development. In this regard, it is necessary to highlight the works of many prominent scientists [6-10].

The electric power spatial infrastructure system of the region is the key element in forming the digital energy production based on the new technological paradigm, which is defined in the strategic documents [11].

The purpose of the study is to analyze the formation of the digital environment for developing the electric power spatial infrastructure system in the region. The research area was the Sverdlovsk region and its electric power spatial infrastructure system. The logic of the research provides for solving the following tasks: first, determining the hierarchical levels of forming the digital environment to develop the spatial infrastructure systems in the region; second, analyzing and evaluating the priorities, the parameters of forming the digital environment to develop the spatial infrastructure system in the region, seeing the balance between them.

2 Materials and methods

The formation of the digital economy in the Russian Federation as a priority direction of development and the condition for increasing the economic efficiency based on the introduction of information technologies, systems and services was given a start relatively recently. In 2016 V. V. Putin addressed his message to the Federal Assembly of the Russian Federation. Since then, the formation of the digital environment for the development of the national and regional economy has begun, which reflected in the adoption and implementation of a number of strategic documents, among which special importance is given to the Program "Digital Economy of the Russian Federation" (approved by the Decree of the Government of the Russian Federation of July 28, 2017, No 1632-r). The strategic goal of the digital economy development at the national and regional levels is "the transition to the economy of a new technological generation, the "digital economy" based on the creation of a sustainable and secure information and telecommunications infrastructure". This determines the vector of development of both the regions and the objects in tem including the infrastructure systems

The development of spatial infrastructure systems and the development of the economic space of the region are interconnected and interdependent processes. Ensuring sustainable development is largely connected with the functioning of the electric power spatial infrastructure system in the region, it depends on the state of this system, its supplying, networking and connecting functions. The processes of neo-industrial development and digitalization cause a change in the role of the electric power spatial infrastructure system of the region plays in the economic space. Digitalization of spatial infrastructure systems is determined in departmental documents. The departmental project "Digital Energy" [12] aims to develop the energy production infrastructure of the Russian Federation by implementing digital technologies and platform solutions through digital services and solutions development. At the corporate level, the direction of digitalization of the electric power spatial infrastructure system of the Russian Federation is determined by the Concept of Digital Transformation 2030 of "Rosseti" PJSC [13], according to which the digital transformation of the networks is based on digital technologies and big data analysis. Thus, at the federal level, the formation of a digital environment for developing an electric power spatial infrastructure system is determined by national, departmental and corporate documents, agreed upon concerning the targets.

3 Empirical model

In terms of regional development, the goal of digitalization is to transform the objects of the regional economy and social sphere based on accelerated implementation of domestic digital technologies and platform solutions [15]. The Sverdlovsk region shows high readiness for digitalization in both ICT and non-ICT industries (Table 1, 2).

Table 1. Analysis of the dynamics of the use of information and communication technologies by organizations of the Sverdlovsk region (excluding small businesses), in% compared to the previous period [14]

Indicators	2016	2017	2018	2019	Growth rate for the period, %
Organizations' expenditures on information and communication technologies, mln RUB.	72,6	125,5	105,7	108,5	149,4
Number of organizations belonging to the ICT sector, units.	127,5	98,5	113,5	91,3	71,6
Headcount in organizations belonging to the ICT sector, thousand people.	99,3	57,9	108,8	104,1	104,8
Shipped own-production goods and services related to information and communication technologies, mln RUB	94,3	93,4	108,7	116,7	123,8

For the period of 2016-2019 organizations' expenditures on information and communication technologies increased by 49.4% (growth rate 149.4%), since 2018 the indicator has been increasing annually. There is also a 23.8% increase in shipped goods and services) of own production related to information and communication technologies. The number of employees on the payroll in ICT-related organizations increased by 4.8% compared to 2016. At the same time, the number of organizations belonging to the ICT sector decreased by 29.4%.

Table 2. The proportion of organizations in the Sverdlovsk region that used global information networks (as a percentage of the total number of surveyed organizations of the respective type of economic activity) [14]

	Organizations using						Organizations having a website		
	Global information networks			Including using the Internet					
	2017	2018	2019	2017	2018	2019	2017	2018	2019
Total	89,7	92,0	92,0	88,9	91,1	91,2	47,4	50,9	51,9
Supply of electricity, gas and steam; air conditioning	92,3	91,5	93,5	89,6	89,1	92,5	47,5	47,7	51,1

In 2017-2019 the dynamics of the share of organizations using global information networks in the type of activity "Supply of electricity, gas and steam ; air conditioning" generally corresponds to the dynamics of the total number of organizations. During the study period, more than 89.0% of the total number of organizations and more than 91.5% by the type of activity used global information networks. In 2017-2019 more than 47.0% of the total number and more than 47.5% of organizations by the type of activity had a website.

As part of implementing the national program "Digital Economy of the Russian Federation" at the regional level in the Sverdlovsk region, the following regional projects are being implemented "Information Infrastructure", "Personnel for the Digital Economy", "Information Security", "Digital Technologies", "Digital Public Administration", the regional program for the development of the digital economy [16]. In accordance with the national program "Digital Economy" and the departmental project of the Ministry of Construction, Housing and Utilities of the Russian Federation on digitalization of the urban economy "Smart City", the passport of the regional program "Smart Cities of the Sverdlovsk region" (hereinafter referred to as the "Smart City" program) was formed, which also includes a list of projects and a number of activities for the implementation of digital platform solutions and smart services

The "Smart City" program includes the following projects: city management; smart housing and utilities; innovations for the urban environment; smart city transport; intelligent public security systems; communication network infrastructure. The projects contain a number of objectives and measures to implement digital platform solutions and smart services. In 2019, the Government of the Sverdlovsk region received an appeal from Smart Light Solutions LLC regarding the implementation of the Smart City program projects in the Sverdlovsk Region using public-private and municipal-private partnership mechanisms involving concession and energy service instruments, as well as the possibility of investing about 10 billion rubles. One of the key tasks of the "Smart City" standard is implementing the "Energy Efficient Smart City Lighting" service (section "Innovations for the Urban Environment").

In 2018, in Ekaterinburg, to promote the Agreement between the Government of the Sverdlovsk region and the "ER-Telecom Holding" company of July 10, 2018 No. 46, the first Internet of Things (IoT) network based on the LPWAN standard was launched, which can be the basis for implementing the "Smart City" projects at a modern level.

Thus, in the region, work is underway to introduce an energy efficient intelligent urban lighting system in the city of Yekaterinburg using the existing urban Internet of Things (IoT) network, the investor of which is "ER-Telecom Holding" JSC.

The owners of the infrastructure – the corporate sector is also developing and implementing programs for digitalization of the regional spatial infrastructure systems. The development of the electric grid infrastructure of the region is carried out as part of the activities of OAO "IDGC of Urals". The Board of Directors of the Company (Minutes of 28.04.2017 No. 224) approved the Program of Innovative Development of OAO "IDGC of Urals" for the period 2016-2020. with a perspective up to 2025.

The main directions of innovative development are new technologies and solutions, digitalization of management and technological processes. In accordance with the Executive Orders of the President of the Russian Federation "On the 2017-2030 Strategy for the Development of an Information Society in the Russian Federation " of 09.05.2017 No 203, "On National Goals and Strategic Objectives of the Development of the Russian Federation through to 2024" of 07.05.2018 No 204, Orders of the Government of the Russian Federation of July 28, 2017 No 1632r to implement the Concept of PJSC ROSSETI "Digital Transformation - 2030", approved by the Board of Directors of PJSC "ROSSETI" (Minutes of the meeting of 21.12.2018 No 336), it is assumed that the main principles and development directions of the power grid complex of OAO "IDGC of Urals" will be reconsidered.

At the moment, the program "Digital Transformation of OAO "IDGC of Urals" for 2020-2030" has been adopted and approved, the purpose of which is "to change the logic of the processes and the transition of the company to risk-oriented management based on digital technologies and of big data analysis" [17] The program contains pilot projects for the digital transformation of the power grid company in the territory of its presence, aimed at improving the quality of power supply to consumers by reducing power losses, increasing the observability and controllability of distribution networks. An integral part of the company's digital transformation program is the Program for the Development of Smart Metering of Electricity in the Company's Power Grids for 2020-2030. The development of the company is also carried out within the agreements with the Government of the Sverdlovsk region on innovative development of the power grid complex. The total costs of the implementation of the digital transformation program will amount to 53 billion rubles, the sources of which are the company's own and borrowed funds.

4 Results and discussions

To summarize the results of the study on the digital environment for the development of the electric power spatial infrastructure system in the region, the levels and target priorities should be highlighted (Table 3).

Table 3. The levels of forming the digital environment to develop the electric power space infrastructure system in the region

Level	Document	Targets
Federal	Program "Digital Economy of the Russian Federation" (approved by the Government of the Russian Federation of July 28, 2017 No. 1632-r); Departmental project " Digital Energy", Package plan for modernization and expansion of the main infrastructure for the period to 2024 (Decree of the Government of the Russian Federation of September 30, 2018 N 2101-r, amended on March 13, 2020); program "Digital Transformation of the Russian Electric Power Industry"; concept of digital transformation 2030 of PJSC "ROSSETI"; "On National Goals and Strategic Objectives of the Development of the Russian Federation through to 2024" (Executive Order of the President of the Russian Federation No. 204 of 07.05.2018)	Increasing the availability and quality of goods and services produced in the digital economy using modern digital technologies; introduction of digital technologies in the energy sector, transition to intelligent energy, demand management and price-dependent consumption reduction, transformation of the Russian energy infrastructure through the introduction of digital technologies and platform solutions to improve its efficiency and safety
Regional	Regional projects of the Sverdlovsk region: "Information Infrastructure", "Personnel for the Digital Economy", "Information Security", "Digital Technologies", "Digital Public Administration", "Smart Region" concept	Systemic implementation of information services and solutions based on the latest achievements in the field of digital technologies for the development of human capital, improving the quality of life and increasing the competitiveness of the economy
Corporate	Concept "Digital transformation of PJSC "ROSSETI" 2030" (approved by the Board of Directors of PJSC "ROSSETI", minutes of December 21, 2018) No. 336; The program of innovative development of OAO	Changing the logic of the processes and transition of the company to risk-oriented management by introduction of

	"IDGC of Urals" for 2016-2020 with a prospect until 2025 (approved by the Board of Directors of the Company, minutes of 28.04.2017 No. 224); program "Digital Transformation of OAO "IDGC of Urals" for 2020-2030".	digital technologies and big data analysis

Thus, the formation of a digital environment for the development of the electric power spatial infrastructure system of the region is represented by several levels: federal, regional, corporate. At each level, a system of documents has been formed that defines the goals and objectives of digitalization. Content analysis of the documents in Table 3 shows that the general target is to use digital technologies for reliable and accessible [18] electricity supply. At the same time, particular tasks are determined for introducing digital technologies into business processes and public administration. However, the documents are not fully interrelated, there is discrepancy in timescales, priorities and planning parameters, which complicates the process of strategic planning and achieving the goals set.

5 Conclusions

The study of the formation of a digital environment for forming spatial infrastructure systems in the region (the case of the electric power industry) allows drawing the following conclusions:

- the region has a digital environment for developing the electric power spatial infrastructure system, which is reflected by federal, regional, corporate documents both complex and by type of activity. However, there is not enough interconnection and balance between them.
- there is hierarchical order of developing the priorities and parameters of digitalization of electric power infrastructure;
- the target parameters of digitalization in the electric power spatial infrastructure system are increased reliability and accessibility for the public and business, effective electricity supply for the regional development;
- there are potential risks of a technological and macroeconomic nature, different understanding of digitalization in the industry companies due to inequality, differences in the access to the use of technologies;
- mechanisms are not defined to coordinate the tasks and digitalization results in horizontal and vertical management.

References

1. I.L. Avdeeva, T.A. Golovina, L.V. Parakhina. Management issues, **6 (49)** (2017)
2. A.N. Kozyrev, *http://digital-economy.ru/* (2021)
3. N.M. Surnina, E.A. Shishkina, Journal of New Economy, **5 (61)**, 69 (2015)
4. V.G. Khalin, G.V. Chernova, Administrative Consulting, **(10)**, 46 (2018)
5. G. Abdrakhmanova, K. Vishnevskiy, L. Gokhberg et al. *Digital Economy Indicators in the Russian Federation* (2020)
6. A.I. Tatarkin, V.V. Akberdina, N.Yu. Buhvalov, Proceedings of 2nd International conference: Strategies for Developing Social Groups, Institutes, and Territories, 41 (2016)
7. E.A. Istomina, Bulletin of Chelyabinsk State University. Economic sciences, **12 (422)**, 108 (2018)

8. N.M. Surnina, E.A. Shishkina, Scientific Publications of the Free Economic Society of Russia, **3 (223),** 251 (2020)
9. L.V. Lapidus, L.S. Leontieva, A.O. Gostilovich, Public administration (E-Journal), **77**, 212 (2019)
10. V.V. Akberdina, Journal of New Economy, **19 (3),** 82 (2018)
11. N.D. Rogalev, Digital energy: a new paradigm of functioning and development, 300 (2019)
12. Departmental project «Digital Energy», https://minenergo.gov.ru/
13. Digital transformation 2030 concept, PJSC ROSSETI, 2018
14. Science, innovation, and the Information Society, Federal State Statistic Service, *https://sverdl.gks.ru/folder/30403* (2021)
15. ANO "Digital economy", https://data-economy.ru/ (2021)
16. Digital economy of the Sverdlovsk oblast', *https://све.рф/* (2021)
17. 2020-2030 Digital Transformation of OAO IDGC of Urals Program, *https://www.mrsk-ural.ru/* (2021)
18. A. Dyachkov, N. Surnina, E. Shishkina, Proceedings of the International Scientific Conference on Competitive, Sustainable and Secure Development of the Regional Economy: Re-sponse to Global Challenges (CSSDRE 2018), Advances in Economics, Business and Management Research, **39**, 485 (2018)

A systematic literature review of smart grids

Ihor Vakulenko[1], *Liudmyla* Saher[2,*], *Oleksii Lyulyov*[2], and *Tetyana* Pimonenko[2]

[1]Oleg Balatskyi Department of Management, Academic and Scientific Institute of Business, Economics and Management, Sumy State University, 2, Rymskogo-Korsakova st., 40007 Sumy, Ukraine
[2]Department of Marketing, Academic and Scientific Institute of Business, Economics and Management, Sumy State University, 2, Rymskogo-Korsakova st., 40007 Sumy, Ukraine

Abstract. The development and implementation of smart grids involve developing new and improvements in existing energy technologies, introducing information systems to manage the smart grid, monitoring and controlling energy consumption, and closely related to alternative energy and decarbonization of the economy. Scientific research of smart grids differs significantly in terms of topics because they aim to solve problems in each of these areas. Thus, this research aims to present a bibliometric overview to define the current scientific production state regarding "Smart Grid." A review of 1359 publications from the Scopus database (2008–2020) was conducted. The "Title, abstract, keywords" field of search in the Scopus database was done. The visualization of the results was made using VOSviewer program to map the material graphically. The study used the co-occurrence of keywords and co-authorship (country) analyzes. As a result, the most productive authors and journals were defined. The most cited studies were determined. Country clusters and keywords (co-occurrence) clusters were represented. The obtained results of the analysis and graphical presentations are relevant, and they form the basis for a better understanding of Smart Grid's concept.

1 Introduction

The development of smart grids is one of the main trends in global energy development. The use of fundamentally new methods in energy and combining them into a holistic, interconnected and independent infrastructure allows to maximize the capabilities of the energy system, implement the principle of distributed energy generation, increase the efficiency of load balancing in the network. In addition, the use of smart technologies in the power grid helps to better meet the needs of consumers.

The global concern about energy issues keeps growing in recent years. Thus, the problems energy efficiency of the country, economic, ecological and energy security is shown by scientists in papers [1-11]. Issues of marketing, communication, environmental management that accompanies the process of building smart grids is discussed in the following studies [12-15]. Issues of sustainable development of society and economy, what are the reason for using smart grids as a means of transition from a traditional energy system to environmentally

*Corresponding author: lyudmila.sager@gmail.com

friendly, based on alternative energy, distributed energy production, and intelligent technologies, considered in [16-21].

The authors in the papers [22-33] presented the results of the study of clean energy sources, and ways to convert existing energy networks into renewable ones. The key issues and features of improving the energy efficiency of households are considered in the works [24-25].

The researches [26-32] analyzed current development trends of smart grids, in particular, issues of network management and infrastructure development, the role of users in accelerating / slowing down the pace of smart grids, motivating stakeholders, research on the state of smart grids deployment in the global dimension. The application of information technologies in energy is considered in [33].

Despite the significant work of scientists in the field of smart grids, it is necessary to systematize the directions of their research. This will identify new areas of research, as well as those areas that require further attention of the scientific community. Fundamental and practical approaches to the systematization of scientific knowledge based on the use of advanced tools of bibliometric analysis are studied in the works of [35-38]. Issues of sustainable development of society and economy, what are the reason for using smart grids as a means of transition from a traditional energy system to environmentally friendly, based on alternative energy, distributed energy production, and intelligent technologies, considered in [16-21].

2 Materials and methods

For a comprehensive understanding of the research problem and in-depth analysis of the peculiarities of the development of scientific research in this area, a bibliometric analysis was carried out to study the bibliographic material in detail and determine the main scientific directions research area. Bibliometric analysis is an advanced tool for identifying existing "gaps" in the research topic, identifying areas of research that are most relevant and in line with current trends. Bibliometric analysis was performed according to the sequence shown in Table 1.

Table 1. Stages of bibliometric analysis

Stage		Filters	Result
1	Choice of suitable information sources	Scopus database	
2	Identification of search field in the database	Title, abstract, keywords	
3	Identification of search keywords	Smart grid	55 015 publications
4	Identification of publication type	Journal articles only; conference papers, books, and chapters of books excluded	16 037 publications
5	Choice of the language	English	14 154 publications
6	Choice of the field of publication	Social Sciences, Business, Management and Accounting, Economics, Econometrics and Finance, Decision Sciences	1 378 publications
7	Identification of the publications time limits	2008-2020 (since the beginning of the growth of the number of publications on the subject)	1 359 publications
8	Manual check	The analysis of the paper in terms of its relevance	1 359 publications

The reliability and certainty of bibliometric analysis largely depend on the rating and transparency of information collection sources. Accordingly, one of the most influential and authoritative scientific databases, Scopus, was chosen for bibliometric analysis. To study the problem as a keyword was chosen "smart grid," this term is the main one and is officially used to describe the modernization of the energy sector based on intelligent technologies. To ensure the study's complexity and integrity, the search field "title, abstract, keywords" was used.

As the research topic is popular and widely researched, restrictions on selecting relevant materials for bibliometric analysis have been introduced. Thus, only articles written in English were selected for further consideration. Besides, restrictions were introduced on the research field (the focus is on publications that contain an economic component) and the period of publication. Also, the inspection showed that all publications correspond to the research problem. Thus, taking into account all the limitations, 1359 publications were selected for further consideration.

3 Results and discussions

Figure 1 shows the dominance of technical publications. In particular, in the field of engineering, informatics, and energy. The share of publications related to economic sectors remains low, but there is a tendency to increase their number.

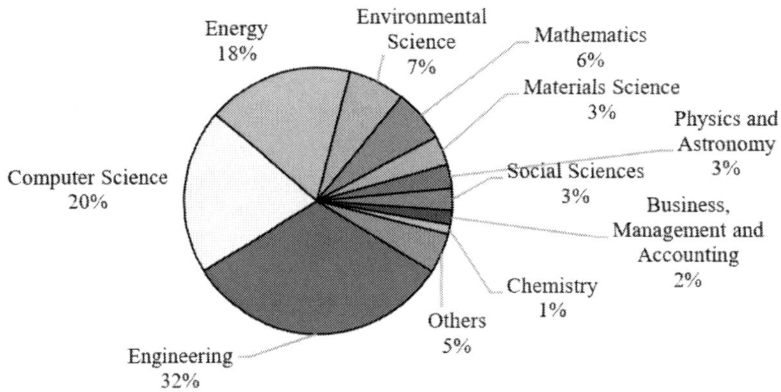

Fig. 1. Publications in the field of smart grids, selected by research areas (based on the Scopus database)

The distribution of scientific works by the studied years is given in Fig. 2. The first publication on the research subject, containing the economic component, dates to 1995 [33], but until 2008 the number of publications did not exceed one per year.

The period of active interest and strengthening of publishing activity in smart grids began in 2008. Since then, there has been a steady trend towards increasing the number of publications (as evidenced by the trend line) with a peak in 2019 – 245 publications. Figure 2 also shows over the last five years (from 2016 to 2020), more than 72.3% of all research materials have been published.

The highest citation rate – 2282 – was achieved in 2017. However, the largest number of citations per publication was in 2009 (81 citations per 1 publication). In 2009, two articles were published, ranking 1st and 3rd among the most cited, respectively.

Analysis of the publications and citations number correlates with trends in the development of smart grids. It allows us to analyze the evolutionary development of research on smart grids. The basis for determining scientific interest is the number of published scientific papers on this topic. Public interest in the subject is determined based on Google Trends, which allows you to track the number of queries related to smart grids in the Google search engine.

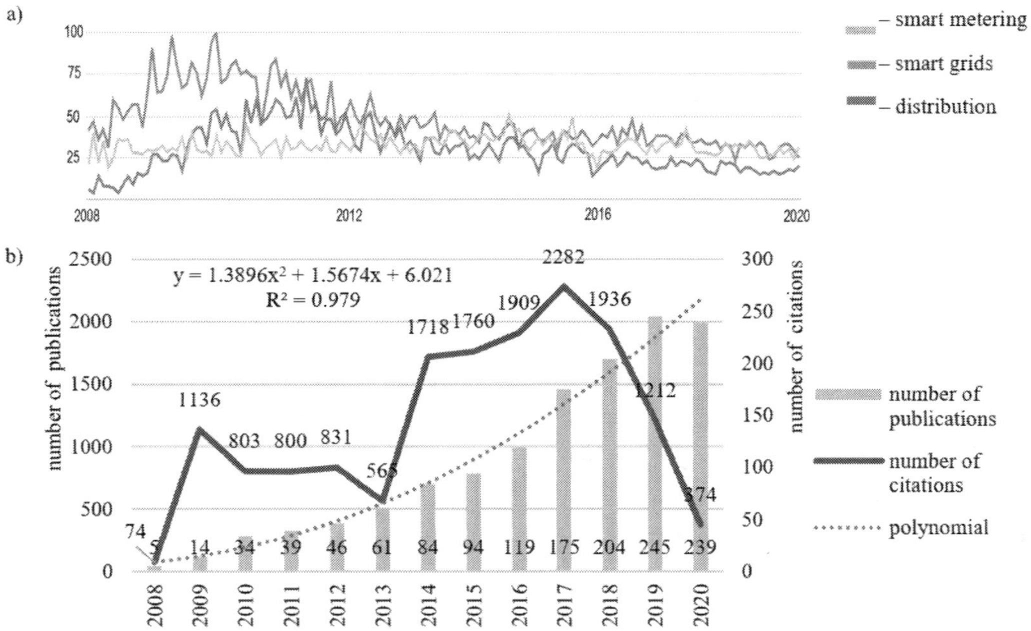

Fig. 2. Comparison of the dynamics of Google search queries on smart grids, smart metering and distributed power generation (a), and scientific publications on smart grids' development, indexed by the database Scopus (b) for 2008-2020.

It is expedient to distinguish four stages in the evolution of scientific and public interest in intelligent power grids. The first stage fell in 2008-2009. This period is characterized by searching for ways to level technical and technological constraints for large-scale implementation of the first stage of smart grid development – forming a basis for full-fledged smart grid projects by deploying smart metering infrastructure. A small number of scientific papers characterizes this period. The shortage of scientific research, which considers the latest technical and technological advances in the development of smart grids, leads to a considerable increase in citations of existing scientific papers. The second stage – 2009-2013 – is the transition from smart metering to the creation of full-featured smart grids. During this period, against the background of the concept's development for introducing smart grids in the energy sector by leading countries (in the introduction of smart grids), researchers focus not on economic but on technical issues of smart grid configuration. The increase in the number of relevant scientific studies is accompanied by an almost twofold decrease in their citations of scientific papers. The results of the analysis in Google Trends show a significant increase in public interest in the topic of smart grids. The third stage – 2014-2017 – is associated with the emergence of new technological constraints in developing smart grids. This period is characterized by a rapid increase in the number of scientific publications and their citation dynamics while reducing public interest in this issue. The fourth stage – from 2017 – is the search for breakthrough technologies that will determine the vector of development of smart grids in the future. During this period, it is worth noting the growing number of scientific papers. Simultaneously, there are multiple reductions of citations of existing developments and also reduction of public interest in this issue due to unresolved technical and technological problems that do not increase smart grid projects` technical and economic efficiency. It is holding back investment in scaling up smart grids.

To identify existing research trends and identify current areas of further investigation, we analyzed the keywords of publications using VOSviewer, a reliable and effective tool to visualize the relationship between the main keywords in research. Three hundred seventeen keywords were selected for analysis (frequency of occurrence more than seven times), after checking the keywords for further consideration, repeated and irrelevant words (for example, "scheduling," "China," and others). A total of 280 keywords were analyzed. Accordingly, the areas of cross-sectoral research of smart grids are shown in Figure 3. Thus, the generated network visualization between 280 keywords includes 6 clusters. The first (most powerful) cluster (red cluster) is formed around the concept of smart grid (the concept establishes 643 connections, the strength of connections – 3529, links – 279). This cluster includes the following main keywords: “energy efficiency” (occurrences (occ.) – 98, total link strength – 743, links – 199), “sustainability energy” (occ. – 81, total link strength – 607, links – 179), “alternative energy” (occ. – 76, total link strength – 584, links – 158), “sustainable development” (occ. – 66, total link strength – 508, links – 175), “renewable energy” (occ. – 63, total link strength – 397, links – 148), “electricity” (occ. – 48, total link strength – 323, links – 136).

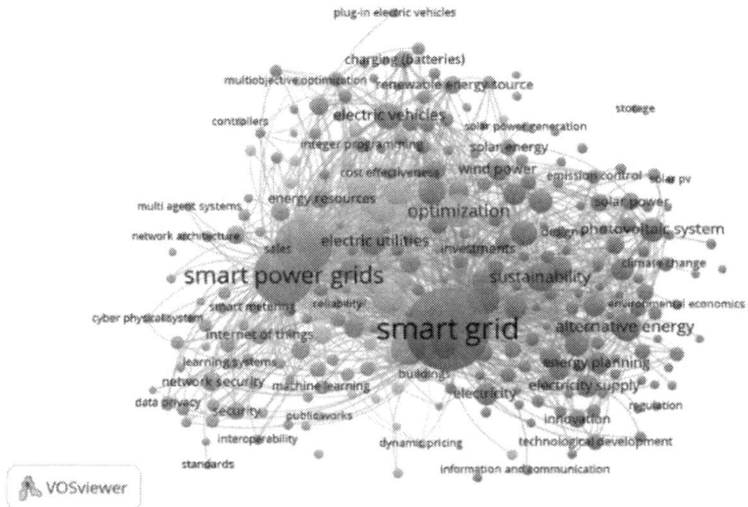

Fig. 3. Identifying areas of cross-sectoral research on smart grids (larger circle diameter means the greater frequency of mention of the concept as a keyword along with smart grids in scientific articles indexed by the scientometric database Scopus for 2008-2020)

The second (green) cluster covers keywords such as “electric power transmission networks” (occ. – 302, total link strength – 2310, links – 255), “smart grids” (occ. – 107, total link strength – 570, links – 190), “network security” (occ. – 34, total link strength – 200, links – 190), “electric power distribution” (occ. – 31, total link strength – 248, links – 116), “electric power system control” (occ. – 31, total link strength – 239, links – 111), “smart meters” (occ. – 32, total link strength – 178, links – 92) and others, and is formed around the concept “smart power grids” (occ. – 331, total link strength – 2388, links – 253).

The third (blue) cluster is tied to the concept “renewable energy resources” (occ.– 73, total link strength – 707, links – 73) and covers keywords “electric vehicles” (occ. – 64, total link strength – 452, links – 139), “wind power” (occ. – 46, total link strength – 333, links – 141), “renewable energy” (occ. – 41, total link strength – 367, links – 142), micro-grid (occ. – 34, total link strength – 339, links – 130), “charging (batteries)” (occ. – 28, total link strength – 269, links – 93).

At the heart of the fourth (yellow) cluster is the concept “optimization” (occ.– 106, total link strength – 813, links – 212). The following main keywords belong to this cluster:

"demand response" (occ. – 90, total link strength – 592, links – 177), "energy management" (occ. – 79, total link strength – 629, links – 170), "costs" (occ. – 72, total link strength – 678, links – 174), "energy utilization" (occ. – 61, total link strength – 541, links – 165).

The last fifth (purple) cluster is formed around the concept "electric utilities" (occ. – 28, total link strength – 269, links – 93), it is advisable to include keywords in this cluster "demand-side management" (occ. – 45, total link strength – 355, links – 135), "energy resources" (occ.– 32, total link strength – 255, links – 114), "economics" (occ.– 35, total link strength – 314, links – 147).

The bibliometric analysis shows that until 2017, research by economists focused mainly on finding ways to commercialize smart grids, including tariff issues. In 2017–2018, the vector of economic analysis of smart grids changed: the focus of scientific attention was on technical and economic optimization of the creation of new and development of existing smart grids projects. In 2018-2019, the emphasis on research was shifted to studying smart grid development's global economic model, their mutual integration, and technical and economic efficiency based on energy consumption patterns (Fig. 4).

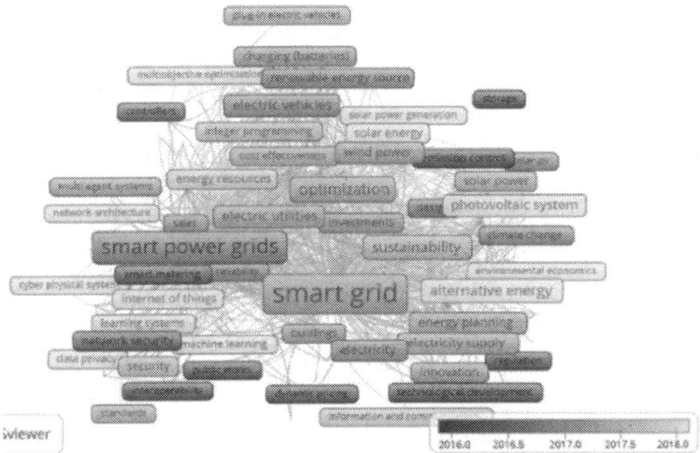

Fig. 4. Visualization map of changing trends in economic research on the development of smart grids, published in 2008-2020 in the publications of the Scopus database

The ten most-cited publications in the "smart grid" in the Scopus database are given in Table 2. All ten articles were cited more than 100 times. The presence of such a tendency to cite indicates that the publications were perceived and highly appreciated by the world scientific community. There is a scientific discussion that emphasizes the relevance of the research topic.

The top 10 most cited publications can be divided into the following research topics: security (items 1, 2), renewable energy and sustainable development (items 3, 8), energy efficiency (items 9, 10), micro-networks (items 5), energy management, including pricing, and demand management (items 4, 6, 7).

Thus, the most cited (734 citations) is the publication «Security and privacy challenges in the smart grid» authors – McDaniel P., McLaughlin S. [31]. In this study, the authors considered the state of the spread of smart grids globally and operational, environmental, and financial factors motivating this process in grid security and storage of personal data. In second place is the publication «Smart-grid security issues,» authors Khurana H., Hadley M., Lu N., Frincke D.A. [30], which was cited 392 times. In this publication, the authors have discussed in detail issues related to the security of smart grids.

The third place is taken by the publication «Cleaner energy for sustainable future» з кількістю цитувань 231, автори with citations of 231, authors – Dovì V.G., Friedler F., Huisingh D., Klemeš J.J. [22].

Table 2. Top 10 most cited studies in the field of smart grids from 2008 to 2020 (based on the Scopus database)

No	Total citations	Article title	Author	Number of universities/ institutions	Number of countries	Journal/Year of publication
1	2	3	4	5	6	7
1	761	Security and privacy challenges in the smart grid [31]	McDaniel P., McLaughlin S.	2	1 (USA)	IEEE Security and Privacy, 2009
2	402	Smart-grid security issues [30]	Khurana H., Hadley M., Lu N., Frincke D.A.	4	1 (USA)	IEEE Security and Privacy, 2010
3	236	Cleaner energy for sustainable future [22]	Dovì V.G., Friedler F., Huisingh D., Klemeš J.J.	4	3 (Hungary, Austria, Italy)	Journal of Cleaner Production, 2009
4	202	The US Department of Energy's Microgrid Initiative [32]	Ton D.T., Smith M.A.	2	1 (USA)	Electricity Journal, 2012
5	160	Smart grids, smart users? the role of the user in demand side management [29]	Goulden M., Bedwell B., Rennick-Egglestone S., Rodden T., Spence A.	1	1 (Great Britain)	Energy Research and Social Science, 2014
6	118	A technical and economic analysis of one potential pathway to a 100% renewable energy system [23]	Connolly D., Mathiesen B.V.	1	1 (Denmark)	International Journal of Sustainable Energy Planning and Management, 2014
7	117	Ten questions concerning model predictive control for energy efficient buildings [24]	Killian M., Kozek M.	1	1 (Austria)	Building and Environment, 2016
8	112	An unsupervised training method for non-intrusive appliance load monitoring [25]	Parson O., Ghosh S., Weal M., Rogers A.	4	1 (Great Britain)	Artificial Intelligence, 2014

The authors who have the most significant number of publications in smart grids in the Scopus database are listed in Table 3.

Thus, the top three authors with the highest publishing activity in the field included Javaid N., Ketter W., Tuite D., Senjyu T. Besides, many articles have been published in leading scientific journals with high SNIP and included in quartiles Q1 and Q2.

The most significant number of citations are published in the journal "Sustainable Cities and Society" (1369), in second place in the number of references is the journal "Journal Of Cleaner Production" (1346), the top three is closed by "Energy Research and Social Science" (1220). The scientific journal "Sustainability" (Switzerland), which has the most publications, is in fourth place in terms of the number of citations (1024 citations). At the same time, per one publication, the most significant number of citations falls on the journal "Energy Research and Social Science" (23.0 citations).

Table 3. Top authors by the number of publications in the field of smart grids during 2008-2020 (based on the Scopus database)

Author	Number of published materials	Author's h-index in the Scopus database
Javaid N.	12	41
Ketter W.	7	19
Tuite D.	7	2
Senjyu, T.	7	54
Skjølsvold T.M.	6	9
Afonso J.L.	5	20
Collins J.	5	18
Ryghaug M.	5	14

Particular attention should be paid to the study of geographical coverage in the study of specific issues both in the country and the authors' affiliation (university/institution).

VosViewer toolkit was used to visualize the obtained data. Thus, Fig. 5 presents the relationships between scientists and researchers of smart grids from different countries.

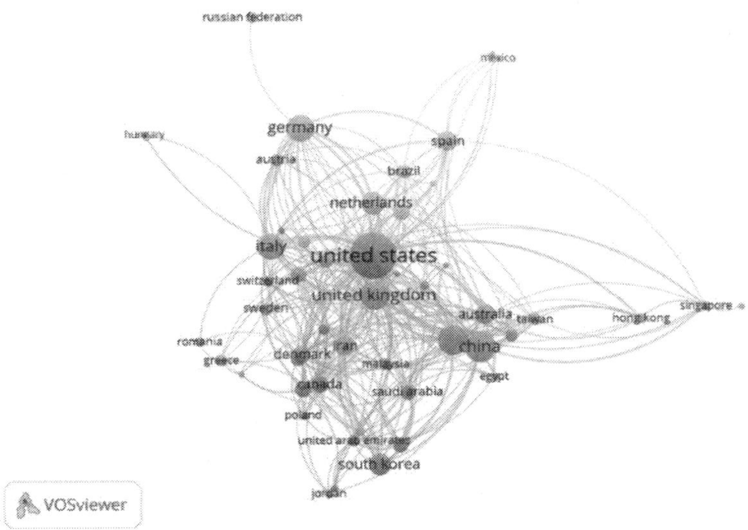

Fig. 5. Visualization map co-authored by scientists (criterion - the country specified in the affiliation), whose joint publications on smart grids were indexed by the Scopus database in 2008–2020

According to the analysis results, eight countries were identified that occupy leading positions in the study. These are USA (277 publications, 5104 citations), China (122 publications, 876 citations), India (117 publications, 500 citations), Great Britain (106 publications, 1823 citations), Germany (89 publications, 860 citations), Italy (85 publications), 1235 citations), the Netherlands (64 publications, 956 citations) and South Korea (63 publications, 424 citations).

Scientists from EU countries have the most significant number of joint publications. In contrast, American researchers' publications are mostly mono-national (prepared without the participation of foreign scientists), although the absolute number of studies by US scientists are undisputed leaders.

An essential step in the analysis of scientific publications is determining the bibliographic relationships related to matches in the lists of references to publications (Fig. 6). This

 https://doi.org/10.1051/e3sconf/202125008006

bibliographic analysis component is used to determine the relationship between two or more publications when they use references to the same document.

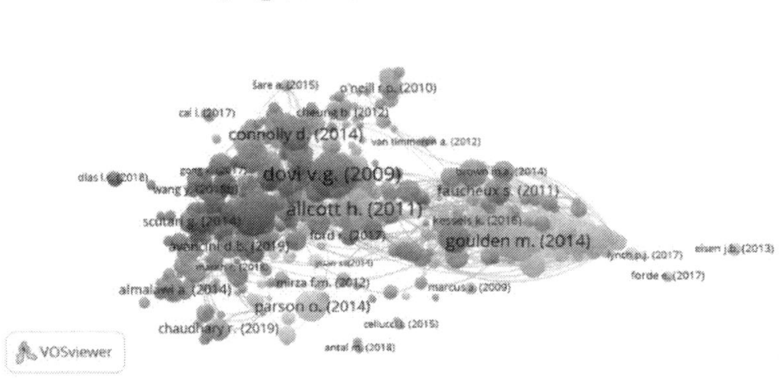

a)

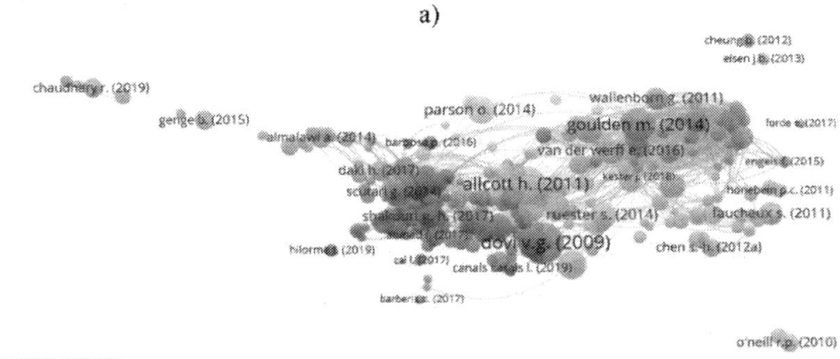

b)

Fig. 6. Visualization map of bibliographic links of publications on smart grids indexed by the Scopus scientometric database in 2008–2020: a) criterion - 10 mutual citations; b) 5 citations

The nodes' size and color represent the number of links made by the authors of the analyzed articles on the development of smart grids. The analysis shows the strength of the relationship, making it possible to state a significant number of cited links common between the two publications. However, the comparison of these results with those obtained during geographical and evolutionary research on the development of the theory of smart grids and publishing activity on this topic allows us to conclude that scientists' level of cooperation to solve problems of smart grids.

4 Conclusions

Thus, our exhaustive analysis of the smart grids that employed a combination of VOSViewer v.1.6.10 and Google Trends allowed more comprehensive and thorough in a formalized by several essential parameters (evolutionary, meaningful, geographical) to generalize the theoretical aspects of the introduction of smart grids in the energy sector.

The analysis presented in this paper creates a basis for further empirical research in the direction of identifying the potential of the smart grids, the interest to the topic and its coverage in the research and scientific literature. In addition, it indicates the prospects and future directions of this area, which can be associated with many unsolved scientific problems that hinder the development of smart grids and the formation of an integrated energy-centered model of the economy.

Acknowledgement

This work was supported by the Ministry of Education and Science of Ukraine (the project No. 0119U100766 "The optimization model of smart and secure energy grids building: an innovative technologies of enterprises and regions ecologisation")

References

1. S. Kolosok, I. Myroshnychenko, L. Zakharkina, CEUR Workshop Proceedings, **2732,** 270-281 (2020)
2. Yu. Harust, V. Melnyk. Marketing and Management of Innovations, **4,** 373-382 (2019).
3. Y. Samusevych, A. Vysochyna, T. Vasylieva, S. Lyeonov, S. Pokhylko, E3S Web of Conferences, **234**, 00012 (2021)
4. V. Levchenko, A. Boyko, T. Savchenko, V. Bozhenko, Yu. Humenna, R. Pilin, Marketing and Management of Innovations, **4,** 364-372 (2019)
5. V. Pavlyk. SocioEconomic Challenges, **4 (1),** 122-128 (2020)
6. S. Kolosok, I. Myroshnychenko, H. Mishenina, I. Yarova, E3S Web of Conferences, **234**, 00021 (2021)
7. Ye. Ziabina, T. Pimonenko, L. Starchenko, SocioEconomic Challenges, **4 (4),** 160-174 (2020)
8. Y. Us, T. Pimonenko, O. Lyulyov, Polityka Energetyczna, **23 (4),** 49-66 (2021)
9. V. Panchenko, Yu. Harust, Ya. Us, O. Korobets, V. Pavlyk, Marketing and Management of Innovations, 1, 256-264 (2020)
10. V. Pavlyk, Financial Markets, Institutions and Risks, 4 (1), 117-123 (2020)
11. L. Starchenko, S. Lyeonov, T. Vasylieva, T. Pimonenko, O. Lyulyov, E3S Web of Conferences, **234**, 00015 (2021)
12. O. Chygryn, A. Rosokhata, O. Rybina, N. Stoyanets, E3S Web of Conferences, **234,** 00004 (2021)
13. R. Boutti, Ad. El. Amri, F. Rodhain, Financial Markets, Institutions and Risks, **3(1),** 18-29 (2019).
14. N. Letunovska, O. Lyuolyov, T. Pimonenko, V. Aleksandrov, E3S Web of Conferences **234**, 00008 (2021)
15. R. Vanickova, Marketing and Management of Innovations, **2,** 56-67 (2020)
16. I. Makarenko, N. Sirkovska. Business Ethics and Leadership, **1(1),** 16-24 (2017)
17. B. George, Business Ethics and Leadership, **4 (3),** 119-126 (2020)
18. E. M. Nelson, SocioEconomic Challenges, **1(3),** 116-119 (2017)
19. M. Kostel, D. Leus, A. Cebotarenco, A. Mokrushina, SocioEconomic Challenges, **1 (3),** 79-90 (2017)
20. R. Gupta, SocioEconomic Challenges, **1 (1),** 81-85 (2017)
21. H. Dave, Financial Markets, Institutions and Risks, **3 (4),** 89-93 (2019)
22. V. G. Dovì, F. Friedler, D. Huisingh, J. J. Klemeš, Journal of Cleaner Production, **17(10)**, 889–895 (2009)
23. D. Connolly, B. V. Mathiesen, International Journal of Sustainable Energy Planning and Management, **1**, 7–28 (2014)
24. M. Killian, M. Kozek, Building and Environment, **105**, 403–412 (2016)
25. O. Parson, S. Ghosh, M. Weal, A. Rogers. Artificial Intelligence, **217,** 1–19 (2014)
26. R. Miskiewicz, Polityka Energetyczna, **21(2)**, 49-62 (2018)

27. H. Dzwigoł, M. Dzwigoł–Barosz, Z. Zhyvko, R. Miskiewicz, H. Pushak, Journal of Security and Sustainability Issues, **8(3)**, 307-317 (2019)
28. I. Vakulenko, L. Saher, L. Syhyda, S. Kolosok, A. Yevdokymova, E3S Web of Conferences, **234**, 00020 (2021)
29. M. Goulden, B. Bedwell, S. Rennick-Egglestone, T. Rodden, A. Spence, Energy Research and Social Science, **2**, 21–29 (2014)
30. H. Khurana, K. M. Rogers, R. Klump, A. A. Aquino-Lugo, T. J. Overbye, IEEE Transactions on Smart Grid, **1 (1)**, 40–47 (2010)
31. P. McDaniel, S. McLaughlin, Security and privacy challenges in the smart grid, IEEE Security and Privacy, **7(3)**, 75–77 (2009)
32. D. T. Ton, M. A. Smith, Electricity Journal, **25 (8)**, 84–94 (2012)
33. P. Waddell, D. Seraphim, Circuit Design, **12 (3),** 1-21 (1995)
34. L. Saher, L. Syhyda, O. Korobets, T. Berezianko, E3S Web of Conferences, **234**, 00011 (2021)
35. G. A. Duffy, V. G. Duffy, International Conference on Human-Computer Interaction, 228-241 (2020)
36. A. Zolkover, V. Terziev, Business Ethics and Leadership, **4 (3),** 107-118 (2020)
37. T. Vasylieva, V. Kasyanenko, Actual Problems of Economics, **144(6)**, 50-59 (2013)
38. K. Djalilov, S. Lyeonov, A. Buriak, Financial Markets and Institutions, 5(4CONT1), 178-187 (2015)

AUTHOR INDEX

AUTHOR INDEX

EDP Sciences – Web of Conferences
17, Avenue du Hoggar
Parc d'Activité de Courtabouf
BP 112
F-91944 Les Ulis Cedex A
France

ISBN 978-1-7138-2870-9